Lecture Notes
in Business Information Processing   214

Series Editors

Wil van der Aalst
 Eindhoven Technical University, Eindhoven, The Netherlands
John Mylopoulos
 University of Trento, Povo, Italy
Michael Rosemann
 Queensland University of Technology, Brisbane, QLD, Australia
Michael J. Shaw
 University of Illinois, Urbana-Champaign, IL, USA
Clemens Szyperski
 Microsoft Research, Redmond, WA, USA

More information about this series at http://www.springer.com/series/7911

Khaled Gaaloul · Rainer Schmidt
Selmin Nurcan · Sérgio Guerreiro
Qin Ma (Eds.)

Enterprise, Business-Process and Information Systems Modeling

16th International Conference, BPMDS 2015
20th International Conference, EMMSAD 2015
Held at CAiSE 2015
Stockholm, Sweden, June 8–9, 2015
Proceedings

Editors
Khaled Gaaloul
Luxembourg Institute of Science
　and Technology
Luxembourg
Luxembourg

Rainer Schmidt
Munich University of Applied Sciences
Munich
Germany

Selmin Nurcan
Université Paris 1 Panthéon - Sorbonne
Paris
France

Sérgio Guerreiro
Lusofona University
Lisbon
Portugal

Qin Ma
Luxembourg Institute of Science
　and Technology
Luxembourg
Luxembourg

ISSN 1865-1348　　　　　　　　ISSN 1865-1356　(electronic)
Lecture Notes in Business Information Processing
ISBN 978-3-319-19236-9　　　ISBN 978-3-319-19237-6　(eBook)
DOI 10.1007/978-3-319-19237-6

Library of Congress Control Number: 2015939169

Springer Cham Heidelberg New York Dordrecht London
© Springer International Publishing Switzerland 2015
This work is subject to copyright. All rights are reserved by the Publisher, whether the whole or part of the material is concerned, specifically the rights of translation, reprinting, reuse of illustrations, recitation, broadcasting, reproduction on microfilms or in any other physical way, and transmission or information storage and retrieval, electronic adaptation, computer software, or by similar or dissimilar methodology now known or hereafter developed.
The use of general descriptive names, registered names, trademarks, service marks, etc. in this publication does not imply, even in the absence of a specific statement, that such names are exempt from the relevant protective laws and regulations and therefore free for general use.
The publisher, the authors and the editors are safe to assume that the advice and information in this book are believed to be true and accurate at the date of publication. Neither the publisher nor the authors or the editors give a warranty, express or implied, with respect to the material contained herein or for any errors or omissions that may have been made.

Printed on acid-free paper

Springer International Publishing AG Switzerland is part of Springer Science+Business Media (www.springer.com)

Preface

This book contains the proceedings of two long-running events held along with the CAiSE conferences relating to the areas of enterprise, business-process and information systems modeling: the 16th International Conference on Business Process Modeling, Development and Support (BPMDS 2015), the 20th International Conference on Exploring Modeling Methods for Systems Analysis and Design (EMMSAD 2015). The two working conferences are introduced below.

BPMDS 2015

BPMDS has been held as a series of workshops devoted to business process modeling, development and support since 1998. During this period, business process analysis and design has been recognized as a central issue in the area of information systems (IS) engineering. The continued interest in these topics on behalf of the IS community is reflected by the success of the last BPMDS events and the recent emergence of new conferences and workshops devoted to the theme. In 2011, BPMDS became a two-day working conference attached to CAiSE (Conference on Advanced Information Systems Engineering). The basic principles of the BPMDS series are:

1. BPMDS serves as a meeting place for researchers and practitioners in the areas of business development and business applications (software) development.
2. The aim of the event is mainly discussions, rather than presentations.
3. Each event has a theme that is mandatory for idea papers.
4. Each event's results are, usually, published in a special issue of an international journal.

The goals, format, and history of BPMDS can be found on the website: http://www.bpmds.org/.

The intention of BPMDS is to solicit papers related to business process modeling, development and support (BPMDS) in general, using quality as a main selection criterion. As a working conference, we aim to attract papers describing mature research, but we still give place to industrial reports and visionary idea papers. To encourage new and emerging challenges and research directions in the area of business process modeling, development and support, we have a unique focus theme every year. Papers submitted as idea papers are required to be of relevance to the focus theme, thus providing a mass of new ideas around a relatively narrow but emerging research area. Full research papers and experience reports do not necessarily need to be directly connected to this theme (they still needed to be explicitly relevant to BPMDS though).

The focus theme for BPMDS 2015 idea papers was "Enabling value creation via business process modeling, development and support." More than two decades after Hammer and Champy made the explicit link between business processes and value, this relationship is still unclear. For the 16th edition of the BPMDS conference, we invited the interested authors to engage, through their idea papers and the discussions during the two days of BPMDS 2015 in Stockholm, in a deep discussion with all participants about what is value, how it is provided, and how it is captured through business process modeling, development and support.

BPMDS 2015 received 42 submissions from 25 countries (Argentina, Austria, Belgium, Brazil, Cameroon, Canada, France, Germany, Hong Kong, Iran, Israel, Italy, The Netherlands, Poland, Portugal, Russia, Spain, Sweden, Switzerland, Tunisia, Turkey, UK, USA, Uruguay, Vietnam). The management of paper submission and reviews was supported by the EasyChair conference system. Each paper received at least three reviews. Eventually, 17 high-quality papers were selected; among them two experience reports and three idea papers. The accepted papers cover a wide spectrum of issues related to business process development, modeling, and support. They are organized under the following section headings:

- Enabling value creation
- Human centric paradigms
- Mining for processes
- Declarative approaches
- Understanding and sharing
- Quality and security issues
- New Areas for BPMDS

We wish to thank all the people who submitted papers to BPMDS 2015 for having shared their work with us, as well as the members of the BPMDS 2015 Program Committee, who made a remarkable effort in reviewing submissions. We also thank the organizers of CAiSE 2015 for their help with the organization of the event, and IFIP WG8.1 for the support.

April 2015

Selmin Nurcan
Rainer Schmidt

EMMSAD 2015

The field of information systems analysis and design includes numerous information modeling methods and notations (e.g., ER, ORM, UML, ArchiMate, EPC, BPMN, DEMO) that are typically evolving. Even with some attempts to standardize (e.g., UML for object-oriented software design), new modeling methods are constantly being introduced, many of which differ only marginally from previous approaches. These ongoing changes significantly impact the way information systems, enterprises, and business processes are being analyzed and designed in practice.

The EMMSAD conference focuses on exploring, evaluating, and enhancing modeling methods and methodologies for the analysis and design of information systems, enterprises, and business processes. Though the need for such studies is well recognized, there is a paucity of such research in the literature. The objective of the EMMSAD conference series is to provide a forum for researchers and practitioners interested in modeling methods for systems analysis and design to meet, and exchange research ideas and results. It also provides the participants an opportunity to present their research papers and experience reports, and to take part in open discussions.

Whereas modeling techniques traditionally have been used to create intermediate artifacts in systems analysis and design, modern modeling methodologies take a more active approach. For instance in Business Process Management (BPM), Model-driven Software Engineering, Domain-specific modeling (DSM), Enterprise Architecture (EA), Enterprise modeling (EM), Interactive Models and Active Knowledge Modelling, the models are used directly as part of the information system of the organization. At the same time, similar modeling techniques are also used for sense-making and communication, model simulation, quality assurance, and requirements specification in connection with more traditional forms of information systems and enterprise development. Since modeling techniques are used in such a large variety of tasks with different goals, it is hard to assess whether a model is sufficiently good to achieve the goals. To provide guidance in this process, knowledge for understanding quality of models and modeling languages is needed.

June 2015

Khaled Gaaloul
Sérgio Guerreiro
Ma Qin

BPMDS 2015 Organization

Organizers

Selmin Nurcan   Université Paris 1 Panthéon - Sorbonne, France
Rainer Schmidt   Munich University of Applied Sciences, Germany

Steering Committee

Ilia Bider   Stockholm University and IbisSoft, Sweden
Selmin Nurcan   Université Paris 1 Panthéon - Sorbonne, France
Rainer Schmidt   Munich University of Applied Sciences, Germany
Pnina Soffer   University of Haifa, Israel

Industrial Advisory Board

Ilia Bider   Stockholm University and IbisSoft, Sweden
Pascal Negros   Arch4IE, France
Gil Regev   EPFL and Itecor, Switzerland

Program Committee

Joao Paulo Almeida   Federal University of Espirito Santo, Brazil
Eric Andonoff   IRIT/UT1, France
Judith Barrios Albornoz   University of Los Andes, Colombia
Ilia Bider   Stockholm University/IbisSoft, Sweden
Karsten Boehm   FH KufsteinTirol - University of Applied Science, Austria
Lars Brehm   Munich University of Applied Science, Germany
Dirk Fahland   Technische Universiteit Eindhoven, The Netherlands
Claude Godart   Loria, France
Renata Guizzardi   Universidade Federal do Espirito Santo, Brazil
Paul Johannesson   Royal Institute of Technology, Sweden
Marite Kirikova   Riga Technical University, Latvia
Agnes Koschmider   Karlsruhe Institute of Technology, Germany
Marcello La Rosa   Queensland University of Technology, Australia
Jan Mendling   Wirtschaftsuniversität Wien, Austria
Bela Mutschler   University of Applied Sciences Ravensburg-Weingarten, Germany
Michael Möhring   Aalen University, Germany

Pascal Negros   Arch4IE, France
Jens Nimis   University of Applied Sciences Karlsruhe,
                                      Germany
Selmin Nurcan   Université Paris 1 Panthéon - Sorbonne, France
Oscar Pastor Lopez   Universitat Politécnica de Valencia, Spain
Elias Pimenidis   University of the West of England, UK
Gil Regev   Ecole Polytechnique Fédérale de Lausanne,
                                      Switzerland
Manfred Reichert   University of Ulm, Germany
Hajo A. Reijers   Eindhoven University of Technology,
                                      The Netherlands
Iris Reinhartz-Berger   University of Haifa, Israel
Stefanie Rinderle-Ma   University of Vienna, Austria
Colette Rolland   Université Paris 1 Panthéon - Sorbonne, France
Michael Rosemann   Queensland University of Technology, Australia
Shazia Sadiq   The University of Queensland, Australia
Rainer Schmidt   Munich University of Applied Sciences, Germany
Samira Si-Said Cherfi   CEDRIC - Conservatoire National des
                                      Arts et Métiers, France
Pnina Soffer   University of Haifa, Israel
Lars Taxén   Linköping University, Sweden
Roland Ukor   FirstLinq Ltd, UK
Barbara Weber   University of Innsbruck, Austria
Matthias Weidlich   Imperial College London, UK
Jelena Zdravkovic   Stockholm University, Sweden

Additional Reviewers

Fdhila, Walid   Popescu, George
Hildebrandt, Tobias   Tapandjieva, Gorica
Knuplesch, David   Teixeira, Maria Das Graças
Kolb, Jens   Tiedeken, Julian
Lanz, Andreas   van der Aa, Han
Moraes Nicola, João Rafael

EMMSAD 2015 Organization

Co-chairs

Khaled Gaaloul	Luxembourg Institute of Science & Technology - LIST, Luxembourg
Sérgio Guerreiro	Universidade Lusófona de Humanidades e Tecnologias, Lisbon, Portugal
Qin Ma	Luxembourg Institute of Science & Technology - LIST, Luxembourg

Advisory Committee

Terry Halpin	INTI International University, Malaysia
John Krogstie	Norwegian University of Science and Technology (NTNU), Norway
Henderik A. Proper	Luxembourg Institute of Science & Technology - LIST, Luxembourg, and Radboud University Nijmegen, The Netherlands
Keng Siau	Missouri University of Science and Technology, USA

Program Committee

Stephan Aier	University of St. Gallen, Switzerland
Antonia Albani	Delft University of Technology, The Netherlands
Raian Ali	Lero, University of Limerick, Ireland
Saïd ASSAR	Telecom Ecole de management, Paris, France
David Aveiro	Madeira University, Portugal
Eduard Babkin	Higher School of Economics, Moscow, Russia
Herman Balsters	University of Groningen, The Netherlands
Annie Becker	Florida Institute of Technology, USA
Giuseppe Berio	University of Torino, Italy
Nacer Boudjlida	Lorraine University, France
Andy Carver	INTI International University, Malaysia
François Charoy	Lorraine University, France
Olfa chourabi	Telecom Ecole de management, Paris, France
Olga De Troyer	Vrije Universiteit Brussel, Belgium
Marwane El Kharbili	Accenture GmbH, Germany
John Erickson	University of Nebraska-Omaha, USA
Peter Fettke	Institute for Information Systems (IWi) at the DFKI, Germany

Wided Guedria	Luxembourg Institute of Science & Technology - LIST, Luxembourg
Remigijus Gustas	Karlstad University, Sweden
Wolfgang Hesse	University of Marburg, Germany
Stijn Hoppenbrouwers	Radboud University Nijmegen, The Netherlands
Philip Huysmans	University of Antwerp
Jon Iden	Norges Handelshøyskole, Bergen, Norway
Marite Kirikova	Riga Technical University, Latvia
Bogdan Lent	University of Applied Sciences, Zurich, Switzerland
Pericles Loucopoulos	Loughborough University, UK
Kalle Lyytinen	Case Western Reserve University, US
Leszek Maciaszek	Wroclaw University of Economics, Poland
Florian Matthes	Technical University München, Germany
Raimundas Matulevičius	University of Tartu, Estonia
Graham McLeod	University of Cape Town, South Africa
Rui Pedro Figueiredo Marques	University of Aveiro, Portugal
Jan Mendling	Humboldt University, Berlin
Tony Morgan	INTI International University, Malaysia
Haralambos Mouratidis	University of East London, UK
Andreas L. Opdahl	University of Bergen, Norway
Sietse Overbeek	Delft University of Technology, The Netherlands
Hervé Panetto	Lorraine University, France
Barbara Pernici	Politecnico di Milano, Italy
Anne Persson	University of Skövde, Sweden
Michaël Petit	University of Namur, Belgium
Nuno Pombo	Universidade da Beira Interior, Portugal
Paul Ralph	Lancaster University, UK
Jolita Ralyté	University of Geneva, Switzerland
Sudha Ram	University of Arizona, USA
Michael Rosemann	Queensland University of Technology, Australia
Matti Rossi	Helsinki School of Economics, Finland
Kurt Sandkuhl	Jönköping University, Sweden
Zohra Sbai	National Engineering School of Tunis, Tunisia
Peretz Shoval	Ben-Gurion University of the Negev, Israel
Piotr Soja	Cracow Economic University, Poland
Janis Stirna	Royal Institute of Technology, Sweden
Inge van de Weerd	Utrecht University, The Netherlands
Dirk van der Linden	Luxembourg Institute of Science & Technology - LIST, Luxembourg
Johan Versendaal	University of Utrecht, The Netherlands
Carson Woo	University of British Columbia, Canada
Jelena Zdravkovic	Stockholm University
Iryna Zolotaryova	Kharkiv National University of Economics, Ukraine
Pär Ågerfalk	Uppsala University, Sweden

Contents

Enabling Value Creation

On the Fragmentation of Process Information: Challenges, Solutions, and Outlook 3
Han van der Aa, Henrik Leopold, Felix Mannhardt, and Hajo A. Reijers

Creating Self-Managed Cross-Professional Teams with Metaphoric Business Process Support Systems 19
Ilia Bider, Amin Jalali, and David Söderström

Human Centric Paradigms

Mining the Organisational Perspective in Agile Business Processes 37
Stefan Schönig, Cristina Cabanillas, Stefan Jablonski, and Jan Mendling

Changing the Focus of an Organization: From Information Systems to Process Aware Information Systems 53
Andrea Delgado and Daniel Calegari

Role and Task Recommendation and Social Tagging to Enable Social Business Process Management 68
Mohammad Ehson Rangiha, Marco Comuzzi, and Bill Karakostas

Mining for Processes

Scalable Process Discovery with Guarantees 85
Sander J.J. Leemans, Dirk Fahland, and Wil M.P. van der Aalst

Multidimensional Process Mining Using Process Cubes 102
Alfredo Bolt and Wil M.P. van der Aalst

Declarative Approaches

Matching of Events and Activities - An Approach Using Declarative Modeling Constraints 119
Thomas Baier, Claudio Di Ciccio, Jan Mendling, and Mathias Weske

PQL - A Descriptive Language for Querying, Abstracting and Changing Process Models. 135
Klaus Kammerer, Jens Kolb, and Manfred Reichert

Enhancing Declarative Process Models with DMN Decision Logic........ 151
Steven Mertens, Frederik Gailly, and Geert Poels

Understanding and Sharing

Using the Process-Assets Framework for Creating a Holistic View Over
Process Documentation... 169
Magnus Josefsson, Kim Widman, and Ilia Bider

Process Fragmentation: An Ontological Perspective 184
Asef Pourmasoumi, Mohsen Kahani, Ebrahim Bagheri, and Mohsen Asadi

Identifying and Quantifying Visual Layout Features of Business
Process Models... 200
Vered Bernstein and Pnina Soffer

Quality and Security Issues

Identifying Quality Issues in BPMN Models: An Exploratory Study........ 217
*Cornelia Haisjackl, Jakob Pinggera, Pnina Soffer, Stefan Zugal,
Shao Yi Lim, and Barbara Weber*

Modeling and Reasoning About Information Quality Requirements
in Business Processes 231
Mohamad Gharib and Paolo Giorgini

From Secure Business Process Models to Secure Artifact-Centric
Specifications... 246
Mattia Salnitri, Achim D. Brucker, and Paolo Giorgini

New Areas for BPMDS

PROtEUS: An Integrated System for Process Execution
in Cyber-Physical Systems.. 265
Ronny Seiger, Steffen Huber, and Thomas Schlegel

Fundamental Issues in Modeling

Applying Predicate Abstraction to Abstract State Machines............. 283
Alessandro Bianchi, Sebastiano Pizzutilo, and Gennaro Vessio

Implementation and First Evaluation of a Molecular Modeling Language ... 293
Alexander Andersson and John Krogstie

Requirements and Regulations

Analyzing Variability of Cloned Artifacts: Formal Framework
and Its Application to Requirements 311
   Iris Reinhartz-Berger, Anna Zamansky, and Mark Kemelman

Solving Semantic Disparity and Explanation Problems in Regulatory
Compliance- A Research-In-Progress Report with Design Science
Research Perspective... 326
   Sagar Sunkle, Deepali Kholkar, and Vinay Kulkarni

Enterprise and Software Ecosystem Modelling

On the Support of Automated Analysis Chains on Enterprise Models 345
   Andrés Ramos, Juan Pablo Sáenz, Mario Sánchez, and Jorge Villalobos

Designing Software Ecosystems: How Can Modeling Techniques Help? 360
   Mahsa H. Sadi and Eric Yu

Information and Process Model Quality

Dealing with Information Quality Requirements....................... 379
   Mohamad Gharib and Paolo Giorgini

Understanding Model Quality Concerns When Using Process Models
in an Industrial Company... 395
   Merethe Heggset, John Krogstie, and Harald Wesenberg

Meta-modeling and Domain Specific Modeling and Model Composition

Towards Metamodelling-In-The-Large: Interface-Based Composition
for Modular Metamodel Development................................. 413
   Srđan Živković and Dimitris Karagiannis

Towards Static Analysis of Executable DSMLs Using Model Typing 429
   Reza Gorgan Mohammadi and Ahmad Abdollahzadeh Barforoush

Modeling of Architecture and Design

Real-Time Design Patterns: Architectural Designs for Automatic
Semi-Partitioned and Global Scheduling 447
   *Amina Magdich, Yessine Hadj Kacem, Adel Mahfoudhi,
   Mickaël Kerboeuf, and Mohamed Abid*

A Generic Traceability Framework for Model Composition Operation 461
Youness Laghouaouta, Adil Anwar, Mahmoud Nassar,
and Jean-Michel Bruel

Novel Applications of Modeling (Short Papers)

Applying Predicate Abstraction to Abstract State Machines 479
Alessandro Bianchi, Sebastiano Pizzutilo, and Gennaro Vessio

A Workaround Design System for Anticipating, Designing,
and/or Preventing Workarounds 489
Steven Alter

An Evaluation of an Enhanced Model Driven Approach
for Computer Game Creation 499
Hong Guo, Shang Gao, John Krogstie, and Hallvard Trætteberg

Author Index 509

Enabling Value Creation

On the Fragmentation of Process Information: Challenges, Solutions, and Outlook

Han van der Aa[1]([✉]), Henrik Leopold[1], Felix Mannhardt[2,3], and Hajo A. Reijers[1,2]

[1] Department of Computer Science, Faculty of Sciences, VU University Amsterdam, De Boelelaan 1081, 1081HV Amsterdam, The Netherlands
j.h.vander.aa@vu.nl
[2] Department of Mathematics and Computer Science, Eindhoven University of Technology, PO Box 513, 5600MB Eindhoven, The Netherlands
[3] Perceptive Software, Gooimeer 12, 1411DE Naarden, The Netherlands

Abstract. An organization's knowledge on its business processes represents valuable corporate knowledge because it can be used to enhance the performance of these processes. In many organizations, documentation of process knowledge is scattered around various process information sources. Such *information fragmentation* poses considerable problems if, for example, stakeholders wish to develop a comprehensive understanding of their operations. The existence of efficient techniques to combine and integrate process information from different sources can therefore provide much value to an organization. In this work, we identify the general challenges that must be overcome to develop such techniques. This paper illustrates how these challenges should be and, to some extent, are being met in research. Based on these insights, we present three main frontiers that must be further expanded to successfully counter the fragmentation of process information in organizations.

1 Introduction

Corporate knowledge is recognized as a principle source of sustainable competitive advantage and value creation [3]. *Process knowledge* is a particular type of corporate knowledge that relates to processes, their context, and their execution. It is regarded as valuable corporate knowledge, because it can be used to enhance the performance of business processes and, hence, also of organizations [23]. In order to meet this value proposition, it is vital that process knowledge is available when needed. A significant threat to this required availability is that process knowledge is often fragmented throughout an organization. Some is only available in the form of tacit knowledge held by specialized process participants or domain experts [13]. Other knowledge is explicitly captured as *process information*, but scattered over a plethora of sources, such as documents, models, and systems. This paper focuses on the latter category of process knowledge.

The problems resulting from the *fragmentation of process information* are considerable. First, information contained in different sources potentially contradicts each other. This can result, for instance, in situations where different stakeholders hold different expectations on what a process aims to establish and take mutually counter-productive measures. Second, even if sources do not contradict each other, insights from multiple sources must be combined to obtain a complete understanding of any given process. This may be quite tedious, depending on the types of sources that are attempted to be combined. Imagine, for example, the task of combining the insights for the same process from its work instruction, a listing of business rules, and a graphical model. So-called *process information defragmentation techniques* counter these problems. They automatically detect and resolve conflicts between various sources, and also integrate information from different sources in order to provide more comprehensive process insights.

At this point, we argue that the development of process information defragmentation techniques takes place in a haphazard way. A number of sources, such as *event logs*, are being studied and harvested intensively; others are virtually being ignored, such as *policies* and *spreadsheets*. Moreover, researchers that look into different information sources are hardly aware of each other's work, while that would actually be highly beneficial to further the defragmentation process. A more thorough understanding of the challenges that are involved with the development of process information defragmentation techniques as well as the opportunities seem called for. Hence, it is the goal of this paper to present a systematic view on this field. We achieve this through the identification of the major challenges that process information defragmentation techniques must meet, a description of the state of the art, and a way forward for the advancement of further techniques.

The remainder of this paper is organized as follows. In Section 2, we first consider why information fragmentation exists and the problems that it causes for organizations. Section 3 then elaborates on the types of heterogeneity across information sources that defragmentation techniques must deal with. In Section 4, we illustrate existing approaches that consider these challenges in order to present insights into the current state of the field. Section 5 discusses the main shortcomings of the existing methods and considers future research directions. Finally, Section 6 concludes the paper.

2 Fragmented Information

The fragmentation of process-related information poses considerable problems for organizations, yet its causes are well-understood. In Section 2.1, we consider the factors that drive information fragmentation. Because it may be difficult or even undesirable to root out these factors, we argue that that the better approach is to repair the resulting defragmentation with efficient techniques. We identify two main streams of such techniques in Section 2.2.

2.1 Causes

There are two ways in which information fragmentation can manifest itself: (i) sources contain different information, and (ii) sources use different representation formats.

Different Information. First, information fragmentation exists because different documents/records/models cover different perspectives on processes. Each type of process documentation can be considered as a simplified description of a process from a certain vantage point: It emphasizes particular aspects and omits details that are not relevant to the party that created it [28]. As such, the process description by the head of the sales department, for instance, may significantly differ from the technical specification used by system administrators. In addition to differing individual perspectives, there are also external sources of process-related information that exist independently from how a company itself characterizes its business processes [38]. Important examples include *policies* or *regulations*, such as Basel III. Clearly, such sources often also contain important pieces of information about business processes, in particular in normative terms [35].

Different Representation Formats. Second, information fragmentation manifests itself through the existence of a variety of different *types of information sources*. This primarily follows from differing preferences of stakeholders about the representation of process-related information. For instance, business professionals, i.e. those who actually conduct the work in the processes, are known not to feel very confident in reading and interpreting *process models* [11]. They rather prefer a verbal description, like a *work instruction*. Some business professionals also use *spreadsheets* for documenting entire business processes [29]. The reason for such a choice is often given by the general familiarity with general purpose tools such as Excel. Business analysts, by contrast, typically do prefer graphical descriptions in the form of processes models [13].

Different representation formats also exist because different types of process information come into existence through different triggers. Information sources such as process models and text documents mainly relate to *build-time* activities of process models, i.e. they are created to characterize business processes. By contrast, the information contained in other sources is created during the *run-time* of a business process. For instance, *event logs* capture execution data generated by users interacting with information systems. Similarly, information recorded in *spreadsheets* support business process execution by providing means to compute certain process steps. Varying information needs and the existence of different types of sources thus result in process information being fragmented over a variety of artifacts, comprising different representation formats.

2.2 Defragmentation Tasks

Two major problems result from process information fragmentation that organizations must deal with. First, information captured in different sources can contradict each other, potentially resulting in highly problematic situations. Second, to obtain all available information on a given process, it does not suffice to consult a single source. To counter these issues, we identify two tasks that process information defragmentation techniques must fulfill: (1) detecting and resolving inconsistencies between information sources, and (2) integrating information captured in different sources.

Inconsistency Detection and Resolution. When different information sources relate to the same process, it is crucial to prevent conflicts in terms of contradicting information. Inconsistencies occur especially when documents are being developed independently from each other [40]. For instance, it has been found that different domain experts often operate with diverging assumptions on how a process is running [13, p.158]. This results in information sources that represent multiple, possibly different expectations. The detection of differences is also crucial for comparisons of the desirable behavior of a process with its actual behavior. i.e. through checking how requirements or regulations imposed by regulatory authorities relate to execution behavior, e.g. captured in event logs. Thus inconsistencies must be detected and resolved to prevent possible misinterpretation of process-related information.

Information Integration. As a result of process information fragmentation, it is not sufficient to consult a single document in order to obtain all information available on a business process. Consider, for instance, two process models related to the same business process. A process model used by managers emphasizes operational details, i.e. *what* has to be done. By contrast, a workflow model representing the actual system implementation focuses on technical details, i.e. *how* the process should be executed [46]. When analyzing this process, it is important to consider the details of both information sources. Otherwise analyses are performed based on incomplete information This can result in sub-optimal decision making and, ultimately, evaporation of valuable corporate knowledge. To ensure that organizations can make optimal decisions, it is therefore worthwhile to integrate the information captured in multiple sources.

3 Challenges

Any process information defragmentation task depends on the ability to relate information from different sources to each other. Without such a relation, it is not possible to check if the information contained in different sources contradicts each other, or to integrate this information into an all-encompassing view. To obtain such relations, we identify two subsequent steps that represent the two main challenges that information defragmentation techniques must always overcome. Figure 1 visualizes these. To automatically compute a relation between process-related information captured in different sources, this information must first be transformed into an interpretable format. In the remainder, we refer to such

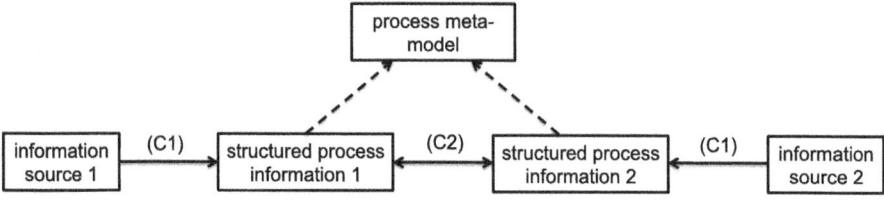

Fig. 1. Two challenges of process information defragmentation

distilled information as *structured process information*. Second, once structured process information has been obtained from various sources, it must be aligned with each other. Due to differences in terminology and level of detail that may exist between various sources, this often poses a considerable challenge. In the remainder of this section we elaborate on these two challenges.

3.1 Challenge 1: Extracting Structured Process Information

In order to integrate or compare process information from multiple sources, it is essential that the information is available in compatible or equal formats. This means that information from any source must be distilled into an interpretable and process-oriented format. While several process meta-models exist (see e.g. [30,34]), we do not argue for the selection of a specific one. This is because the exact requirements differ per situation. We therefore rather say that the distilled information should adhere to a given process meta-model that contains sufficient information for the purpose of a particular defragmentation task. For instance, if a defragmentation technique only considers the behavioral or control-flow perspective of a process, it suffices to be able to express the flow of process elements that are to be executed through, e.g. sequencing, feedback loops, and iteration [34]. In the remainder, we refer to information that adheres to such a process meta-model as *structured process information*.

Extracting structured process information is relatively straightforward for source types with an explicit, process-oriented structure such as *process models* and *business rules*. However, for information sources with a less apparent structuring of its constituting elements, i.e. sources containing *semi-structured data*, this extraction represents a significant challenge. The problem is that semi-structured data can be difficult to interpret in an automated manner [6]. This follows from the lack of an explicit, known, internal representation [8] and because it often cannot be exploited with *traditional* means [19].

Because each type of source has different internal representation of information, all types present their own challenges with regards to the extraction of structured process information. For instance, for textual documents the main challenges revolve around the identification of process model elements and their interrelations contained in a larger text. Which phrases actually refer to the steps that are being taken in a process and which to the actors involved in it?

Answering such questions can only be achieved by acquiring some understanding of the textual semantics. For spreadsheets, on the other hand, it is crucial to correctly determine inter-relations between the information stored in different cells. Some cells may refer to the various phases that can be generally distinguished in a process, where other data may refer to execution data for specific cases. Because of the broad variation in type-specific challenges, dedicated extraction techniques for different types of sources must be developed. Yet, it seems crucial to exploit advances that have been made for one source to the fullest extent for harvesting others.

3.2 Challenge 2: Aligning Extracted Process Information

Correlating process information captured from different sources with each other represents the second challenge for defragmentation approaches. *Correspondence links* or *alignments* associate one or more elements from an information source with its corresponding elements in another source [44]. Any technique that checks for inconsistencies between sources, or attempts to integrate information captured in them, depends on such alignments. However, even if process information has been extracted into a structured form from both sources, creating this alignment is often not trivial. Although process analysis techniques often assume or require that extracted process model elements from multiple sources can be directly linked [5], these assumptions are generally not applicable in practice. Especially when information sources have been created by different stakeholders, it is highly likely that they differ with regard to terminology, structure, and abstraction level.

Differing Terminology. The usage of different vocabularies or naming schemes often leads to terminological differences between sources [43]. As an example, consider the two activities *Evaluate invoice* and *Check bill*. In fact, these labels may refer to the very same task. Both *invoice* and *bill*, as well as *check* and *evaluate*, are synonyms. In order to successfully recognize this, a technique would be required to identify semantic relationships between words.

Differing Grammatical Structures. The previously discussed reasons for terminological variety may also lead to varying grammatical structures. These pertain to sentences in documents as well as the short text labels in process models or spreadsheets, for instance *Invoice creation* and *Create invoice*. In order to successfully relate process concepts with such differing grammatical styles, the different part of speeches such as verbs, nouns, and adjectives must be automatically recognized.

Differing Abstraction Levels. Aside from differing terminology and structure, differences may also arise due to the usage of differing abstraction levels between sources. Figure 2 presents an example of different abstraction levels that

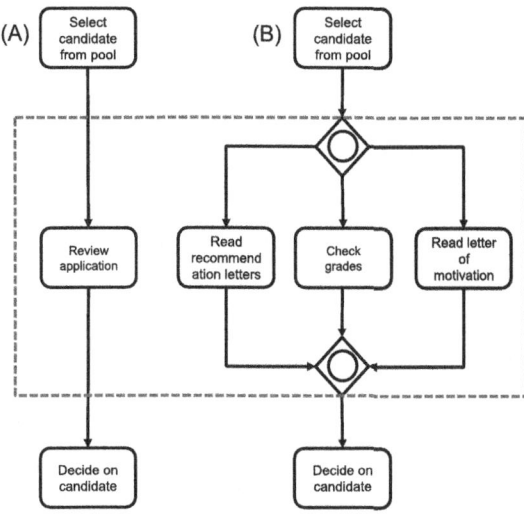

Fig. 2. Processes with different abstraction levels, from [44]

must be resolved. While the review of an application comprises a single activity in model (A), it consists of three different activities in model (B). The detection of such one-to-many relationships still represents a considerable challenge for alignment techniques [10].

These three types of differences, together with the extraction of structured process information must be addressed for any information defragmentation task to succeed. Section 4 presents existing techniques that tackle these challenges.

4 Overview

A variety of existing approaches already address the two main challenges of process information defragmentation techniques. In Section 4.1 we consider challenges related to the extraction of structured process information specific to several types of sources, as well as a number of techniques that deal with them. Section 4.2 then considers how existing techniques tackle the issues related to the alignment of structured process information obtained from different sources.

4.1 Extracting Structured Process Information

In this section we illustrate the extraction of structured process information from a selection of source types. The selected sources greatly differ in terms of their structure (or lack thereof) and type of information they can contain. This selection is not intended to be exhaustive. We rather aim to illustrate the broad variety of sources that can contain valuable process information and the particular challenges that must be dealt with in order to extract this information. For each of these source types we first consider the process-relevant information they potentially contain. Second, we consider the major extraction challenges specific to these sources. Finally, we introduce techniques that tackle or attempt to tackle these challenges. Note that the vast majority of these techniques extract structured process information in order to ultimately generate *process models*. This is simply one of the most popular formats to represent process information. Because of their general adherence to a certain process meta-model it is straightforward to conceive of the use of a comparable technique to other formats, e.g. a textual list of the various steps the process is composed of.

Business Rules. Business rules represent a crucial and widely used information source for constraint-based process information. They are formal expressions of business policies, regulations and common-sense constraints that make business concerns explicit and traceable [17]. A plethora of different business rules representations formats exist, including the natural language SBVR (Semantics of Business vocabularies and Business rules) format [42], decision tables [12], attribute-relationship diagrams [26], and formal logics such as LTL (Linear Temporal Logic) and first order logic [9]. Business rules management (BRM) can be regarded as complementary to BPM [48]. This is because business rules and processes differ in abstraction levels [26]. Rules are also more practical to be updated, which makes them more suitable for the modeling of dynamic process aspects. For these reasons, business rules are sometimes regarded as the integration link between business and process modeling [28]. Their close relation to business processes make business rules important containers of process information. Recognizing this, several works extract structured process information in the form of process models from various business rule formats.

Existing work translates a variety of business rule representations into process models. Methods that achieve this include ones that convert Attribute Relationship Diagrams [26], decision tables [12], and SBVR rules [17]. Because of their formal nature and close relation to business processes, the distillation of business process elements from business rules to is relatively straightforward. The considered approaches therefore generate high quality results that can be confidently used for defragmentation tasks.

Spreadsheets. Spreadsheets are interactive documents for the organization and analysis of data stored in tabular form. Their use is diverse, ranging from inventory administration, to educational applications, to financial modeling [21]. Due to their intuitive usage, they can be created by a large fraction of the working population. For example, Winston [47] estimates that 90% of all analysts in industry perform calculations in spreadsheets, while Panko found that 95% of US firms used them for financial reports [39]. Despite their extensive usage for business critical applications [21], spreadsheets are often ignored in business process management contexts [29]. For example, the business process modeling taxonomy of [16] does not include spreadsheets as a way of modeling processes. Admittedly, Krumnow et al. [29] briefly consider the possibilities of using spreadsheets for modeling, analysis, controlling, and simulation of business processes. Nevertheless, due to the lack of attention for spreadsheets, comprehensive insights into the kind of relevant process information spreadsheets contain and ways to distill this, are missing.

A downside of spreadsheets is that their contents are often not intuitive to comprehend. Users indicate that they experience difficulty understanding spreadsheets created by others [21]. This is problematic, because spreadsheet development and usage often includes multiple people [39]. This results in a situation where many persons involved in the maintenance and usage of spreadsheets do not fully understand the contents. A related result is that spreadsheets have a high risk of undetected errors [22,39]. These issues call for methods that allow users to visualize the contents of spreadsheets in a process-oriented manner. Due to the broad range of usage possibilities of spreadsheets, obtaining process information from them in an automated manner can be considered challenging.

To extract information from spreadsheets, the underlying semantics of the cells and their inter-relations must be understood. For this, there are two main difficulties to overcome: (1) cell type classification, and (2) name resolution. Different cell types exists within spreadsheets. For instance [2] identifies header, footer, data, and filler cells, whereas [20] differentiates between label, data, formula, and empty cells. Once cell types have been successfully identified, appropriate labels or headers must be attached to cells capturing values or formulas. For instance, in the data captured in Figure 3, it is important to recognize that the label "monthly payment" relates to the values € 101,25 and € 472,56. Although they are popular in practice, approaches that relate spreadsheets to business process concepts are scarce.

	A	B	C	D
1	loan amount	interest (%)	term (years)	monthly payment
2	€ 10.000,00	5	10	€ 101,25
3	€ 56.000,00	6	15	€ 472,56

Fig. 3. Excerpt of tabular data typically captured in spreadsheets

A notable exception is the method of [21], which visualizes computations in Excel spreadsheets. The resulting data-flow diagrams can be used to visualize data manipulating processes, and dependencies between data elements. While this work is certainly promising, it focuses on a single type of process information. Due to the wide application range of spreadsheets, it seems paramount to consider the extraction of process information from them in a broader fashion.

Text Documents. Text documents are widely used throughout organizations for a variety of purposes. On the one hand, this results from the potential of natural language to freely express almost any type of information. On the other hand, text documents are popular because they can be created and understood by all stakeholders of an organization [4]. In the context of business processes, text documents may include verbal process descriptions and protocols, laws and regulations as well as annotations that provide additional information about a process and its steps.

Despite their wide usage, text documents have an important downside from an analytic point of view: Natural language can be highly ambiguous. Hence, text documents are not always completely clear and easy to understand. As a result, process information captured in more structured representations, such as process models, can be easier and faster interpreted [13]. The problem of ambiguity is amplified when texts are analyzed in an automatic manner. To handle this, recent research has turned towards the use of *natural language process* (NLP) techniques to extract structured process information from text documents.

The issues that must be addressed when extracting structured process information from natural language text are considerable. [15] presents the most important issues found in scientific literature. A seminal problem is the mismatch between syntax and semantics in natural language. This mismatch follows because the syntactic structure of a sentence is not directly linked to its semantics [14]. Another issue that is typical for natural language texts is the usage of anaphoric references or *anaphoras*. Anaphoras are usually pronouns ("he", "her", "it") or determiners ("this", "that") that refer to a previously introduced unit. For instance, the word "it" in the sentence "If it is not available, it is backordered", only receives it meaning if we consider the sentence that precedes it: "If a part is available, it is reserved." Dealing with these and other challenges related to natural language text is complex and prone to errors.

Nevertheless, a variety of methods exist that generate process models from different types of text documents. Examples include methods for generating models from textual process descriptions [15], group stories [18], and use cases [41]. These techniques differ greatly in their applicability. Some (e.g. [18]) require specific domain knowledge, for instance in the form of ontologies, while others, such as [15], are more generally applicable.

Event Data. Information systems store data related to the execution of business processes. Contrast to information captured in e.g. process models, business rules, and text documents, which represents desired or supposed process behavior, so-called *event data* represents information about the reality of a process. *Process mining* techniques analyze event data, captured in *event logs*, to discover, monitor, and improve processes [1]. An important use case of process mining is the generation of a process model from an event log. The challenges that these techniques must deal with are very different from the issues faced when extraction process information from other considered sources.

An event log stores information on a case level, where every case represents a single instance of a process execution. To distill event data into the same abstraction level as information obtained from other types of sources, process mining techniques must generalize instance level behavior captured in event logs. This is achieved by *process discovery* techniques. The main issue that process discovery techniques must deal with is the balance between over- and underfitting of process models to the behavior captured in event logs. This is necessary, because event logs can often contain noisy or exceptional behavior, that may obscure the understandability of a generated process model [7].

A plethora of process discovery techniques have been developed that address this problem in many different ways. Due to the widespread attention that process mining has received over the past decade, these techniques can be considered to be far more mature than extraction techniques of other types of information sources. We therefore refer the interested reader to e.g. [1] for broader insights into the different types of process discovery algorithms.

4.2 Aligning Extracted Process Information

Challenge 2 recognizes that to compare or combine process information from different sources, alignments must be created between the process concepts stored in them. So-called *matchers* aim to achieve this in an automated manner. In order to succeed, matchers must address the issues introduced in Section 3.2: differences in terminology, grammatical structures, and abstraction levels.

To deal with terminological differences, e.g. *Evaluate invoice* and *Check bill*, semantic similarity (i.e. synonymous relations) between words must be identified. In prior work, this turned out to be a major challenge in business process contexts despite the existence of frequently employed tools that support the identification of synonyms [31]. This is because popular tools, such as WordNet [37], do not contain the full range of vocabulary that occurs in organizational documents. Hence, many terms cannot be related simply because there are not part of the available taxonomy. Therefore, better strategies to measure semantic similarity have to be identified in the future. Promising directions may, for instance, include co-occurrence based techniques such as DISCO [27]. They are based on text corpora and, hence, can be trained on any large text collection.

Business process concepts captured in different information sources may also exhibit differences in labeling or grammatical styles. To recognize that labels such as *Invoice creation* and *Create invoice*, relate to the same activity, *parts-of-speech*, e.g. verbs and objects, must be identified in these labels. While there are precise techniques available to analyze full natural language sentences (see e.g. [25]), techniques for analyzing short process model text labels have not yet reached the same level of maturity [32]. Current solutions are based on rules and heuristics, rather than statistical models. These models, however, are notably successful in generic natural language parsers [24]. Hence, it might be a promising direction to transfer those concepts from standard natural language processing tools to the analysis of short text labels.

Differences in abstraction levels used to describe process concepts pose the third class of problems for the creation of alignments. To take on these situations, matchers must recognize situations where a concept in one information source represents a sub-part of a larger concept contained in another source. Although there are taxonomies such as the MIT process handbook [36] or WordNet [37] that define part-of-relationships, their applicability is again limited by the low coverage of the variety of terms that are contained in business documents.

Hence, also in this direction novel solutions are required. To work around the lack of appropriate taxonomies, some matchers, e.g. [31,33,45], turn towards the context in which activities occur. Namely, they exploit the structural relations between process concepts in order to match activities based on similar neighboring elements. Consider, for instance, the two models contained in Figure 2, the relation between the middle activity in model (A) and the three activities in model (B), becomes very clear when the context of the activities is considered. This is because both sets of activities are preceded and succeeded by similar (even equal) activities.

Despite the incorporation of aforementioned, sophisticated techniques to deal with the challenges of process alignment, the quality of matchers still leaves room for improvement. This is illustrated by the results of the *Process Model Matching Contest 2013*, in which seven matchers from different research groups tried to solve the same matching problem [10]. The best technique only achieved an f-measure of 0.45. This highlights that the alignment of process concepts is a very relevant and open challenge that impedes successful information defragmentation.

5 Discussion

The previous sections outlined the major challenges to be met, and how existing techniques are meeting these to support information defragmentation. To advance the state of these techniques, we identify three main frontiers where further developments are needed.

First, many of the extraction methods considered in Section 4.1 represent preliminary approaches that leave plenty of room for improvement. On the one hand, this results from the limited scope that methods take into consideration. The majority of methods focus only on control-flow information, while other perspectives, e.g. organizational and data perspectives, are largely ignored. On the other hand, the quality of the results obtained through existing techniques is in some cases low. This is especially the case for sources with a less explicit structure, such as text documents and spreadsheets, that pose considerable information extraction challenges. As a result, existing techniques yield sub-optimal results. We therefore argue that in many cases it is necessary to improve upon the quality of existing methods for extracting structured process information from semi-structured sources.

Quality issues also manifest themselves with respect to approaches to align extracted process information from various sources. As seen in Section 4.2, even matchers designed to deal with information from homogeneous source types, i.e. process model matchers, achieve far from perfect results. We therefore argue to further develop the quality of existing matchers as a second frontier, despite the sophisticated techniques already incorporated in them.

Finally, we note the lack of comprehensive insights into process information sources that are available within organizations. To overcome this, it is important to identify which sources are available within organizations, and how these can be used for process information defragmentation. The information sources considered in Section 4.1 represent a set of important and widely available source types, but they merely provide an illustration of the variety of sources that contain process information. Furthermore, even for these known sources, comprehensive insights into their usability for information defragmentation are missing. Particularly, source types with a broad range of uses, namely text documents and spreadsheets, can contain process-related information in a large variety of forms. This lack of insights impedes the results achievable by process information defragmentation since important information may remain unconsidered.

6 Conclusions

The fragmentation of process information is a major issue that impedes the ability of organizations to obtain value from their documented process knowledge. Through the insights presented in this work, we have demonstrated the relevance of extraction and matching techniques to tackle problems caused by process information fragmentation. Despite the existence of various techniques to tackle these issues, there is a window of opportunity to improve upon the state of art.

We have noted that the problems related to information fragmentation in business process contexts have not yet been investigated in a systematic manner. Therefore, with this work we aimed to structure the challenges that impede process information defragmentation techniques. This resulted in actionable steps that any defragmentation technique faces. To further advance this field, we identified several major development frontiers. First, we argue for the importance of improved and new techniques for the extraction and alignment of process information from various sources. Aside from those directions, an important driver for future research is the lack of a comprehensive overview of relevant information sources. In the end we hope that this work sets the stage for future efforts to resolve the fragmentation of documented knowledge on business processes.

References

1. Van der Aalst, W.: Discovery, conformance and enhancement of business processes. Springer (2011)
2. Abraham, R., Erwig, M.: Header and unit inference for spreadsheets through spatial analyses. In: 2004 IEEE Symposium on Visual Languages and Human Centric Computing, pp. 165–172. IEEE (2004)
3. Alavi, M., Leidner, D.E.: Review: Knowledge management and knowledge management systems: Conceptual foundations and research issues. MIS quarterly, 107–136 (2001)
4. Allweyer, T.: BPMN 2.0: introduction to the standard for business process modeling. BoD-Books on Demand (2010)
5. Baier, T., Mendling, J.: Bridging abstraction layers in process mining by automated matching of events and activities. In: Daniel, F., Wang, J., Weber, B. (eds.) BPM 2013. LNCS, vol. 8094, pp. 17–32. Springer, Heidelberg (2013)

6. Blumberg, R., Atre, S.: The problem with unstructured data. DM REVIEW **13**, 42–49 (2003)
7. Buijs, J., Van Dongen, B., Van der Aalst, W.: Quality dimensions in process discovery: The importance of fitness, precision, generalization and simplicity. International Journal of Cooperative Information Systems **23**(01) (2014)
8. Buneman, P., Davidson, S., Fernandez, M., Suciu, D.: Adding structure to unstructured data. In: Afrati, F.N., Kolaitis, P.G. (eds.) ICDT 1997. LNCS, vol. 1186, pp. 336–350. Springer, Heidelberg (1996)
9. Caron, F., Vanthienen, J., Baesens, B.: Comprehensive rule-based compliance checking and risk management with process mining. Decision Support Systems **54**(3), 1357–1369 (2013)
10. Cayoglu, U., Dijkman, R., Dumas, M., Fettke, P., García-Bañuelos, L., Hake, P., Klinkmüller, C., Leopold, H., Ludwig, A., Loos, P., Mendling, J., Oberweis, A., Schoknecht, A., Sheetrit, E., Thaler, T., Ullrich, M., Weber, I., Weidlich, M.: Report: the process model matching contest 2013. In: Lohmann, N., Song, M., Wohed, P. (eds.) BPM 2013 Workshops. LNBIP, vol. 171, pp. 442–464. Springer, Heidelberg (2014)
11. Chakraborty, S., Sarker, S., Sarker, S.: An exploration into the process of requirements elicitation: a grounded approach. Journal of the Association for Information Systems **11**(4), 1 (2010)
12. De Roover, W., Vanthienen, J.: On the relation between decision structures, tables and processes. In: Meersman, R., Dillon, T., Herrero, P. (eds.) OTM-WS 2011. LNCS, vol. 7046, pp. 591–598. Springer, Heidelberg (2011)
13. Dumas, M., La Rosa, M., Mendling, J., Reijers, H.A.: Fundamentals of business process management. Springer (2013)
14. Fillmore, C.J.: The case for case. (1967)
15. Friedrich, F., Mendling, J., Puhlmann, F.: Process model generation from natural language text. In: Mouratidis, H., Rolland, C. (eds.) CAiSE 2011. LNCS, vol. 6741, pp. 482–496. Springer, Heidelberg (2011)
16. Giaglis, G.M.: A taxonomy of business process modeling and information systems modeling techniques. International Journal of Flexible Manufacturing Systems **13**(2), 209–228 (2001)
17. Goedertier, S., Vanthienen, J.: Declarative process modeling with business vocabulary and business rules. In: Meersman, R., Tari, Z. (eds.) OTM-WS 2007, Part I. LNCS, vol. 4805, pp. 603–612. Springer, Heidelberg (2007)
18. Gonçalves, J., Santoro, F.M., Baiao, F.A.: Business process mining from group stories. In: 13th International Conference on Computer Supported Cooperative Work in Design, pp. 161–166. IEEE (2009)
19. Herbst, J., Karagiannis, D.: An inductive approach to the acquisition and adaptation of workflow models. In: Proceedings of the IJCAI, vol. 99, pp. 52–57 (1999)
20. Hermans, F., Pinzger, M., van Deursen, A.: Automatically extracting class diagrams from spreadsheets. In: D'Hondt, T. (ed.) ECOOP 2010. LNCS, vol. 6183, pp. 52–75. Springer, Heidelberg (2010)
21. Hermans, F., Pinzger, M., van Deursen, A.: Supporting professional spreadsheet users by generating leveled dataflow diagrams. In: Proceedings of the 33rd International Conference on Software Engineering, pp. 451–460. ACM (2011)
22. Hesse, R., Scerno, D.H.: How electronic spreadsheets changed the world. Interfaces **39**(2), 159–167 (2009)
23. Jung, J., Choi, I., Song, M.: An integration architecture for knowledge management systems and business process management systems. Computers in Industry **58**(1), 21–34 (2007)

24. Jurafsky, D., Martin, J.H.: Speech & language processing. Pearson Education India (2000)
25. Klein, D., Manning, C.D.: Accurate unlexicalized parsing. In: 41st Meeting of the Association for Computational Linguistics, pp. 423–430 (2003)
26. Kluza, K., Nalepa, G.J.: Automatic generation of business process models based on attribute relationship diagrams. In: Lohmann, N., Song, M., Wohed, P. (eds.) BPM 2013 Workshops. LNBIP, vol. 171, pp. 185–197. Springer, Heidelberg (2014)
27. Kolb, P.: Disco: A multilingual database of distributionally similar words. In: Proceedings of KONVENS-2008, Berlin (2008)
28. Kovacic, A.: Business renovation: business rules (still) the missing link. Business Process Management Journal 10(2), 158–170 (2004)
29. Krumnow, S., Decker, G.: A concept for spreadsheet-based process modeling. In: Mendling, J., Weidlich, M., Weske, M. (eds.) BPMN 2010. LNBIP, vol. 67, pp. 63–77. Springer, Heidelberg (2010)
30. La Rosa, M., Reijers, H.A., Van Der Aalst, W., Dijkman, R.M., Mendling, J., Dumas, M., García-Bañuelos, L.: Apromore: An advanced process model repository. Expert Systems with Applications 38(6), 7029–7040 (2011)
31. Leopold, H., Niepert, M., Weidlich, M., Mendling, J., Dijkman, R., Stuckenschmidt, H.: Probabilistic optimization of semantic process model matching. In: Barros, A., Gal, A., Kindler, E. (eds.) BPM 2012. LNCS, vol. 7481, pp. 319–334. Springer, Heidelberg (2012)
32. Leopold, H., Smirnov, S., Mendling, J.: On the refactoring of activity labels in business process models. Information Systems 37(5), 443–459 (2012)
33. Ling, J., Zhang, L., Feng, Q.: Business process model alignment: an approach to support fast discovering complex matches. In: Enterprise Interoperability VI, pp. 41–51. Springer (2014)
34. List, B., Korherr, B.: An evaluation of conceptual business process modelling languages. In: Proceedings of the 2006 ACM symposium on Applied computing, pp. 1532–1539. ACM (2006)
35. Ly, L.T., Rinderle-Ma, S., Knuplesch, D., Dadam, P.: Monitoring business process compliance using compliance rule graphs. In: Meersman, R., Dillon, T., Herrero, P., Kumar, A., Reichert, M., Qing, L., Ooi, B.-C., Damiani, E., Schmidt, D.C., White, J., Hauswirth, M., Hitzler, P., Mohania, M. (eds.) OTM 2011, Part I. LNCS, vol. 7044, pp. 82–99. Springer, Heidelberg (2011)
36. Malone, T.W., Crowston, K., Herman, G.A.: Organizing business knowledge: the MIT process handbook. MIT press (2003)
37. Miller, G.A.: Wordnet: a lexical database for english. Communications of the ACM 38(11), 39–41 (1995)
38. Möhring, M., Schmidt, R., Härting, R.-C., Bär, F., Zimmermann, A.: Classification framework for context data from business processes. In: Fournier, F., Mendling, J. (eds.) BPM 2014 Workshops. LNBIP, vol. 202, pp. 440–445. Springer, Heidelberg (2015)
39. Panko, R.R.: What we know about spreadsheet errors. Journal of Organizational and End User Computing (JOEUC) 10(2), 15–21 (1998)
40. Rahm, E., Bernstein, P.A.: A survey of approaches to automatic schema matching. The VLDB Journal 10(4), 334–350 (2001)
41. Sinha, A., Paradkar, A.: Use cases to process specifications in business process modeling notation. In: 2010 IEEE International Conference on Web Services (ICWS), pp. 473–480. IEEE (2010)
42. Team, S., et al.: Semantics of business vocabulary and rules (sbvr). Tech. rep., Technical Report dtc/06-03-02 (2006)

43. Wache, H., Voegele, T., Visser, U., Stuckenschmidt, H., Schuster, G., Neumann, H., Hübner, S.: Ontology-based integration of information-a survey of existing approaches. In: IJCAI-01 workshop: Ontologies and Information Sharing, vol. 2001, pp. 108–117. Citeseer (2001)
44. Weidlich, M., Barros, A., Mendling, J., Weske, M.: Vertical alignment of process models – how can we get there? In: Halpin, T., Krogstie, J., Nurcan, S., Proper, E., Schmidt, R., Soffer, P., Ukor, R. (eds.) Enterprise, Business-Process and Information Systems Modeling. LNBIP, vol. 29, pp. 71–84. Springer, Heidelberg (2009)
45. Weidlich, M., Dijkman, R., Mendling, J.: The ICoP framework: identification of correspondences between process models. In: Pernici, B. (ed.) CAiSE 2010. LNCS, vol. 6051, pp. 483–498. Springer, Heidelberg (2010)
46. Weidlich, M., Weske, M., Mendling, J.: Change propagation in process models using behavioural profiles. In: IEEE International Conference on Services Computing, SCC 2009, pp. 33–40. IEEE (2009)
47. Winston, W.: Executive education opportunities millions of analysts need training in spreadsheet modeling, optimization, monte carlo simulation and data analysis. OR MS TODAY **28**(4), 36–39 (2001)
48. Zoet, M., Versendaal, J., Ravesteyn, P., Welke, R.: Alignment of business process management and business rules (2011)

Creating Self-managed Cross-Professional Teams with Metaphoric Business Process Support Systems

Ilia Bider[1,2(✉)], Amin Jalali[1], and David Söderström[1]

[1] DSV - Stockholm University, Stockholm, Sweden
{ilia,aj,ids}@dsv.su.se
[2] IbisSoft AB, Stockholm, Sweden

Abstract. The typical values that can be obtained by employing a business process support (BPS) system are optimization, compliance to rules and regulations, leanness, etc. These are obtained by introducing a kind of "conveyor belt" in business processes. While these values can be important for some processes at some phases of the enterprise development, they can be counter-productive for other processes and other phases. This idea paper investigates a possibility of obtaining completely different value from a BPS system, namely it serving as a means of creating self-managed/self-directed cross-professional teams of process participants. The paper suggests using visualized folk and fairy tales and interactive game technology as a basis for building BPS systems that can bring this value. The discussion is done based on the example of representing a consultative sales process using a treasure hunting plot.

Keywords: Business process · Cross-professional · Cooperation · Sales process · Treasure hunting · Interactive game · Metaphor · Alternative reality game

1 Introduction

From the beginning [1], the Business Process Management (BPM) movement was aimed at optimizing business activities of an enterprise or organization by streamlining the operations completed in the frame of its business processes and balancing resources engaged. The first Business Process Support (BPS) systems were designed to support this aim of optimization. They were based on the operational view on business processes – workflow – and were introducing a kind of "conveyor belt" principle in non-manufacturing processes, which became the primary value that was brought by BPS systems[1].

One of the original ideas with BPM was to destroy the walls between organizational units (silos), as many business processes run across multiple units. Employing a "conveyor belt" seemingly can help in attaining this goal. However, there is a risk that instead of walls between the departments/units, it might create invisible walls

---

[1] This is not the only value that can be achieved from a BPS system of the workflow type, for example, such a system makes it also possible imposing compliance, unification of work, gathering data and making measurements, etc.

between people participating in the processes. They will be executing activities that a BPS system "moves" to them along with the data to be processed without feeling the needs to cooperate with others and/or understand what is happening in other parts of the process instances in which they participate. While the conveyor belt principle can bring optimization of resource usage, its area of application is limited. Even in production, there is a tendency to use self-managed teams instead of a conveyor belt as soon as there is no possibility to fully standardize the work. The possibility to work in the conveyor belt fashion in the frame of more soft (nonproduction) business processes is even more limited. For example, data, e.g. documents, received as an input for some activity may require clarification, which in turn leads to communication acts with the people who produced these documents thus breaking the pattern of the conveyor belt.

In cases where the unsuitability of the workflow ("conveyor belt") is obvious, instead of operational view on the processes, the data-centric view is used for building BPS systems of Case Management (CM) and Adaptive Case Management types (ACM) [2]. Such a system, usually have some kind of shared space, e.g. in form of folders, where all relevant information for each business process instance/case is stored. This shared space minimized the needs to send data between the participants of the same case as all information is available for everybody.

Employing a BPS system with shared spaces substitutes the "conveyor belt" principle of workflow systems with the "construction site" principle [3,4]. The shared space in such a system serves as a "construction site" where every participant comes to do his part of the job, the results of which all other participants can see, while continuing building upon these results.

Provided that the shared space clearly identifies the progress in the given process instance/case, a BPS system based on shared spaces could bring a new type of value. Namely, it can serve as a means of making the participants of the given process work as a self-managed, or even as self-directed team, even when the members of the team are not working at the same location. The main characteristics of such a team [5], in our view, is the minimum level of explicit management, be it from the team leader or from the upper management. The members of the team help each other when seeing that such help required even when nobody asks for help explicitly.

To bring the value of facilitator of a self-managed/directed (often virtual) process team, the shared space of the BPS system should be highly visual, so that the team members could see themselves where the help is needed without being asked by the colleagues, or the team leader. Such a visualization, most probably is not too difficult to find if the team is homogeneous, i.e. the members belong to the same profession, or have long experience of working with the people of other professions and have detailed understanding of their professional jargon, and what kind of activities they perform and why. In this case, the visualization consists of finding visual representation for an already existing written or oral professional jargon. This type of visualization can be in a form of complex graphs, which can be mystical for the outsiders, but be perfectly clear for the members of the process team.

If the process team, however, is cross-professional and is not working together all the time, finding a common visualization might be a problem. There is a risk that the shared space, though common as far as access rights are concerned, is divided into isolated islands as far as understanding the content of the shared space is concerned. In this situation, explicit management will be required to coordinate the work of the team.

This idea paper is looking for a solution for using a BPS system for bridging linguistic and cultural differences of cross-professional teams so that they can become self-managed/directed with the minimum explicit management. For this end, we investigate the idea of using visualized metaphors based on folk and/or fairy tales well known to all participants. More specifically, we suggest using metaphors that realize the quest plot [6] as they are consistent with the goal orientation of business processes.

Assume that we can translate the happenings inside a business process instance into the analogous happening in a story with a quest plot familiar to all participants. Assume further that we can visualize the tale in the same manner and using the same technique as is used for interactive game creation. Then, it would be easier for all engaged to understand what is happening in the process, even if each participant does not understand all technical details of actions completed by others.

The rest of the paper is structured in the following way. In section 2, we discuss the advantages and problems of having cross-professional teams. In Section 3, we discuss the challenges to overcome to make a cross-professional team self-managed. In Section 4, we describe our idea of overcoming these challenges using an example of consultative sales processes represented with a metaphor of treasure hunting. Section 5 is devoted to related research. Section 6 contains concluding remarks and plans for future research.

2 Why Cross-Professional Business Process Teams Are Needed

As we analyzed in [7], the flexibility required from a business process often depends on the Marketing Position (MP) in which an enterprise operates, see Fig. 1. In MP1: *Entering an emergent market*, the maximum flexibility is required. This is a situation of start-up when a team works in close cooperation trying to figure out how to survive and win the new emergent market. This situation can be visualized as in Fig. 2, which represents a software product development company. The diagram features the enterprise boundary, its assets, from which we will consider only the workforce assets, and business processes. The latter can be internal (like product development) and boundary (like sales, or field services). The boundary processes run partly inside the organization and partly outside it (in the surrounding environment). It is in the frame of boundary processes the organization has a contact with the external world – its environment - and gets information needed for decision making, e.g. what product to develop.

This information is crucial if the company product (software system) is fairly complex, and it takes long time, say a year, to produce a new version. While the developers design, code and test, the needs in the emerging dynamic market would continue to evolve creating a risk that the new version will be outdated when it hits the market.

In this situation, there is a need to provide the development team with information on the changes in the market demands so that they can adjust the product to the evolving needs even when the process is in the design and coding phases.

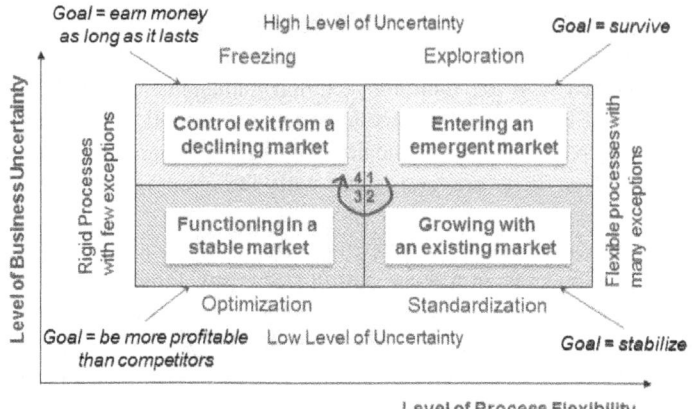

Fig. 1. 4 Market positions (MP), borrowed from [7]

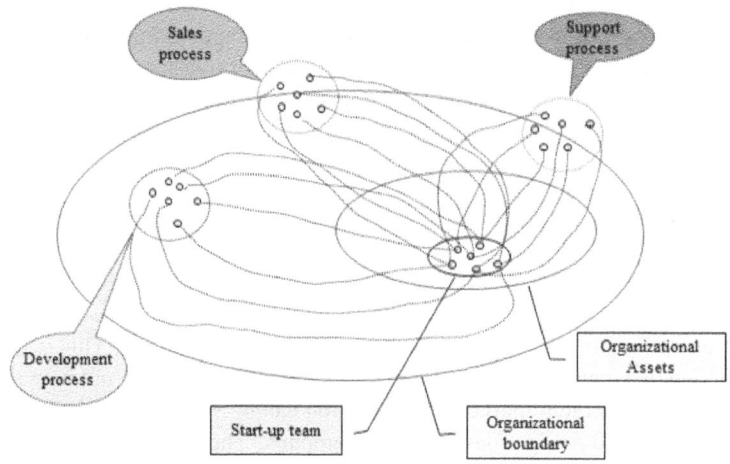

Fig. 2. Business processes in MP1 "Entering an emerging market" – startup organization

In the startup organization the needs of obtaining information from the outer world is solved in a relatively simple way. In this case, there is only one team and all members of the team participate in all business processes. This means that developers are visiting existing and potential customers with the sales personal to help to understand the needs, and they also provide support to the customers themselves at this phase. Also, sales people participate in the development processes and contribute to their understanding of the emerging market. At that stage, based on trials and errors when probing the new market, the team, naturally, works in self-managed/self-directed fashion possessing a high-level of motivation and creativity. At this phase, most processes are ad-hoc (loose in the terminology of [7]), and they cannot be supported by a workflow BPS system. According to [7], a BPS system that suit the startup setting should be some kind of social software, e.g. wiki, forum, etc.

After surviving in an emerging market, the company goes through the phase of growth and arrives to MP3: *Functioning in a stable market* in Fig. 1, the goal of which is to be more profitable than the competitors. At this point, usually, the enterprise structure changes so that each process is manned by people from a corresponding department as it is depicted in Fig. 3. Thus sales processes are conducted by the staff from the sales department, support processes are conducted by the staff from the support department, product development is conducted by the staff from the engineering department, etc. In such a situation, people who man the boundary processes and have a chance of acquiring the tacit knowledge (in terms of Polanyi [8]) that might be needed for the development do not directly participate in the development process. Also people who sell, or support the existing customers do not have intimate knowledge on the new features under development, and thus might miss a potential customer, or a solution soon to be available.

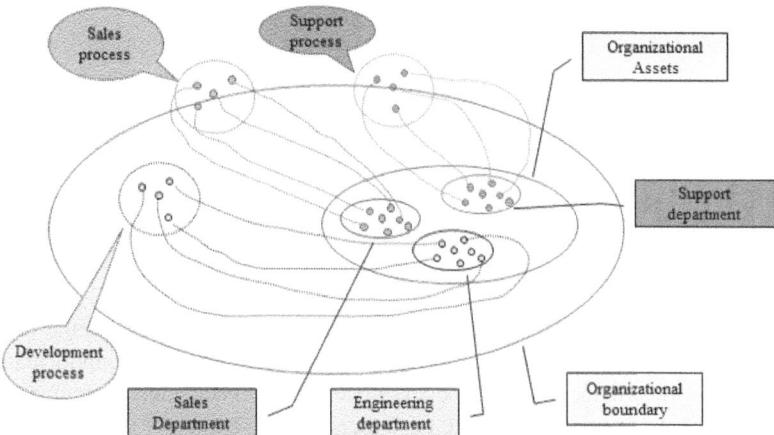

Fig. 3. Business processes in MP3 "Functioning in a stable market" – traditional organization

There are two options to solve the problem above. One is to arrange internal processes for transferring the knowledge on changes in the outer world. This will, require converting the tacit knowledge into an explicit form, e.g. analysis, recommendations, etc. before sending it to the development team. In addition, there should be some incentive schemes to encourage such transfer. Both add overheads and make the organizational structure more bureaucratic and complicated. This scheme has also risks that the explicit information may not be properly interpreted by the receiving side, as there are professional and cultural differences that could affect how one sees and express the reality, and interpret the messages received.

The other solution is cross-manning of business processes as it is represented in Fig. 4. Cross-manning means that people participating in the development process participate also in the boundary processes (e.g. sales and/or field services). It can be sales people, or support engineers that participate in the development process, or (and) the developers that also participate in the sales and (or) support processes. The advantage of cross-manning is that there is no need for additional internal processes for knowledge transfer. Development team gets the knowledge directly through participating in the boundary processes. The disadvantages are that business processes should be adjusted to allow participation of the "foreigners". This might lead to each process becoming less "optimal" on its own. There can be lost off efficiency in each of the processes. However, this loss might be a small penalty for ensuring the shortest time to the market for the product that will also be "right" for the market.

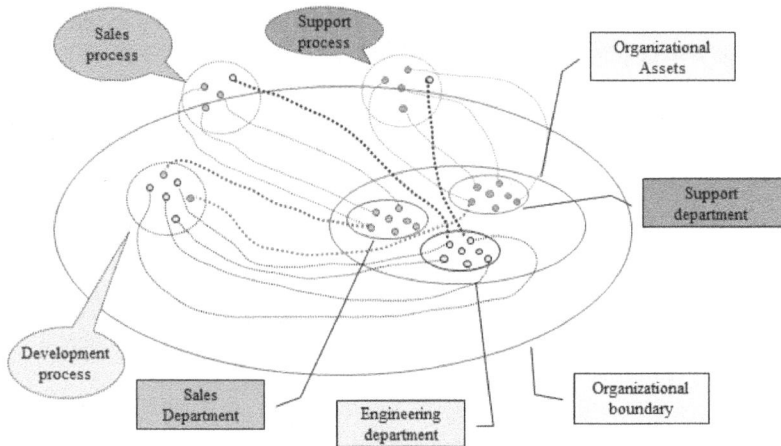

Fig. 4. Cross-manning of business processes

Effective transfer of knowledge on the current needs and problems to the development team is not the only effect of cross-manning. The sales, or support team that possesses the knowledge on the products under development may start transferring this knowledge to existing and future customers far in advance of the product reaching the market. This will create expectations and ensure smoother transferring of knowledge on how to use the new product in practice.

3 Challenges to Overcome

Cross-manning of business processes as depicted in Fig. 4 have serious implications on the way the processes are to be developed, maintained, and managed, including the following two:

1. The process team becomes heterogeneous, i.e. it includes people with different backgrounds, professional jargons, experiences and professional cultures.
2. Each cross-manned process has multiple goals to achieve. For example, a sales process besides its main goal to sell a product (or service), has a goal of giving the development team knowledge on the current customer's needs/problems, even if

they cannot be satisfied by the existing assortment of products/services. Having different goals may in addition to (1) above create discordance between the members of the process team.

The differences in the professional language, culture goals, etc., create a challenge in a cross-manned process team becoming self-managed/self-directed. If the members do not understand the language, activities and goals of others, how can they determine when their help is needed? Thus such team requires some explicit management from a team leader/manager to determine what to do next and who need to be engaged. Alternatively, an intelligent BPS system can take the role of such management based on some rules like when a certain milestone in the process, has been reached, it is time a person of another profession to be engaged.

Both solutions above have disadvantages against having self-managed/self-directed process team. Having explicit management requires an experience leader/manager who understands all details of the process and having time to react on the development in it. Even when there are no practical problems to engage such a person, it will constitute a significant overhead. Creating a highly automated intelligent BPS system that would substitute an experienced team leader might be possible in the future, but so far this goal has not been achieved in practice. In both cases, the team will, most probably, not have the same high level motivation and creativeness of the start-up team discussed in the previous section. The third option of having the team learn in details the language, culture, goals of each other may not be working in an environment of a bigger enterprise. Most employees may prefer to stick to the profession they have chosen and consider learning the other language, ways of thinking, etc. as excessive, and not in line with their chosen career path. In other words to reproduce the atmosphere of a start-up in a big company could be quite difficult.

So, could we build a BPS system that would help in creating of self-managed cross-professional teams in a situation as expressed in the diagram of Fig. 4? We believe that it might be possible. Such a BPS system should be based on the shared spaces architecture as discussed in [9]. However, as was discussed in Section 1, having the content expressed in the professional language, even if visualized, would create islands of the shared space that are understandable only to some members of the process team. Thus, the challenge that needs to be overcome is how to create a common and understandable for all participants visual presentation of the process instance goals, progress, plans, etc. This presentation should show the state of the affairs in the given instance/case so that it is clear what is happening and where a given member can contribute directly, or soon. This common view does not need to show all professional details, they can still be represented in professional languages, but should show sufficient information for a team become self-managed with the minimum level of explicit management.

4 A Proposed Solution

4.1 Game Technology as a Help

To reduce the language and cultural barriers for the cross-professional cooperation in the frame of business processes, we suggest using visualized metaphors based on folk and/or fairy tales well known to all participants. More specifically, we suggest using metaphors that realize the quest plot as they are consistent with the goal orientation of business processes. To such a plot belong, for example, treasure hunting, or sleeping beauty/sleeping princess tales.

Assume that we can translate the happenings inside a business process instance into the analogous happening in a story with a quest plot familiar to all participants. Assume further that we can visualize the tale in the same manner and using the same technique as is used for interactive game creation. Then, it would be easier for all engaged to understand what is happening in the process, even if each participant does not understand all technical details of actions completed by others, which will not be presented in the fairy tale plot visualization. Such visualization will also appeal to the next generation of workers coming to the working life, as they have been growing up with interactive games and team spirit of playing them.

4.2 Example: Consultative Sales Process vs. Treasure Hunting

We will use an example to clarify further our suggested solution. As an example of a business process, we have chosen a consultative sales process that requires the cross-professional cooperation. As an example of a fairy tale plot, we will use treasure hunting.

Consultative sales is a well-known process for which there exist many publications of both literature for practitioners and academic works, see for example [10]. A consultative sales process is aimed at closing a deal that concern a complex customized solution to be delivered by a vendor to a potential customer. It includes finding where in the customer organization there are needs that can be satisfied by a vendor solution, identifying the needs, sketching a solution and negotiating a deal. This process needs cooperation of people of different professions, sales people, business analysts, solution engineers, lawyers.

Treasure hunting is a popular plot that is used in books, films, interactive games, etc. It may take different forms: concrete - when the treasure is a chest with gold coins or gems, or abstract - when the treasure is something valuable but not converted to money. The plot may include travel, for example, by a ship to a treasure island, locating the possible place for digging, digging to the treasure, lifting it up, and safely coming home with it.

If we compare the general outline of the consultative sales and treasure hunting, it is easy to see the similarity and find correspondence between the activities in both. On a high level, the correspondence can be expressed as in Fig. 5. The similarity can be found not only on the highest level, but also in the details. For example, both processes are not sequential, as could be imagined from the simplified diagram of Fig. 5. The treasure hunters may do a lot of digging before the treasure could be located. For example, this may happen when they get information on the location of the treasure that is somewhat vague. The same is true for the consulting sales. First, there could be only a suspicion on that some particular customer department needs a product/service that could be delivered by the vendor. This suspicion needs to be proved or disproved by more thorough investigation whether the needs really exist and are not satisfied by already existing products and services.

Consultative Sales Process

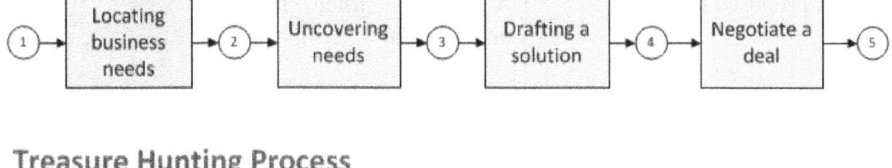

Treasure Hunting Process

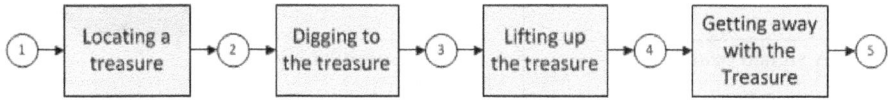

Fig. 5. Correspondence between Consulting Sales and Treasure hunting

Complementing Fig. 5 is the correspondence between the sub-goals achieved in each phase presented in Table 1, where numbers in the first column correspond to the numbers in circles of Fig. 5. The third column gives an example how the progress in achieving a sub-goal can be visualized. The process starts with getting a map of the island, and not a very good one. Locating a treasure can include making the map more detailed and ringing the areas of potential treasure hiding places. Digging to the treasure could be represented as the growing depth of the shaft down to the treasure. The progress in lifting the treasure can be represented by the position of the chest with treasure in the shaft. Getting away with the treasure can be represented as a position of the ship on the way home.

Table 1. Correspondence between the sub-goals in consultative sales and treasure hunting

#	Consultative sales	Treasure hunting	Representing the progress
1	Info about the customer planning to invest in a solution of a given type	A mystical map of the island, a rumor about pirates' treasure hidden in an island	
2	Department/unit/person in the company who owns the problem to be solved	A cross on the map where to start digging for a treasure	

Table 2. (*Continued*)

3	A document describing the needs/problem/requirements on the solution	A shaft going down to the treasure
4	A proposed solution	A treasure lifted from underground
5	A signed deal	A treasure safely at home

In the same way as Fig. 5, it is possible to establish the correspondence between other concepts of both processes. Table 2 shows an example of correspondence between the professions of a consultative sales process team and a treasure hunting team. Table 3 shows an example of correspondence between the actors at the potential customer site and inhabitants of the treasury island. Besides these two groups of people related to the process, there are a number of others, for example competitors. The latter can be fair play competitors, or the ones that can be represented by bandits in the treasury island story (though such an allusion may need to be considered from the ethical point of view).

Table 2. Professions involved in consultative sales and treasure hunting

Professions in Consultative Sales	Activities in Consultative Sales	Professions in Treasure Hunting	Activities in Hunting
Sales person	Charting a map of the organization and locating business needs	Surveyor, Prospector	Detailing the map and locating the treasure
Business analyst	Understanding business needs discovering requirements	Digger, Driller	Digging/drilling a shaft to the treasure
Solution Engineer	Drafting a solution to offer	Lifter	Lifting the treasure
Negotiator (lawyer)	Negotiating a deal	Captain of the ship	Navigating the ship home

Table 3. Counter partners

Customer Actors		Inhabitant of the treasure island
CEO	⇔	King of the island
Middle Manager	⇔	Local tribe chief
Field employees	⇔	Inhabitants

Besides mapping the activities and professions, roles related to both process we can relate other things, e.g., tools - by mapping methods of uncovering customer needs to digging tools, a spade or excavator, or drilling machine. The tool to be used depends on the ground, the more difficult the ground is the heavier, or more precise the tools should be, see Table 4 for an example. In the same way methods of negotiating contracts could be mapped to the types of ships to take the treasure home safely.

Table 4. Mapping business analysis methods to shaft digging/drilling tools

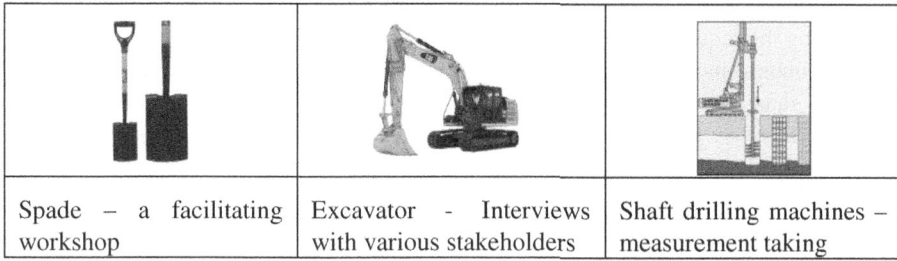

Spade – a facilitating workshop	Excavator - Interviews with various stakeholders	Shaft drilling machines – measurement taking

Based on the limited investigation above, we have shown that, in principle, the consultative sales business process can be mapped into a treasure hunting story. This allows to represent the state of the affairs, the progress, speed, achievements in the sales through visualizing the correspondent development in treasure hunting. Such visualization includes maps, progress in digging shafts, speed in which the digging goes, tools involved, etc. Such visualization would be easy to understand for any member of the process team. From it, it could be clear where the progress goes too slow and help is needed, and where a sub-goal could be reached soon, which requires planning of the next step. The process team of consultative sales could use this visualization to become self-manage/self-directed. This, naturally, will not happen automatically. A BPS system with the visualization as described is only a means to attain the goal, there need to be the will of the process participants to function as a self-managed team.

4.3   Summary of the Idea

Summarizing the above, let us consider the main characteristics of a tale with the quest plot [6]:

- Moving in a physical space in order to reach a goal
- Hinders that comes and goes when moving in the space towards the goal
- Cooperation between different roles

- Presence of third-parties that can be helpers or obstructers

If we look at a business process from the perspective of the state-oriented view [11], we get the same characteristics, except that the movement is not in the physical space, but in an abstract multidimensional state space. Therefore, by mapping (a) the abstract process state space into physical space of a tale and (b) different roles related to the process into the roles of the story, we can create a metaphoric representation of a business process. By visualizing this representation using the game technology we can create a BPS system of a completely different kind than a current generation of BPS systems.

There are major advantages of using the idea as above:

1. More engaging game-like computer environment instead of somewhat boring text-based representation of the progress in the process. This would appeal to the coming generation of workers.
2. The quest plot is much used in the interactive multi-user games, including treasure hunting games. This allows exploiting existing technology, e.g. game engines, for building the required visualization
3. More understandable presentation of the goals, state of the affairs and the work of others than the one done in the professional jargon of others professions. This would increase the team spirit and help to overcome cross-professional language and cultural barriers, and thus make it possible for a business process team to become self-managed/self-directed.

5 Related Research

The use of game technology for other purposes than entertainment has long been exploited by both researchers and practitioners. This area is called gamification, and it is defined as "the use of design elements characteristic for games in non-game contexts" [12]. This concept has been applied in different domains such as productivity, finance, health, education, sustainability, news and entertainment media [12], and recently it started to get more attention from the researchers, see, for example, [13,14].

There are a number of works that aim at using the gaming technology in BPM in different ways. For example, [15] enhances a process elicitation methodology using 3D simulation techniques of the workplace. [16] focuses on Games for Health by explaining the areas and approaches to developing these games. [17] explores the use of game elements to present orientation information to new students at the university. [18] shows that using the gamification approaches can yield significant improvements such as enjoyment, flow or perceived ease of use through a SAP ERP case. However, to the best of our knowledge, none of the existing works suggests using gaming technology in the way presented in this paper.

The term gamification is being used differently by different researchers, and it is hard to determine whether a work belongs to this category or not. [12] suggests a simple categorization of the works related to gaming to help positioning different works and products. This categorization is illustrated in Fig. 6. In it, different works and products can be categorized based on two dimensions, i.e. *Whole/Parts* and *Playing/Gaming*. We consider that our approach lies in the left upper corner *Whole/Gaming*, though it cannot be called serious game, as the plot is developing in the real world and time. In the world of entertainment, our approach corresponds to the Alternative Reality Games (ARG) [19].

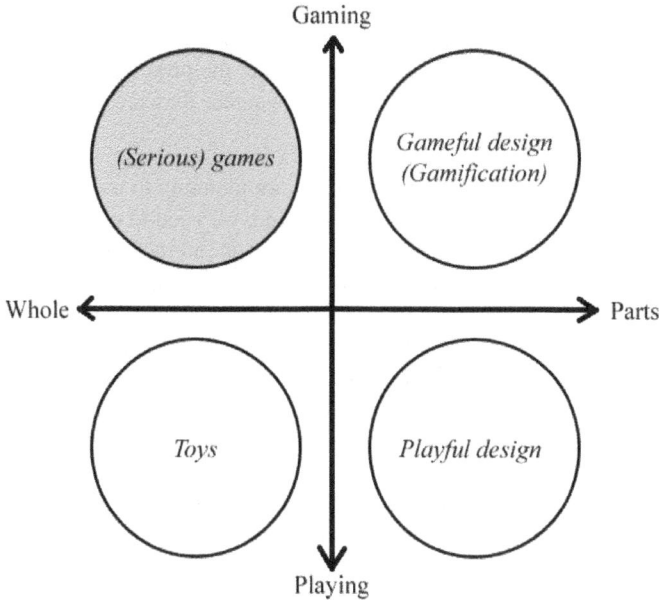

Fig. 6. Categories of approaches related to gaming [12]

6 Conclusion

This paper exploits the idea of attaining a new kind of value from BPS systems – bringing to the mature organizations the spirit of the start-ups. This would be done by bringing the game technology in the area of BPM. This technology however is suggested to be used in a different way than the one already under exploitation. The current attempts either use serious games for education, or gaming elements in ordinary BPS systems. Our suggestion is about representing the whole business process as a folk or fairy tale story as it develops in real time and space. The nearest correspondence to our approach in the world of entertainment is not an ordinary game with a built-in fixed or half-fixed plot, but an alternative reality game, where the plot is self-developing.

The described in this paper idea is only a hypothesis, which could be proved to be wrong, but it is of the nature of worth testing. In case of success, it can be become a foundation of the "next generation" of BPS systems that could appeal to the next generation of professionals, i.e. so-called digital natives who grew up with interactive gaming. It is highly improbable that this generation will silently accept the boring old-fashion BPS systems currently in use when they become the majority.

To test the hypothesis, we are working on:

1. Completing the mapping of the world of consultative sales into the world of treasure hunting.
2. Building and evaluating a full or partial prototype of a new kind of BPS using an existing game, or virtual reality engine. The evaluation is planned to be of the expert evaluation type. We will present the idea and the prototype or parts of it to the people engaged in consultative selling to see whether they are prepared to use such a system, if it has been built.

As far as building a prototype is concerned, we are planning to use one of the gaming engines, like [20]. To be able to use such an engine, we need to translate the terminology of the business processes in the terminology of interactive games. One of the main terms that is used in gaming engine is "quest", which identify a task completing of which would progress the game. Examples of quests include: *Deliver Items*, *Gather or produce Items*, *Defeat monsters* (to obtain items), *Reach a place*, *Guard an area*, *Survive for a time limit*, *Answer a question*, *Put an item on a place*, etc. [21].

In a single player game, the quest is received from Non-Player Character (NPC), which can be treated as an analogous for a manager controlling the given business process instance. In a multiplayers game, the quest can be obtained not only from NPC, but also from other players that can cooperate, compete, or trade [21]. These techniques are of special interest for our research as they can be used as communication mechanisms to coordinate the efforts of the process participants in attaining the common goal. Thus our additional goal is investigating the mechanisms of arranging interactions in the game world, especially the cooperative ones, to see which of them could be suitable for our purpose.

References

1. Hammer, M., Champy, J.: Reengineering the corporation: A manifesto for business revolution. HarperBusiness, New York (1993)
2. Swenson, K. (ed.): Mastering the Unpredictable: How Adaptive Case Management Will Revolutionize the Way That Knowledge Workers Get Things Done. Meghan-Kiffer Press, Tampa, USA (2010)
3. Bider, I., Johannesson, P., Perjons, E.: A strategy for merging social software with business process support. In: Muehlen, M., Su, J. (eds.) BPM 2010 Workshops. LNBIP, vol. 66, pp. 372–383. Springer, Heidelberg (2011)
4. Bider, I., Johannesson, P., Schmidt, R.: Experiences of using different communication styles in business process support systems with the shared spaces architecture. In: Mouratidis, H., Rolland, C. (eds.) CAiSE 2011. LNCS, vol. 6741, pp. 299–313. Springer, Heidelberg (2011)
5. Duimering, P., Robinson, R.: Situational influences on team helping norms: Case study of a self-directed team. Journal of Behavioral Management 9(1), 62–87 (2007)
6. Booker, C.: The Seven Basic Plots: Why We Tell Stories. Continuum, London (2004)

7. Bider, I., Kowalski, S.: A framework for synchronizing human behavior, processes and support systems using a socio-technical approach. In: Bider, I., Gaaloul, K., Krogstie, J., Nurcan, S., Proper, H.A., Schmidt, R., Soffer, P. (eds.) BPMDS 2014 and EMMSAD 2014. LNBIP, vol. 175, pp. 109–123. Springer, Heidelberg (2014)
8. Polanyi, M.S.: Knowing and Being. University of Chicago, Chicago (1969)
9. Bider, I., Johannesson, P., Perjons, E.: Justifying ACM: Why do we need a paradigm shift in BPM. In: Empowering Knowledge Workers. New ways to leverage CAse Management. Future strategies, Lighthous Point, Florida, US, pp. 67–78 (2013)
10. Moncrief, W.C., Marshall, G.W.: The evolution of the seven steps of selling. Industrial Marketing Management **34**, 13–22 (2005)
11. Khomyakov, M., Bider, I.: Achieving Workflow Flexibility through Taming the Chaos. In: OOIS 2000 - 6th International Conference on Object Oriented Information Systems, pp. 85–92 (2000)
12. Deterding, S., Dixon, D., Khaled, R., Nacke, L.: From game design elements to gamefulness: defining "Gamification". In: MindTrek, Tampere, Finland (2011)
13. Deterding, S., O'Hara, K., Sicart, M., Dixon, D., Nacke, L.: Gamification: using game design elements in non-gaming contexts. In: CHI EA (2011)
14. McGonigal, J.: Reality Is Broken: Why Games Make Us Better and How They Can Change the World. The Penguin Press (2011)
15. Brown, R., Rinderle-Ma, S., Kriglstein, S., Kabicher-Fuchs, S.: Augmenting and assisting model elicitation tasks with 3D virtual world context metadata. In: Meersman, R., Panetto, H., Dillon, T., Missikoff, M., Liu, L., Pastor, O., Cuzzocrea, A., Sellis, T. (eds.) OTM 2014. LNCS, vol. 8841, pp. 39–56. Springer, Heidelberg (2014)
16. McCallum, S.: Gamification and Serious Games for Personalized Health. In: pHealth (2012)
17. Fitz-Walter, Z., Tjondronegoro, D., Wyeth, P.: Orientation passport: using gamification to engage university students. In: 23rd Australian Computer-Human Interaction Conference (2011)
18. Herzig, P., Strahringer, S., Ameling, M.: Gamification of ERP Systems - Exploring Gamification Effects on User Acceptance Constructs (2012)
19. Szulborski, D.: This is not a game: a guide to alternate reality gaming. New Fiction Publishing (2005)
20. Unity Technologies: Unity - Game Engine. http://unity3d.com/
21. Oh, G., Y., K.: "Effective quest design in MMORPG environment" - presentation at the Game Developers Conference 2005, March 2015. http://ielab.ajou.ac.kr/ielab/professor/gdc2005

Human Centric Paradigms

Mining the Organisational Perspective in Agile Business Processes

Stefan Schönig[1(✉)], Cristina Cabanillas[2], Stefan Jablonski[1], and Jan Mendling[2]

[1] University of Bayreuth, Bayreuth, Germany
{stefan.schoenig,stefan.jablonski}@uni-bayreuth.de
[2] Vienna University of Economics and Business, Vienna, Austria
{cristina.cabanillas,jan.mendling}@wu.ac.at

Abstract. Agile processes depend on human resources, decisions and expert knowledge, and they are especially versatile and comprise rather complex scenarios. Declarative, i.e., rule-based, process models are well-suited for modelling these processes. Although there are several mining techniques to discover such declarative process models from event logs, they put less emphasis on the organisational perspective, which specifies how resources are involved in the activities. As a consequence, the resulting models do not specify who should execute which task and with which constraint (like separation of duties) in mind. In this paper, we propose a process mining approach to discover resource-aware, declarative process models. Our specific contribution is the extraction of complex rules for resource assignment that integrate the control-flow and organisational perspectives. Our experiments demonstrate the expressiveness of the mined rules with a reference to the Workflow Resource Patterns and a real-world use case.

Keywords: Declarative process mining · Organisational perspective · Resource perspective · Event log analysis

1 Introduction

The success of an organisation primarily depends upon its ability to accomplish its tasks in a structured and reliable manner. A well accepted method for structuring the activities carried out in an organisation is business process management (BPM). BPM usually involves modelling, executing and analysing processes [1]. In this context, two different types of processes can be distinguished [2]: well-structured routine processes with exactly predescribed control flow and agile processes with control flow that evolves at run time without being fully predefined a priori. Agile processes are common in healthcare where, e.g., patient diagnosis and treatment processes require flexibility to cope with unanticipated circumstances.

This work is funded by the "Europäischer Fonds für regionale Entwicklung" (EFRE) and the Austrian Research Promotion Agency (FFG) under grant 845638 (SHAPE).

In a similar way, two different representational paradigms can be distinguished: procedural models describe which activities can be executed next and declarative models define execution constraints that the process has to satisfy. The more constraints are added to the model, the less possible execution alternatives remain. As agile processes may not be completely known a priori, they can often be captured more easily using a declarative rather than a procedural modelling approach [3–5].

For purposes of compliance and process improvement, organisations are interested in the way their processes are actually executed [1]. Process mining aims at discovering processes by extracting knowledge from event logs, e.g., by generating a process model reflecting the behaviour recorded in the logs [6]. Declarative languages like Declare [3] or DPIL [7] can be used to represent these models, and tools like DeclareMiner [8] or MINERful [9] to generate them, often with a focus on control flow and data [10,11]. Agile processes, however, need to explicitly integrate the organisational perspective due to the importance of human decision-making and expert knowledge [12]. Recent research has identified the potential of role mining [13,14] and process mining of the organisational perspective [15]. However, these results have not yet been integrated with declarative process models.

In this paper, we fill this research gap by proposing a process mining approach to discover resource-aware, declarative process models. In particular, we are able to extract complex allocation rules, such as binding of duties between activities, as well as cross-perspective rules involving the control-flow and the organisational perspectives together. The latter consider the influence of resource allocation on the control flow of the process, e.g., an activity can only be executed by a specific role if a specific activity was performed previously. The expressiveness of the extracted rules has been evaluated using the Workflow Resource Patterns [16]. The applicability of the approach has been validated with a real-world event log from the university domain.

The remainder of this paper is structured as follows: In Section 2 we explain background upon which our approach is built. In Section 3 we describe our approach to extract resource-aware, declarative process models. In Section 4 we provide details about the implementation as well as experimental results. In Section 5 we present the results of the evaluations performed. In Section 6 we describe the related work and Section 7 concludes this paper.

2 Background

This section describes the language that we will use as a basis for mining models. We choose *Declarative Process Intermediate Language (DPIL)* [7] for this purpose due to several reasons. First, it is multi-perspective, i.e., it allows representing several business process perspectives, namely, control flow, data and resources. Since we want to extract resource-aware process models, the modelling language needs to support the modelling of rules related to the organisational perspective. The expressiveness of DPIL and its suitability for business process modelling have been evaluated [7] with respect to the well-known Workflow

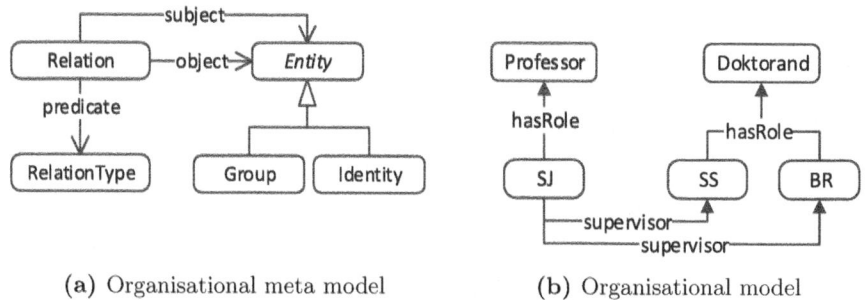

Fig. 1. Organisational meta model and example organisational model

Patterns [16]. Second, it is multi-modal, meaning that it allows defining two different types of rules: rules representing mandatory relations (called *ensure* in DPIL) and rules representing recommended relations (called *advice* in DPIL). The latter are useful, e.g., to reflect good practices.

In order to express organisational relations, DPIL builds upon a generic organisational meta model that has been described in [17] and is depicted in Fig. 1a. It comprises the following elements: *Identity* represents agents that can be directly assigned to activities, i.e., both human and non-human resources. *Group* represents abstract agents that may describe several identities as a whole, e.g., roles or groups. *Relation* represents the different relations (*RelationType*) that may exist between these elements. It is suitable to define, e.g., that an identity has a specific role, that a person is the boss of another person, or that a person belongs to a certain department. Fig. 1b illustrates an exemplary organisational model of a university research group. It is composed of two roles (Professor, Student) assigned to three people (SJ, SS, BR) and several relations between them indicating who is supervised by whom.

DPIL provides a textual notation based on the use of *macros* to define reusable rules. For instance, the *sequence* macro (*sequence(a,b)*) states that the existence of a *start event* of task b implies the previous occurrence of a *complete event* of task a; and the *role* macro (*role(a,r)*) states that an activity a is assigned to a role r. Fig. 2 shows an example of a process for trip management modelled with DPIL that uses the organisational model defined in Fig. 1b. It states that it is mandatory to approve a business trip before a flight can be booked. Moreover, it is recommended but not necessary that the approval be carried out by a resource with the role *Professor*.

In the work at hand, we use DPIL and build upon the described organisational meta model. However, please note that any other declarative process modelling language like Declare [3] or Dynamic Condition Response graphs [18] in combination with a suitable organisational meta model that fulfils the identified requirements can be used.

Applying process mining techniques it is possible to generate process models like the one depicted in Fig. 2, as long as the event logs contain the required

```
use group Professor

process BusinessTrip {
   task Book flight
   task Approve Application

   advice role(Approve Application, Professor)
   ensure sequence(Approve Application, Book flight)
}
```

Fig. 2. Process for trip management modelled with DPIL

information. Each event in a log refers to an activity, i.e., a well-defined step in the process, and is related to a particular process instance. We refer to the ordered set of events of a particular process instance as a *trace*. Event logs usually contain information about the resource performing an activity [6], as well as additional information that may be useful for subsequent analysis purposes. For instance, the following excerpt of a business trip process event log encoded in the XES logging format [19] shows the recorded information of the *start event* of an activity *Apply for trip* performed by a resource *SS*.

```
<string key="concept:name" value="SS_Riga2013"/>
<event>
   <string key="org:resource" value="SS"/>
   <date key="time:timestamp" value="2013-08-06T14:58:00.000+01:00"/>
   <string key="concept:name" value="Apply for trip"/>
   <string key="lifecycle:transition" value="start"/>
</event>
```

3   Mining Resource-Aware Declarative Process Models

In this section, we describe our approach to automatically discover resource-aware, multi-modal, declarative process models from event logs. First, we describe rule candidates and a support and confidence framework. Finally, we cover rule templates along with an improvement based on pre- and post-processing.

3.1   Generation and Checking of Rule Candidates

Declarative process modelling languages like DPIL are based on so-called *rule templates*. A rule template captures frequently needed relations and defines a particular type of rules. Templates have formal semantics specified through logical formulae and are equipped either with user-friendly graphical representations (e.g., in Declare) or with macros in textual languages (e.g., in DPIL) that make the model easier to understand. In contrast to concrete rules, a rule template consists of placeholders, i.e., typed variables. It is instantiated by providing concrete values for these placeholders. For instance, the model described in Section 2 makes use of two rule templates represented by the macros sequence(T_1,T_2) and role(T,G). These templates comprise placeholders of type *Task T* as well as *Group G*. In all well-known declarative process mining approaches, rule templates are used for querying the provided event log and to find solutions to the placeholders. A solution to the query is any combination of concrete values for the placeholders that yields a concrete rule that is satisfied in the event log. First, all possible rules need to be constructed by instantiating the given set of rule templates with all possible combinations of occurring process elements provided in the event log. The *sequence* template, e.g., consists of 2 placeholders of type *Task*. Assuming that $|T|$ different tasks occur in the event log, $|T|^2$ *rule candidates* are generated. All the resulting rule candidates are subsequently checked w.r.t. the event log.

In many cases, a rule candidate can be trivially valid. Consider, e.g., the rule candidate direct(t_1,i_1), i.e., start(of t_1) implies start(of t_1 by i_1), which holds when task t_1 is performed by an identity i_1, and the example event log of Table 1. The provided event log notation (first column) encodes the recorded start and complete events of a specific task t performed by an identity i with s(t,i) and c(t,i), respectively. The given events are ordered temporally so that timestamps are not encoded explicitly. In the first trace the rule holds trivially because t_1 never happens. Using the terminology of [20], we say that the rule is vacuously satisfied. It is necessary to discriminate between traces where a rule is trivially true and traces in which the rule is non-vacuously satisfied. Only traces in which a rule candidate non-trivially holds are considered interesting [8]. For first order logic rules that depict implications of the form $A \rightarrow B$ like in DPIL, trivially and non-vacuously valid rules can be discriminated by additionally checking the condition A of the rule separately.

Table 1 also shows the results of checking the non-vacuous satisfaction of the direct(t_1,i_1) rule (third column) as well as its condition (second column) for each trace of the example event log. In the first trace the rule is not (non-vacuously) satisfied because t_1 is never started, i.e., the condition is $false$. The rule holds non-vacuously in the traces two to four while it is violated in trace five.

3.2 Support and Confidence Framework to Classify Rules

Checking rule candidates as described above provides for every candidate the number of instances, i.e., traces in the event log where it non-vacuously holds.

Table 1. Event log and satisfaction of exemplary rule and its condition

Trace	start(of t_1)	direct(t_1,i_1)
{s(t_2,i_1), c(t_2,i_2), s(t_3,i_1), c(t_3,i_1)}	false	false
{s(t_1,i_1), c(t_1,i_1), s(t_2,i_2), c(t_2,i_2), s(t_3,i_1), c(t_3,i_1)}	true	true
{s(t_1,i_1), c(t_1,i_1), s(t_3,i_3), c(t_3,i_3), s(l_2,i_2), c(t_2,i_2)}	true	true
{s(t_1,i_1), c(t_1,i_1), s(t_3,i_3), c(t_3,i_3), s(t_2,i_2), c(t_2,i_2)}	true	true
{s(t_1,i_4), c(t_1,i_4), s(t_3,i_1), c(t_3,i_1)}	true	false

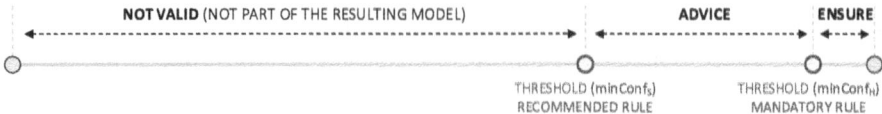

Fig. 3. Classification of rule candidates based on the confidence value

Based on these values it is possible to classify rules and to separate non-valid from valid ones. Therefore, [20] adopted a support and confidence framework proposed by association rule mining methods to evaluate the relevance of rule candidates.

Definition 1 (Support and Confidence). *Let $|\Phi|$ be the number of traces in an event log Φ. Let $|\sigma_{nv}(r)|$ be the number of traces in which a rule $r : A \rightarrow B$ is non-vacuously satisfied. The support $supp(r)$ and confidence $conf(r)$[1] values of a rule r are defined as:*

$$supp(r) := \frac{|\sigma_{nv}(r)|}{|\Phi|}, \qquad conf(r) := \frac{supp(r)}{supp(A)} \qquad (1)$$

Considering the event log of Table 1 and the $direct(t_1,i_1)$ rule, its support evaluates to $supp(r) = 0.6$ and its confidence to $conf(r) = 0.75$. The support value is used for pre-processing, as described in Section 3.4. We make use of the confidence value in order to classify a rule candidate r as (i) a *mandatory* rule, i.e., satisfied in almost all traces; (ii) a *recommended* rule, i.e., not always satisfied but with a tendency to be satisfied; or (iii) a *non-valid* rule, i.e., violated in most of the recorded traces. As visualized in Fig. 3, two thresholds $minConf_S$ and $minConf_H$ are introduced to classify rule candidates. Candidates r with $conf(r) > minConf_H$ are classified as mandatory (*ensure*) and $minConf_S < conf(r) < minConf_H$ as recommended (*advice*). All rule candidates r with $conf(r) < minConf_S$ are non-valid rules and are not part of the resulting process model. Using the confidence values of rule candidates it is directly possible to generate a DPIL process model reflecting the recorded behaviour.

3.3 Rule Templates for Analysing the Organisational Perspective

The previous section showed how it is possible to automatically generate a multi modal, declarative process model by checking a set of rule candidates whose structure is defined by rule templates w.r.t. a given event log. Since DPIL builds upon a flexible organisational meta model, it is possible to define rule templates that define the structure of organisational relations. By instantiating these resource-aware rule templates with all possible parameter combinations of

---

[1] In case of rules that do not depict implications, the condition is satisfied in every trace. Here $supp(A) = 1$ and $conf(r) = supp(r)$.

defined resources, groups and relation types, it is possible to generate rule candidates that focus on the organisational perspective of the process to be analysed. These candidates can then be checked under consideration of the corresponding organisational model. Based on the resulting confidence values, a resource-aware process model can be generated automatically.

We now define rule templates and their macros that can be used to mine the organisational perspective and to generate a resource-aware, declarative process model. We identified three different groups of organisational rule templates: resource allocation templates related to a single task, resource allocation templates related to more than one task, and cross-perspective rule templates, i.e., templates that express the influence of resources on the execution order of tasks. For every group we provide at least two representative examples that cover frequently needed organisational relations, according to the Workflow Resource Patterns [16]. Nonetheless, further rule templates can be defined individually according to the analyst's needs.

We first focus on rule templates that define *resource allocation* patterns, i.e., rules that specify the resources which are allowed to perform a certain task. The *direct allocation* of resources to a task can be extracted by analysing the *direct(T,I)* template. Given the free variables T and I and an event log with $|T|$ distinct tasks and $|I|$ distinct resources, there are $|T| \cdot |I|$ candidates to be checked.

```
direct(T,I) iff start(of T) implies start(of T by I)
```

Role-based allocation of resources can be identified with the *role(T,G)* template. Here, rule candidates for every task and group combination are generated, i.e., $|T| \cdot |G|$ rule candidates need to be checked.

```
role(T,G) iff start(of T by :p) implies
             relation(subject p predicate hasRole object G)
```

Organisational patterns can also relate to more than one task at the same time. The *binding(T_1,T_2)* template, e.g., can be used to discover if a task is always (mandatory) or should (recommended) be performed by the same resource as another task. With the *separate(T_1,T_2)* template, on the contrary, it is possible to discover task combinations that need to be or should be performed by different resources. For both templates, $|T|^2$ candidates need to be checked.

```
binding(T1,T2) iff start(of T1 by :p) and start(of T2) implies
                start(of T2 by p)

separate(T1,T2) iff start(of T1 by :p) and start(of T2) implies
                 start(of T2 by not p)
```

The *orgDist(T_1,T_2,RT)* template is used to discover relations that are defined in the organisational model between the resources that performed two different tasks. This template incorporates a variable RT for the different relation types in the considered organisational model. Like this, $|T|^2 \cdot |RT|$ rule candidates exist.

```
orgDist(T1,T2,RT) iff start(of T1 by :p1) and start(of T2 by :p2) implies
                       relation(subject p1 predicate RT object p2)
```

Moreover, the organisational perspective can affect the execution order of tasks, i.e., the control flow of the process. There may be processes in which a sequence between tasks only holds for specific resources or for resources with a specific role. These patterns can be discovered by the *roleSequence(T_1,T_2,G)* and the *resourceSequence(T_1,T_2,I)* templates, respectively. For these templates, $|T|^2 \cdot |G|$ and $|T|^2 \cdot |I|$ candidates need to be checked.

```
roleSequence(T1,T2,G) iff start(of T2 by :p at :t) and
                          relation(subject p predicate hasRole object G)
                          implies complete(of T1 at < t)

resourceSequence(T1,T2,I) iff start(of T2 by I at :t) implies
                              complete(of T1 at < t)
```

3.4 Pre- and Post-Processing

Real-life event logs and organisational models potentially contain a big set of distinct tasks, resources and groups. For instance, the BPI challenge 2011 event log [21] of a hospital information system contains 623 different tasks and 42 organisational groups. By only considering the *role* template, this already leads to $623 \cdot 42 = 26166$ candidates to be checked. Although many of these parameter combinations never occur together in the same trace, the corresponding rules need to be checked. This problem also becomes obvious when considering task/resource combinations of the event log in Table 1. The resource i_4 only occurs together with task t_1. Hence, candidates of the *direct* template where $I = i_4$ and $T \neq t_1$ are trivially true in all traces and can be neglected without checking. The method proposed by Maggi et al. [20] uses the well-known Apriori algorithm to pre-process the log and to extract *task combinations* that frequently occur together. A task combination is considered to be relevant if it occurs in a sufficient number of traces, i.e., above a given threshold *minSupp*. A *minSupp* of 5%, e.g., claims that only rule candidates are considered whose parameter combinations occur in at least 5% of the recorded traces. We extended this method in [22] to also extract *task/resource* and *task/group* combinations that frequently occur together. This way, it is also possible to dramatically reduce the number of organisational rule candidates by abstracting from infrequent parameter combinations. Hence, for the example log, only 1 out of 3 *direct(T,i_4)* candidates are generated and checked.

Furthermore, when automatically generating a declarative process model, there are potentially extracted rules that are redundant. Consider, e.g., that a specific task t_1 has always been performed by a resource i_1 who has a role g_1 according to the organisational model. Then, the proposed method will (inevitably) discover a *role(t_1,g_1)* rule. This rule is redundant, since a direct allocation rule *direct(t_1,i_1)* will also be discovered. In case of i_1 *hasRole* g_1 the *role* rule is already implied in the *direct* rule. Redundant rules complicate the understandability of discovered

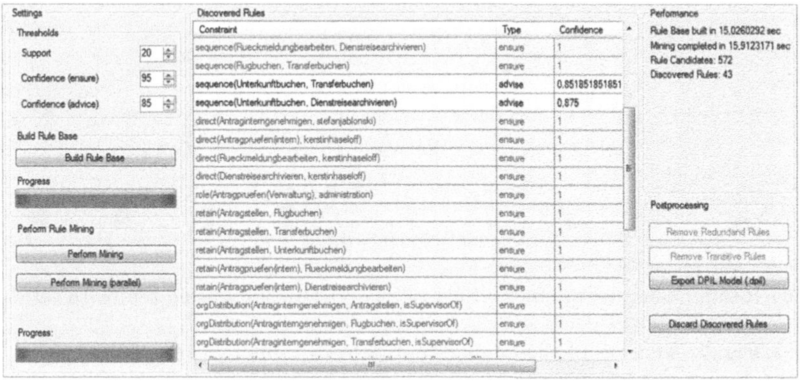

Fig. 4. User-interface of the DpilMiner during the analysis of an event log

models. Maggi et al. [23] proposed a technique to post-process a discovered model and to remove redundant, *weaker* rules if they are already implied in *stronger* rules only focusing on the hierarchy of control flow perspective templates. We extended this method to also consider the rule hierarchies of organisational rules. Redundancy may also be caused by the interplay of three or more organisational rules. Consider, e.g., a set of discovered *binding* rules, such as $binding(t_1,t_2)$, $binding(t_2,t_3)$ and $binding(t_1,t_3)$. Here, the rule between t_1 and t_3 is redundant because it belongs to the transitive closure of the other rules. In other words, if task t_1 has always been performed by the same resource as t_2, and t_3 has always been performed by the same resource as t_2, then also t_1 and t_3 have been performed by the same resource. Not all rule types can be reduced using transitive reduction. *Separate* rules, e.g., are not transitive, i.e., if t_1 is not performed by the same resource as t_2, and t_2 is not executed by the same resource as t_3, then we cannot conclude automatically that t_1 is also not performed by the same resource as t_3.

4 Implementation and Experiments

The problem of checking a large set of rule candidates can be solved by efficient pattern matching methods like the *rete algorithm* [24]. Instead of checking each rule separately, the rete algorithm first identifies common parts of the provided set of rules and constructs a *rete network*. Based on this decision network, common rule parts just need to be checked once. The JBoss Drools platform [25] provides a current implementation of this method. In order to check rule candidates with Drools, they are translated to the Drools Rule Language (DRL).

The described approach has been implemented in the *DpilMiner* application. Fig. 4 shows the DpilMiner user-interface and some discovered rules when analysing the application to an event log. To analyse performance and applicability, we applied the DpilMiner with different configurations using an event log

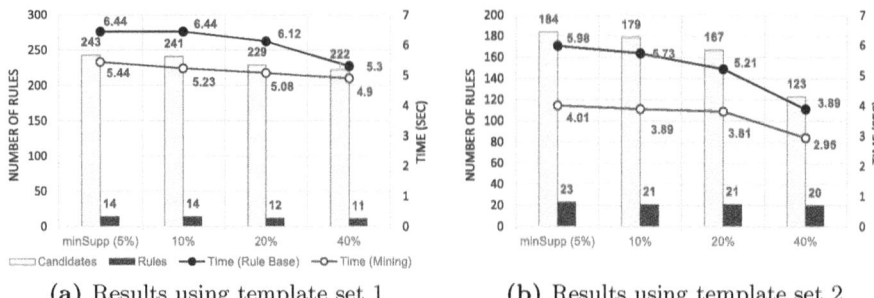

(a) Results using template set 1   (b) Results using template set 2

Fig. 5. Analysis of the mining approach with different template sets

of a university business trip management system[2] The log contains 2104 events of 10 different activities related to the application and the approval of university business trips as well as the management of accommodations and transfers, i.e., booking hotels, flights or trains. The system has been used for 6 months by 10 employees. In order to show the complete functionality of our approach, the underlying organisational model of the domain is required. The given organisational model assigns these identities to 4 distinct roles, specifically, 6 phd students, 1 professor, 1 secretary as well as 2 administration employees. In total, there are 128 business trip cases recorded in the log. All the computation times reported in this section are measured on a Core i7 CPU @2.80 GHz with 8 GB RAM. Our approach has been tested with two different sets of rule templates. Fig. 5 outlines the results. Fig. 5a shows the results of applying the approach with template *set* 1, which contains four different rule templates purely focusing on the organisational perspective, i.e., *direct*, *role*, *binding* as well as the *orgDist* template. The diagram in Fig. 5b shows the results for template *set* 2, which analyses the execution order of tasks under consideration of the organisational perspective, i.e., the templates *sequence* and *roleSequence*. We analysed the time to build the rete network, i.e., the rule base, as well as the time to perform the actual mining process under consideration of a different number of rule candidates. This was achieved by considering different *minSupp* values during the preprocessing phase. The analysis shows the feasibility of our approach since in both tests, despite a big amount of candidates, only a manageable number of rules has been discovered. Especially the diagram of *set* 2 additionally highlights the benefit of the pre-processing approach. With increasing *minSupp*, the number of rule candidates to check considerably decreases. However, almost the same number of rules has been discovered.

In order to evaluate and compare the performance of our approach, we also applied the ProM implementation of the DeclareMiner [26] by only analysing the *precedence* template of Declare [3], which equates to the *sequence* template of DPIL. With standard settings, the DeclareMiner needed 14.85 sec to analyse

---

[2] The event log is available for download at *workbench.kppq.de*.

Table 2. Overview of Workflow Resource Patterns that can be discovered

	Resource Pattern	DPIL Rule Template
✓	Direct Allocation	start(of T) implies start(of T by :p)
✓*	Role-Based Allocation	start(of T by :p) implies
		relation(subject p predicate hasRole object G)
⊘	Deferred Allocation	—
⊘	Authorization	—
✓	Separation of Duties	start(of T_1 by :p) and start(of T_2) implies
		start(of T_2 by not p)
✓	Case Handling	forall(task A start(of T) implies start(of T by :p)
✓	Retain Familiar	start(of T_1 by :p) and start(of T_2) implies
		start(of T_2 by p)
✓*	Capability-Based Allocation	start(of T by :p) implies
		relation(subject p predicate RT object G)
⊘	History-Based Allocation	—
✓*	Organisational Allocation	start(of T_1 by :p_1) and start(of T_2 by :p_2) implies
		relation(subject p_1 predicate RT object p_2)
★	Automatic Execution	invoke(T)

✓=Supported ✓*=Organisational model required ⊘=Non supported

the provided event log with the *precedence* template. Even if we analysed the example log with 2 respectively 4 rule templates, our approach was still faster in any case, as depicted in Fig. 5. Note that the rule base only needs to be built once for several different mining applications.

5 Evaluation

In the following section, we evaluate the expressiveness and the applicability of the described approach.

5.1 Discovering Creation Workflow Resource Patterns

We use the group of so-called *creation patterns* of the well-known Workflow Resource Patterns [16] to evaluate which of the resource allocation patterns can be discovered by our mining approach. The Workflow Resource Patterns have been used in several modelling approaches as a reference model to specify how resources take part in process activities. As described in Section 3, our approach is able to discover a particular resource pattern if it is possible to define a parametrized DPIL rule template that represents the pattern.

Table 2 shows an overview of the different resource patterns as well as the corresponding rule templates, when possible. As seen, the patterns *Direct Allocation*, *Role-Based Allocation*, *Separation of Duties*, *Binding of Duties* and *Organisational Allocation* are extractable with the rule templates *direct*, *role*, *separate*, *binding* and *orgDist*, respectively. Since event logs do not usually contain

```
ensure direct(Approve Application, SJ)
ensure role(Check Application, Administration)
ensure binding(Apply for trip, Book flight)
ensure binding(Apply for trip, Book accommodation)
ensure binding(Apply for trip, Book transfer)
ensure orgDist(Approve Application, Apply for trip, supervisor)
advice sequence(Apply for trip, Book flight)
ensure roleSequence(Apply for trip, Book flight, Student)
```

Fig. 6. Excerpt of the discovered business trip process model

information about the resource assignment mechanism used, it is not possible to extract if the resource assignment has been deferred to run time, i.e., the *Deferred Allocation* pattern cannot be discovered. Since in DPIL it is not possible to specify relations that relate to other process instances, the *History-Based Allocation* pattern cannot be discovered. The *Automatic Execution* pattern can only be discovered if the event log makes use of certain event types which indicate an automatic task processing. DPIL, e.g., defines invoke(T) events to call an automatic non-human service T. Even without considering invoke events our approach is able to discover 7 out of 11 resource creation patterns.

5.2 Case Study

In this section, we describe our findings when applying the approach to the university business trip event log that has already been described in Section 4. We analysed the event log with 6 different organisational rule templates comprising resource allocation patterns as well as resource influence on task execution order, i.e., cross-perspective patterns. With $minSupp = 10\%$ in the pre-processing phase and after removing redundant rules in the post-processing phase, we extracted 34 rules in total. The extracted resource allocation patterns are composed of 4 *direct*, 1 *role*, 5 *binding* and 4 *orgDist* rules. The control flow related pattern set is composed of 14 *sequence* and 6 *roleSequence* rules. For the classification in mandatory and recommended rules, we used $minConf_H = 95\%$ and $minConf_S = 85\%$. For space reasons, we only describe some interesting parts of the resulting model in Fig. 6.

We first focus on interesting resource allocation patterns. The discovered model shows that task "Approve Application" has always been performed by the identity "SJ" (mandatory direct allocation). Furthermore, "Check Application" has always been performed by a resource with the role "Administration" (mandatory role-based allocation). The three binding of duties rules show that the resource who booked the flight, the accommodation as well as the transfer service has to be the applicant herself (mandatory binding of duties). The model additionally shows that the resource who approved the trip application has been the supervisor of the applicant (mandatory organisational distribution). Moreover, we also discovered the influence of resources on the execution order of tasks.

Although employees usually applied for the trip before they booked the corresponding flight, it is not mandatory (recommended task sequence). Nonetheless, there are cases in which certain employees already booked the flight without applying for the trip. However, when analysing the ordering of tasks under consideration of performing resources, we extracted that students always applied for the trip before they booked a flight (mandatory role-based sequence). While professors are free to book a flight without an approved application, students mandatorily have to stick to a certain order of tasks.

In order to evaluate the precision of the mining results, the model has finally been discussed in a workshop with process particpants. Here, 29 out of 34 rules have been identified as relevant while 5 rules have been classified as non-relevant. This leads to a precision value of 0.85.

6 Related Work

The work presented in this paper relates to two streams of research: declarative versus procedural process mining and mining of the organisational perspective of a process. Recently, several techniques for the automated discovery of declarative process models from event logs have been proposed. The DeclareMiner [8] and its enhancements aim to improve the mining performance [20] as well as the readability of discovered models [10,11,23]. Furthermore, efficient algorithms to discover Declare models are presented in [9,27,28]. The work on Dynamic Condition Response Graphs [18] proposes an alternative formalism. In essence, the focus of these approaches is control flow with extensions to cover data. Complementary to these papers are approaches on role mining [13,14] and process mining of the organisational perspective [15] that aim to make use of the rich information on who has been executing a particular task [11]. Mining methods for analysing event logs with respect to resource information are mainly focused on enriching a given procedural model with resource information [15], on extracting an underlying organisational model [29] or social network [30], or on analysing the influence of resources on process performance [31]. Approaches on role mining [13,14] are interested in separation of duty constraints. The research reported in this paper takes a step towards the integration of both streams and the mining of constraints that express resource assignments depending on control flow and vice versa. In this way, it has the potential for evolving to a useful tool for compliance management of agile processes.

7 Conclusions and Future Work

In this paper, we presented a process mining method to discover resource-aware and multi-modal, declarative process models. Our approach is based upon the textual Declarative Process Intermediate Language (DPIL), in which organisational relations of processes can be modelled. We proposed a set of rule templates that can be used to mine the organisational perspective and to generate a resource-aware declarative model. Our approach has been implemented in the

DpilMiner application. We analysed performance and tested its applicability using an event log of a university business trip management system. Moreover, we evaluated the expressiveness against the Workflow Resource Patterns.

Since our approach is based upon DPIL, the mining capabilities are limited to the expressiveness of the language. Hence, e.g., inter-case dependencies such as those in the History-Based Allocation pattern cannot be discovered. Furthermore, the analysis of certain rule templates leads to many discovered rules. The *separate* template, e.g., discovers all combinations of tasks, which are performed by different resources. It is in some cases difficult to manage the amount of rules. Hence, mechanisms to reduce the rule set must be explored. Finally, we are currently evaluating possibilities to map the DPIL to existing graphical process modelling notations, such as RALph [32]. This will increase the understandability of the resulting process models for systems analysts.

References

1. Dumas, M., Rosa, M.L., Mendling, J., Reijers, H.A.: Fundamentals of Business Process Management. Springer (2013)
2. Jablonski, S.: MOBILE: a modular workflow model and architecture. In: Working Conference on Dynamic Modelling and Information Systems (1994)
3. van der Aalst, W., Pesic, M., Schonenberg, H.: Declarative workflows: Balancing between flexibility and support. Computer Science - Research and Development **23**(2), 99–113 (2009)
4. Pichler, P., Weber, B., Zugal, S., Pinggera, J., Mendling, J., Reijers, H.A.: Imperative versus declarative process modeling languages: an empirical investigation. In: Daniel, F., Barkaoui, K., Dustdar, S. (eds.) BPM Workshops 2011, Part I. LNBIP, vol. 99, pp. 383–394. Springer, Heidelberg (2012)
5. Vaculín, R., Hull, R., Heath, T., Cochran, C., Nigam, A., Sukaviriya, P.: Declarative business artifact centric modeling of decision and knowledge intensive business processes. In: EDOC, pp. 151–160 (2011)
6. van der Aalst, W.: Process mining: discovery, conformance and enhancement of business processes. (2011)
7. Zeising, M., Schönig, S. Jablonski, S.: Towards a common platform for the support of routine and agile business processes. In: Collaborative Computing: Networking, Applications and Worksharing (2014)
8. Maggi, F.M., Mooij, A., van der Aalst, W.: User-guided discovery of declarative process models. In: Computational Intelligence and Data Mining, pp. 192–199 (2011)
9. Di Ciccio, C., Mecella, M.: A two-step fast algorithm for the automated discovery of declarative workflows. In: Computational Intelligence and Data Mining, pp. 135–142 (2013)
10. Maggi, F.M., Dumas, M., García-Bañuelos, L., Montali, M.: Discovering data-aware declarative process models from event logs. In: Daniel, F., Wang, J., Weber, B. (eds.) BPM 2013. LNCS, vol. 8094, pp. 81–96. Springer, Heidelberg (2013)
11. Bose, R.P.J.C., Maggi, F.M., van der Aalst, W.M.P.: Enhancing declare maps based on event correlations. In: Daniel, F., Wang, J., Weber, B. (eds.) BPM 2013. LNCS, vol. 8094, pp. 97–112. Springer, Heidelberg (2013)

12. Marin, M., Hull, R., Vaculín, R.: Data Centric BPM and the Emerging Case Management Standard: A Short Survey Case Management, vol. 257593
13. Leitner, M., Baumgrass, A., Schefer-Wenzl, S., Rinderle-Ma, S., Strembeck, M.: A case study on the suitability of process mining to produce current-state RBAC models. In: La Rosa, M., Soffer, P. (eds.) BPM Workshops 2012. LNBIP, vol. 132, pp. 719–724. Springer, Heidelberg (2013)
14. Baumgrass, A., Strembeck, M.: Bridging the gap between role mining and role engineering via migration guides. Inf. Sec. Techn. Report **17**(4), 148–172 (2013)
15. Zhao, W., Zhao, X.: Process mining from the organizational perspective. In: Foundations of Intelligent Systems, pp. 701–708 (2014)
16. Russell, N., van der Aalst, W.M.P., ter Hofstede, A.H.M., Edmond, D.: Workflow resource patterns: identification, representation and tool support. In: Pastor, Ó., Falcão e Cunha, J. (eds.) CAiSE 2005. LNCS, vol. 3520, pp. 216–232. Springer, Heidelberg (2005)
17. Bussler, C.: Organisationsverwaltung in Workflow-Management-Systemen. Dt. Univ.-Verlag (1998)
18. Hildebrandt, T.T., Mukkamala, R.R., Slaats, T., Zanitti, F.: Contracts for cross-organizational workflows as timed dynamic condition response graphs. J. Log. Algebr. Program. **82**(5–7), 164–185 (2013)
19. Verbeek, H.M.W., Buijs, J.C.A.M., van Dongen, B.F., van der Aalst, W.M.P.: XES, xESame, and ProM 6. In: Soffer, P., Proper, E. (eds.) CAiSE Forum 2010. LNBIP, vol. 72, pp. 60–75. Springer, Heidelberg (2011)
20. Maggi, F.M., Bose, R.P.J.C., van der Aalst, W.M.P.: Efficient discovery of understandable declarative process models from event logs. In: Ralyté, J., Franch, X., Brinkkemper, S., Wrycza, S. (eds.) CAiSE 2012. LNCS, vol. 7328, pp. 270–285. Springer, Heidelberg (2012)
21. Bose, R.P.J.C., van der Aalst, W.M.P.: Analysis of patient treatment procedures. In: Daniel, F., Barkaoui, K., Dustdar, S. (eds.) BPM Workshops 2011, Part I. LNBIP, vol. 99, pp. 165–166. Springer, Heidelberg (2012)
22. Schönig, S., Gillitzer, F., Zeising, M., Jablonski, S.: Supporting rule-based process mining by user-guided discovery of resource-aware frequent patterns. In: ICSOC 2014 Workshops (2014) (in press)
23. Maggi, F.M., Bose, R.P.J.C., van der Aalst, W.M.P.: A knowledge-based integrated approach for discovering and repairing declare maps. In: Salinesi, C., Norrie, M.C., Pastor, Ó. (eds.) CAiSE 2013. LNCS, vol. 7908, pp. 433–448. Springer, Heidelberg (2013)
24. Forgy, C.: Rete: A fast algorithm for the many pattern/many object pattern match problem. Artificial Intelligence **19**(1), 17–37 (1982)
25. JBoss Drools, T.: JBoss Drools Documentation - Chapter 7: Rule Language Reference (2013)
26. Maggi, F.M.: Declarative process mining with the declare component of ProM. In: BPM (Demos) (2013)
27. Westergaard, M., Stahl, C., Reijers, H.: UnconstrainedMiner: Efficient Discovery of Generalized Declarative Process Models. BPM Center Report, No. BPM-13-28 (2013)
28. Di Ciccio, C., Maggi, F.M., Mendling, J.: Discovering target-branched declare constraints. In: Sadiq, S., Soffer, P., Völzer, H. (eds.) BPM 2014. LNCS, vol. 8659, pp. 34–50. Springer, Heidelberg (2014)
29. Song, M., van der Aalst, W.: Towards comprehensive support for organizational mining. Decision Support Systems (2008)

30. Van Der Aalst, W.M., Reijers, H.A., Song, M.: Discovering Social Networks from Event Logs. CSCW **14**(6), 549–593 (2005)
31. Nakatumba, J., van der Aalst, W.M.P.: Analyzing resource behavior using process mining. In: Rinderle-Ma, S., Sadiq, S., Leymann, F. (eds.) BPM 2009. LNBIP, vol. 43, pp. 69–80. Springer, Heidelberg (2010)
32. Cabanillas, C., Knuplesch, D., Resinas, M., Reichert, M., Mendling, J., Ruiz-Cortés, A.: RALph: A Graphical Notation for Resource Assignments in Business Processes. In: CAiSE 2015, LNCS, vol. 9097, pp. 53–68. Springer, Heidelberg (2015)

Changing the Focus of an Organization: From Information Systems to Process Aware Information Systems

Andrea Delgado[✉] and Daniel Calegari[✉]

Instituto de Computación, Facultad de Ingeniería, Universidad de la República,
Julio Herrera y Reissig 565 11300, Montevideo, Uruguay
{adelgado,dcalegar}@fing.edu.uy

Abstract. Organizations that manage their Business Processes (BPs) poorly -or that not manage them at all- have well-known problems regarding their operation, both at the BPs and the Information Systems (IS) levels. Some of these problems are due to a vertical and functional vision of the organization, without any global BP vision. In this context, similar decentralized organizational units very often perform the same BPs sometimes in different ways. Moreover, some BPs are implicit in the IS supporting them. In this article we present an experience report of a BPM pilot project we have carried out within our university, as an initiative to improve BPs management and corresponding IS support. We started specifying a process map for management support BPs, and then we selected key BPs which where specified, modeled and implemented using BPMN 2.0 and Bonita BPMS, to shift the organization focus from traditional IS to Process Aware IS (PAIS).

Keywords: Business Process Management (BPM) · Process Aware Information Systems (PAIS) · Business Process Management Systems (BPMS) · BPMN 2.0

1 Introduction

Organizations that manage their Business Processes (BPs) poorly -or that not manage them at all- have well-known problems regarding their operation, both at the BPs and the Information Systems (IS) levels. Some of these problems are due to the size of these organizations, which have many organizational units, some of them decentralized and sometimes with duplicated responsibilities, carrying out the same BP in different units in a different manner and maybe with a different name. BP variants are mostly unknown to the organization, not only regarding the activities each one entitles, but also the conditions or variation points that determine the different variants. Also, the fact that most BPs are implicit in IS, supported by menu options with no sequence defined among them and no explicit conditions defined to perform one or another, make it difficult to identify the control flow of the BPs. In most cases, these organizations rely on people knowledge about BPs or at least regarding the activities of the organizational unit to which they belong (without complete knowledge of the

BP in the organization). This lack of global vision and explicit knowledge of the BPs throughout the organization prevents it to assess its daily operation, and to identify improvement opportunities, since BP measures and objectives are also missing.

The university of which we are part of, Universidad de la República (Udelar) from Uruguay, suffers from most of these problems, due to characteristics such as its size regarding faculty members and students, its organizational structure and the fact of having organizational units decentralized, among others. Despite having an Management Improvement Group aimed at defining and taking actions to improve management throughout the university, there is no deep knowledge of Business Process Management (BPM) [1,2] nor of technological support such as BPM Systems (BPMS) at the university general level. Although BPM is recognized as a way of working with explicitly defined BPs, there is no knowledge of even simplest definitions such as what a BP is, e.g. they identify common functions such as accounting and human resources management as processes, which are clearly areas composed of many different processes. In terms of an organizational maturity model such as the Business Process Maturity Model (BPMM) [3], the university can be seen as mainly at Level 1: Initial - where BPs are performed in inconsistent and sometimes ad hoc ways, with results that are difficult to predict. Having detecting the many problems that the university has regarding its characteristics and context presented above, and with the main goal of improving managerial efforts, the Management Improvement Group defined several lines of action in 2012, regarding existing human resources, the current organizational structure and the BPs that are carried out within the university.

In this article we present an experience report from a process management improvement project we have carried out as part of these lines of action, regarding the management support BPs carried out within the university. For doing so an interdisciplinary group was created integrating business and software visions and knowledge. The project was recently finished and comprises two years with focus on: firstly the definition of a process map of the university management area to have a global view of their BPs (Phase 1), and secondly a methodological experimentation with an organizational unit in order to identify and specify selected key BPs as well as to show technological options to support them (Phase 2). Both phases were steps for leading to a real adoption of BPM in the university, as part of a more general and longer project which includes the adoption of a methodology for BPM, the evaluation and acquisition of a BPMS, the specification and implementation of all identified BPs and the replication of the experience in other organizational units.

As support methodology we used the Business Process Continuous Improvement Process (BPCIP) [4] which provides a guide for carrying out and integrating improvement efforts in the organization, which was defined within our research group. It extends the traditional BP lifecycle [2] with specific measurement and improvement elements. We selected this methodology since it was proposed within our research group so we were familiar with the disciplines, activities and roles it defines, and as it was already validated in the context of a PhD thesis by means of a case study within a hospital in Spain, which is also an institution which presents the characteristics mentioned above. Main organizational elements we analyze are: (1) although BPCIP was validated within a similar institution, it has never been used before in an educational institution and this one has no previous knowledge of BPM and its technological support (2) since the authors are part of the university (not external consultors) and were also part of the team which carried out the experience, we worked together with business people providing our own knowledge of the institution, and (3) the institution cultural context is a singular one since it is autonomous (from the government) and co-managed by its teachers, students and alumni.

The rest of this paper is structured as follows. In Section 2 we introduce the business process improvement project and its context, and in Section 3 we present its actual execution. In Section 4 we discuss lessons learned and reflections on the results we have obtained. Finally, in Section 5 we draw some conclusions and future work.

2 The Business Process Improvement Initiative

The Udelar university employs near ten thousands teachers (professors and lecturers) from all academic areas (Social, Medicine, Engineering, Architecture, Chemistry, Arts, etc.), more than six thousands of non-teaching staff, and serves near one hundred thousand students. It is composed of a central unit and near twenty decentralized schools which are grouped into conceptual areas (Health, Science and Technologies, Social and Humanities, etc.). Each school is in turn organized with its own structure regarding academic departments (i.e. Engineering School has institutes such as Computer Science, Electrical Engineering, Civil Engineering, etc.). Each school also has its own management structure composed of many administrative units such as admission office, human resources, accounting, building maintenance, library, etc., some of them coordinated by specific offices at the central unit. Udelar is the only public university of the country and its main educational institution regarding not only grade and postgraduate degrees, but also research. It is autonomous and co-managed by its teachers, students and alumni, a political system which although defended by all interested parties, adds many levels of discussion and delays in making decisions.

As a starting point of the BP improvement initiative, the main concern of this project was the definition of a process architecture [5] with respect to management support BPs. A process architecture is a conceptual model that shows the BPs of the organization and makes their relationships explicit. It has several levels of detail, as depicted in Fig. 1. Level 1 shows the main BPs at a very abstract level whereas Level 2 is composed of a refined version of such BPs, but still in an abstract way. In Level 3 we have a detailed version of those BPs, as for example using the Business Process Modeling Notation (BPMN) [6]. Level 1 presents the most important challenge for the definition of a process architecture since it must be understandable and sufficiently complete to be accepted as a description of the organization. In addition, three different types of BPs can be identified: strategy BPs which are performed by the organizations head to define objectives and plans, key BPs which contributes to the organization's mission regarding the defined objectives, and management support BPs which are those carried out to support the key ones, and subject of this work.

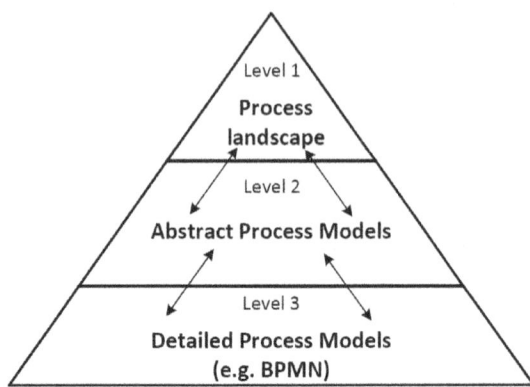

Fig. 1. The different levels of detail in a process architecture [5]

For the definition of such an architecture, we have followed the BPCIP [4] methodology, which provides a guide for carrying out improvement efforts in the organization, as mentioned above. It extends the traditional BP lifecycle [2] with measurement and improvement activities which involves the definition of a BP Execution Measurement Model (BPEMM) [4], a model that integrates execution measures for BPs realized by services in a comprehensive way. BPCIP consists of the same four phases as [2], from modeling a new BP or redesigning an existing one, to the evaluation of its real execution to identify improvement opportunities. We partially addressed the Design&Analysis phase with respect to the Business Modeling (BM) and BP Execution Measurement (EM) disciplines. The BM discipline aims to obtain a map of the organization and its BPs and to gain a better understanding of the business by representing their BPs explicitly as models. Moreover, the EM discipline sets out to show explicitly the execution measurement activities to perform in the extended BP lifecycle of BPCIP. In particular, we addressed the following activities within these disciplines.

- **BM1 - Asses the Organization**. The current state of the organization is described in terms of their current BPs, tools, people skills, customers, competitors, technological challenges, problems and areas of improvement, among others.
- **BM2 - Identify Business Processes**. To understand and describe the BPs in the organization, mainly those related to the application being developed, specifying the BPs models using BPMN 2.0.
- **EM1 - Select Execution Measures**. The execution measures are selected from BPEMM in order to define which data will be registered from the BP execution, to be able to analyze the execution.

2.1 The Design and Analysis Project

The project was divided in two one-year consecutive phases which are described next and detailed in Section 3:

- **Phase 1: Process map**. We mainly developed a process map identifying every management support BPs in the organization and categorizing them in conceptual areas, i.e. abstract categories relating BPs without any direct relation with a section of the organization to avoid ownership by definition. For each BP we also describe a very abstract data sheet including the process owner, participants, objectives, existent software support tools, among other aspects. This phase corresponds to the BPCIP's activity "BM1 - Asses the Organization" and provide us with the Level 1 and Level 2 description of our process architecture. The process map is intended to be a strong communication artifact for supporting the BPM initiative.
- **Phase 2: Specification of BPs**. We focused on a detailed specification of BPs (Level 3 models based on BPCIP's activity "BM2 - Identify Business Processes"). However, since the organization is not currently set for the adoption of BPM, we selected some priority BPs and conducted a pilot project. We included not only the detailed description of BPs but also the definition of execution measures (EM1 - Select Execution Measures) for the continuous improvement of those BPs and the development of functional prototypes as a way of showing existing technological alternatives. We also worked on a methodological guide, based on BPCIP, which allows replicating the pilot experience with other BPs.

2.2 Upcoming projects

There are three other phases projected for the BP improvement initiative:

- **Phase 3: Evaluation of Business Process Management Systems (BPMS)**. We need to evaluate different BPMS solutions fit to the organization. We plan to follow our systematic approach for evaluating BPMS [7]. This approach provides a list of key characteristics of BPMS which are ranked by the organization and evaluated using test cases and quantitative criteria. A mandatory requirement is to establish centralized BPM support and avoid the proliferation of management systems in such distributed organization. In this sense, the evaluation must be conducted together with software professionals from the University Central Informatics Service ((Servicio Central de Informática Universitaria, SeCIU). Although some software systems are locally used in the many divisions, SeCIU provides centralized software infrastructure and support to several management systems. Thus, their participation is a must for making key decisions for the whole Udelar.
- **Phase 4: BPMS acquisition and configuration**. The BPMS identified in Phase 3 as best suited for the organization must be acquired. Moreover, we plan to configure those BPs from Phase 2 and execute them. This will be the first complete application of BPCIP in Udelar which will allow us to adjust the methodology and take it as a basis for future projects.
- **Phase 5: BPCIP iteration**. This is a never ending phase in which concrete projects must be defined for the full application of BPCIP to other conceptual areas of the process map. This project will allow us to spread the BPM vision. Individual projects would be managed by the central group of people in charge of BPM.

3 Sowing the Seeds of BPM

The two-year project we present here comprises the first two phases defined which started in the second quarter of 2013. As mentioned above, it was sponsored by the Management Improvement Group and the main management authority within Udelar (Pro Rector de Gestión Administrativa), as well as other authorities. In addition to our participation providing the BPM methodological and software vision, the project team was also integrated with a researcher from the Faculty of Economics and Business Administration (Facultad de Ciencias Económicas y Administración) providing a purely management vision of BPM and BPs, a software professional from SeCIU providing the technological vision of the university's capacities; and two members from the management team of the main sponsor of the project (Pro Rector de Gestión Administrativa) providing the global vision for the results of the project. In the following we describe the execution of the project from the point of view of each BPCIP activity performed and the corresponding project phase.

3.1 Assessing the Organization

The focus of the first phase was set on defining a process map for the management support BPs of the Udelar, since the sponsor of the project was the management area they wanted to help identifying and organizing the BPs for the management structure. The steps defined in BPCIP to carry out this activity include understanding and specifying several aspects of the organization such as BPs context, technological context and human context, problems and improvement opportunities, and stakeholders for the business modeling effort. To do so, as the first activity of the project we carried out a workshop with administrative directors and organizational units responsible, presenting the general theory of BPM and the initial proposal of the project. Then working in groups of five or six people, we asked them to identify and named as many BPs as they recognize as possible, their responsibles, organizational units involved and the existence or not of software support. The result of this activity was a first coarse-grained identification of conceptual areas and their main mangement support BPs, which we wrap up together with the working groups.

To further detail and identify BPs with a finer granularity, we conducted several two hour interviews with the workshop participants at they workplaces, digging into the work each organizational unit performs within each defined BP, and identifying new ones. For doing so, we divided the project group into two interviewers groups, scheduling two interviews by week (one by each group on Tuesdays and Wednesdays) and putting together the results in a general meeting of the project group on Thursdays, where we analyzed and discussed the advances. Although the interviews were planned to be conducted within two months, they were actually carried out within the period July to October 2013, mainly due to difficulties in scheduling them with the required participants. As the main product of this activity and of Phase 1 of the project, a process map for the management support BPs of the university was generated, validated with the participants and the sponsors, and diffused within the organization, including a presentation in an internal management working day.

As mentioned before, the structure of the process map was defined in two levels: a first level in which we defined conceptual areas such as human resources, planning and accounting, student's admission and registry, identifying the management support BPs in each one, and a second level in which we assigned each BP from each conceptual area to an organizational unit. We also identified two points of view for the execution of the management support BPs: centralized within the central unit of the university and the central offices, which coordinates with organizational units within each faculty, and decentralized within each faculty, which are executed solely or mainly within each faculty without centralized intervention. We did not detail these BPs as the main focus of the project was the ones including centralized stakeholders. Fig. 2 presents an example of the process map and the two defined levels, where the conceptual areas are identified by different colors and numbering, with decentralized (e.g. those from each Library) and centralized BPs (e.g. those from SeCIU).

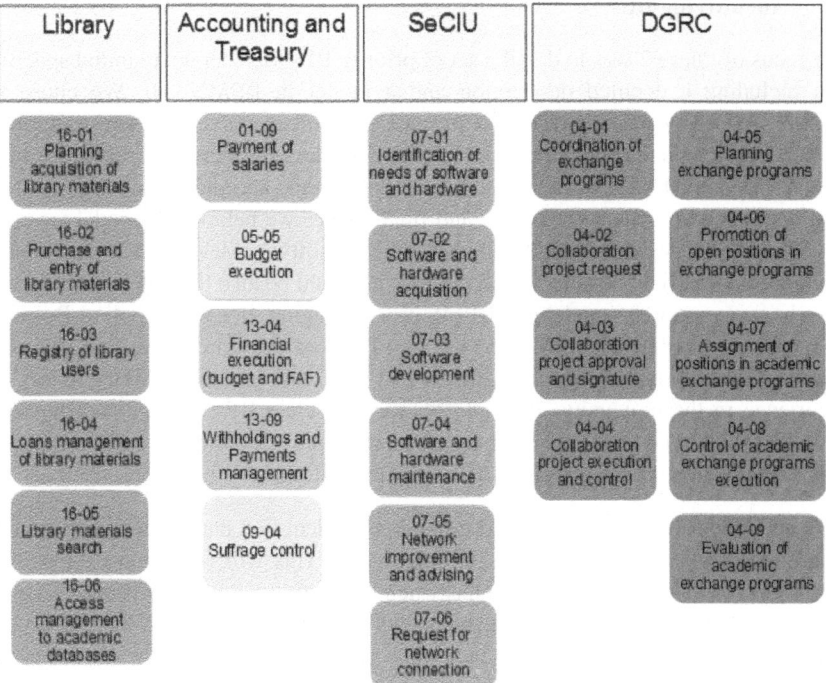

Fig. 2. Process map example with centralized and decentralized BPs

For each BP defined in the process map we also generated an identification data sheet with a high level description of the BP, its objective and owner, its participants in three levels: initiator, intermediate and final, the main artifacts it manages and the existence of support software. Fig. 3 shows the data sheet for the "Assignment of positions in academic exchange programs" BP.

Process Name:	Assignment of positions in academic exchange programs	
Objectives:	To assign the available positions in academic exchange programs for students	
Participants:	Student, Faculty, Dirección General de Relaciones y Cooperación (DGRC), Comisión de Relaciones Internacionales y Cooperación Regional e Internacional (CRI y CRI)	
Owner:	Dirección General de Relaciones y Cooperación (DGRC)	
Description:	It includes the reception the application forms for academic exchange programs from students, the evaluation of those students and the final assignment of avilable positions.	
Software:	No	

Inputs		
Items	Initiator	Comments
Application forms	Student	Open positions are defined by foreign universities
Intermediate Products		
Items	Participant	Comments
Preliminary priority list	Faculty, DGRC, CRI y CRI	Each Faculty analyzes their correspondent application forms
Outputs		
Items	Receiver	Comments
Assignment list	Student, CRI, CRI	

Fig. 3. BP data sheet for the "Assignment of positions in academic exchange programs" BP

3.2 Identifying BPs

The focus of Phase 2 was to detail a set of priority BPs to generate an initial specification including a detailed description and a model in BPMN 2.0. We chose this notation for many reasons: firstly it provides several elements and constructions to support most modeling situations including workflow patterns [8], secondly in last years it has gained acceptance in the business, academic and industrial world, and it is well understood by business people; and finally since we believe the model we specify with business people should be the basis for the software development, it should be changed minimally to be able to execute it as it should be also the one business people see when executing the model in a BPMS process engine such as Activiti or Bonita.

To select the BPs to work with, we first detailed the criteria we had drafted in Phase 1, trying to balance several aspects that we found interesting to try in this first experience in the university. We did not want to select BPs that will not be representative of the type or size of the management support BPs the university deals with. Based on this objective we defined the following criteria:

- Users: BPs should cover the largest number of identified participants, i.e. teachers, staff, students, other institutions, among others. This will cover most types of requirements that each type of user has when interacting with the university.
- Impact: BPs should navigate between many organizational units. This will cover different types of internal interactions that occurs inside the university to perform the BPs, highlighting communication issues that should be addressed.
- IS support: BPs should not have or have minimal software support to be performed since many IS were developed with a vertical vision focused on the organizational unit requirements and will limit the global vision of BPs.
- Definition: BPs should have been identified in the first phase of the project, and they should be well defined although they are not specified or modeled. The organizational unit responsible for the BPs should be able to detail them.
- Number of BP instances: BPs should have a high number of instances. This represents how much the BP is performed within the organization and it is desirable to deal with BPs demanding everyday work.

Based on the defined criteria and the process map we have identified, we selected a total of six BPs corresponding to the central unit General Direction of Cooperation and International Relations (Dirección General de Cooperación y Relaciones Internacionales, DGRC), regarding academic exchange programs for graduate and postgraduate students and teachers, and collaboration projects between the university and external institutions (national and international). They meet the criteria we have defined since they involve several types of users (students, faculty, and staff) and they impact in many organizational units with internal and external interaction going from different university schools to external institutions and internal administrative units. The responsible unit is centralized and BPs definition is adequate (there is documentation and previous experiences that allow a good approximation). Furthermore, they can be extended to specific schools on a future testing phase, which makes the volume of instances manageable. The selected BPs are as follows:

- Academic exchange programs
  — 04-07 Assignment of positions in academic exchange programs
  — 04-08 Control of academic exchange programs execution
  — 04-09 Evaluation of academic exchange programs
- Collaboration projects
  — 04-02 Collaboration project request
  — 04-03 Collaboration project approval and signature
  — 04-04 Collaboration project execution and control

An advantage of working with this counterpart was that they were really excited about participating in the Phase 2 of the project and to try and model their own BPs. We agreed on a weekly schedule of work, with meetings with the DGRC staff on Tuesdays afternoon at their workplace, and general meetings of the project group on Thursdays to analyze and process the data. We worked within the period August to December 2014 in the selection, specification and modeling of the selected BPs.

We started with the "Assignment of positions in academic exchange programs" BP, in which for every exchange program in which the university participates (Erasmus, Santander, Fulbright, etc.), positions are announced and interested people (students, faculty, staff) apply for them. The DGRC and other authorities analyzed applicant's merits and after some steps and meetings define the assignments, which are then notified to the applicants. Following the BPCIP steps of the Identify BPs activity, we started with the identification of participants and actors involved, generating a table with each exchange program in the rows side and each participant in the column side, and marking each corresponding cell whenever a participant is involved in an exchange program. We defined participant names as general roles e.g. Evaluator, External Counterpart, etc. and assigned the actual organizations to them based on what they do within each exchange program.Then we carried out the following steps as defined by BPCIP: identify activities to be performed andthe sequence of activities realization, identify decision nodes in the flow, message interaction with other participants (if any), and business rules. It is worth mentioning that when modeling the BP we take into account the workflow patterns [8] and best modeling practices in [9] such as to use verbs to name activities so instead of having "Candidates evaluation" we use

"Evaluate candidates". In Fig. 4 we present the complete model of the "Assignment of positions in academic exchange programs" BP, specified in BPMN 2.0 and with all possible participants and defined interactions.

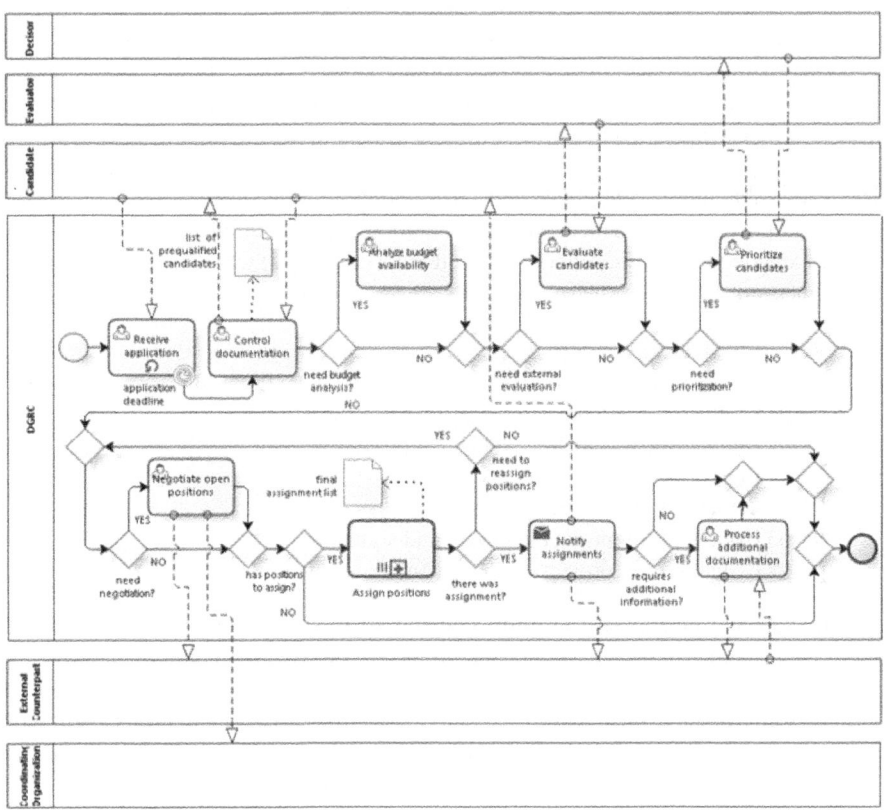

Fig. 4. "Assignment of positions in academic exchange programs" BP model

The first problem we found for specifying the BP control flow was that they visualized one BP for each program, since different people were in charge of different programs and they believed they performed different activities as program requirements are different. We, on the other hand, clearly identified a BP variants problem were all programs share a common set of activities, and based on specific conditions of each program (variation points) different paths are executed, probably involving different participants [10-12]. Although there are many ways to approach the variation modeling of a BP, we selected the definition of blocks based on XOR gateways that will be executed or not depending on the programs settings. We selected this option since it was the easiest way we find to immediately solve the problem at hand and also to provide a simpler way for business people to understand the modeling of all exchange programs within a unique model. To do so for each program we present the corresponding variant with its execution path colored so they can visualize for each program which is the corresponding control flow.

In Fig. 5 we show and example of the model for the Erasmus Mundus program, only with the DGRC pool since it is where the variability is defined depending on the exchange program. It can be seen that when executing the Erasmus Mundus exchange program, the activities that are not colored will not be executed since the condition will be set on "NO" in the configuration of the exchange program. We did the same for the rest of the selected BP so we also provide a unique model for collaboration projects with variants regarding the type of collaboration e.g. national or international, central or decentralized, among other conditions.

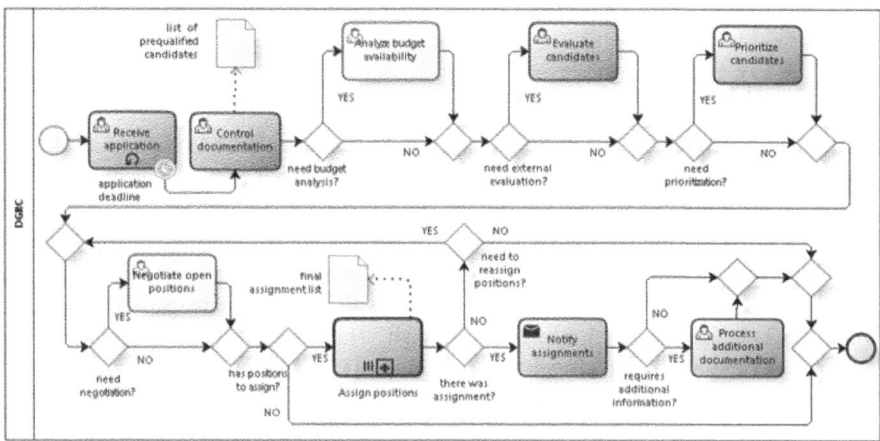

Fig. 5. Variant for the Erasmus Mundus exchange program with colored path

At the same time that we were modeling the selected BPs, we defined a set of requirements for the implementation of a prototype using Bonita BPMS. We developed the prototype as a way of showing to business and IT people how technology supports BPM, how BPs execution is addressed in practice, and how a BPMS integrates many existing technologies for supporting daily operations. We decided to use Bonita since it provide aid for a rapid development and deployment of a BP and we used it (along with Activiti) in graduate and postgraduate courses regarding BPM we teach at the university. We defined to develop two versions of the prototype for the "Assignment of positions in academic exchange programs" BP: the first with a functional focus and the second with technological focus. For the first prototype the requirements were to use as BD postgreSQL (with separated schemas for the BPMS and the organization), Tomcat as web server and JSF for the web front end to interact with the BP from outside the organization (i.e. without having a user and password to execute the intranct web portal), and specific tables for setting and loading the configuration of each exchange program for the execution of corresponding defined variants. In this tables we record for each XOR condition defining a selection block for a variant, whether a program executes it or not. In Table 1 we show how the data for the XOR blocks and exchange programs is registered. When executing the BP, the first selection the user makes is the exchange program, and based on that process variable, the variables of each XOR condition are set from the data in the DB.

Table 1. Example of configuration of XOR blocks for exchange program for the Fig.5 variant

Program Id	XOR1	XOR2	XOR3	XOR4	XOR5	XOR6	XOR7
Erasmus M.	NO	YES	YES	NO	YES	YES	YES

For the second prototype the requirements were to integrate to Bonita the following technologies: Alfresco as a document management system, Apache LDAP as an authentication system for the BPMS users, a Web Services invocation to check with the students admission and registry system whether the student is active and able to apply in an exchange program, and a mail server to send notifications to all applicants.

3.3 Select Execution Measures

As defined by BPCIP this activity is executed in the Analysis&Design phase to determine which execution measures will be calculated when the BP is executed. BPCIP also provides the BPEMM execution measurement model to aid the selection of key measures for BPs from the ones integrated in the model. We presented them with the execution measures defined in BPEMM [4] and helped to select a few to start, since we believe is better to start with a few measures that can be managed and analyzed when the organization have no measures at all. The selected measures included the throughput time of the BP (regarding activities, BP cases and average times for the BP), capacity of the BP (regarding defined resources), cost of the BP, and specific KPI defined with them for the exchange programs within a period, as follows:

- number of exchange programs executed
- number of applications received for all exchanged programs and for each program, and for each type of applicant (student, faculty, staff)
- number of positions granted for all exchange programs and for each program
- number of applications received and positions granted for each university school
- number of origin and target universities and associated countries for all programs and for each program

4 Lessons Learned and Reflection

From an organizational and cultural perspective, we found that a key driver for the success or failure of BPM in an organization is to establish a culture that supports the enactment of efficient and effective BPs. In this project we did not focused on cultural aspects, but on the practical application of some BPM activities in order to promote BPM within the organization. However, we think that the results of the whole BPM initiative can be improved by assessing with some accuracy the cultural level of our organization with respect to BPM, e.g. using some tool like [13]. We know that we are working in a very immature organization. Nevertheless, more information about the sources of immaturity could help to define concrete strategies for improving our organizational context in parallel to the other phases of the initiative. An indicator of a growing culture is the definition of a BPM group formally established within the organizational structure that is responsible for any BPM initiative giving support to all organizational units with a unified methodological approach. This group must be composed of both management and software engineering professionals. Similarly, BPs themselves need a stable organizational structure where participants and especially process owners are well identified. In this project we did not found an updated and formally approved functional organization chart. This entails several disadvantages that prevents the BPM group establishment as well as may hinder process management when there are no clear formal relationships among participants.

We also claim that the process map is a strong communication artifact for supporting the BPM initiative. In our experience, we request its approval from the highest authorities. This process had some opponents to the extent that one of the organizational units defined without notice another process map focused on their BPs. However, this result has the same vertical view of the organization as ever, forgetting about other organizational units. We found some causes for this opposition. Some organizational units felt not adequately covered either because our methodology was insufficient when extracting information, or there was no careful review from their own side. Moreover, not everyone understood the process map as an evolving artifact which must be reviewed and improved in specific projects such as the one of Phase 2. Even with this opposition, the process map promoted interesting discussions and led to many units to rethink their roles, which is another mandatory step for setting a BPM culture. In other projects we found two different kinds of opposition: to the results and to the people. The one we described before was to the resulting process map. In fact, it was not focused on the process map itself, whose content was praised by everyone, but on the formal approval of the process map, i.e. its content is useful but not as comfortable in the eyes of the whole organization. With respect to opposition to our participation in the project, we found that our belonging to the univeristy generates two different reactions. On the one hand we found that people is more open to share experiences with the team and trust in the activities we propose, especially when there is no want-to-sell-something behavior. On the other hand we found that there is some resistance from managers, mistakenly seeing endanger their roles by the project.

From a conceptual perspective, since the knowledge of BPM and BPs in the organization was minimal, we had to explain several concepts and the vision we had taken to develop the project. A hard concept to transmit was about the arranging of the process map into several conceptual areas which are then assigned to the organizational units responsible for their execution, but which are not necessarily in a one-to-one relation. The global vision beyond the work that is performed within each organizational unit was very difficult to understand, in particular with cross cutting BPs in many organizational units. Another problem related to this was the granularity of the BPs identified. We tried to balance it so each BP identified is at the same level of abstraction, but sometimes this was difficult since the stakeholders did not know the complete detailed BP or confused the part in which they participated with all the BP. The process map shows some disparity in existing BPs. Many BPs are chained and therefore its definition was given in terms of the four typical phases of an administrative BP: detection of needs, planning, execution and control. In some cases it was found that this chain lacks some parts, particularly those related to detection of needs and control. Furthermore, we observed that in some cases the BPs are owned by a centralized unit that coordinates whereas others are distributed and enacted differently in many units (mainly in decentralized schools). We think that a uniform view of BPs in those four phases helps in detecting potential flaws. In the long-term a periodical revision of the process map must be led by the BPM group using the BPCIP-based methodology we developed, working together with BPs owners and main participants.

Also, in each reunion we found ourselves continually repeating that BPM is much more than modeling BPs or installing a BPMS to execute them, BPM is to provide an organic support to the entire BPs life cycle and to justify both business and technical decisions, using analytical results of their enactment. In this way, beyond that we did not focused on cultural aspects, we believe that short projects let build that culture when main business and technical stakeholders are involved from the beginning, and projects also include BPM skills training activities useful for the development of such project. As an example of this, in Phase 2 we found that a minimum training in BPMN 2.0, as well as the discussion about variability, changed the way the organizational unit sees its BPs. Regarding measures about BPs, we detected that the DGRC lacks some basic information for a detailed report of its daily work, e.g. the return date of the exchange students. The early selection of BP measures as proposed by BPCIP and the execution measures provided by BPEMM generated much interest, helping to define new KPIs to measure specific BPs aspects.

From a technological perspective, we are convinced that it is necessary to address platform-independent initiatives in order to perceive BPM as a wider paradigm and avoid bias in the acquisition of a BPMS. However, since it is desirable to support BPs life-cycle using a BPMS instead of many adhoc software systems, it was necessary to balance people's expectations, pointing out the future value of the activities we were doing with a prototype of the final system. Although we consider that this action was useful in terms of BPM cultural growing, it naturally built the expectation of having a functional running system in the short term. Once again, without a BPM group and a centralized BPMS this false expectation can be dangerous in terms of the interest of people for developing partial BPM projects as the one in Phase 2.

Finally, from a methodological perspective, we confirmed that it is very important to guide the BPM effort with a systematic approach to identify and specify PBs, and that the BPCIP methodology provided a useful template that can be adapted to specific aspects of the organization, in particular with respect to the people involved in a project and the interaction mechanisms the project needs.

5 Conclusions and Future Work

We have presented an experience report corresponding to a BPM improvement initiative we are carrying out within Udelar. This two-year project involved the definition of a process map with respect to mangement support BPs and a pilot phase of BP modeling and prototyping. The project was carried out by an interdisciplinary group of both business and technical stakeholders. Although the organization has a very immature level with respect to BPM, we found much interest in the ideas we have proposed and an interesting engagement of stakeholders. Some ideas were very hard to transmit and the results were not easily accepted by everyone. However, the project led to very interesting discussions. In fact, the management authorities perceived the value of this project and support its continuation. Beyond the refered value for the organization, we identify open research fields with respect to BP variability and we expect that in the short-term this ongoing work could feed back this initiative.

The BPCIP methodology was a useful guide and helped us to carry out the work in a systematic way. The use of BPMN 2.0 promotes the participation of non-technical stakeholders, as well as the discussion around execution measures, and the development of a prototype builds a long-term vision. Nevertheless, we need to continue with the upcoming projects in order to strengthen our conclusions.

Acknowledgements. This work was funded by the Prorectorado de Gestión Administrativa, Udelar, Uruguay. Res. N° 3 CDGAP 01/07/13 .

References

1. van der Aalst, W.M., ter Hofstede, A.H., Weske, M.: Business process management: a survey. In: van der Aalst, W.M., ter Hofstede, A.H., Weske, M. (eds.) BPM 2003. LNCS, vol. 2678, pp. 1–12. Springer, Heidelberg (2003)
2. Weske, M.: Business Process Management - Concepts, Languages, Architectures, 2nd Edition. Springer (2012)
3. Object Management Group. Business Process Maturity Model (BPMM) (2008)
4. Delgado, A., Weber, B., Ruiz, F., de Guzmán, I.G.R., Piattini, M.: An integrated approach based on execution measures for the continuous improvement of business processes realized by services. Information & Software Technology 56(2) (2014). Web. http://alarcos.esi.uclm.es/MINERVA/BPCIP/Published/
5. Dumas, M., Rosa, M.L., Mendling, J., Reijers, H.A.: Fundamentals of Business Process Management. Springer (2013)
6. Object Management Group. Business Process Model And Notation (BPMN) v2.0 (2011)
7. Delgado, Andrea, Calegari, Daniel, Milanese, Pablo, Falcon, Renatta, García, Esteban: A systematic approach for evaluating BPM systems: case studies on open source and proprietary tools. In: Damiani, Ernesto, Frati, Fulvio, Riehle, Dirk, Wasserman, Anthony I. (eds.) OSS 2015. IFIP AICT, vol. 451, pp. 81–90. Springer, Heidelberg (2015)
8. van der Aalst, W., ter Hofstede, A., Kiepuszewski, B., Barros, A. Workflow patterns. Distributed & Parallel Databases 14(3) (2003)
9. Mendling, J., Reijers, H.A., van der Aalst, W.M.P,: Seven process modeling guidelines (7PMG). Information and Software Technology 52(2) (2010)
10. Rosa, M.L., van der Aalst, W.M.P., Dumas, M., Milani, F.: Business process variability modeling: A survey. ACM Computing Surveys (2013)
11. Valenca, G., Alves, C., Alves, V., Niu, N.: A systematic mapping study on business process variability. Int. Journal of Computer Science & Inf. Technology (IJCSIT) 5(1) (2013)
12. Ayora, C., Torres, V., Weber, B., Reichert, M., Pelechano, V.: VIVACE: A framework for the systematic evaluation of variability support in process-aware information systems. Information and Software Technology 57(1) (2015)
13. Schmiedel, T., vom Brocke, J., Recker, J.: Development and Validation of an Instrument to Measure Organizational Cultures' Support of Business Process Management. Information & Management 51(1) (2014)

Role and Task Recommendation and Social Tagging to Enable Social Business Process Management

Mohammad Ehson Rangiha[1(✉)], Marco Comuzzi[1], and Bill Karakostas[2]

[1] Centre on Adaptive Computing Systems, City University London, London, UK
{Mohammad.Rangiha.2,Marco.Comuzzi.1}@city.ac.uk
[2] VLTN GCV, Terninckstraat 13/208 2000, Antwerp, Belgium
bill.karakostas@vltn.be

Abstract. Traditional Business Process Management (BPM) poses a number of limitations for the management of ad-hoc processes, where the execution paths are not designed a priori and evolve during enactment. Social BPM, which predicates to integrate social software into the BPM lifecycle, has emerged as an answer to such limitations. This paper presents a framework for social BPM in which social tagging is used to capture process knowledge emerging during the enactment and design of the processes. Process knowledge concerns both the type of activities chosen to fulfil a certain goal and the skills and experience of users in executing specific tasks. Such knowledge is exploited by recommendation tools to support the design and enactment of future process instances. We first provide an overview of our framework, introducing the concepts of role and task recommendations, which are supported by social tagging. These mechanisms are then elaborated further by an example. Eventually, we discuss a prototype of our framework enabling collaborative process design and execution.

Keywords: Social BPM · Social tagging · BPM · Task recommendation · Process knowledge

1 Introduction

Business Process Management (BPM) is a discipline widely used across almost all large corporations. According to [1], BPM is defined as a field for involving any combination of modelling, automation, execution, control, measurement and optimization of business activity flows, in support of enterprise goals, spanning systems, employees, customers and partners within and beyond the enterprise boundaries.

Traditional approaches to BPM consider a traditional BPM "lifecycle" comprising process design, deployment, enactment, monitoring, and improvement [2]. Each phase is supported by different components of a Process-Aware Information System (PAIS) [3] and involves a specific set of different stakeholders.

Such a traditional approach to BPM presents the following limitations:

Lack of Information Fusion: BPM normally follows a top-down approach, where processes are designed by a group of individuals and passed on end users to follow [4][5].

Model-Reality Divide: The top-down approach of traditional BPM drastically limits the participation of end users in the design of processes. This often results in the processes not being followed correctly, which consequently creates a gap between the designed process and the process which is executed [6][4][5].

Information Pass-On Threshold and Lost Innovation: Useful feedback from end users is not captured in process design due to rigid hierarchical controls in the design and deployment phases. As a result, valuable first-hand knowledge to improve processes may remain unused [6][4].

Strict Access-Control: This is present in most traditional BPM approaches, that is, only actors which have been selected and given specific access are allowed to execute them [7]. This will limit the users who are able to participate into the business process life-cycle.

These properties of the standard BPM cycle make it unsuitable for so-called ad-hoc processes which are under the consideration in this paper.

To address the above limitations of traditional BPM, over the recent years there has been much research done around the emerging idea of Social BPM (SBPM). Social BPM can be defined as an approach to enhance organizational performance through a controlled participation of external stakeholders, in order to improve process design and execution [8]. Offering BPM as social software is a promising approach that fosters improved communication and collection of knowledge by allowing multiple users to work on the design, operation and improvement of a business process simultaneously and without many of the access control restrictions typical of traditional BPM [9].

Overall Social BPM can have a number of benefits, such as exploitation of weak ties and implicit knowledge [10][11], increased transparency of information sharing [11][13] and decision distribution [11] [12], and improved knowledge sharing [8][12][13]. At the same time, a number of potential limitations of Social BPM have also been identified, including the steep learning effort [14][10], security [8][13], lower quality of input in process design and enactment [8], and difficulties in evaluation [13].

Researchers have proposed an extension of BPMN to include features of social software, such as setting up polls [15]. Others [16] have also discussed agile and flexible business process development to overcome some of the limitation of the traditional BPM systems and discussed different types of process flexibilities (*by change, deviation, underspecification, and design*) [14]. However, such approaches do not change the rigid sequential nature of the traditional BPM lifecycle, as they only increase the number of features available in process design and enactment.

Social tagging in the context of social media networks has increased immensely over the recent years, as powerful and effective tools to capture and share user knowledge. However, these have not been used in the context of BPM so far. In this regard, only [17] has touched upon this briefly discussing where social networks are used to support process models by providing recommendations to people and supporting collaboration.

This paper builds on the existing works in the area of Social BPM [18][17][4] and aims at proposing an innovative framework for Social BPM. Our framework exploits the benefits of social tagging to capture process knowledge from users during the enactment and design of the processes. Such knowledge may refer to either design-time concerns, such as the type of tasks successfully considered in the past to fulfil a certain goal, or run-time concerns, such as the experience and skills of users demonstrated in previous execution of the processes. Process knowledge is exploited to support role and task recommendation in the phases of process design and enactment. In other words, task fit for purpose and users' experience and skills as emerged during the design and execution of processes is captured via social tagging to be reused in current or future enactments of processes.

Overall, our framework is more geared towards 'ad-hoc processes' [19], as opposed to structured processes where the process steps are pre-determined and remain unchanged, or even semi-structured processes. According to [19] in ad-hoc processes the execution path is defined during the enactment of the processes and the participants are free to choose the course of action they wish to follow. In terms of the ratio of flexibility and rigidness of the processes as introduced by [20], our approach starts from the exploration of the processes and then potentially moves to the process standardisation, once the structure of processes has converged to a stable state. Exploration requires flexible processes and collaboration inside loosely structured teams whilst standardization requires division of responsibilities [20]. Organisation of social activities in organisational contexts or unstructured problem solving involving a large number of potential participants with different expertise are typical examples of ad-hoc processes.

The paper is structured as follows: Section 2 introduces our framework discussing the static and the behavioural. This section also provides background about social tagging and then moves on to explaining the recommendation components of our framework. Section 3 illustrates with an example the proposed design, while Section 4 presents highlights from the proof of concept implementation of the design. Conclusions and future work are presented in Section 5.

2 Social Business Process Management Framework

Our framework is defined by a static 'conceptual' view and a dynamic 'behavioural' view.

2.1 The Static View

Figure 1 shows the conceptual model underpinning our framework. A *process* in our framework may have one or multiple *tasks* that need to be fulfilled in order to be completed. As far as actors are concerned, we distinguish between community members and process owners. The *community members* participate in the design and execution of processes. The community determines when new instances of *processes* should be started. Community members collaboratively discuss to design processes, i.e. deciding which tasks should be part of a process, to assign community members to tasks, and to capture the knowledge emerging from process execution, through tagging. The *process owner* is responsible for the overall design and execution of the process, to ensure deadlines are met and that the discussions among community members related to the tasks come to a decision and conclusion. Generally, the process owners have more experience and expertise for one or more given processes.

T*ags* are keywords used by community members to capture certain segments of the discussion among community members during the design and execution of process instances. There are two categories of tags, the *system-defined* tags are used for tasks that are populated by the system and refer to the process the tasks are related to, while *user-defined* tags are added by the community members or process owners to the tasks to specify the skill-set the task refers to. The *tag cloud* is a method of presenting tags where the more frequently used tags are presented and re-emphasized, to facilitate reuse [21].

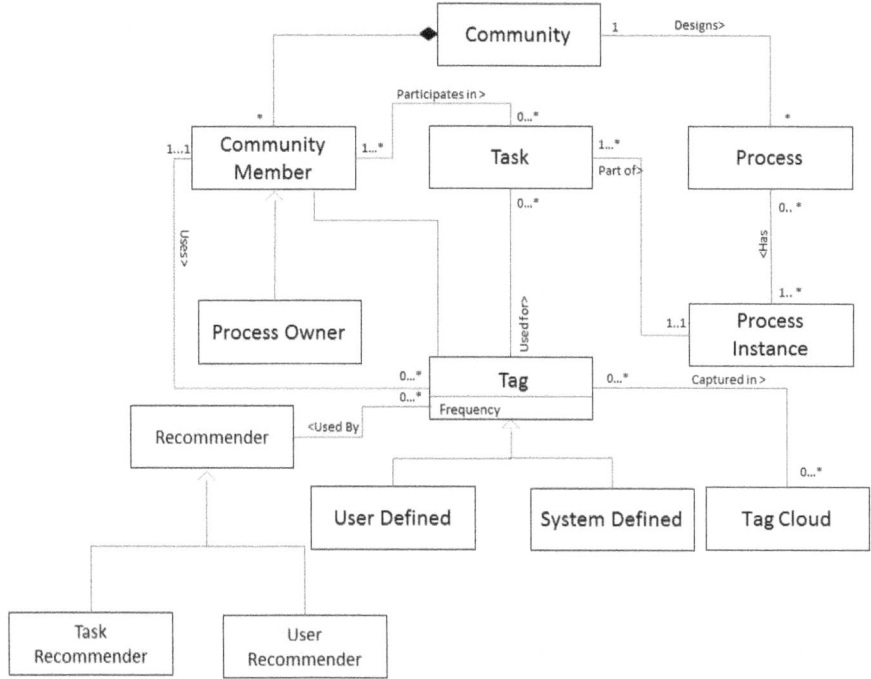

Fig. 1. Social BPM Framework

Tags represent the basis of the *task* and *role recommender*. For any given process instance, the framework recommends to the community all the tasks performed in previous instances of similar processes (task recommendation). The role recommender suggests a list of community members with their expertise which is based on their performance, as well as the tasks they have previously participated in that specific process. The two main mechanisms of the model assist the members in reusing previously captured process knowledge and recommend tasks, as well as present community members with their skill set which have been recommended and tagged by previous process owners.

Having presented a static conceptual model of our framework, the following section presents a more dynamic view of the framework.

2.2 The Behavioural View

The process model in Fig. 2 presents the behavioural view of our framework, illustrating the different activities from the outset of process initiation until the end of its execution that are supported by our framework.

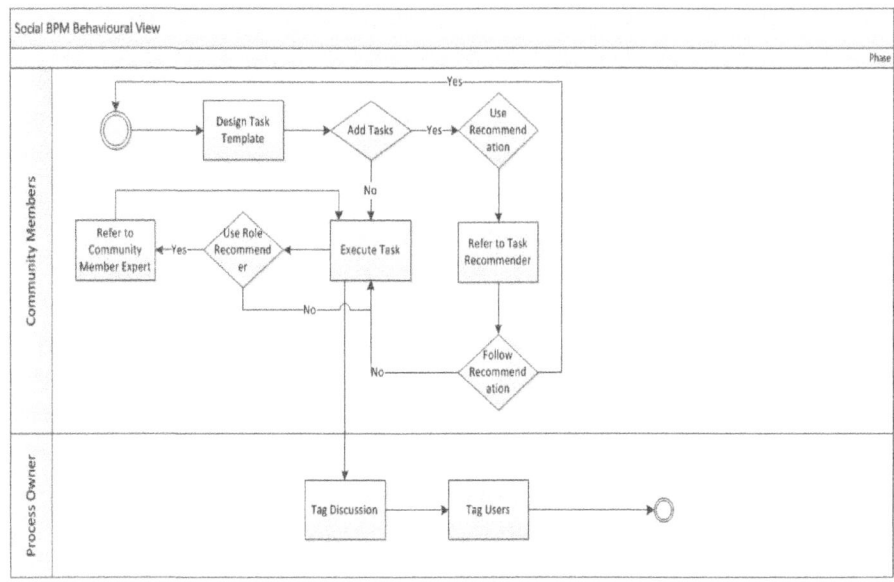

Fig. 2. Social BPM Behavioural View

Fig. 2 shows that the task design and execution in our framework are blended, as typical of ad-hoc processes. While defining and add tasks to the process, the community members may use the recommendation provided by our framework. During the execution of task or after those have been executing, process owners are responsible for making the knowledge captured in terms of tasks and processes and skills set of community members to fulfil such tasks through tagging.

2.3 Social Tagging for Capturing Process Related Knowledge

Tagging is the assignment of unrestricted keywords to all kinds of content and it becomes social when tags are shared among users and different users are allowed to tag the same content unit [22]. Social tagging has become part and parcel of most of the social networking sites over the past few years. In other contexts social tagging assists in integrating models into knowledge management systems [23]. Existing models consider tagging as an activity where an individual user assigns a set of tags to a resource [24], however so far this has not been applied in the context of business processes in order to exploit process knowledge.

In our research we are incorporating and utilising social tagging into the context of social BPM for the purpose of process knowledge discovery. In the proposed social BPM model as discussed below, there are two types of roles, the community member who contributes towards accomplishing the tasks, and the process owner who is a community member which in addition to contributing is responsible for the guiding of the process to completion and also to tag the discussions.

The user interaction model during execution has been discussed previously in [25]; in social tagging we are focusing on the post-execution phase. After or during the execution of the tasks related to the process, the process owner (or any community member) is responsible for going through the discussions and tag the segments which are useful process knowledge for future executions. The process owner tags the discussion based on his judgment and usefulness of the process knowledge captured in the discussions.

The following two sections expand on the role and task recommendation elements of our model.

2.4 Task Recommender

The task recommender in our framework presents all the tasks which have been executed in previous executions of a specific process. This is in order to benefit from previous process knowledge that has been accumulated. This process knowledge is captured with the content that it was executed in as well in order for the community member to benefit from the experience.

There have been various recommendation techniques for business process models that have inspired our task recommender approach. Attachment, structural, and textual recommendations [26] are examples of this. Specifically, our framework exploits the attachment recommendation, that supports designers during modelling tasks by finding appropriate services which are meaningful to the designer [27]. Furthermore, [28] discusses the same approach which helping process designers in modelling by providing a list of related services to the current designed model. In the task recommender component of our framework, relevant tasks are suggested to support community members during process design and execution.

The tasks are recommended for each process based on the tasks created in the past for the same type of process. There can be cases where a task is captured within another task. These tasks are those that have emerged throughout the discussion which has been taken as part of another task. In such cases it is primarily the responsibility of the process owner to create an independent task, so this can also be suggested as part of the task recommendation. For instance (see Fig. 3), task D has been completed and community members have provided their input, at this point the process owner has realised that, as part of task B, there is another part of the discussion in which a different task has emerged, which would be useful in the future to do as part of process X. This emerged task is captured and a new task (i.e, D in this case) is created. Task D will then be recommended going forward for anyone who is going to be running a process of type X in the future.

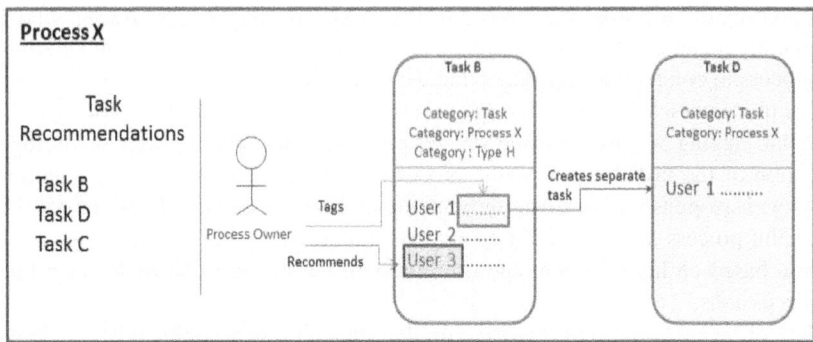

Fig. 3. Task Recommender in Social BPM

Therefore, the community sets the 'agenda', that is, it determines the list of tasks that need to be carried out in the context of a specific process (this could be added as they go along and might differ in different instances of the same process). The community decides the initial list of tasks and creates a new instance of the process in the system.

In order to get support from the system, a community member who participates in the process, e.g. a process owner, uses tags to describe the type of the process according to some taxonomy agreed within the community. This is in order to get assistance from the task recommender component. Going forward, every time the community is running a task, the task recommender will display a list of relevant tasks that were carried out in previous instances of the process. When the community decides that a task has been completed, someone from the community (the process owner for instance) needs to tag the task so that it can be reused in the future. This tag specifies under which process this task took place (System defined tag), and then the process owner (or/and other community members) classifies the main topic of the task (User defined tag). Since there is no standard classification agreed by everyone else, the one tagging is free to use any tag they like. The 'tag cloud' provides support to this phase, by showing the tags used in the past and their frequency of use.

2.5 Role Recommender

After the execution of tasks, the process owner is responsible to go through the execution log and recommend the community members who have offered valuable contribution to the tasks that have been executed. The process owner's two main primary roles, therefore, are *firstly* to ensure the discussion in fulfilling a specific task is followed up till completion and not left abandoned. *Secondly* the process owner goes through the discussions after the completion of the process and tags segments that would be useful process knowledge to be utilised in future executions. This process is illustrated in Fig. 4. The process owner has recommended community member 1 and 2 on Tasks A and N respectively. This is the decision of the individual who is tagging based on the community members' contribution.

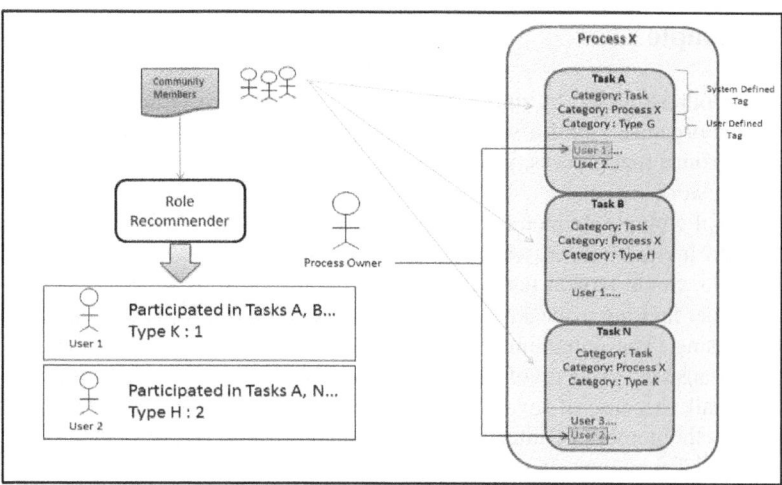

Fig. 4. Role Recommendation in Social BPM

Community members are listed with the previous tasks they have participated in, and also the number of times they have been recommended for having contributed in a task with a specific skill area. Skill sets are assigned to tasks by process owners as they see suitable and recommend community members who have contributed positively in the specific task. The list of community members, the number of times they have been recommended (tagged), as well as the tasks they have participated in are listed to allow community members to see who has expertise and experience and in which areas. This can be utilised either as taking up the role of being a process owner, or just a resource centre which community members can contact to benefit from the process knowledge other community members have.

After the execution of a process, the process owners (and the community) decide who is going to be recommended for their contribution to the tasks. This can be a collective decision so the recommendation is not biased. For instance assume James was of great help in the task that dealt with financial matters of a given process. If the community decided to recommend James for his contribution, this would add to James' profile rating by increasing his rating for 'financial expertise'. In the future when looking at member's profile, the accumulated scores for the different categories of tasks would be shown against each of the community members as illustrated in Fig. 4. This would help the community to select members for a particular task based on their previous contributions.

To conclude, the role recommender does not suggest specific community members for the task, because the knowledge of who has been good at what could be biased, incomplete or inaccurate, thus, such recommendations need to be used only as an approximate indication and not accurate measurements of members' skill sets.

3 Example

We clarify our approach by applying it in the context of a typical social business process, the organisation of a study circle in a non-profit organisation. This example is adapted from a real process, which is the object of the ongoing empirical evaluation of our framework.

A group of community members would like to run a process called "Study Circle", a type of an invited talk, in which a specific subject is discussed by an expert in a specific field. Some typical tasks involved in organising a study circle are setting a date and time, inviting a speaker, booking a venue, ordering food, designing a poster and advertising. The community does not have a standard way of executing such process, because the tasks involved will change every time, often during the organisation of the talk, because of several factors, such as the availability and preferences of the speaker, the number of interested participants, or the scheduled date. To support this in a social way, any of the community members should be able to initiate the process, setting an agenda as to what tasks have initially to be achieved. Our framework supports the definition of the process by capturing the knowledge of the community members via tagging and making it available to improve the design and execution of the process.

3.1 Task Recommendation

When a new process instance is enacted (for example a "Study circle in 2015"), the task recommender suggests all the tasks executed in previous instances for this process type of process, i.e. "Study Circle" (see under Task Recommendations in Fig. 5). This allows the community members to get ideas about potential tasks they could also be considering for the current execution.

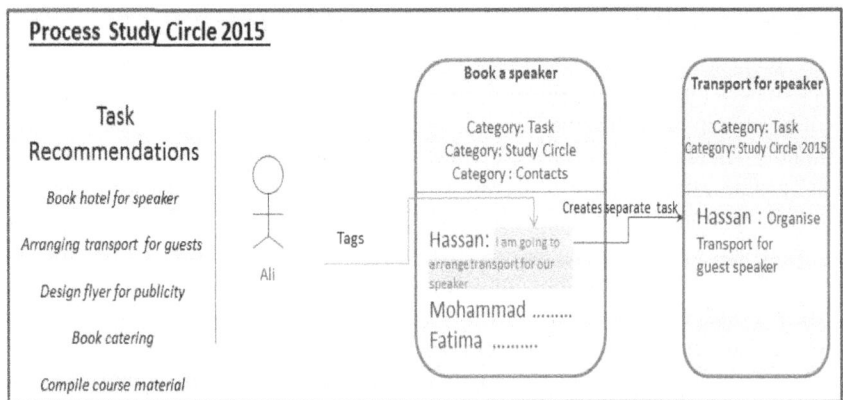

Fig. 5. Task-Recommendation in a Study Circle Process

The tags of tasks can belong to different categories (see Fig. 5). First, a category is used to identify tasks as opposed to other elements in our framework, e.g. the user profiles community member or process owner. The second category of tags captures the process to which the task belongs to, which is Study Circle 2015 in this specific case. In the third category, task can be tagged to explain their type and the expertise that is needed by community members for their execution. For instance, Book a Speaker can be tagged with contacts to identify the skill set related to this task. The knowledge captured by this last category of tags will be used in the role recommender to identify the strengths of the community member who should execute tasks.

Additionally, if as part of the definition of a task, for example "book a speaker", a new task has emerged (e.g "organising transport for speaker") then the community members are able to tag this and create a separate task for this. The emerged task will then appear in the list of recommended tasks in the future to suggest to other community members to consider also arranging transport for their invited speaker when organising a study circle.

3.2 Role Recommendation

The knowledge captured through tags in the task definition (see Fig. 5) is also exploited to assign tasks to the community members who have the most experience and/or expertise to execute them.

In Fig. 6, Fatima participated in two tasks in the past ("Book a Speaker" and "Design Flyer") and was rated by Ali, the process owner of previous editions of the Study Circle, in the 'design flyer' task, for instance because she has volunteered to design a poster for the event and did that successfully and on time. This task has been tagged with the task 'publicity', which shows the skills of Fatima in the area of publicity. 'Publicity' is part of a specific set of tags used by the community interested in the processes "Study Circle" to specify competencies related to the types of tasks that are likely to be performed in the process.

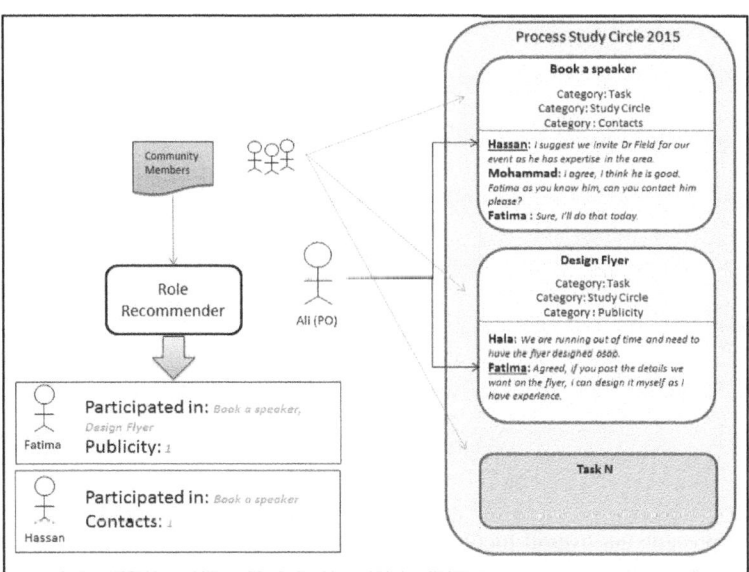

Fig. 6. Role-Recommendation in a Study Circle Process

The role recommender presents the list of community members with their skill sets, expressed as the number of times they have been rated by process owners for the execution of a given task. This helps community members to identify who in the community have experience in the required tasks. In our Study Circle example, this list is used to find out who has experience in finding and contacting speakers and designing a flyer, e.g. Hassan and Fatima, respectively, in Fig. 6, or has experience finding a venue and arranging transport for the speaker.

Note that in complex processes involving large communities machine learning techniques and algorithms [29] could potentially be adapted to assist data classification and tagging.

4 Proof of Concept Implementation

In order to validate our framework, we have produced a proof of concept prototype. Pimki, a Personal Information Manager based on Wiki technology, has been chosen as the basis for our implementation. The main reason for choosing Wiki technology is that it clearly supports community collaboration and it provides native mechanisms to capture tags. The prototype, however, required also ad-hoc extensions for specific aspects of our framework.

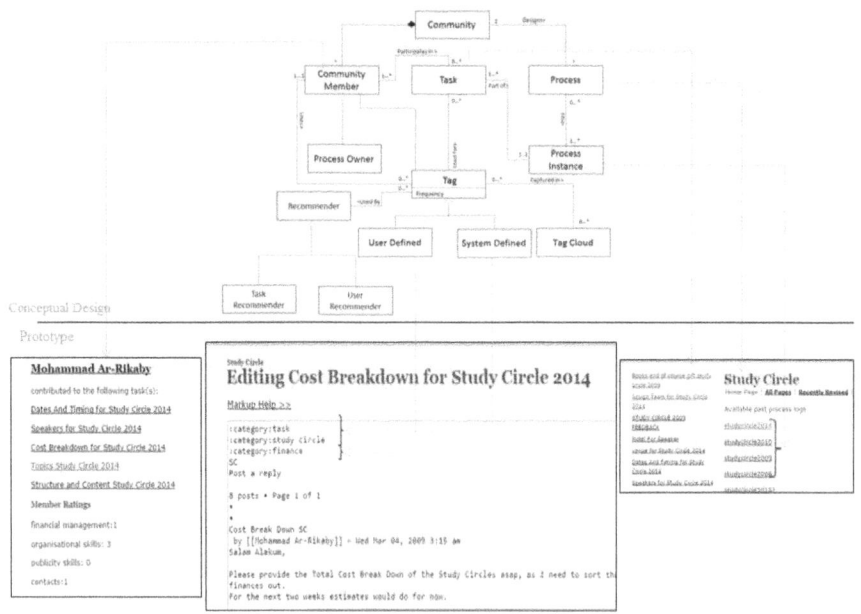

Fig. 7. Design - Prototype Mapping

Fig. 8 presents an overall picture of the mapping between elements of our framework to some screens from the prototype.

In our prototype tasks and processes are created as pages, which the community members can contribute to. Pages are the building block of a Wiki and can be uniquely named, stored and searched. The community decides the initial list of tasks and creates a new instance of the process in the system (in our prototype this refers to a new page for the process with links to each page that corresponds to a task). Every task is a discrete entity in the prototype, so that the system can store and index it (using tags) and retrieve them later on.

Fig. 9 shows an example of how a community member can be tagged for their valuable contribution in suggesting an innovative task, i.e. Suggesting present a gift to the study circle participants. The user-defined tag that most suitably explains the type of the task is populated by the process owner (or any member of the community) as shown below, i.e "publicity".

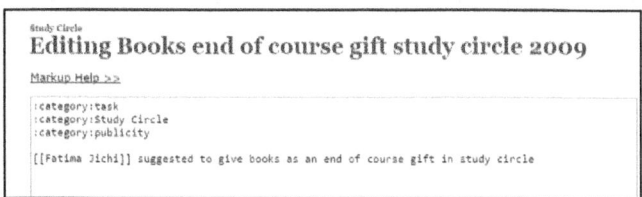

Fig. 8. Example of Recommending Role in the prototype

The community members' page in our prototype shows the tasks to which the member has contributed to and also what skill they have been rated for (see Fig. 10).

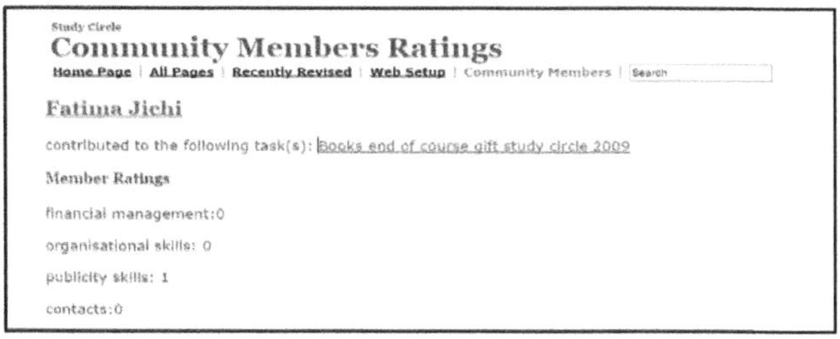

Fig. 9. Example of a Community Member Profile in the prototype

5   Conclusion

This paper presented a framework for social BPM using social tagging to exploit emergent process knowledge about user experience and skills to support the design and execution of future process instances. Our framework particularly addresses the limitations of traditional BPM, such as the model-reality divide and lack of information fusion. In our framework, process design and execution become blended, thus there is no or limited gap between the designed processes and what is actually executed. The utilisation of social tagging in order to benefit from previously captured process knowledge in other occurrences of the same process is also an attempt to overcome the gap related to information pass-on threshold and lost innovation.

The applicability of our framework is limited to ad-hoc business processes, and it does not extend to highly procedural and codified processes.

In the framework proposed by [4], in the proposed SBPM model, during the initial runs of process instance execution, there is a loosely structured team which are free to participate in any process instance, however as process knowledge is accumulated and members are recommended for their positive contributions, the processes move towards stabilisation. Firstly, in fact, there are more tasks recommended from previous executions which would broadly consist of similar tasks, and secondly individuals with more experience and expertise are ranked, which can be utilised by other community members to assist in process execution.

Our framework allows for a large degree of flexibility in processes where the design and execution of each instance can benefit from the knowledge generated by participants in previous executions. In principle, each instance of the same process may deviate from the others and the community members have more possibilities to take a decision on how and in which order to act in a specific process instance.

There are generally two avenues to balance flexibility and rigidness in business processes. One approach is to start with flexible process and then add controls and restrictions in place. The second is to start with a rigid sequential workflow system and then add elements of flexibility to it [16]. Our framework is closer to the former approach, in the sense that social tagging and collaborative participation of users restricts the freedom of community members, by making knowledge generated in previous executions of the process emerge. Control factors in our approach are the process tasks which need to be achieved in order to completed the overall processes, and also the presence of a process owner who guides the process enactment to ensure the process execution is completed.

As far as future work is concerned, our work can be extended by integrating automated semantic tagging of tasks and roles and by providing more explicit support to process execution at the implementation level. An empirical evaluation of our framework is ongoing.

References

1. Palmer, N., Schooff, P., Dugan, L., Farina, C.: Passports to Success in BPM; Real-World, Theory and Applications. Future Strategies Inc. Publication (2014)
2. Lee, R.G., Dale, B.G.: Business process management: a review and evaluation. Business Process Management Journal 4(3), 214–225 (1998). MCB UP Ltd
3. Dumas, M., van der Aalst, W., Hofstede, A.: Process-Aware Information Systems: Bridging People and Software Through Process Technology. Elsevier Science & Technology Books (2007)
4. Schmidt, R., Nurcan, S.: BPM and social software. In: Ardagna, D., Mecella, M., Yang, J. (eds.) Business Process Management Workshops. LNBIP, vol. 17, pp. 649–658. Springer, Heidelberg (2009)

5. Palmer, N.: The role of trust and reputation in social BPM. In: Swenson, KD., Palmer, N., et al. (eds.) Social BPM Work, Planning and Social Collaboration Under the Impact of Social Technology, BPM and Workflow Handbook Series, pp. 35–43 (2011)
6. Filipowska, A., Kaczmarek, M., Koschmider, A., Stein, S., Wecel, K., Abramowicz, W.: Social Software and Semantics for Business Process Management - Alternative or Synergy? Journal of Systems Integration 2(3), 54–69 (2011)
7. Wohed, P., Henkel, M., Andersson, B., Johannesson, P.: A new paradigm for work organization. In: Business Process Management Workshops, LNBIP, pp. 659–665. Springer, Heidelberg (2009)
8. Brambilla, M., Fraternali, P., Vaca Ruiz, C.: A model-driven approach to social BPM applications. In: Social BPM Handbook, BPM and Workflow Handbook series, Future Strategies, USA, 95–112 (2012)
9. Duipmans, E.: Business Process Management in the cloud: Business Process as a Service (BPaaS). University of Twente (2012). http://tinyurl.com/8f44g5l
10. Kemsley, S.: Leveraging social BPM for enterprise transformation. In: Swenson, K.D., Palmer, N., et al. (eds.) Social BPM Work, Planning and Social Collaboration Under the Impact of Social Technology, BPM and Workflow Handbook Series, pp. 77–83 (2011)
11. Brambilla, M., Fraternali, P.: Combining social web and BPM for improving enterprise performances: the BPM4People approach to social BPM. In: Proceedings of the 21st International Conference Companion on World Wide Web, pp. 223–226 (2012)
12. Richardson, C.: Is social BPM a Methodology, A Technology, Or just a lot of Hype? May 20, 2010 [Blog Entry]. http://blogs.forrester.com/clay_richardson/10-05-20-social_bpm_methodology_technology_or_just_lot_hype (accessed April 2015)
13. Erol, S., Granitzer, M., Happ, S., Jantunen, S., Jennings, B., Koschmider, A., Nurcan, S., Rossi, D., Schmidt, R., Johannesson, P.: Combining BPM and Social Software: Contradiction or Chance? Journal of Software Maintenance and Evolution: Research and Practice 22, 449–476 (2010). John Wiley & Sons, Ltd
14. van der Aalst, W.M.P., ter Hofstede, A.H.M., Kiepuszewski, B., Barros, A.P.: Workflow patterns. Distrib Parallel Databases 14(1), 5 (2003)
15. Brambilla, M., Fraternali, P., Vaca, C.: BPMN and design patterns for engineering social BPM solutions. In: Daniel, F., Barkaoui, K., Dustdar, S. (eds.) BPM Workshops 2011, Part I. LNBIP, vol. 99, pp. 219–230. Springer, Heidelberg (2012)
16. Bider I., Jalali A.: Agile business process development: why, how and when - applying Nonaka's theory of knowledge transformation to business process development. Information Systems and e-Business Management. http://link.springer.com/article/10.1007/s10257-014-0256-1
17. Rangiha, M.E., Karakostas, B.: A socially driven, goal-oriented approach to business process management. International Journal of Advanced Computer Science and Applications, IJACSA (2013)
18. Burkhart, T., Loos, P.: Flexible Business Processes - Evaluation of Current Approaches: 2010, MKWI 2010, Institute for Information Systems (IWi) at the German Research Center for Artificial Intelligence (DFKI). Göttingen (2010)
19. Bider, I., Kowalski, S.: A framework for synchronizing human behavior, processes and support systems using a socio-technical approach. In: Bider, I., Gaaloul, K., Krogstie, J., Nurcan, S., Proper, H.A., Schmidt, R., Soffer, P. (eds.) BPMDS 2014 and EMMSAD 2014. LNBIP, vol. 175, pp. 109–123. Springer, Heidelberg (2014)
20. Smith, G.: Tagging people – powered metadata for the social web. Pearson Education, San Francisco (2008)

21. Cantador, I., Konstas, I., Jose, J.: Categorising Social Tags to Improve Folksonomy-based Recommendations. Journal Web Semantics: Science, Services and Agents on the World Wide Web **9**(1), 1–15 (2011)
22. Bruno, G., Dengler, F., Jennings, B., Khalaf, R., Nurcan, S., Prilla, M., Sarini, M., Schmidt, R., Silva, R.: Key challenges for enabling agile BPM with social software. J. Softw. Maint. Evol.: Res. Pract. **23**, 297–326 (2011)
23. Kim, Scerri, Decker, Breslin: The State of the Art in Tag Ontologies: A Semantic Model for Tagging and Folksonomies. In: Proc. Int'l Conf. on Dublin Core and Metadata Applications (2008)
24. Rangiha, ME., Karakostas, B.: A goal-oriented social business process management framework. In: Conference Proceedings The International Conference Business Process Management, ICBPM 2014. London, UK, September 26–27, 2014
25. Kluza, K., Baran, M., Bobek, S., Nalepa, G.: Overview of recommendation techniques in business process modeling. In: Proc. KESE. AGH University of Science and Technology (2013)
26. Born, M., Brelage, C., Markovic, I., Pfeiffer, D., Weber, I.: Auto-completion for executable business process models. In: Ardagna, D., Mecella, M., Yang, J. (eds.) Business Process Management Workshops. Lecture Notes in Business Information Processing, vol. 17, pp. 510–515. Springer, Heidelberg (2009)
27. Chan, N., Gaaloul, W., Tata, S.: Context-based service recommendation for assisting business process design. In: Huemer, C., Setzer, T. (eds.) E-Commerce and Web Technologies. Lecture Notes in Business Information Processing, vol. 85, pp. 39–51. Springer, Heidelberg (2011)

Mining for Processes

Scalable Process Discovery with Guarantees

Sander J.J. Leemans[✉], Dirk Fahland, and Wil M.P. van der Aalst

Eindhoven University of Technology, Eindhoven, The Netherlands
{s.j.j.leemans,d.fahland,w.m.p.v.d.aalst}@tue.nl

Abstract. Considerable amounts of data, including process event data, are collected and stored by organisations nowadays. Discovering a process model from recorded process event data is the aim of process discovery algorithms. Many techniques have been proposed, but none combines scalability with quality guarantees, e.g. can handle billions of events or thousands of activities, and produces sound models (without deadlocks and other anomalies), and guarantees to rediscover the underlying process in some cases. In this paper, we introduce a framework for process discovery that computes a directly-follows graph by passing over the log once, and applying a divide-and-conquer strategy. Moreover, we introduce three algorithms using the framework. We experimentally show that it sacrifices little compared to algorithms that use the full event log, while it gains the ability to cope with event logs of 100,000,000 traces and processes of 10,000 activities.

Keywords: Big data · Scalable process mining · Block-structured process discovery · Directly-follows graphs · Rediscoverability

1 Introduction

Considerable amounts of data are collected and stored by organisations nowadays. For instance, ERP systems log business transaction events; high tech systems such as X-ray machines record software and hardware events; and web servers log page visits. These event logs are often very large [5], i.e. contain billions of events. From these event logs, process mining aims to extract information, such as compliance to rules and regulations; and performance information, e.g. bottlenecks in the process [3].

Figure 1 shows a typical process mining workflow. The system executes and produces an event log. From this log, a *process discovery technique* obtains a model of the process. As a next step, the log is filtered: the behaviour in the log that does not match with the model (typically 20%) is analysed manually, while the behaviour that matches with the log is analysed with respect to the model, for instance by selecting interesting parts and drilling down on it using filtering, after which a new model is discovered.

Process mining is usually studied in small settings, for instance on processes containing a few activities (i.e. process steps) and a few thousand traces (i.e. cases; customers), e.g., the BPI challenge log of 2012 [14] contains 36 activities

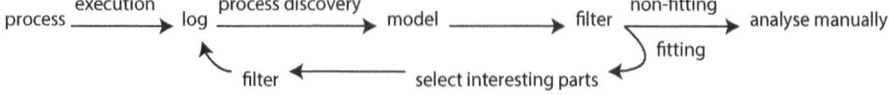

Fig. 1. A typical process mining workflow

and around 13,000 traces. In this paper, we explore a first step towards process mining on bigger scales: we experimented using logs containing 100,000,000 traces and processes containing 10,000 activities. While such numbers seem large for a complaint-handling process in an airline, even much larger processes exist, for instance in control systems of the Large Hadron Collider [18]. Current process discovery techniques *do not scale well* when facing big event logs (i.e. containing billions of events), when facing event logs of complex systems (i.e. containing thousands of activities), or *do not provide basic, essential guarantees*, such as deadlock-freedom or the ability to rediscover the original process.

Given a big or complex log, several strategies could be of help: 1) sampling or drilling down beforehand using process cubes might reduce logs to sizes manageable for existing discovery techniques; 2) restrictions or guarantees might be dropped; 3) taylor-made algorithms could be used. In this paper, we focus on situations in which the first two options are not possible, such as if the process consists of 10,000 activities, (because of which sampling to manageable sizes would throw away too much information); or if the use case prefers taking the full log into account, such as in auditing [2]. In this paper, we push the boundary of how far we can get while taking all recorded behaviour into account.

Event logs with billions of events challenge current discovery algorithms, as most techniques require that the event log is present in main memory. If an event log is too big to fit into main memory of a single machine, few algorithms can be used. Ideally, such a discovery algorithm should require only a *single pass through the event log*, and its run time after this pass should be constant and not depend on the number of events and traces in the event log. The single-pass property would cancel the need for the event log to be in main memory; the independence of run time to the number of events and traces would ensure that logs containing billions of events can still be handled. Independence of the number of activities cannot be achieved, as activities are part of the output of any technique.

Single-pass algorithms should produce models that are as close as possible to other techniques in terms of quality criteria. In the context of process mining, several criteria exist that describe the quality of a process model, such as fitness, precision, generalisation and simplicity, of which the first three are measured with respect to an event log. Fitness describes what part of the event log is represented by the model, precision describes what part of the model is present in the log, generalisation expresses a confidence that future behaviour of the process will be representable by the model, and simplicity expresses whether a process model requires few constructs to express its behaviour. Any discovery algorithm needs to balance these sometimes conflicting four criteria [10]. However, a more basic quality criteria is whether the model is *sound*, which means it contains neither deadlocks nor other anomalies. While an unsound model can be useful for human interpretation, it is often unsuitable for further (automated) analysis, including the computation of fitness, precision and generalisation [20].

A desirable property of discovery techniques is *rediscoverability*: Rediscoverability measures the ability of a technique to find the actual process that produced the partial observations in the event log. Typically, rediscoverability is proven assuming that the log contains enough information (is *complete*), and assuming that the process adheres to some restrictions [8,19].

In this paper, we investigate how current techniques perform in big-data settings and show what happens to several discovery algorithms if they are adapted to the single-pass property. We adapt the Inductive Miner framework (IM) [19] to recurse on directly-follows graphs rather than on logs. Directly-follows graphs can be computed in a single pass over the event log, and their computation can even be parallelised, for instance using highly-scalable map-reduce techniques [15]. Moreover, the size of a directly-follows graph only depends quadratically on the number of activities (i.e, types of process steps) in a log, and is independent of the number of traces or events. We show that this adaptation combines the scalability of directly-follows graphs, as used by the Heuristics Miner (HM) [25] and the α-algorithm (α) [6], with guarantees such as soundness and rediscoverability as provided by the IM framework. The adapted framework is called the *Inductive Miner - directly-follows based* (IMD framework).

We use the new framework in three algorithms: we introduce a basic algorithm, an algorithm focused on infrequent behaviour handling and an algorithm focused on handling incomplete behaviour, similar to the algorithms developed for the IM framework. These algorithms are implemented as plug-ins of the ProM framework, and are available for download at http://www.promtools.org.

To show how the use of directly-follows graphs influences the adapted framework and its algorithms, we test their suitability in a big data context where we deliberately create larger and larger logs, until reaching a log size that cannot be handled anymore, or it becomes clear it poses no restrictions on the log size (at 100 million traces). As the new algorithms have less information available than in the IM framework, we expect that they will need more information for rediscoverability and infrequent behaviour handling. Therefore, we evaluate this trade-off as well. It turns out we can get significant performance and scalability improvements, i.e. the feasibility of handling event logs with over 3.000.000.000 events (>200GB), while producing similar process models as existing non-scalable techniques.

Outline. First, related work is discussed in Section 2. Second, process trees, directly-follows graphs and cuts are introduced in Section 3. In Section 4, the framework and three algorithms using it are introduced. The algorithms using the framework are evaluated in Section 5. Section 6 concludes the paper.

2 Related Work

Process discovery and process discovery in highly scalable environments have been studied before. In this section, we discuss related process discovery techniques and their application in scalable environments.

Process Discovery. Process discovery techniques such as the Evolutionary Tree Miner (ETM) [9], the Constructs Competition Miner (CCM) [23] and Inductive Miner (IM) [19] provide several quality guarantees, in particular soundness and some offer rediscoverability, but do not manage to discover a model in a single pass. ETM applies a genetic strategy, i.e. generates an initial population, and then applies random crossover steps, selects the 'best' individuals from the population and repeats. While ETM is very flexible towards the desired quality criteria to which respect the model should be 'best' and guarantees soundness, it requires multiple passes over the event log and does not provide rediscoverability.

CCM and IM use a divide-and-conquer strategy on event logs. In the Inductive Miner IM framework, first an appropriate cut of the process activities is selected; second, that cut is used to split the event log into sub logs; third, these sub logs are recursed on, until a base case is encountered. If no appropriate cut can be found, a fall-through ('anything can happen') is returned. CCM works similarly by having several process constructs compete with one another. While both CCM and the IM framework guarantee soundness and IM guarantees rediscoverability, both require multiple passes through the event log (the event log is being split and recursed on).

Scalability. Several techniques exist that satisfy the single-pass requirement, for instance the α-algorithm (α) and its derivatives [6,26,27], and the Heuristic Miner (HM) [25]. These algorithms first obtain an abstraction from the log, which denotes what activities directly follow one another; in HM, this abstraction is filtered. Second, from this abstraction a process model is constructed. Both α and HM have been demonstrated to be applicable in highly-scalable environments: event logs of 5 million traces have been processed using map-reduce techniques [15]. Moreover, α guarantees rediscoverability, but neither α nor HM guarantees soundness. We show that our approach offers the same scalability as HM and α, but provides both soundness and rediscoverability.

Some commercial tools such as Fluxicon Disco (FD) [16] and Perceptive Process Mining (PM) offer high scalability, but have no executive semantics (FD) or do not support parallelism (PM) [22].

Other well-known discovery techniques such as the ILP miner [28] satisfy neither of the two requirements.

Streams. Another set of approaches that aims to handle even bigger (i.e. unbounded) logs assumes that the event log is an infinite stream of events. Some approaches such as [13,17] work on click-stream data, i.e. the sequence of web pages users visit, to extract for instance clusters of similar users or web pages. However, we aim to extract end-to-end process models, in particular containing parallelism. HM, α and CCM have been shown to be applicable in streaming environments [11,24]. While streaming algorithms could handle event logs containing billions of events by feeding them as streams, these algorithms assume the log can never be examined completely and, as of the unbounded stream, eventually cannot store information for an event without throwing away information about an earlier seen event. In this paper, we assume the log is bounded and we investigate how far we can get using all information in it.

3 Process Trees, Directly-Follows Graphs

Our approach combines the single-pass property of directly-follows graphs with a divide-and-conquer strategy. This section recalls these existing concepts.

An *event log* is a multiset of *traces* that denote process executions. For instance, the event log $[\langle a,b,c \rangle, \langle b,d \rangle^2]$ denotes the event log in which the trace consisting of a followed by b followed by c was executed once, and the trace consisting of b followed by d was executed twice.

A *process tree* is an abstract representation of a block-structured hierarchical process model, in which the leaves represent the *activities*, i.e. the basic process steps, and the *operators* describe how their children are to be combined [9]. τ denotes the activity which execution is not visible in the event log. We consider four operators: $\times$, $\rightarrow$, $\wedge$ and $\circlearrowleft$. $\times$ describes the exclusive choice between its children, $\rightarrow$ the sequential composition and $\wedge$ the parallel composition. The first child of a loop $\circlearrowleft$ is the *body* of the loop, all other children are *redo* children. First, the body must be executed, followed by zero-or-more times a redo child and the body again. For instance, the language of the process tree $\times(\circlearrowleft(a,b), \rightarrow(\wedge(c,d),e))$ is $\{\langle a \rangle, \langle a,b,a \rangle, \langle a,b,a,b,a \rangle \ldots \langle c,d,e \rangle, \langle d,c,e \rangle\}$. Process trees are inherently sound.

A *directly-follows graph* can be derived from a log and describes what activities follow one another directly, and with which activities a trace starts or ends. In a directly-follows graph, there is an edge from an activity a to an activity b if a is followed directly by b. The weight of an edge denotes how often that happened. For instance, the directly-follows graph of our example log $[\langle a,b,c \rangle, \langle b,d \rangle^2]$ is shown in figure 2. Notice that the multiset of start activities is $[a,b^2]$ and the multiset of end activities is $[c,d^2]$. A directly-follows graph can be obtained in a single pass over the event log with minimal memory requirements [15].

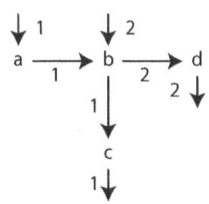

Fig. 2. Example of a directly-follows graph

A *partition* is a non-overlapping division of the activities of a directly-follows graph. For instance, $(\{a,b\},\{c,d\})$ is a binary partition of the directly-follows graph in Figure 2. A *cut* is a partition combined with a process tree operator, for instance $(\rightarrow,\{a,b\},\{c,d\})$. In the IM framework, finding a cut is an essential step: its operator becomes the root of the process tree, and its partition determines how the log is split.

Suppose that the log is produced by a process which can be represented by a process tree T. Then, the root of T leaves certain characteristics in the log and in the directly-follows graph. The most basic algorithm that uses the IM

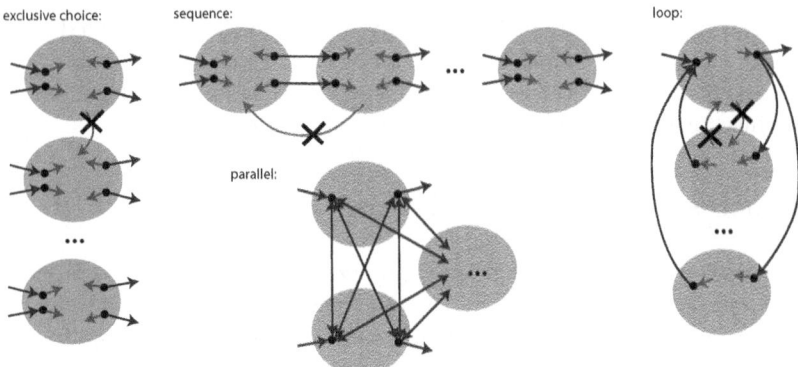

Fig. 3. Cut characteristics

framework, i.e. IM [19], searches for a cut that matches these characteristics perfectly.

Each of the four process tree operators ×, →, ∧ and ↻ leaves a different characteristic footprint in the directly-follows graph. Figure 3 visualises these characteristics: for exclusive choice, each trace will be generated by one child; so we expect several unconnected clusters in the directly-follows graph. For sequential behaviour, each child generates a trace; in the directly-follows graph we expect to see a chain of clusters without edges going back. For parallelism, each child generates a trace and these traces can occur in any intertwined order; we expect all possible connections to be present between the child-clusters in the directly-follows graph. In a loop, the directly-follows graph must contain a clear set of start and end activities; all connections between clusters must go through these activities. For more details, please refer to [19].

4 Process Discovery Using a Directly-Follows Graph

Algorithms using the IM framework guarantee soundness, and some even rediscoverability, but do not satisfy the single-pass property, as the log is traversed and even copied during each recursive step. Therefore, we introduce an adapted framework: *Inductive Miner - directly-follows based* (IM framework) that applies the recusion on the directly-follows graph directly. In this section, we first introduce the IMD framework and a basic algorithm using it. Second, we introduce two more algorithms: one to handle infrequent behaviour; another one that handles incompleteness.

The input of the IMD framework is a directly-follows graph, and it applies a divide-and-conquer strategy: first, a cut of the directly-follows graph is selected. Second, using this cut the directly-follows graph is split, which yields several smaller directly-follows graphs. Third, the framework recurses on these smaller directly-follows graphs until a base case is encountered. If no cut can be found, a fall-through, i.e. a model that allows for all behaviour, is applied. The returned

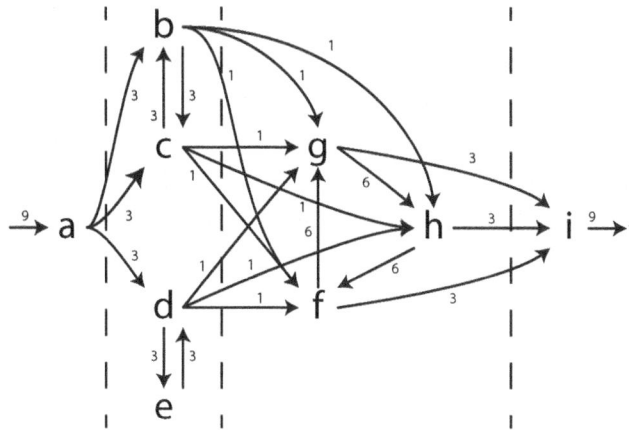

Fig. 4. Directly-follows graph D_1 of L. In a next step, the partition $(\{a\}, \{b, c, d, e\}, \{f, g, h\}, \{i\})$, denoted by the dashed lines, will be used.

process model is the hierarchical composition of operators, base cases and fall-throughs.

4.1 Inductive Miner - Directly-Follows Based

As a first algorithm that uses the framework, we introduce Inductive Miner - directly-follows based (IMD). We explain the stages of IMD in more detail by means of an example: Let L be $[\langle a, b, c, f, g, h, i\rangle, \langle a, b, c, g, h, f, i\rangle, \langle a, b, c, h, f, g, i\rangle, \langle a, c, b, f, g, h, i\rangle, \langle a, c, b, g, h, f, i\rangle, \langle a, c, b, h, f, g, i\rangle, \langle a, d, f, g, h, i\rangle, \langle a, d, e, d, g, h, f, i\rangle, \langle a, d, e, d, e, d, h, f, g, i\rangle]$. The directly-follows graph D_1 of L is shown in Figure 4.

Cut Detection. IMD searches for a cut that perfectly matches the characteristics mentioned in Section 3. Cut detection has been implemented using standard graph algorithms (connected components, strongly connected components) [19], which run in linear time, given the number of activities and directly-follows edges in the graph.

In our example, the cut $(\rightarrow, \{a\}, \{b, c, d, e\}, \{f, g, h\}, \{i\})$ is selected: as shown in Figure 3, every edge crosses the cut lines from left to right. Therefore, it perfectly matches the sequence cut characteristic. Using this cut, the sequence is recorded and the directly-follows graph can be split.

Directly-Follows Graph Splitting. Given a cut, the IMD framework splits the directly-follows-graph in disjoint subgraphs. The idea is to keep the internal structure of each of the clusters of the cut by simply projecting a graph on the cluster. Figure 5 gives an example of how D_1 (Figure 4) is split using the sequence cut that was discovered in our example. If the operator of the cut is $\rightarrow$ or ↺, the start and end activities of child might be different from the start and end activities of its parent. Therefore, every edge that enters a cluster is counted as a start activity, and edges leaving a cluster are counted as end activities.

In our example, the cluster $\{f, g, h\}$ gets as start activities all edges that are entering the cluster $\{f, g, h\}$ in D_1; similar for all end activities. The result is shown in Figure 5a. In case the operator of the cut is $\times$ or $\wedge$, traces that are crossing cluster boundaries do not cause a child to start or end.

The choice for a sequence cut and the split directly-follows graphs are recorded in an intermediate tree: $\rightarrow((D_2), (D_3), (D_4), (D_5))$, denoting a sequence operator with 4 unknown sub-trees that are to be derived from 4 directly-follows graphs.

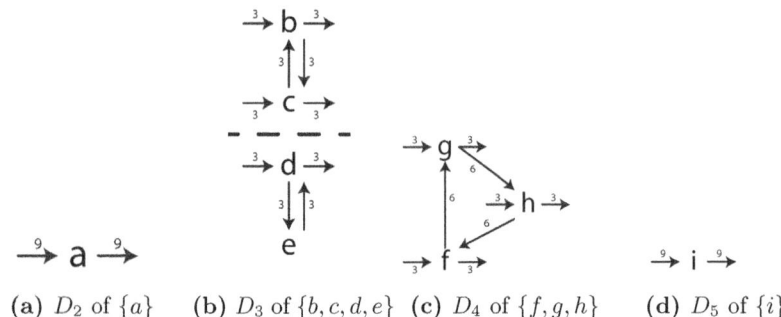

(a) D_2 of $\{a\}$ (b) D_3 of $\{b, c, d, e\}$ (c) D_4 of $\{f, g, h\}$ (d) D_5 of $\{i\}$

Fig. 5. Split directly-follows graphs of D_1. The dashed line is used in a next step and denotes another partition.

Recursion. Next, IMD recurses on each of the new directly-follows graphs (find cut, split, ...) until a base case (see below) is reached or no perfectly matching cut can be found. Each of these recursions returns a process tree that is inserted as a child.

Base Case. Directly-follows graphs D_2 (Figure 5a) and D_5 (Figure 5d) contain base cases: in both graphs, only a single activity is left. The algorithm turns these into leaves of the process tree and inserts them at the respective spot of the parent operator. In our example, detecting the base cases of D_2 and D_5 yields the intermediate tree $\rightarrow(a, (D_3), (D_4), i)$, in which D_3 and D_4 indicate directly-follows graphs that are not base cases and will be recursed on later.

Fall-Through. Consider D_4 as, shown in Figure 5c. D_4 does not contain unconnected parts, so does not contain an exclusive choice cut. There is no sequence cut possible, as f, g and h form a strongly connected component. There is no parallel cut as there are no dually connected parts, and no loop cut as all activities are start and end activities. Hence, IMD selects a fall-through, being a process tree that allows for any behaviour consisting of f, g and h. The intermediate tree of our example up till now becomes $\rightarrow(a, (D_3), \circlearrowleft(\tau, f, g, h), i)$ (remember that τ denotes the activity which execution is invisible).

Example Continued. In D_3, shown in Figure 5b, a cut is present: $(\{b, c\}, \{d, e\})$. No edge in D_3 crosses this cut, hence this is an exclusive choice cut. The directly-follows graphs D_6 and D_7, shown in Figures 6a and 6b, result after splitting D_3. The tree of our example up till now becomes $\rightarrow(a, \times((D_6), (D_7)), \circlearrowleft(\tau, f, g, h), i)$.

In D_6, shown in Figure 6a, a parallel cut is present, as all possible edges cross the cut, i.e. the dashed line, in both ways. The dashed line in D_7 (Figure 6b) denotes a loop cut, as all connections between $\{d\}$ and $\{e\}$ go via the set of start and end activities $\{d\}$. Four more base cases give us the complete process tree $\rightarrow(a, \times(\wedge(b,c), \circlearrowleft(d,e)))$, $\circlearrowleft(\tau, f, g, h), i)$.

(a) D_6 of $\{b,c\}$ in D_3 (b) D_7 of $\{d,e\}$ in D_3

Fig. 6. Split directly-follows graphs. Dashed lines denote cuts, which are used in the next steps.

To summarise: IMD selects a cut, splits the directly-follows graph and recurses until a base case is encountered or a fall-through is necessary. As each recursion removes at least one activity from the graph and cut detection is $O(n^2)$, IMD runs in $O(n^3)$, in which n is the number of activities in the directly-follows graph.

By the nature of process trees, the returned model is sound. By reasoning similar to IM [19], IMD guarantees rediscoverability on the same class of models, i.e. assuming that the model is representable by a process tree without using duplicate activities, and it is not possible to start loops with an activity they can also end with [19]. This makes IMD the first single-pass algorithm to offer these guarantees.

4.2 Handling Infrequency and Incompleteness

The basic algorithm IMD guarantees rediscoverability, but, as will be shown in this section, is sensitive to both infrequent and incomplete behaviour. To solve this, we introduce two more algorithms using the IMD framework.

Infrequent Behaviour. Infrequent behaviour in an event log is behaviour that occurs less frequent than 'normal' behaviour, i.e. the exceptional cases. For instance, most complaints sent to an airline are handled according to a model, but a few complaints are so complicated that they require ad-hoc solutions. This behaviour could be of interest or not, which depends on the use case.

Consider again directly-follows graph D_3, shown in Figure 5b, and suppose that there is a single directly-follows edge added, from c to d. Then, $(\times, \{b,c\}, \{d,e\})$ is not a perfectly matching cut, as with the addition of this edge the two parts $\{b,c\}$ and $\{d,e\}$ became connected. Nevertheless, as 9 traces showed exclusive-choice behaviour and only one did not, this single trace is probably an outlier and in most cases, a model ignoring this trace would be preferable.

To handle these infrequent cases, we apply a strategy similar to IMi [20] and introduce an algorithm using the IMd framework: Inductive Miner - infrequent - directly-follows based (IMiD). Infrequent behaviour introduces edges in the directly-follows graph that violate cut requirements. As a result, a single edge makes it impossible to detect an otherwise very strong cut. To handle this, IMiD first searches for existing cuts as described in Section 4.1. If none is found (when IMd would select a fall through), the graph is filtered by removing edges which are infrequent with respect to their neighbours. Technically, for a parameter $0 \leq h \leq 1$, we keep the edges (a, b) that occur more than $h \times \max_c(|(a, c)|)$, i.e. occur frequently enough compared to the most occurring outgoing edge of a. Start and end activities are filtered similarly.

Incompleteness. A log in a "big-data setting" can be assumed to contain lots of behaviour. However, we only see example behaviour and we cannot assume to have seen all possible traces, even if we use the rather weak notion of directly-follows completeness [21] as we do here. Moreover, sometimes smaller subsets of the log are considered, for instance when performing slicing and dicing in the context of process cubes [4]. For instance, an airline might be interested in comparing the complaint handling process for several groups of customers, to gain insight in how the process relates to age, city and frequent-flyer level of the customer. Then, there might be combinations of age, city and frequent-flyer level that rarely occur and the log for these customers might contain too little information.

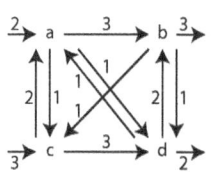

Fig. 7. An incomplete directly-follows graph

If the log contains little information, edges might be missing from the directly-follows graph and the underlying real process might not be rediscovered. Figure 7 shows an example: the cut $(\{a, b\}, \{c, d\})$ is not a parallel cut as the edge (c, b) is missing. As the event log only provides example behaviour, it could be that this edge is possible in the process, but has not been seen yet. Given this directly-follows graph, IMd can only give up and return a fall-through flower model, which yields a very imprecise model. However, choosing the parallel cut $(\{a, b\}, \{c, d\})$ would obviously be a better choice here, providing a better precision.

To handle incompleteness, we introduce Inductive Miner - incompleteness - directly-follows based (IMinD), which adopts ideas of IMin [21] into the IMd framework. IMinD first applies the cut detection of IMd and searches for a cut that perfectly matches a characteristic. If that fails, instead of a perfectly matching cut, IMinD searches for the most probable cut of the directly-follows graph at hand.

IMinD does so by first estimating the most probable behavioural relation between any two activities in the directly-follows graph. In Figure 7, a and b are most likely in a sequential relation as there is an edge from a to b. a and c are most likely in parallel as there is a forth and a back edge. Loops and choices have similar local characteristics. For each pair of activities x and y the probability $P_r(x, y)$ that x and y are in relation R is determined. The best cut is then a partition into sets of activities X and Y such that the average probabilities that $x \in X$ and $y \in Y$ are in relation R is maximal. For more details, please refer to [21].

In our example, the probability of cut $(\wedge, \{a, b\}, \{c, d\})$ is the average probability that (a, c), (a, d), (b, c) and (b, d) are parallel. IMINDD chooses the cut with highest probability, using optimisation techniques. This approach gives IMINDD a run time exponential in the number of activities, but still requires a single pass over the event log.

5 Evaluation

In this section, we illustrate handling of big event logs and complex systems, and we study the possible losses of quality of the resulting model. The new algorithms were compared to various existing discovery algorithms in two experiments: First, we test their scalability in handling big event logs and complex systems, and second we test their ability to handle infrequent behaviour. The new algorithms were implemented as plug-ins of the ProM framework, taking as input a directly-follows graph. In order to obtain these, we used an ad-hoc script that builds a directly-follows graph from an event log incrementally.

5.1 Scalability

First, we compare the IMD algorithms with several process discovery tools and algorithms in their ability to handle big event logs and complex systems using limited main memory.

Table 1. Scalability results. #: max traces/events/activities an algorithm could handle.

	A: tree of 40 activities		B: tree of 10,000 activities		
	#traces	#events	#traces	#events	#activities
α	10,000	522,626	1	420	141
HM	100,000	5,218,594	1	420	141
ILP	1,000	51,983	1	420	141
P-IF	10,000	522,626	*1	420	141
P-PT	10,000	522,626	*1	420	141
IM	100,000	5,218,594	100	82689	6941
IMD	100,000,000	3,499,987,460	100,000	76,793,937	10,000
IMI	100,000	5,218,594	10	5886	2053
IMID	100,000,000	3,499,987,460	100,000	76,793,937	10,000
IMIN	100,000	5,218,594	*10	5886	2053
IMIND	100,000,000	3,499,987,460	*10	5886	2053

Setup. This experiment searches for the largest event log that a process discovery algorithm currently can handle. All algorithms are tested on the same set of XES event logs, which are created randomly from a process tree of 40 activities. First, each algorithm gets as input a log of t traces generated from the model. The algorithm can handle this log if it returns a model using only the allocated memory of 2GB, i.e. it terminates and does not crash. The ad-hoc pre-processing step was not exempted from this restriction. If the algorithm is successful, t is multiplied by 10 and the procedure is repeated. The maximum number of traces that an algorithm can handle is recorded. This procedure (A) is repeated for a process tree of 10,000 activities (B). For the algorithms implemented as ProM plug-ins, the logs are imported using a disk-buffered import plug-in; enough SSD disk space is available for this plug-in. Algorithms are restarted between all runs to release all memory.

Besides the new algorithms introduced in this paper, from the ProM 6.4.1 framework we included IM, IMi, IMin, α, HM and ILP in the comparison. From the PMLAB suite [12], we included the IMMEDIATELY_FOLLOWS_CNET_FROM_LOG (P-IF) and the PN_FROM_TS (P-PT) functions. To obtain the directly-follows graphs, a one-pass pre-processing step is executed before applying the actual discovery.

Results. Table 1 shows the results. Some results could not be obtained; they are denoted with *; (for instance, IMin and IMinD ran for over a week). The largest log we could generate was 217GB (100,000,000 traces) for A, limited by disk space; and 100,000 traces for B, limited by RAM needed for log-generation.

For A, many implementations (HM, IM, IMi, IMin, P-IF and P-PT) are limited by their requirement to have the complete log in main memory: the ProM disk-caching importer (HM, IM, IMi, IMin) could handle 100,000 traces, the PMLAB importer (P-IF, P-PT) 10,000. This shows the value of the single-pass property: for a single-pass algorithm, there is no need to import the log. In [15], a single-pass version of HM and α are described. We believe such a version of HM could be memory-restricted, but still this would not offer any of the guarantees described. Of the IM framework, single-pass versions cannot exist due to the necessary log splitting. The IMd framework algorithms are clearly not limited by log size.

For B, log importers were not a problem. Algorithms that are exponential in the number of activities (IMin, IMinD, α, HM) clearly show their limitations here: none of them could handle our log of 100 traces/6941 activities. This experiment clearly shows the limitiations of sampling: by sampling a log to a size manageable for other algorithms, many activities are removed, making the log incomplete. In fact, IMd and IMiD were the only algorithms that could handle the 10,000 activities.

Timewise, our ad-hoc pre-processing step on the log of 100,000,000 traces (A) took a few days, after that mining was a matter of seconds; P-IF, α, HM, IM, IMi and IMin took a few minutes; P-PT took days, ILP a few hours.

This experiment clearly shows the scalability of the IMD framework. A manual inspection revealed that IMD and IM algorithms always returned the same model for A once the log was complete.

5.2 Infrequent Behaviour

The goal of the second experiment is investigate the loss of quality faced by the algorithms of the IMD framework compared to the IM framework. We do this with two use cases in mind. First use case is to obtain a model describing almost all behaviour, i.e. having a fitness close to 1.0, preferably with good precision and generalisation. Second use case is, if precision and generalisation make the user consider all found 100% models to be unacceptable, to obtain a model that is accurate for 80% of the log, i.e. having a fitness of around 0.8, that represents the main flow of the process and allows for classification and separate analysis of outliers. Current evaluation techniques force us to perform this experiment on rather small event logs.

Setup. First, a random tree of 40 activities is generated. Second, from this tree a log of 1000 random traces is generated. Third, we vary the number t of infrequent behaviour traces that are added to the log and do not fit the model. Each infrequent trace is generated by inserting errors at certain decisions points with a probability of 0.2; as a result the infrequent trace does not fit the model. The total number of deviations in the log is recorded. Fourth, a discovery algorithm is applied, and conformance checking is measured using fitness [1], precision and generalisation [7]. This process is repeated for the same algorithms as the infrequency experiments, and increasing t. In this experiment, we compared only algorithms that return process trees, as only on sound models fitness, precision and generalisation can be measured reliably [20]. Simplicity is not measured as by nature of these algorithms all trees contain each activity once.

We performed a similar experiment on a real-life log of a financial institution of the Netherlands [14]. This log, consisting of 36 activities, was split in three parts, containing the activities prefixed by respectively A, O or W. On these logs, fitness, precision and generalisation were measured.

Results. The results of the infrequency experiments are shown in Table 2. For some measurements, mining finished but computation of fitness, precision and generalisation failed; these measurements are denoted by empty cells. (As shown in the scalability experiment, discovery algorithms easily handle much larger logs.) For each deviation level, there is an algorithm in both groups that gives a high fitness, i.e. use case 1. As to be expected, IMi and IMiD with higher parameter settings are needed when the log has more deviations. IMD usually scores worse in precision than the IM counterparts: IM clearly benefits from the full information in the log. However, IMD sometimes scores slightly better in generalisation.

Using parameter setting .8, IMiD returns models with fairly high precision and generalisation, even reaching levels similar to the best IMi parameter settings, although not achieving similar fitness. Therefore, IMiD at .8 seems to be

Table 2. Results of infrequent behaviour on a tree of 40 activities. f: fitness, p: precision, g: generalisation. t: # deviating traces. Model most suitable for use case $\underline{1}$ and $\utilde{2}$.

t \ log size deviations	0 \ 1000 0	10 \ 1010 30	100 \ 1100 450	1000 \ 2000 4472
IM		f $\underline{1.00}$ p $\underline{0.90}$ g $\underline{0.92}$	f 1.00 p 0.62 g 0.94	f 1.00 p 0.39 g 0.95	f $\underline{1.00}$ p $\underline{0.16}$ g $\underline{0.95}$
IMD		f 1.00 p 0.81 g 0.93	f $\underline{0.94}$ p $\underline{0.53}$ g $\underline{0.93}$	f 1.00 p 0.15 g 0.93	f 1.00 p 0.15 g 0.95
IMi	0.01	f $\underline{1.00}$ p $\underline{0.90}$ g $\underline{0.92}$	f 1.00 p 0.79 g 0.93	f 0.99 p 0.44 g 0.93	
IMiD	0.01	f 1.00 p 0.81 g 0.93	f 0.93 p 0.58 g 0.93	f 1.00 p 0.35 g 0.94	f 1.00 p 0.15 g 0.95
IMi	0.05	f $\underline{1.00}$ p $\underline{0.90}$ g $\underline{0.92}$	f $\underline{1.00}$ p $\underline{0.83}$ g $\underline{0.92}$	f $\underline{0.96}$ p $\underline{0.83}$ g $\underline{0.93}$	
IMiD	0.05	f 1.00 p 0.81 g 0.93	f 0.93 p 0.72 g 0.93	f $\underline{0.96}$ p $\underline{0.57}$ g $\underline{0.94}$	
IMi	0.20	f $\underline{1.00}$ p $\underline{0.90}$ g $\underline{0.92}$	f 0.98 p 0.81 g 0.92	f 0.94 p 0.82 g 0.93	
IMiD	0.20	f 0.98 p 0.90 g 0.92	f 0.89 p 0.62 g 0.93	f 0.95 p 0.51 g 0.94	f $\underline{0.94}$ p $\underline{0.46}$ g $\underline{0.95}$
IMi	0.80	f $\underline{1.00}$ p $\underline{0.90}$ g $\underline{0.92}$	f $\utilde{0.56}$ p $\utilde{0.88}$ g $\utilde{0.86}$	f $\utilde{0.61}$ p $\utilde{0.91}$ g $\utilde{0.87}$	f $\utilde{0.55}$ p $\utilde{0.69}$ g $\utilde{0.93}$
IMiD	0.80	f $\utilde{0.97}$ p $\utilde{0.91}$ g $\utilde{0.92}$	f $\utilde{0.70}$ p $\utilde{0.82}$ g $\utilde{0.91}$	f $\utilde{0.66}$ p $\utilde{0.68}$ g $\utilde{0.90}$	f $\utilde{0.67}$ p $\utilde{0.62}$ g $\utilde{0.90}$
IMin		f $\underline{1.00}$ p $\underline{0.90}$ g $\underline{0.92}$	f 1.00 p 0.55 g 0.94	f 1.00 p 0.29 g 0.96	
IMinD		f $\underline{1.00}$ p $\underline{0.90}$ g $\underline{0.92}$	f 0.86 p 0.74 g 0.93		

a good candidate default algorithm to get an 80% model from a log with infrequent behaviour. In a practical use case, this 80% model can be used to separate outliers from main flow, i.e. traces that fit the log and traces that do not. This classification can be used to investigate main flow and outliers using separate techniques, possibly using IMinD to achieve robust results on small logs.

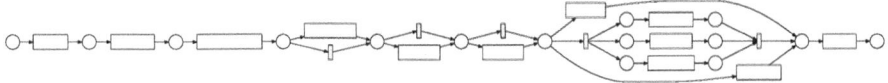

Fig. 8. Result without activity names of IMi 0.2 applied to BPIC12_A. f1.00, p0.89, g0.99.

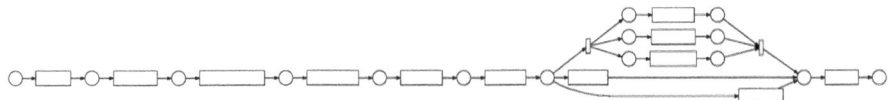

Fig. 9. Result without activity names of IMiD 0.2 applied to BPIC12_A. f0.93, p1.00, g0.99.

We illustrate the results of the experiments on the real-life logs using two models: Figure 8 shows the Petri net returned by IMi 0.2; Figure 9 the model by IMiD 0.2. This illustrates the different trade-offs made between IM and IMD: these models are very similar, except that for IMi, three activities can be skipped. Without the log, IMiD could not decide to make these activities skippable, which lowers fitness a bit (0.93 vs 1) but increases precision (1 vs 0.89).

To summarise, these experiments show that the IMD family is able to adequately handle logs with infrequent behaviour in different use cases, and can achieve results similar to IM with only minor losses in quality (except for the 100 deviations log).

6 Conclusion

Process discovery aims to obtain process models from event logs. Currently, there is no process discovery technique that works on logs containing billions of events or thousands of activities, and that guarantees both soundness and rediscoverability. In this paper, we pushed the boundary on what can be done with logs containing billions of events.

We introduced the *Inductive Miner - directly-follows based* (IMD) framework and three algorithms using it. The input of the framework is a directly-follows graph, which has been shown to be obtainable in highly-scalable environments, for instance using Map-Reduce style log analyses. The framework uses a divide-and-conquer strategy that splits this graph and recurses on the sub-graphs, until a base case is encountered.

We showed that the memory usage of all algorithms of the IMD framework is independent of the number of traces in the event log considered. In our experiments, the scalability was only limited by the logs we could generate. The IMD framework managed to handle over five billion events, while using 2GB of RAM; some other techniques required the event log to be in main memory and therefore could handle up to 10,000-100,000 traces. Besides performance, we also investigated how the new algorithms compare qualitatively to existing techniques that use more knowledge, but also have higher memory requirements, in a number of practical process mining use cases. We showed that the IMiD algorithms allow to handle infrequent behaviour adequately. In particular, parameter setting .8 allows to obtain an "80% model" of a process which helps to investigate main flow and outliers in more detail, an operation often needed for detailed analysis in process mining projects. Altogether, these results suggest that the IMD family not only handles big event logs, but that its algorithms can be used in a variety of use cases, eliminating the need to operate with different classes of algorithms.

Future Work. In the typical processs mining workflow as described in Section 1, process discovery is just the first step. Further analysis steps can be mostly manual, which makes them hard to perform on logs containing millions of traces. As we encountered during the evaluation, automatic steps are not guaranteed to work on these logs either. We envision further steps in the analysis of models, such as using alignments [7], in contexts of big data.

In streaming environments, it is assumed that events arrive in small intervals and that the log is too big to store, so the run time for each event should be short and ideally constant. The α algorithm and the Heuristics Miner have been applied in streaming environments [11] by using directly-follows graphs; the IMD framework might be applicable as well.

References

1. van der Aalst, W.M.P., Adriansyah, A., van Dongen, B.: Replaying history on process models for conformance checking and performance analysis. Wiley Interdisciplinary Reviews: Data Mining and Knowledge Discovery **2**(2), 182–192 (2012)
2. van der Aalst, W.M.P., van Hee, K.M., van der Werf, J.M.E.M., Verdonk, M.: Auditing 2.0: Using process mining to support tomorrow's auditor. IEEE Computer **43**(3), 90–93 (2010)
3. van der Aalst, W.M.P.: Process Mining: Discovery, Conformance and Enhancement of Business Processes. Springer (2011)
4. van der Aalst, W.M.P.: Process cubes: slicing, dicing, rolling up and drilling down event data for process mining. In: Song, M., Wynn, M.T., Liu, J. (eds.) AP-BPM 2013. LNBIP, vol. 159, pp. 1–22. Springer, Heidelberg (2013)
5. van der Aalst, W.M.P.. In: Data Scientist: Enigneer of the Future. I-ESA, vol. 7, pp. 13–26 (2014)
6. van der Aalst, W.M.P., Weijters, A., Maruster, L.: Workflow mining: Discovering process models from event logs. IEEE Trans. Knowl. Data Eng. **16**(9), 1128–1142 (2004)
7. Adriansyah, A., Munoz-Gama, J., Carmona, J., van Dongen, B.F., van der Aalst, W.M.P.: Alignment based precision checking. In: La Rosa, M., Soffer, P. (eds.) BPM Workshops 2012. LNBIP, vol. 132, pp. 137–149. Springer, Heidelberg (2013)
8. Badouel, E.: On the α-reconstructibility of workflow nets. In: Haddad, S., Pomello, L. (eds.) PETRI NETS 2012. LNCS, vol. 7347, pp. 128–147. Springer, Heidelberg (2012)
9. Buijs, J., van Dongen, B., van der Aalst, W.: A genetic algorithm for discovering process trees. In: IEEE Congress on Evolutionary Computation, pp. 1–8. IEEE (2012)
10. Buijs, J.C.A.M., van Dongen, B.F., van der Aalst, W.M.P.: On the role of fitness, precision, generalization and simplicity in process discovery. In: Meersman, R., Panetto, H., Dillon, T., Rinderle-Ma, S., Dadam, P., Zhou, X., Pearson, S., Ferscha, A., Bergamaschi, S., Cruz, I.F. (eds.) OTM 2012, Part I. LNCS, vol. 7565, pp. 305–322. Springer, Heidelberg (2012)
11. Burattin, A., Sperduti, A., van der Aalst, W.M.P.: Control-flow discovery from event streams. In: IEEE Congress on Evolutionary Computation, pp. 2420–2427 (2014)
12. Carmona, J., Solé, M.: PMLAB: an scripting environment for process mining. In: BPM Demos. CEUR-WP, vol. 1295, p. 16 (2014)
13. Datta, S., Bhaduri, K., Giannella, C., Wolff, R., Kargupta, H.: Distributed data mining in peer-to-peer networks. IEEE Internet Computing **10**(4), 18–26 (2006)
14. van Dongen, B.: BPI Challenge 2012 Dataset (2012). http://dx.doi.org/10.4121/uuid:3926db30-f712-4394-aebc-75976070e91f
15. Evermann, J.: Scalable process discovery using map-reduce. In: IEEE Transactions on Services Computing (2014, to appear)
16. Günther, C., Rozinat, A.: Disco: Discover your processes. In: BPM (Demos). CEUR Workshop Proceedings, vol. 940, pp. 40–44. CEUR-WS.org (2012)
17. Hay, B., Wets, G., Vanhoof, K.: Mining navigation patterns using a sequence alignment method. Knowl. Inf. Syst. **6**(2), 150–163 (2004)
18. Hwong, Y., Keiren, J.J.A., Kusters, V.J.J., Leemans, S.J.J., Willemse, T.A.C.: Formalising and analysing the control software of the compact muon solenoid experiment at the large hadron collider. Sci. Comput. Program. **78**(12), 2435–2452 (2013)

19. Leemans, S.J.J., Fahland, D., van der Aalst, W.M.P.: Discovering block-structured process models from event logs - a constructive approach. In: Colom, J.-M., Desel, J. (eds.) PETRI NETS 2013. LNCS, vol. 7927, pp. 311–329. Springer, Heidelberg (2013)
20. Leemans, S.J.J., Fahland, D., van der Aalst, W.M.P.: Discovering block-structured process models from event logs containing infrequent behaviour. In: Lohmann, N., Song, M., Wohed, P. (eds.) BPM 2013 Workshops. LNBIP, vol. 171, pp. 66–78. Springer, Heidelberg (2014)
21. Leemans, S.J.J., Fahland, D., van der Aalst, W.M.P.: Discovering block-structured process models from incomplete event logs. In: Ciardo, G., Kindler, E. (eds.) PETRI NETS 2014. LNCS, vol. 8489, pp. 91–110. Springer, Heidelberg (2014)
22. Leemans, S.J.J., Fahland, D., van der Aalst, W.M.P.: Exploring processes and deviations. In: Fournier, F., Mendling, J. (eds.) BPM 2014 Workshops. LNBIP, vol. 202, pp. 304–316. Springer, Heidelberg (2015)
23. Redlich, D., Molka, T., Gilani, W., Blair, G., Rashid, A.: Constructs competition miner: process control-flow discovery of bp-domain constructs. In: Sadiq, S., Soffer, P., Völzer, H. (eds.) BPM 2014. LNCS, vol. 8659, pp. 134–150. Springer, Heidelberg (2014)
24. Redlich, D., Molka, T., Gilani, W., Blair, G.S., Rashid, A.: Scalable dynamic business process discovery with the constructs competition miner. In: SIMPDA 2014. CEUR-WP, vol. 1293, pp. 91–107 (2014)
25. Weijters, A., van der Aalst, W., de Medeiros, A.: Process mining with the heuristics miner-algorithm. BETA Working Paper series 166, Eindhoven University of Technology (2006)
26. Wen, L., van der Aalst, W., Wang, J., Sun, J.: Mining process models with non-free-choice constructs. Data Mining and Knowledge Discovery **15**(2), 145–180 (2007)
27. Wen, L., Wang, J., Sun, J.: Mining invisible tasks from event logs. In: Dong, G., Lin, X., Wang, W., Yang, Y., Yu, J.X. (eds.) APWeb/WAIM 2007. LNCS, vol. 4505, pp. 358–365. Springer, Heidelberg (2007)
28. van der Werf, J., van Dongen, B., Hurkens, C., Serebrenik, A.: Process discovery using integer linear programming. Fundam. Inform. **94**(3-4), 387–412 (2009)

Multidimensional Process Mining Using Process Cubes

Alfredo Bolt[✉] and Wil M.P. van der Aalst

Department of Mathematics and Computer Science,
Eindhoven University of Technology, Eindhoven, The Netherlands
{a.bolt,w.m.p.v.d.aalst}@tue.nl

Abstract. Process mining techniques enable the analysis of processes using event data. For structured processes without too many variations, it is possible to show a relative simple model and project performance and conformance information on it. However, if there are multiple classes of cases exhibiting markedly different behaviors, then the overall process will be too complex to interpret. Moreover, it will be impossible to see differences in performance and conformance for the different process variants. The different process variations should be analysed separately and compared to each other from different perspectives to obtain meaningful insights about the different behaviors embedded in the process. This paper formalizes the notion of *process cubes* where the event data is presented and organized using different dimensions. Each cell in the cube corresponds to a set of events which can be used as an input by any process mining technique. This notion is related to the well-known OLAP (Online Analytical Processing) data cubes, adapting the OLAP paradigm to event data through *multidimensional process mining*. This adaptation is far from trivial given the nature of event data which cannot be easily summarized or aggregated, conflicting with classical OLAP assumptions. For example, multidimensional process mining can be used to analyze the different versions of a sales processes, where each version can be defined according to different dimensions such as location or time, and then the different results can be compared. This new way of looking at processes may provide valuable insights for process optimization.

Keywords: Process cube · Process mining · OLAP · Comparative process mining

1 Introduction

Process Mining can be seen as the missing link between model-based process analysis (e.g., simulation and verification) and data-oriented analysis techniques such as machine learning and data mining [1]. It seeks the "confrontation" between real event data and process models (automatically discovered or handmade). Classical process mining techniques focus on analysing a process as a whole, but in this paper we focus on isolating different *process behaviors* (versions) and present them in a way that facilitates their *comparison* by approaching process mining in a multidimensional perspective.

Multidimensional process mining has been approached recently by some authors. The *event cube* approach described in [2] presents an exploratory view on the applications of OLAP operations using *events*. The *process cube* approach is introduced by the second author in [3] with an initial prototype implementation [4]. The process cube notion was proven useful in case studies [5,6]. These approaches have established a *conceptual* framework for process cubes, however, they still present some conceptual limitations. One of the limitations of [3] is related to concurrency issues (e.g. derived properties are created on the *event base* which may be used with many *process cube structures*, which would force all the dimensions that correspond to a specific property to have exactly the same meaning and value set. This is an undesired behavior when for example, calculating in different *process cube structures* a dimension *customer type* according to different criteria). Other limitations are the structure within dimensions (e.g. there is no composition of attributes and no hierarchies of aggregation, therefore no *roll-up* and *drill-down* directions) and the (lack of) granularity-level definitions (used for defining the cube cells distribution and filter the events in each cell). In this paper we provide an improved formalization of the *process cube* conceptual framework.

The idea is related to the well-known OLAP multidimensional paradigm [7]. OLAP techniques organize the data under multiple combinations of dimensions and typically numerical measures, and accessing the data through different OLAP operations such as slicing, dicing, rolling up and drilling down. Lots of research have been conducted to deal with OLAP technical issues such as the materialization process. An extensive overview of such approaches can be found in [8]. The application of OLAP on non-numerical data is increasingly being explored. Temporal series, graphs and complex event sequences are possible applications [9–11]. However, there are two significant differences between OLAP and Process Cubes: Summarizability and Representation. The first refers to the classic OLAP cubes assumption on the summarizability of facts. This allows for pre-computations of the different multidimensional perspectives of the cube, which provides real-time (On-Line) analysis capabilities. Some authors have studied summarizability issues in OLAP [15,16] and attempt to solve it by introducing rules and constraints to the data model. In Process Cubes, summarizability is not guaranteed because of the process-oriented nature of the event data used. In Process Mining, each event is related to one or more traces, and the relevance of an event as data is given mostly by its relations with other events within those traces. One cannot simply merge or split Process Cube cells as summarizable OLAP cells because events are ordered, and any slight change in that ordering may change the whole representation of the cell where that event is being contained. The second refers to classical OLAP relying on the aggregation of facts for reducing a set of values into a single value that can be represented in many ways. On the other hand, Process Cubes have to deal with a much more complex representation of data. Process Cube cells are associated to process models and not just event data, and both are directly related. Observed and modeled behavior can be compared, process models can be discovered from events, and events can be used to replay behavior into otherwise static process models.

The remainder is organized as follows. In Sec 2. we define the process cube notion as a means for viewing event data from different perspectives. Sec 3. presents our implementation of process cubes. In Sec 4. we discuss the experiments and benefits that can be achieved through our approach. Finally Sec 5. concludes the paper by discussing some challenges and future work.

2 Process Cubes

In this section we will formalize the notion of a process cube, defining all of its inner components. A process cube is formed by a structure that describes the "shape" of the cube (distribution of cells) and by the real data that will be used as a basis to "fill" those cells.

2.1 Event Base

Normally, *event logs* serve as the starting point for process mining. these logs are created having a particular process and a set of questions in mind. An event log can be viewed as a multiset of *traces*. Each trace describes the life-cycle of a particular *case* (i.e., a *process instance*) in terms of the *activities* executed. Often event logs store additional information about events. For example, many process mining techniques use extra information such as the *resource* (i.e., person or machine) executing or initiating the activity, the *timestamp* of the event, or *data elements* recorded with the event.

An *event collection* is a set of events that have certain properties, but no defined *cases* and *activities*. Table 1 shows a small fragment of some larger event collection. Each event has a unique id and several properties. For example, event 0001 is an instance of action A that occurred on December 28th of 2014 at 6:30 am, was executed by John, and costed 100 euros. An event collection can be transformed into an event log by selecting event properties (or attributes) as *case_id* and *activity_id*. For example, in Table 1, *sales order* could be the *case_id* and *action* could be the *activity_id* of an event log containing all events of the event collection.

For process cubes we consider an *event base*, i.e., a large collection of events not tailored towards a particular process or predefined set of questions. An event base can be seen as an all-encompassing event log or the union of a collection of related event logs. The events in the event base are used to populate the cells in the cube. Throughout the paper we assume the following universes.

Definition 1 (Universes). $\mathcal{U}_V$ *is the universe of all attribute values (e.g., strings, numbers, etc..).* $\mathcal{U}_S = \mathcal{P}(\mathcal{U}_V)$ *is the universe of value sets.* $\mathcal{U}_A$ *is the universe of all attribute names (e.g., year, action, etc...).*

Table 1. A fragment of an event collection: each row corresponds to an event

event_id	sales order	timestamp	action	resource	cost
0001	1	28-12-2014:06.30	A	John	100
0002	1	28-12-2014:07.15	B	Anna	
0003	1	28-12-2014:08.45	C	John	
0004	2	28-12-2014:12.20	A	Peter	150
0005	1	28-12-2014:20.28	D	Mike	
0006	2	28-12-2014:23.30	C	Anna	
...	...	...	...	...	...

Note that $v \in \mathcal{U}_V$ is a single value (e.g., $v = 5$), $V \in \mathcal{U}_S$ is a set of values (e.g., $V = \{Europe, America\}$), $a \in \mathcal{U}_A$ is a single attribute name (e.g., age).

Definition 2 (Event Base). *An event base $EB = (E, P, \pi)$ defines a set of events E, a set of event properties P, and a function $\pi \in P \to (E \nrightarrow \mathcal{U}_V)$. For any property $p \in P$, $\pi(p)$ (denoted π_p) is a partial function mapping events into values. If $\pi_p(e) = v$, then event $e \in E$ has a property $p \in P$ and the value of this property is $v \in \mathcal{U}_V$. If $e \notin dom(\pi_p)$, then event e does not have a property p and we write $\pi_p(e) = \bot$ to indicate this.*

An event base is created from an event collection like the one presented in Table 1. If we transform this table into an EB, then the set of events E consist of all different elements of the *event_id* column of Table 1. In this case, $E = \{0001, 0002, 0003, 0004, 0005, 0006,...\}$. The set of properties P is the set of column headers of Table 1, with the exception of *event_id*. In this case, $P = \{sales\ order, timestamp, action, resource, cost\}$. The function π retrieves the value of each row (event) and column (property) combination (cell) in Table 1. For example, the value of the property *action* for the event 0001 is given by $\pi_{action}(0001) = A$. In the case that this value is empty in the table, we will use $\bot$ to denote it in the EB (e.g., $\pi_{cost}(0002) = \bot$).

Note that an event identifier (*event_id*) $e \in E$ does not have a meaning, but it is unique for each event.

2.2 Process Cube Structure

Independent of the event base EB we define the *structure* of the process cube. A *Process Cube Structure (PCS)* is fully characterized by the set of *dimensions* defined for it, each dimension having its own *hierarchy*.

Before defining the concepts of hierarchy and dimension, we need to define some basic graph properties.

Definition 3 (Directed Acyclic Graph). *A directed acyclic graph (DAG) is a pair $G = (N, E)$ where N is a set of nodes and $E \subseteq N \times N$ a set of edges connecting these nodes, where:*

- $n_1, n_2 \in N, n_1 \neq n_2 : e_1 = (n_1, n_2) \in E$ is a *directed edge* that starts in n_1 and ends in n_2,
- A *walk* $W \in E^*$ with a length of $|W| \geq 1$ is an ordered list of directed edges $W = (e_1, ..., e_k)$ with $e_j \in E : e_j = (n_j, n_{j+1}) \Rightarrow e_{j+1} = (n_{j+1}, n_{j+2})$, $1 \leq j < k \in \mathbb{N}$, and
- $\forall n \in N$: there is no walk $W \in E^*$ that starts and ends in n.

Note that there cannot be any directed cycles of any length in a DAG. For example, part (1) in Fig 1 shows a DAG with nodes: {*City, Country, etc...*}.

Definition 4 (Dimension). *A dimension is a pair* $d = ((A, H), valueset)$ *where the* hierarchy (A, H) *is a DAG with nodes* $A \subseteq \mathcal{U}_A$ *(attributes) and a set of directed edges* $H \subseteq A \times A$, *and valueset* $\in A \to \mathcal{U}_S$ *is a function defining the possible set of values for each attribute.*

The attributes in A are unique. The set of directed edges H defines the navigation directions for exploring the dimension. An edge $(a_1, a_2) \in H$ means that attribute a_1 can be *rolled up* to attribute a_2 (defined in Sec 2.5). A dimension should describe events from a single perspective through any combination of its attributes (e.g., attributes *city* and *country* can describe a *Location*) where attributes describe the perspective from higher or lower levels of detail (e.g., *city* describes a *Location* in a more fine-grained level than *country*). However, this is not strict and users can define dimensions as they want.

An attribute $a \in A$ has a $valueset(a)$ that is the set of possible values and typically only a subset of those values are present in a concrete instance of the process cube. For example, $valueset(age) = \{1, 2, ..., 120\}$ for $age \in A$. Another example: $valueset(cost) = \mathbb{N}$ allows for infinitely many possible values. We introduce the notation A_d to refer to the set of attributes A of the dimension d, and $\mathcal{U}_D$ as the universe of all possible dimensions. Fig 1. shows some examples of dimensions, each containing a DAG and a valueset function.

Definition 5 (Process Cube Structure). *A process cube structure is a set of dimensions* $PCS \subseteq \mathcal{U}_D$, *where for any two dimensions* $d_1, d_2 \in PCS, d_1 \neq d_2 : A_{d_1} \cap A_{d_2} = \emptyset$.

All dimensions in a process cube structure are independent from each other, this means that they do not have any attributes in common, so all attributes are unique, however, their value sets might have common values. We introduce the notation A_{pcs} to refer to the union of all sets of attributes $\bigcup_{d \in PCS} A_d$.

2.3 Compatibility

A process cube structure PCS and an event base EB are independent elements, where the PCS is the structure and the EB is the content of the cube. To make sure that we can use them together, we need to relate them through a mapping function and then check whether they are compatible.

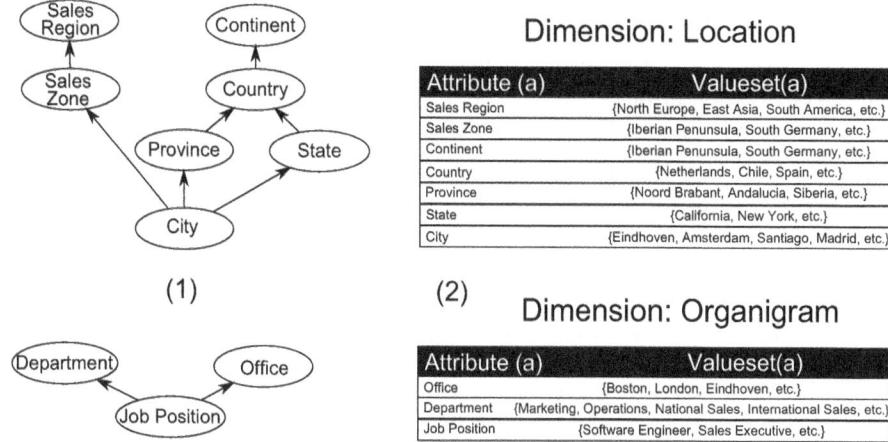

Fig. 1. Example of two dimensions (Location, Organigram), both conformed by a directed acyclic graph (1) and a valueset function (2)

Definition 6 (Mapper). *A mapper is a triplet $M = (PCS, EB, R)$ where PCS is a process cube structure, $EB = (E, P, \pi)$ is an event base and R is a function: $R \in A_{pcs} \to (\mathcal{P}(P) \times (E \nrightarrow \mathcal{U}_V))$. For an attribute $a \in A_{pcs}$, $R(a) = (P', g_a)$ where $P' = \{p_1, ..., p_n\} \in \mathcal{P}(P)$ is a set of properties and g_a is a calculation function mapping events into values used to calculate the value of a, so that for any event $e \in E$ with $\forall p \in P' : e \in dom(\pi_p)$, the value of attribute a for the event e is given by $g_a(e) = f(\pi_{p_1}(e), ..., \pi_{p_n}(e)) = v \in \mathcal{U}_V$. If for any $p \in P' : e \notin dom(\pi_p)$ then event e does not have the property p and the value of the attribute a cannot be calculated, so we write $g_a(e) = \bot$ to indicate this.*

Note that each attribute is related to one set of properties which is used to calculate the value of the attribute for any event through a specific calculation function. For example, an attribute *day type* = {*weekend, weekday*} can be calculated using the event properties {*day, month, year*} according to some specific calendar rules. Another example is an attribute *age* which can be calculated from properties {*birthday, timestamp*} by the function: $g_{age}(e) = \pi_{timestamp}(e) - \pi_{birthday}(e)$.

A set of properties can be used by more than one attributes producing different results if the calculation function is different. For example, in sales one could use the set of properties {*purchase amount, purchase num*} to classify customers into an attribute *customer type* = {*Gold, Silver*} (i.e., if *purchase amount* > 50 and *purchase num* > 10, then *customer type* = *Silver*) and at the same time to detect fraud into an attribute *fraud risk* = {*High, Low*} (i.e., if *purchase amount* > 100000 and *purchase num* = 1, then *fraud risk* = *High*).

Given a mapper $M = (PCS, EB, R)$ we say that PCS and EB are *compatible* through R, making all views of PCS also compatible with the EB.

2.4 Process Cube View

Once a proces cube structure is defined, it does not change. While applying typical OLAP operations such as slice, dice, roll up and drill down (defined in Sec 2.5) we only change the way we are visualizing the cube and its content. A *process cube view* defines the visible part of the process cube structure.

Definition 7 (Process Cube View). *Let PCS be a process cube structure. A process cube view is a triplet* $PCV = (D_{vis}, sel, gran)$ *such that:*

- $D_{vis} \subseteq PCS$ *are the visible dimensions,*
- $sel \in A_{pcs} \to \mathcal{U}_S$ *is a function selecting a part of the value set of the attributes of each dimension, such that for any* $a \in A_{pcs}: sel(a) \subseteq valueset(a)$, *and*
- $gran \in D_{vis} \to \mathcal{U}_A$ *is a function defining the granularity for each one of the visible dimensions.*

The *sel* function selects sets of values per attribute (including attributes in not visible dimensions). For example, in the dimension *Organigram* in Fig 1, one could select the job position *Sales Executive*, but many departments could have that same job position, so we could also select the department *National Sales* to only see the Sales Executives that work in National Sales. On the other hand, if this selection is done incorrectly, it might lead to empty results. For example in the dimension *Location* in Fig 1. one could select the city *Eindhoven* and the country *Spain* and this would produce empty results since no event can have both values. In our approach we made this as flexible as possible, so it is up to the user to check if the selection is done properly.

For each visible dimension, the *gran* function defines one of its attributes as the granularity. This will be used to define the *cell set* of the cube where each value of the granularity attribute corresponds to a cell. For example, in the dimension *Organigram* in Fig 1, one could define the Job Title as granularity.

Many different process cube views can be obtained form the same process cube structure. For example, Fig. 2. shows two process cube views obtained from the same process cube structure.

Definition 8 (Cell Set). *Let PCS be a process cube structure and* $PCV = (D_{vis}, sel, gran)$ *be a view over PCS with* $D_{vis} = \{d_1, ..., d_n\}$. *The* cell set *of PCV is defined as* $CS_{pcv} = AV_{d_1} \times ... \times AV_{d_n}$, *where for any* $d_i \in D_{vis} : AV_{d_i} = gran(d_i) \times sel(gran(d_i))$ *is a set of attribute-value sets.*

Although the term *cube* suggests a three dimensional object a process cube can have any number of visible dimensions.

A cell set CS is the set of visible cells of the process cube view. For example, for a process cube view with visible dimensions *Location* and *Time* with their granularity set to: $gran(Location) = \{City\}$ and $gran(Time) = \{Year\}$ and the selected values of those attributes were: $sel(City) = \{Eindhoven, Amsterdam\}$ and $sel(Year) = \{2013, 2014\}$, the cube would have the following 4 cells: $\{(City, Eindhoven), (Year, 2013)\}$, $\{(City, Eindhoven), (Year, 2014)\}$, $\{(City, Amsterdam), (Year, 2013)\}$, and $\{(City, Amsterdam), (Year, 2014)\}$.

Multidimensional Process Mining 109

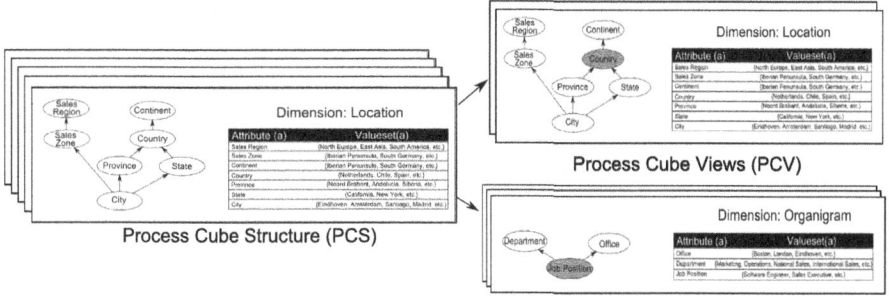

Fig. 2. Example of two PCVs created from the same PCS, both selecting some dimensions, selecting a part of the valuesets, and selecting attributes as granularity for the selected dimensions

2.5 Process Cube Operations

Next we consider the classical OLAP operations in the context of our process cubes.

The *slice operation* produces a new cube by allowing the analyst to filter (pick) specific values for attributes within one of the dimensions, while removing that dimension from the visible part of the cube.

The *dice operation* produces a subcube by allowing the analyst to filter (pick) specific values for one of the dimensions. No dimensions are removed in this case, but only the selected values are considered. Fig 3. illustrates the notions of slicing and dicing. For both operations, the same filtering was applied. In the case of the slice operation, the *Location* dimension is no longer visible, but in dice one could still use that dimension for further operations (i.e., drilling down to *City*) keeping the same dimensions visible.

The *roll up* and *drill down* operations do not remove any dimensions or filter any values, but only change the level of granularity of a specific dimension. Fig 4. shows the concept of drilling down and rolling up. These operations are intended

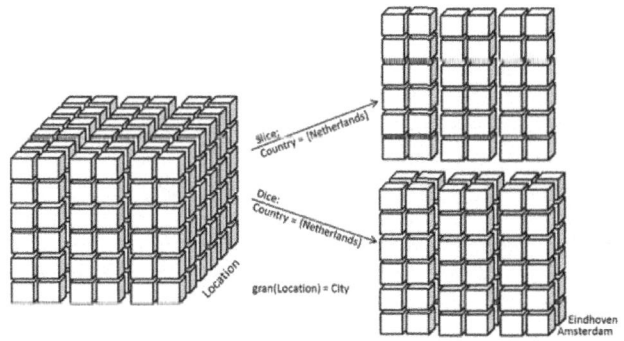

Fig. 3. Slice and Dice operations

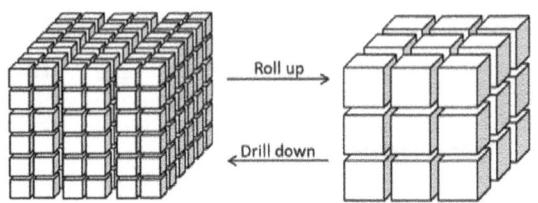

Fig. 4. Roll up and Drill down operations

to show the same data with more or less detail (granularity). However, this is not guaranteed as it depends on the dimension definition.

Definition 9 (Slice). Let PCS be a process cube structure and let $PCV = (D_{vis}, sel, gran)$ be a view of PCS. We define slice for a dimension $d \in D_{vis}$: $d = ((A, H), valueset)$ and a filtering function $fil \in A_d \to \mathcal{U}_S$ where for any $a \in A_d : fil(a) \subseteq valueset(a)$, as: $slice_{d,fil}(PCV) = (D'_{vis}, sel', gran)$, where:

- $D'_{vis} = D_{vis} \setminus \{d\}$ is the new set of visible dimensions, and
- $sel' \in A_{pcs} \to \mathcal{U}_S$ is the new selection function, where:
  - for any $a \in A_d : sel'(a) = fil(a)$, and
  - for any $a \in A_{pcs} \setminus A_d : sel'(a) = sel(a)$.

The *slice* operation produces a new process cube view. Note that d is no longer a visible dimension: $d \notin D'_{vis}$ but it will be used for filtering events. The new sel' function will still be valid as a value set selection function when filtering. Also, note that the *gran* function remains unaffected. For example, for sales data one could slice the cube for a dimension *Location* for City *Eindhoven*, the Location dimension is removed from the cube and only sales of the stores in Eindhoven are considered. One could also do more complex slicing. For example, for a dimension *time*, one could slice that dimension and select years 2013 and 2014 and months *January* and *February*, then the time dimension is removed from the cube and only sales in January or February of years 2013 or 2014 are considered.

Definition 10 (Dice). Let PCS be a process cube structure and let $PCV = (D_{vis}, sel, gran)$ be a view of PCS. We define dice for a dimension $d \in D_{vis}$: $d = ((A, H), valueset)$ and a filtering function $fil \in A_d \to \mathcal{U}_S$ where for any $a \in A_d : fil(a) \subseteq valueset(a)$, as: $slice_{d,fil}(PCV) = (D_{vis}, sel', gran)$, where:

- $sel' \in A_{pcs} \to \mathcal{U}_S$ is the new selection function, where:
  - for any $a \in A_d : sel'(a) = fil(a)$, and
  - for any $a \in A_{pcs} \setminus A_d : sel'(a) = sel(a)$.

The *dice* operation is very similar to the *slice* operation defined previously, where the only difference is that in *dice* the dimension is not removed from D_{vis}.

Definition 11 (Change Granularity). *Let PCS be a process cube structure and $PCV = (D_{vis}, sel, gran)$ a view of PCS. We define* chgr *for a dimension $d \in D_{vis} : d = ((A, H), valueset)$ and an attribute $a \in A_d$ as: $chgr_{d,a}(PCV) = (D_{vis}, sel, gran')$, where $gran'(d) = a$, and for any $d' \in D_{vis} \setminus \{d\} : gran'(d') = gran(d')$.*

This operation produces a new process cube view and allows us to set any attribute of the dimension d as the new granularity for that dimension, leaving any other dimension untouched. However, typical OLAP cubes allow the user to "navigate" through the cube using roll up and drill down operations, changing the granularity in a guided way through the hierarchy of the dimension. Note that D_{vis} and sel always remain unaffected when changing granularity. Now we define the roll up and drill down operation using the previously defined $chgr$ function.

Definition 12 (Roll up and Drill Down). *Let PCS be a process cube structure, $PCV = (D_{vis}, sel, gran)$ a view of PCS with a dimension $d \in D_{vis} : d = ((A, H), valueset)$. We can* roll up *the dimension d if $\exists a \in A_d : (gran(d), a) \in H$. The result is a more coarse-grained cube: $rollup_{d,a}(PCV) = chgr_{d,a}(PCV)$. We can* drill down *the dimension d if $\exists a \in A_d : (a, gran(d)) \in H$. The result is a more fine-grained cube: $drilldown_{d,a}(PCV) = chgr_{d,a}(PCV)$.*

If there is more than one attribute a that the dimension could be *rolled up* or *drilled down* to, then any of those attributes can be a valid target, but we can pick only one each time. For example, for a dimension *Location* described in Fig 1, we could roll up the dimension from *City* to *Province*, *State* or *Sales Zone*.

2.6 Materialized Process Cube View

Once we selected a part of the cube structure, and there is a cell set defined as the visible part of the cube, now we have to add content to those cells. In other words, we have to add *events* to these cells so they can be used by process mining algorithms.

Definition 13 (Materialized Process Cube View). *Let $M = (PCS, EB, R)$ be a mapper with PCS being a process cube structure, $EB = (E, P, \pi)$ being an event base, and let $PCV = (D_{vis}, sel, gran)$ be a view over PCS with a cell set CS_{pcv}. The materialized process cube view for PCV and EB is defined as a function $MPCV_{pcv,eb} \in CS_{pcv} \to \mathcal{P}(E)$ that relates sets of events to cells $c \in CS_{pcv} : MPCV_{pcv,eb}(c) = \{e \in E | (1) \land (2)\}$, where:*

(1) $\forall d \in D_{vis}, R(gran(d)) = (P', g_{gran(d)}) : (gran(d), g_{gran(d)}(e)) \in c$
(2) $\forall a \in A_{pcs}, R(a) = (P', g_a) : g_a(e) \in sel(a)$

The first condition (1) relates an event to a cell if the event property values are related to the (attribute,value) pairs that define that cell. For example, for a cell $c = \{(year,2012),(city,Eindhoven)\}$ one could relate all events that have both attribute values to that cell.

The second condition (2) is to check if the events related to each cell are not filtered out by any other attribute of any dimension of the process cube structure. Note that this condition becomes specially useful when slicing dimensions.

Fig 5 shows an example of a materialized process cube view. Each of the selected dimensions conform the cell distribution of the cube, and the events in the event base are mapped to these cells.

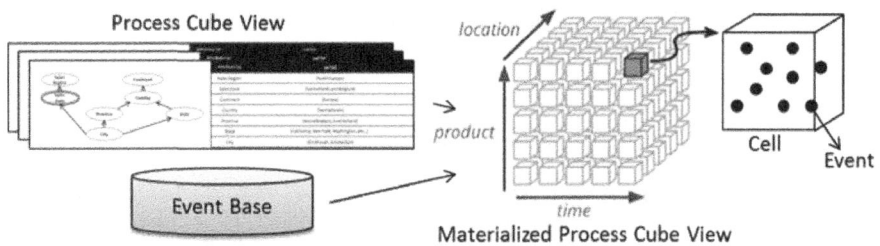

Fig. 5. Example of a materialized process cube view (MPCV) for an event base (EB) and a process cube view (PCV). The cells of MPCV contain events.

Normally events are related to specific activities or facts that happen in a process, and they are grouped in cases. In order to transform the set of events of a cell into an event log, we must define a *case id* to identify cases and an *activity id* to identify activities. The *case id* and *activity id* must be selected from the available attributes $(case\_id, activity\_id \in A_{pcs})$ where an attribute can be directly related to an event property without transformations. Given a set of events $E' \subseteq E$, we can compute a multiset of traces $L \in (valueset(activity\_id))^* \to valueset(case\_id)$ where each trace $\sigma \in L$ corresponds to a case. For example, in Table 1 if we select *sales order* as the *case id*, all events with *sales order* = 1 belong to case 1, which can be presented as $\langle A, B, C, D \rangle$. Similarly, case 2 can be presented as $\langle A, C \rangle$. Most control-flow discovery techniques [12–14] use such a simple representation as input. However, the composition of traces can be done using more attributes of events, such as *timestamp, resource* or any other attribute set $A' \subseteq A_{pcs}$.

3 Implementation

This approach has been implemented as a stand-alone java application (available in http://www.win.tue.nl/~abolt) named *Process Mining Cube (PMC)* (shown in Fig 6) that has 2 groups of functionalities: *log splitting* and *results generation*. The first consists of creating sublogs (cells) from a large event collection using the operations defined in Sec 2.5, allowing the user to interactively explore the data and isolate the desired behavior of an event collection. The second consists of converting each *materialized cell* into a process mining result, obtaining a collection of results visualized as a 2-D grid, facilitating the comparison between cells. For transforming each *materialized cell* into a process mining result we use existing components and plugins from the ProM framework [17] (www.processmining.org) which provides hundreds of plug-ins providing a wide range of analysis techniques.

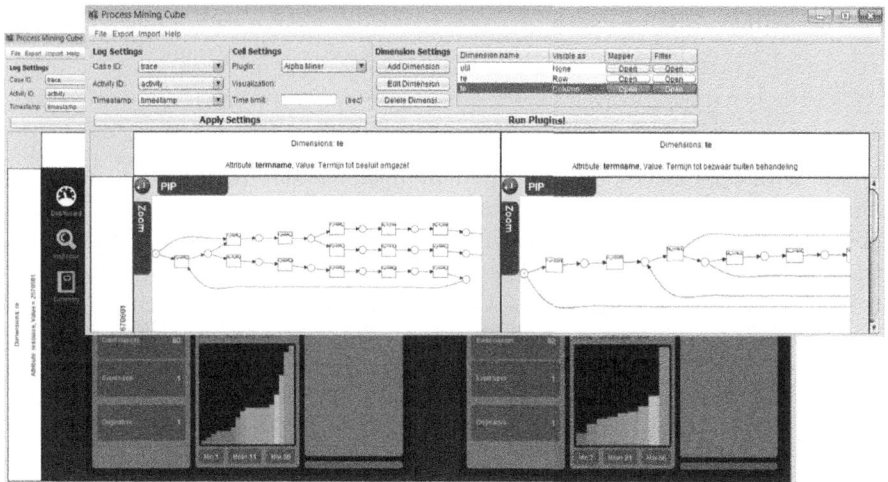

Fig. 6. Process Mining Cube (PMC): Implementation of this approach

The plugins and components used to analyze the cube cells in PMC v1.0 are described in Table 2. This plugin list is extendable. We expect to include more and more plugins for the following versions of PMC.

4 Experiments

In order to compare the performance of our implementation (*PMC*) with the current state of the art (*ProCube*) which was introduced in [4] (also cited in [3]), we designed an experiment using a real life set of events: the WABO1 event log. This log is publicly available in [18] and it is a real-life log that contains 38944 events related to the process of handling environmental permit requests of a Dutch municipality from October 2010 to January 2014. Each event contains more than 20 data properties. From this property set, only two of them were used as dimensions: *Resource* and *(case) termName*, which produces a 2D cube. Both dimensions were drilled down to its finest-grained level, so every different combination of values from these dimensions creates a different cell.

Table 2. Plugins available

Plugin Name	Plugin Description
Alpha Miner	Miner used to build a Petri net from an event log. Fast, but results are not always reliable
Log Visualizer	Visualization that allow us to get a basic understanding of the event log that we are processing
Inductive Miner	Miner that can provide a Petri net or a Process tree as output. Good when dealing with infrequent behavior and large event logs, ensures soundness
Dotted Chart	Visualization that represents the temporal distribution of events
Fast Miner	Miner based on a directly-follows matrix, with a time limit for generating it. The output is a directly-follows graph (Not a ProM plugin)

For both approaches, we compared the *loading time* (e.g. time required to import the events) and *creation time* (e.g. time required to create and materialize all cells and visualize them with the *Log Visualizer* plugin of ProM[17]) using 9 subsets of this log with different number of events. The more events we include in the subset, the larger the value set of a property gets (until the sample is big enough to contain all original values) and more cells are obtained. The experiment results for the 9 subsets are presented in Table 3.

Table 3. Performance benchmark for different-sized subsets of a log

Subset num.		Sub. 1	Sub. 2	Sub. 3	Sub. 4	Sub. 5	Sub. 6	Sub. 7	Sub. 8	Sub. 9
Number of events		1000	5000	10000	15000	20000	25000	30000	35000	38944
Number of cells		48	104	176	187	187	216	216	234	252
ProCube	Load (sec)	2.0	3.0	5.0	8.0	9.0	9.0	9.0	13.0	13.0
	Create (sec)	25.8	106.5	715.3	868.7	1053.2	1220.0	1399.3	1522.5	2279.3
PMC	Load (sec)	0.6	0.9	1.2	1.4	1.6	1.9	2.0	2.1	2.5
	Create (sec)	2.9	6.1	10.1	15.6	21.8	29.5	35.1	41.3	49.6
PMC Load *Speedup*		3.3	3.3	4.1	5.7	5.6	4.7	4.5	6.1	5.2
PMC Create *Speedup*		8.8	17.4	70.8	55.6	48.3	41.3	39.8	36.8	45.9

These results show that *PMC* out-performs the current state of the art in every measured perspective. All *Loading* and *Creation* (Create) times are measured in seconds. Notice that the *Speedup* of *PMC* over *ProCube* is quite considerable, as the average Creation *Speedup* is 40.5 (40 times faster). Also notice that when using the full event log (Sub. 9), *PMC* provides an acceptable response time by creating 252 different process analysis results in less than a minute, something that would take many hours, even days to accomplish if done by hand. This performance improvement makes *PMC* an attractive tool for the academic community and business analysts.

All the above experiments were performed in a laptop PC with an Intel i7-4600U 2.1GHz CPU with 8Gb RAM and Sata III SSD in Windows 7 (x64).

5 Conclusions

As process mining techniques are maturing and more event data becomes available, we no longer want to restrict analysis to a single all-in-one process. We would like to analyse and compare different variants (behaviors) of the process from different perspectives. Organizations are interested in comparative process mining to see how processes can be improved by understanding differences between groups of cases, departments, etc. We propose to use *process cubes* as a way to organize event data in a *multi-dimensional data structure tailored towards process mining*. In this paper, we extended the formalization of process cubes proposed in [3] and provided a working implementation with an adequate performance needed to conduct analysis using large event sets. The new framework gives end users the opportunity to analyze, explore and compare processes interactively on the basis of a multidimensional view on event data. We implemented the ideas proposed in this paper in our *PMC* tool, and we encourage the process mining community to use it. There is a huge interest in tools supporting process cubes and the practical relevance is obvious. However, some of the challenges discussed in [3] still remain unsolved (i.e. comparison of cells and concept drift). We aim to address these challenges using the foundations provided in this paper.

References

1. van der Aalst, W.M.P.: Process Mining: Discovery. Conformance and Enhacement of Business Processes. Springer, Berlin (2011)
2. Ribeiro, J.T.S., Weijters, A.J.M.M.: Event Cube: Another Perspective on Business Processes. In: Meersman, R., et al. (eds.) OTM 2011, Part I. LNCS, vol. 7044, pp. 274–283. Springer, Heidelberg (2011)
3. van der Aalst, W.M.P.: Process Cubes: Slicing, Dicing, Rolling Up and Drilling Down Event Data for Process Mining. In: Song, M., Wynn, M.T., Liu, J. (eds.) AP-BPM 2013. LNBIP, vol. 159, pp. 1–22. Springer, Heidelberg (2013)
4. Mamaliga, T.: Realizing a Process Cube Allowing for the Comparison of Event Data. Master's thesis, Eindhoven University of Technology, Eindhoven (2013)
5. van der Aalst, W.M.P., Guo, S., Gorissen, P.: Comparative Process Mining in Education: An Approach Based on Process Cubes. In Lesage, J.J., Faure, J.M., Cury, J., Lennartson, B. (eds.) 12th IFAC International Workshop on Discrete Event Systems (WODES 2014). IFAC Series, pp. PL1.1–PL1.9. IEEE Computer Society (2014)
6. Vogelgesang, T., Appelrath, H.J.: Multidimensional Process Mining: A Flexible Analysis Approach for Health Services Research. In: Proceedings of the Joint EDBT/ICDT 2013 Workshops (EDBT 2013), pp. 17–22. ACM, New York (2013)
7. Chaudhuri, S., Dayal, U.: An overview of data warehousing and OLAP technology. SIGMOD Rec. 26, pp. 65–74 (1997)
8. Han, J., Kamber, M.: Data mining: concepts and techniques, The Morgan Kaufmann series in data management systems. Elsevier (2006)

9. Chen, C., Yan, X., Zhu, F., Han, J., Yu, P.S.: Graph OLAP: a multi-dimensional framework for graph data analysis. Knowledge and Information Systems 21, 41–63 (2009)
10. Li, X., Han, J.: Mining approximate top-k subspace anomalies in multi-dimensional time-series data. In: Proceedings of the 33rd International Conference on Very Large Data Bases (VLDB), pp. 447–458. VLDB Endowment (2007)
11. Liu, M., Rundensteiner, E., Greenfield, K., Gupta, C., Wang, S., Ari, I., Mehta, A.: E-Cube: Multi-dimensional event sequence processing using concept and pattern hierarchies. In: International Conference on Data Engineering, pp. 1097–1100 (2010)
12. van der Aalst, W.M.P., Weijters, A.J.M.M., Maruster, L.: Workflow Mining: Discovering Process Models from Event Logs. In: IEEE International Enterprise Computing Conference (EDOC 2011), pp. 55–64. IEEE Computer Society (2011)
13. Carmona, J., Cortadella, J.: Process Mining Meets Abstract Interpretation. In: Balcazar, J.L. (ed.) ECML/PKDD 210. Lecture Notes in Artificial Intelligence, vol. 6321, pp. 184–199. Springer-Verlag, Berlin (2010)
14. Cook, J.E., Wolf, A.L.: Discovering Models of Software Processes from Event-Based Data. ACM Transactions on Software Engineering and Methodology **7**(3), 215–249 (1998)
15. Niemi, T., Niinimäki, M., Thanisch, P., Nummenmaa, J.: Detecting summarizability in OLAP. In: Data & Knowledge Engineering, vol. 89, pp. 1–20, Elsevier (2014)
16. Mazón, J., Lechtenbörger, J., Trujillo, J.: A survey on summarizability issues in multidimensional modeling. In Data & Knowledge Engineering, vol 68, pp. 1452–1469. Elsevier (2009)
17. van Dongen, B.F., de Medeiros, A.K.A., Weijters, A.J.M.M., van der Aalst, W.M.P.: The ProM Framework: A New Era in Process Mining Tool Support. In: Ciardo, G., Darondeau, P. (eds.) Applications and Theory of Petri Nets 2005. LNCS, vol. 3536, pp. 444–454. Springer, Berlin (2005)
18. J.C.A.M. Buijs. Environmental permit application process (WABO), CoSeLoG project Municipality 1. Eindhoven University of Technology. Dataset (2014). http://dx.doi.org/10.4121/uuid:c45dcbe9-557b-43ca-b6d0-10561e13dcb5

Declarative Approaches

Matching of Events and Activities - An Approach Using Declarative Modeling Constraints

Thomas Baier[1](✉), Claudio Di Ciccio[2], Jan Mendling[2], and Mathias Weske[1]

[1] Hasso Plattner Institute at the University of Potsdam,
Prof.-Dr.-Helmert-Str. 2-3, D-14482 Potsdam, Germany
{thomas.baier,mathias.weske}@hpi.de
[2] Wirtschaftsuniversität Wien, Welthandelsplatz 1, 1020 Vienna, Austria
{claudio.di.ciccio,jan.mendling}@wu.ac.at

Abstract. Nowadays, business processes are increasingly supported by IT services that produce massive amounts of event data during the execution of a process. This event data can be used to analyze the process using process mining techniques to discover the real process, measure conformance to a given process model, or to enhance existing models with performance information. Mapping the produced events to activities of a given process model is essential for conformance checking, annotation and understanding of process mining results. In order to accomplish this mapping with low manual effort, we developed a semi-automatic approach that maps events to activities using the solution of a corresponding constraint satisfaction problem. The approach extracts Declare constraints from both the log and the model to build matching constraints to efficiently reduce the number of possible mappings. The evaluation with an industry process model collection and simulated event logs demonstrates the effectiveness of the approach and its robustness towards non-conforming execution logs.

Keywords: Process mining · Event mapping · Business process intelligence · Constraint satisfaction

1 Introduction

Organizations often support the execution of business processes with IT systems that log each step of participants or systems. Individual entries in such logs represent the execution of services, the submission of a form, or other related tasks that in combination realize a business process. To improve business processes and to align IT process execution with existing business goals, a precise understanding of processes execution is necessary. Using the event data logged by IT systems, process mining techniques help organizations to have a more profound awareness of their processes, in terms of discovering and enhancing process models, or checking the conformance of the execution to the specification [2]. Yet, these process mining techniques face an important challenge: the mapping of log entries produced by IT systems to the corresponding process activities in the process models has to be known. A discovered process model can only be fully understood when the presented results use the terminology that is known to the business analysts. However, such a mapping is often not existing because (i) the logging mechanism of IT systems captures fine-granular steps on a technical level and (ii) especially with legacy systems, the way in which events are recorded is rarely customizable. In fact, it is often a tedious task to reconstruct a mapping from cryptic names in a database to the activities in a process model.

In this paper, we offer means to help the analyst identify the mapping between a process model and events in an event log, in a semi-automated fashion. Defining such a mapping is generally hard to do manually, due to its combinatorial complexity. While there exist automatic techniques such as [5] or [6], these approaches have limitations on their applicability. For example, [5] requires event names that are processable using linguistic techniques, which are not always provided. The work presented in [6] overcomes this limitation, yet it is able to handle only 1:1 relations between events and activities, and requires pre-processing to handle 1:N relationships. The approach presented in this paper overcomes these limitations by using declarative constraints, in order to turn the matching problem into a constraint satisfaction problem. In this way, we not only lift the limitations of [6], but also drastically narrow down the effort for an analyst. Our approach also informs research into Declare, as it has been mainly used for the modeling of discovered processes from event logs [8,21]. Here, we also devise techniques to derive Declare constraints from an existing imperative process model, in order to reason about possible matches between events and activities, through the comparison of Declare constraints inferred from the event log and the process model.

The remainder of this paper is structured as follows. Section 2 starts by further illustrating the problem with an example and stating the formal definition of the mapping problem and the required formal concepts. Having laid the foundations, the matching technique is introduced in Section 3. In Section 4, the proposed approach is evaluated using an industry process model collection and simulated event logs. Related work is discussed in Section 5 and Section 6 concludes the work.

2 Preliminaries

This section gives a running example to illustrate the problem and introduces the main concepts used for the mapping approach. We will formally introduce the notion of a process model, an event log, and the used Declare rules.

2.1 Illustrating Example

Starting with an example, Table 1 shows an exemplary event log with 5 traces, which have been produced by an IT system supporting the order process depicted in Fig. 1. Obviously, it is not straightforward to interpret the given event log, because the event labels are cryptic database field names, which cannot be easily matched to the names of the activities in the process model. It can be seen that for some of the activities multiple events are being logged, while for others only one type of events can be observed. Looking at the events of the "Change order" activity, not necessarily all events belonging to an activity are generated when the activity is executed. Once the mapping is established as shown in Tab. 2, we can use the event log to check conformance between the model and the log. For example, we are able to detect that there is a case in the log, in which the customer has already been notified before the products were shipped. It is critical for organizations to detect, and accordingly react to such non-conforming behavior [2]. Moreover, using process discovery techniques, a new process model that reflects the actual as-is process, including all deviations, can be automatically created using the known terminology.

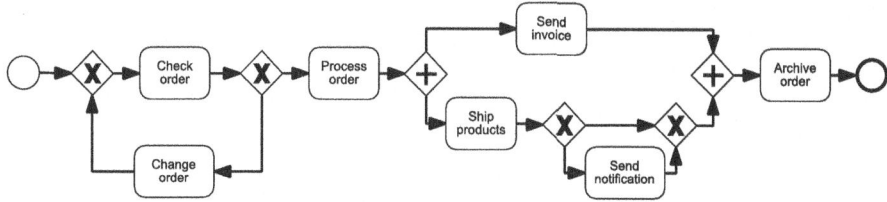

Fig. 1. Order process model in BPMN

Table 1. Event log (L) of order process (M)

	Label sequence
t_1	⟨ O_CHK_S, O_PR_S, O_PR_E, I_SM_E, P_SP_E, O_ARC_S, O_ARC_E ⟩
t_2	⟨ O_CHK_S, O_RC_SB, O_RC_E, O_CHK_S, O_PR_S, O_PR_E, P_SP_E, P_NOT_E, I_SM_E O_ARC_S, O_ARC_E ⟩
t_3	⟨ O_CHK_S, O_PR_S, O_PR_E, P_SP_E, P_NOT_E, I_SM_E, O_ARC_S, O_ARC_E ⟩
t_4	⟨ O_CHK_S, O_RC_SA, O_RC_E, O_CHK_S, O_PR_S, O_PR_E, P_NOT_E, I_SM_E, P_SP_E, O_ARC_S, O_ARC_E ⟩
t_5	⟨ O_CHK_S, O_PR_S, O_PR_E, P_SP_E, P_NOT_E, I_SM_E, O_ARC_S, O_ARC_E ⟩

Table 2. Mapping Map

Activity	Event label
Check order	O_CHK_S
Change order	O_RC_SA, O_RC_SB, O_RC_E
Process order	O_PR_S, O_PR_E
Send invoice	I_SM_E
Ship products	P_SP_E
Send notification	P_NOT_E
Archive order	O_ARC_S, O_ARC_E

2.2 Process Model and Event Log

Let S be a finite set of states, and A be a set of activities. A process model $M = (S, s_I, s_F, A, T)$ is a transition system that defines the allowed sequences of activity executions. Here, $T \in (S \times A \times S)$ is a transition relation modeling the allowed activities in a given state that result in a succeeding state. For example $(s_1, a, s_2) \in T$ implies that we can perform activity a in state s_1 and reach state s_2. A model has an initial state $s_I \in S$ and a final state $s_F \in S$. The function $\tau : M \to \mathcal{P}(A^*)$ captures all execution sequences starting with the initial state s_I and ending in the final state s_F that are allowed in T. Note that the number of execution sequences is infinite if the model contains loops. An execution sequence is also referred to as a process instance. For example, the model $M = (\{s_1, s_2, s_3\}, s_1, s_3, \{a, b, c\}, \{(s_1, a, s_2), (s_2, b, s_2), (s_2, c, s_3)\})$ has the execution sequences $\tau(M) = \{\langle a, c \rangle, \langle a, b, c \rangle, \langle a, b, b, c \rangle, \ldots\}$.

An IT system that supports process executions typically records events for each process instance in an event log [2]. Note that the relation of event instances to process instances might not be trivial in every practical setting. Yet, there exist approaches relating event instances to process instances that use event correlation (see, e.g., [22]). In this work, we therefore assume that this relation is already given. We abstract events as symbols of an alphabet E, which is often referred to as the set of event classes. Each process instance is represented as a sequence of events and also referred to as trace $t \in E^*$. For example, $\langle o, p, o, q \rangle$ is a trace with four consecutive events and three different event classes, $o, p, q \in E$. An event log L is a multiset of traces.

Confronted with a process model M and an event log L, the challenge is to derive the mapping relation between the activities $a \in A$ and the event classes $e \in E$. In this paper, we assume a 1:N relation as events are typically on a more fine-granuar level than activities. Thus, we are looking for the surjective function $Map : E \to A$ that maps event classes to their corresponding activities.

2.3 Declare

Having a process model and an event log, the approach presented in this paper will use Declare to describe their behavior. Declare [3] is natively a declarative process modeling language. It models workflows by means of temporal rules.[1] Such rules are meant to impose specific conditions on the occurrence of tasks in process instances. The rationale is, that every behavior in the process enactment is allowed, as long as it does not violate the specified rules. Due to this, declarative models are said to be "open", in contrast with the "closed" fashion of classical procedural models [21]. The Declare standard provides a predefined library of templates, listing default restrictions that can be imposed on the process control-flow. For instance, *Participation*(a) is a Declare rule expressed on activity a. It states that a must occur in every trace. *NotCoExistence*(a, b) constrains a and b, and imposes that a and b never occur together in the same trace. *Participation*(a) expresses a condition on the execution of a single activity. It is thus said to be an *existence rule*, as opposed to *relation rules*, such

---

[1] In literature, they are called "constraints". Nevertheless, we prefer not to make use of such term, in order to avoid the conflict with "constraints" in the context of constraint satisfaction problems (CSPs).

Table 3. Declare templates

Rule	Explanation	Cat.	Positive and negative examples
$Participation(\mathsf{a})$	a occurs at least *once*	$\mathcal{C}_E$	✓ bcac ✓ bcaac × bcc × c
$Init(\mathsf{a})$	a is the *first* to occur	$\mathcal{C}_E$	✓ acc ✓ abac × cc × bac
$End(\mathsf{a})$	a is the *last* to occur	$\mathcal{C}_E$	✓ bca ✓ baca × bc × bac
$Precedence(\mathsf{a},\mathsf{b})$	b occurs only if preceded by a	$\mathcal{C}_R^{\rightarrow}$	✓ cacbb ✓ acc × ccbb × bacc
$AlternatePrecedence(\mathsf{a},\mathsf{b})$	Each time b occurs, it is preceded by a and no other b can recur in between	$\mathcal{C}_R^{\rightarrow}$	✓ cacba ✓ abcaacb × cacbba × acbb
$ChainPrecedence(\mathsf{a},\mathsf{b})$	Each time b occurs, then a occurs immediately beforehand	$\mathcal{C}_R^{\rightarrow}$	✓ abca ✓ abaabc × bca × bacb
$CoExistence(\mathsf{a},\mathsf{b})$	If b occurs, then a occurs, and viceversa	$\mathcal{C}_R$	✓ cacbb ✓ bcca × cac × bcc
$Succession(\mathsf{a},\mathsf{b})$	a occurs if and only if it is followed by b	$\mathcal{C}_R^{\rightarrow}$	✓ cacbb ✓ accb × bac × bcca
$AlternateSuccession(\mathsf{a},\mathsf{b})$	a and b if and only if the latter follows the former, and they alternate each other in the trace	$\mathcal{C}_R^{\rightarrow}$	✓ cacbab ✓ abcabc × caacbb × bac
$ChainSuccession(\mathsf{a},\mathsf{b})$	a and b occur if and only if the latter immediately follows the former	$\mathcal{C}_R^{\rightarrow}$	✓ cabab ✓ ccc × cacb × cbac
$NotSuccession(\mathsf{a},\mathsf{b})$	a can never occur before b	$\mathcal{C}_R^{\rightarrow}$	✓ bbcaa ✓ cbbca × aacbb × abb
$NotCoExistence(\mathsf{a},\mathsf{b})$	a and b never occur together	$\mathcal{C}_R$	✓ cccbbb ✓ ccac × accbb × bcac

as $NotCoExistence(\mathsf{a},\mathsf{b})$, which indeed constrains pairs of activities. In the following, existence templates will be denoted as $\mathcal{C}_E$, and $\mathcal{C}_E(x)$ is the rule that applies template $\mathcal{C}_E$ to activity $x \in A$. Relation rules will instead be denoted as $\mathcal{C}_R$. $\mathcal{C}_R(x,y)$ applies template $\mathcal{C}_R$ to $x, y \in A$. $Precedence(\mathsf{a},\mathsf{b})$ is the relation rule establishing that, if b occurs in the trace, then it must be *preceded* by at least one occurrence of a. In addition to relation rules, it imposes a condition on the *ordering* in which constrained activities can occur. Therefore, $Precedence(\mathsf{a},\mathsf{b})$ falls under the category of *ordering relation* rules. Templates of such category will be denoted as $\mathcal{C}_R^{\rightarrow}$. $\mathcal{C}_R^{\rightarrow}(x,y)$ indicates an *ordering relation* rule applied to $x, y \in A$. In particular, $\mathcal{C}_R^{\rightarrow}(x,y)$ always specifies the order in which the occurrences of x and y are considered: x first, y afterwards (henceforth, *order direction*).

Table 3 lists the set of Declare rules that are mentioned in the remainder of the paper, along with the *category* (i.e., either $\mathcal{C}_E$, $\mathcal{C}_R$ or $\mathcal{C}_R^{\rightarrow}$) to which they belong. For every rule, two examples of complying traces and two examples of violating traces are provided. The complete list of rules can be found in [3].

Declare rules, when discovered from event logs, are usually associated to a reliability metric, namely *support* [9,21]. Support is a normalized value, ranging from 0 to 1, which measures to what extent traces are compliant with a rule. A support of 0 stands for a rule which is always violated. Conversely, a value of 1 is assigned to the support of rules which always hold true. According to the measurement introduced by the work of [9], the analysis of a trace $t_1 = \langle b, a, c, b, a, b, b, c \rangle$ would, e.g., lead to a support of 1 to $Participation(a)$, 0 to $NotCoExistence(a, b)$, and 0.75 to $Precedence(a, b)$, as 3 b's out of 4 are preceded by an occurrence of a. Considering an event log, which consists of t_1 and $t_2 = \langle c, c, a, c, b \rangle$, the support of $Participation(a)$ and $NotCoExistence(a, b)$ would remain equal to 1 and 0, respectively, whereas the support of $Precedence(a, b)$ would be 0.8 (4 b's out of 5 are preceded by an occurrence of a). [9] provides further details on the computation of support values for each rule. This metric is usually utilized to prune out those rules which are associated to a value below a user-defined threshold.

3 Mapping Event Log and Process Model

This section introduces the approach for the mapping of events to given activities in a process model. The approach consists of three phases. The first one builds and solves a constraint satisfaction problem, to reduce the number of possible mappings between activities and events. The result of this phase is a set of potential event-activity mappings. During the second phase, the analyst is guided to select the correct mapping from the derived potential mappings. Finally, the last phase is used to automatically transform one or many event logs to reflect the activities in the process model. In the following sections, we will elaborate on each of the three phases.

3.1 Reduction of the Potential Set of Event–Activity Mappings

The first phase of our approach deals with the definition of a constraint satisfaction problem (CSP), which is used to restrain the possible mappings of events and activities. A CSP is a triple $CSP = (X, D, C)$ where $X = \langle x_1, x_2, \ldots, x_n \rangle$ is an n-tuple of variables with the corresponding domains specified in the n-tuple $D = \langle D_1, D_2, \ldots, D_n \rangle$ such that $x_i \in D_i$ [13]. $C = \langle c_1, c_2, \ldots, c_t \rangle$ is a t-tuple of constraints. We use predicate logic to express the constraints used in this paper. The set of solutions to a CSP is denoted as $S = \{S_1, S_2, \ldots, S_m\}$, where each solution $S_k = \langle s_1, s_2, \ldots, s_n \rangle$ is an n-tuple with $k \in 1..m$, $s_i \in D_i$ and such that every constraint in C is satisfied.

To build the CSP, first, the activities and event labels need to be mapped to the set of variables and their domains. Therefore, a bijective function $var : E \to X$ is defined, which assigns each event label to a variable with the natural numbers $1..|A|$ as domain. Furthermore, a bijective function $val : A \to 1..|A|$ is defined, which assigns each activity a natural number in the range from 1 to the number of activities. Table 4 and Table 5 show the mapping *var* and the mapping *val* respectively for the example given in Section 2.

Table 4. Mapping *var*

Event $e \in E$	O_CHK_S	O_RC_SA	O_RC_SB	O_RC_E	O_PR_S	O_PR_E	LSM_E	P_SP_E	P_NOT_E	O_ARC_S	O_ARC_E
Variable $var(e) \in X$	x_1	x_2	x_3	x_4	x_5	x_6	x_7	x_8	x_9	x_{10}	x_{11}

Table 5. Mapping *val*

Activity $a \in A$	Check order	Change order	Process order	Send invoice	Ship Products	Send notification	Archive order
Value $val(a) \in 1..|A|$	1	2	3	4	5	6	7

With the variables and domains defined, the solutions to the CSP reflect all possible mappings between events and activities, i.e., for n activities and m events there are potentially n^m solutions. For the example given in section 2.1 these are $7^{11} = 1,977,326,743$ possible mappings. Yet, this also includes solutions where not all activities are assigned to an event or solutions where all events are mapped to one single activity. As these solutions are not desired, we first restrict the set of solutions to those that assign each activity to at least one event. Note that we assume that the execution of each activity in the process model is being logged by the supporting IT system. Thus, those activities that are not recorded, are not considered in the processing. We assume that each event in the given log relates to exactly one activity in the process model, whereas one activity can relate to multiple events. Thus, we are using the NVALUE constraint, available in many constraint problem solvers (cf. [13]). This constraint ensures that each value in the domain of the variables is assigned at least once. Still, the complexity of the matching problem remains very high. In the following, we present an approach to tackle this complexity issue by combining the information available in the log with knowledge on the process model structure.

To be able to reduce the number of possible mappings, we look at Declare rules describing the behavior of event logs and process models as defined in Section 2.3. The techniques described in [9] are utilized to derive the described rules from event logs. In order to infer Declare rules from process models, we build upon the following assumption: if an event log is given, such that at least one trace is recorded for each legal path in the process model, then the Declare rules which are discovered out of such log, reflect the behavior of the original process model.[2] Hence, we can generate an event log from the process model using the simulation techniques described in [23], and thereafter apply the discovery algorithm of [9] to derive the Declare rules. We denote the set of all Declare rules inferred from the event log as $\mathcal{B}_L$, and the set of Declare rules discovered from the process model as $\mathcal{B}_M$. Next, we prune all discovered rules having a support lower than a given minimal threshold β. From our experience, a minimal support of $\beta = 0.9$ has turned out to be the most effective choice. Experimental findings

---

[2] Without loss of generality, loops can be unraveled and treated as an optional path that is traversable multiple times.

reported in the use cases of [8] confirm this assumption. Yet, the value of β can be redefined by the user if needed. From the ordering rules ($\mathcal{C}_R^{\rightarrow}$), only the rule with the highest support for each pair of events / activities is kept. In case there is no ordering rule with a support above β for a given pair of events / activities, we add the pair to the set of interleaving events / activities, denoted as $\mathcal{I}$.

Having the Declare rules from both the model and the event log as well as the set of interleaving pairs of events / activities, we can define a number of constraints to reduce the number of possible solutions of the CSP. These will be introduced in the following equations. For each equation, $e_1, e_2 \in E$ denote two different event classes, i.e., $e_1 \neq e_2$. In the same manner, $a_1, a_2 \in A$ denote two different activities, i.e., $a_1 \neq a_2$. The constraint introduced in Equation (1) ensures that events for which an *Init* rule exists, are only mapped to activities for which an *Init* rule exists. Equation (2) and 3 work in the same manner for *End* and *Participation* rules.

$$Init(e_1) \wedge (Map(e_1) = a_1) \implies Init(a_1) \quad (1)$$

$$End(e_1) \wedge (Map(e_1) = a_1) \implies End(a_1) \quad (2)$$

$$Participation(e_1) \wedge (Map(e_1) = a_1) \implies Participation(a_1) \quad (3)$$

The CSP constraints derived from *CoExistence* and *NotCoExistence* rules found in the event log, are similar to those derived from the existence rules, but look at pairs of events and activities. If two event classes that are co-existing (not co-existing) are matched to two different activities, these activities should also be co-existing (not co-existing).

$$\begin{aligned} NotCoExistence(e_1, e_2) \wedge (Map(e_1) = a_1) \wedge (Map(e_2) = a_2) \\ \implies NotCoExistence(a_1, a_2) \end{aligned} \quad (4)$$

$$\begin{aligned} CoExistence(e_1, e_2) \wedge (Map(e_1) = a_1) \wedge (Map(e_2) = a_2) \\ \implies CoExistence(a_1, a_2) \end{aligned} \quad (5)$$

In contrast to this, it cannot be assumed that *events* in an ordering relation necessarily map to *activities* in the same ordering relation. This is due to the fact that, for a pair of parallel activities, the log may contain a dominant ordering of the corresponding events. For instance in the order process example of Section 2.1, events I_SM and P_NOT_E are in *ChainSuccession* because P_NOT_E always occurs directly before I_SM. Yet, their corresponding activities "Send invoice" and "Send notification" are in interleaving order in the process model. Such a situation is still coherent, with respect to the model. Therefore, we specify in Equation (6) that if two events, for which an ordering rule exists, are mapped to two different activities, then these two activities either have to be in an ordering relation enforcing the same order direction, or this pair of activities has to be in the set of interleaving activities.

$$\mathcal{C}_R^{\rightarrow}(e_1, e_2) \wedge Map(e_1) = a_1 \wedge Map(e_2) = a_2 \implies \mathcal{C}_R^{\rightarrow}(a_1, a_2) \vee (a_1, a_2) \in \mathcal{I} \quad (6)$$

Regarding the pairs of events for which no ordering rule exceeds β, Equation (7) ensures that if a pair of interleaving events is mapped to a pair of activities, these activities also have to be interleaving.

$$(e_1, e_2) \in \mathcal{I} \wedge (Map(e_1) = a_1) \wedge (Map(e_2) = a_2) \implies (a_1, a_2) \in \mathcal{I} \quad (7)$$

Having the constraint definitions in equations 1-7, we add a constraint $c_i, i \in 1..|\mathcal{B}_L|$ for each Declare rule derived from the event log to the CSP. For example, constraint $Init(O\_CHK\_S)$ is derived from the event log. In the set of inferred Declare constraints from the process model there is only one $Init$ rule, namely $Init(Check\ order)$. Using the mappings defined in Table 4 and Table 5 we can derive the corresponding constraint for the CSP: $c_1 \equiv x_1 = 1$. Having defined all constraints, the CSP can be solved to retrieve all possible mappings. If the CSP returns multiple solutions, the analyst has to choose the correct one. The next section shows how the analyst is supported in this selection.

3.2 Selection of the Correct Event–Activity Mapping

The previous section introduced the approach for automatic matching of event labels and activities. This section discusses why there are often multiple solutions to the defined constraint satisfaction problem and introduces means to guide the user through the set of potential mappings returned by the CSP solver.

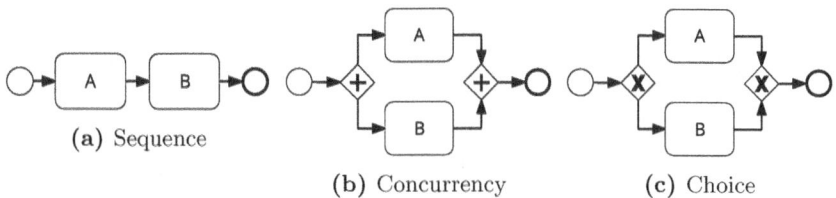

Fig. 2. Process model fragments leading to multiple solutions

Consider the trace $t_1 = \langle k, l, m, n \rangle$ and the simple sequence of activities a and b shown in Fig. 2a. When matching t_1 and the sequence model, the corresponding CSP returns three solutions. In all three solutions k is matched to a, and n is matched to b. For l and m it cannot be said whether they belong to a or b without further knowledge. It may be that both belong to A, or both belong to b, or l belongs to a and m belongs to b. The only mapping that can be excluded, is that l belongs to b and m belongs to a at the same time. If we want to match t_1 to the model shown in Fig. 2b, actually every combination of mappings is possible, besides those where all events are mapped to only one of the activities. For the matching with the process model depicted in Fig. 2c, we add the trace $t_2 = \langle p, q, r, s \rangle$. In this case the CSP returns two solutions: Either k, l, m, n belong to activity a and the rest to b, or the other way around.

Such ambiguous mappings, i.e., cases in which the CSP has multiple solutions, cannot be automatically resolved and require a domain expert to decide the mapping for the concerned events and activities. Nonetheless, this decision can be supported by the mapping approach. To aid the analyst with the disambiguation of multiple potential mappings, we introduce a questioning approach, which is inspired by the work of La Rosa et al. [24], in which the user is guided through the configuration of a process model using a questionnaire procedure. The analyst is presented one event label at a time, along with the possible activities to which this event label can be mapped. Once the analyst decides which of the candidate activities belongs to the event label, this mapping is converted into a new constraint that is added to the CSP. Consecutively, the CSP is solved again. In case there are still multiple solutions, the analyst is asked to make another decision for a different event label. This procedure is repeated until the CSP yields a single solution. The goal is to pose as few questions to the analyst as possible. To achieve this goal, we look into all solutions and choose the event label that is assigned to the highest number of different activities.

3.3 Transformation of the Event Log

Having defined the procedure to build a CSP and iteratively resolved any ambiguities, the next step is to use the selected solution of the CSP as mapping Map to transform the event log. Mapping Map is used to iterate over all traces in the event log and replace each event e_i with the activity returned by $Map(e_i)$. This resulting event log, where each event carries the label of its corresponding activity, is processed using the activity clustering approach described in [4] in order to correctly reflect activity instances. The transformed event log can then be used as input for any process mining technique.

4 Evaluation and Discussion

4.1 Evaluation

For the purpose of evaluation, the approach presented in this paper was implemented as a plug-in in the process mining framework ProM[3]. The Petri net notation has been chosen as modeling language for the implementation of the approach, because it has well-defined semantics and can be verified for correctness [1]. Furthermore, most of the common modeling languages, as e.g. BPMN and EPC, can be transformed into Petri nets [20]. As solver for the constraint satisfaction problem, the java library CHOCO[4] has been used.

To evaluate our approach with real life business processes, we used the *BIT process library, Release 2009*, which has been analyzed by Fahland et al. in [12] and is openly available to academic research. The process model collection contains models of financial services, telecommunications, and other domains. First, the models were transformed into Petri nets and only 1-bounded models that are free of life locks and deadlocks, and do not contain disconnected activities, have

---

[3] See http://processmining.org
[4] See http://www.emn.fr/z-info/choco-solver/

been kept. This restriction is due to the fact that models with such characteristics cannot be simulated. Moreover, some of the larger process models needed to be filtered out, as the resulting CSP could not be solved by the CSP solver due to memory shortage. This is mainly a limitation of the used CSP solver. Yet, for most of these processes, there is little behavioral distinction between activities or pairs of activities, e.g. all activities are interleaving to each other. Thus, it is not possible to match these processes without further knowledge. After the filtering step, 595 models remained with which we tested our approach. For these process models, event logs were generated by simulating the process activities' enactment through event generators. Such event generators followed the patterns (event models) illustrated in Fig. 3. Figure 3a shows a simple model with one start and one end transition, demonstrating a typical pattern found in many systems. For each activity that is assigned to this event model, a start and an end transition are generated for each execution of that activity. The second event model, depicted in Fig. 3b, generates for each execution either an event "Start1" or an event "Start2" and always an end event. Thus, there are two alternative starts for such an activity, e.g. it could be started by an incoming mail or by a telephone call. The event model presented in Fig. 3c also has two different start transitions, but in contrast to the model in Fig. 3b, both start events always occur with no restriction on their order. For the simulation of the process models, each activity is randomly assigned to one of these three event models, or it is left as is, generating only a single event. All generated event logs contain 1.000 traces and are limited to 1.000 events per trace as a stop condition for process models containing loops.

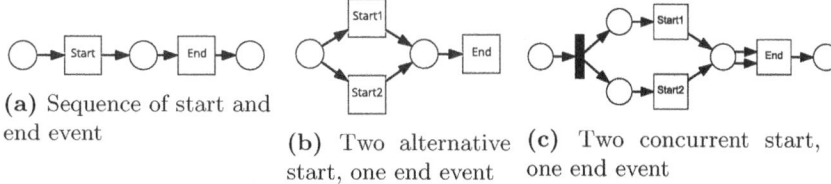

(a) Sequence of start and end event

(b) Two alternative start, one end event

(c) Two concurrent start, one end event

Fig. 3. Different event models used to generate events

In reality, event logs rarely completely comply to the defined process models due to noise and misbehavior. Thus, we generated for each process model five sets of simulated logs, in which we randomly inserted noise by shuffling, duplicating and removing events for a different percentage of traces. Figure 4 shows the results of this experiment. For 22 % of the processes, the mapping between events and activities can be established without asking any question, regardless the amount of noise injected. Looking at logs with no noise or where only 25 % of the traces contain noise, another 17 % of the processes require only one or two questions. With more noise in the event logs, this number continuously shrinks and more questions are needed for these processes. Yet, Fig. 4 shows that even

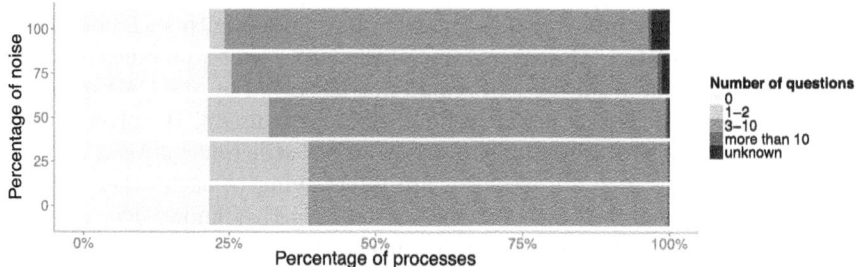

Fig. 4. Number of necessary questions with respect to noise

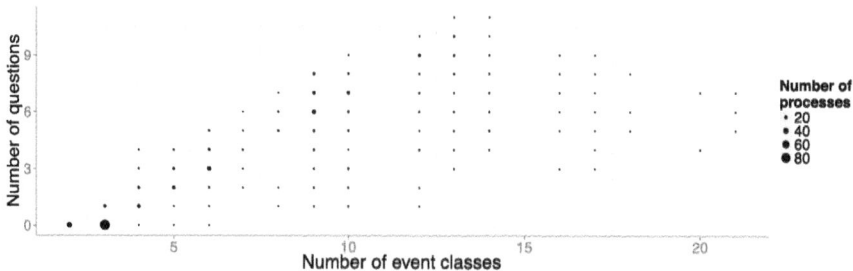

Fig. 5. Number of necessary questions with respect to number of event classes

when all traces contain noise, 97 % of the processes require less than 11 questions to build the correct mapping. Furthermore, it can be seen that with increasing noise a few processes cannot be processed by the CSP solver due to resource problems. This is mainly an implementation issue of the used CSP solver.

Figure 5 depicts the number of necessary questions with respect to the number of different events in a log without any noise. While one can see that there is a slight trend towards more questions with increasing numbers of event classes, it can also be seen that this trend also reverses for larger numbers of event classes. For example, it can be seen that there are processes with 20 different events, requiring only 5 questions. Thus, the number of required questions is rather independent of the number of activities. In fact, it depends on the structure of the process model and how well activities and pairs of activities can be distinguished from each other by their behavior.

4.2 Discussion

In the light of our experimental results, the approach turns out to be promising, especially with regards to resilience to noise. It requires in most of the cases only little manual intervention. Still, there are some processes that could not be handled, mainly due to massive parallelism and resulting memory shortage. Future work should investigate how these processes can be handled or, at least, automatically discovered. Moreover, it might be beneficial to combine this approach with the linguistic matching presented in [5], for cases in which the event labels carry more useful information than cryptic database field names. Finally, it is our plan to investigate, how the approach can be extended to support N:M relations, namely cases in which a single event class can be related to multiple activities – e.g. events representing shared functionalities. In the N:M case, the already very large search space for the matching problem grows drastically and other techniques might be necessary to handle this. Yet, event logs containing shared functionalities could be handled with the approach presented in this paper using preprocessing that essentially removes such events. If applicable, the approach presented in [5] could be used for the detection of shared functionalities.

5 Related Work

Related research can be subdivided into approaches working on event logs and approaches working on process models. Looking at approaches focusing on event logs, there are several approaches aiming at the abstraction of events to activities. Günther et al. introduce in [14] an approach that clusters events to activities using a distance function based on time or sequence position. Due to performance issues with this approach, a new means of abstraction on the level of event classes is introduced by Günther et al. in [16]. These event classes are clustered globally based on co-occurrence of related terms, yielding better performance but lower accuracy. A similar approach introducing semantic relatedness, N:M relations, and context dependence is defined by Li et al. in [19]. Another approach that uses pattern recognition and machine learning techniques for abstraction is introduced by Cook et al. in [7]. Together with the fuzzy miner, Günther and van der Aalst present an approach to abstract a mined process model by removing and clustering less frequent behavior [15]. While all these approaches aim at a mapping of events to activities, they are designed to automatically construct activities and not to match events to activities that have already been defined a-priori. In [5] and [6], we introduced approaches that aim at the mapping of events to pre-defined activities. Nevertheless, the approach in [5] still required more manual work as the precision of matchings is not sufficiently high. In contrast, the approach presented in this paper requires only very little manual effort to match events to pre-defined activities. The approach presented in [6] only works with 1:1 relations between events and activities and requires pre-processing for 1:N relations. Furthermore, it is only able to capture behavior from traces that can be replayed on the model. This is resolved by the work of this paper.

Another branch of related approaches working on event logs are those dealing with event correlation to group events belonging to the same process instance, as e.g. the work by Perez et al. in [22]. Yet, these approaches work on a more coarse-grained level as they focus on the relation to process instances rather than to activities. In fact, we assume that the correlation of events to process instances is either already given, or can be established by an approach like [22].

Our work is also related to automatic matching for process models. While matching has been partially addressed in various works on process similarity [10], there are only a few papers that cover this topic as their major focus. The work on the ICoP framework defines a generic approach for process model matching [25]. This framework is extended with semantic concepts and probabilistic optimization in [17,18]. Further, general concepts from ontology matching are adopted in [11]. The implications of different abstraction levels for finding correspondences is covered in [26]. However, all these works focus on finding matches between two process models, not between events and activities.

Our approach adopts MINERful [9] for the computation of constraints' support. MINERful is a declarative process miner, based on a two-phase technique. During the first step, it creates a so-called knowledge base, containing statistics about the occurrences of events in the log. The second step computes the support for constraints by querying such knowledge base. It is proven to be among the fastest Declare miners [9].

6 Conclusion

In this paper we introduce a novel technique for the mapping of events to activities, which can be used as a preprocessing step to enable business process intelligence techniques (e.g., process mining). The approach uses Declare rules derived from existing business process models and from event logs generated by IT systems, to establish a connection between conceptual process models and operational execution data. The key contribution of this approach is the establishment of a relation between events and a given set of activities in a process model using behavioral knowledge captured by Declare rules. Thereby, also 1:N relations can be handled. As shown in the evaluation in Section 4, the newly introduced matching technique performs well and requires little manual intervention. It also reveals to be robust towards noise.

References

1. van der Aalst, W.M.P.: Verification of workflow nets. In: Azéma, P., Balbo, G. (eds.) ICATPN, LNCS, vol. 1248, pp. 407–426. Springer (1997)
2. van der Aalst, W.M.P.: Process Mining: Discovery, Conformance and Enhancement of Business Processes, 1st edn. Springer (2011)
3. van der Aalst, W.M.P., Pesic, M.: DecSerFlow: Towards a Truly Declarative Service Flow Language. In: Bravetti, M., Núñez, M., Zavattaro, G. (eds.) WS-FM 2006. LNCS, vol. 4184, pp. 1–23. Springer, Heidelberg (2006)
4. Baier, T., Mendling, J.: Bridging Abstraction Layers in Process Mining: Event to Activity Mapping. In: Nurcan, S., Proper, H.A., Soffer, P., Krogstie, J., Schmidt, R., Halpin, T., Bider, I. (eds.) BPMDS 2013 and EMMSAD 2013. LNBIP, vol. 147, pp. 109–123. Springer, Heidelberg (2013)
5. Baier, T., Mendling, J., Weske, M.: Bridging abstraction layers in process mining. Information Systems **46**, 123–139 (2014)
6. Baier, T., Rogge-Solti, A., Weske, M., Mendling, J.: Matching of Events and Activities - An Approach Based on Constraint Satisfaction. In: Frank, U., Loucopoulos, P., Pastor, Ó., Petrounias, I. (eds.) PoEM 2014. LNBIP, vol. 197, pp. 58–72. Springer, Heidelberg (2014)

7. Cook, D.J., Krishnan, N.C., Rashidi, P.: Activity discovery and activity recognition: A new partnership. IEEE T. Cybernetics **43**(3), 820–828 (2013)
8. Di Ciccio, C., Mecella, M.: Mining artful processes from knowledge workers' emails. IEEE Internet Computing **17**(5), 10–20 (2013)
9. Di Ciccio, C., Mecella, M.: On the discovery of declarative control flows for artful processes. ACM Trans. Manage. Inf. Syst. 5(4), 24:1–24:37 (2015)
10. Dijkman, R.M., Dumas, M., van Dongen, B.F., Käärik, R., Mendling, J.: Similarity of Business Process Models: Metrics and Evaluation. Information Systems **36**(2), 498–516 (2011)
11. Euzenat, J., Shvaiko, P.: Ontology Matching. Springer (2007)
12. Fahland, D., Favre, C., Koehler, J., Lohmann, N., Völzer, H., Wolf, K.: Analysis on demand: Instantaneous soundness checking of industrial business process models. Data & Knowledge Engineering **70**(5), 448–466 (2011)
13. Freuder, E., Mackworth, A.: Handbook of Constraint Programming, Foundations of Artificial Intelligence, vol. 2, ch. Constraint satisfaction: An emerging paradigm, pp. 13–27. Elsevier (2006)
14. Günther, C.W., van der Aalst, W.M.P.: Mining activity clusters from low-level event logs. In: BETA Working Paper Series. vol. WP 165. Eindhoven University of Technology (2006)
15. Günther, C.W., van der Aalst, W.M.P.: Fuzzy Mining – Adaptive Process Simplification Based on Multi-perspective Metrics. In: Alonso, G., Dadam, P., Rosemann, M. (eds.) BPM 2007. LNCS, vol. 4714, pp. 328–343. Springer, Heidelberg (2007)
16. Günther, C.W., Rozinat, A., van der Aalst, W.M.P.: Activity Mining by Global Trace Segmentation. In: Rinderle-Ma, S., Sadiq, S., Leymann, F. (eds.) BPM 2009. LNBIP, vol. 43, pp. 128–139. Springer, Heidelberg (2010)
17. Klinkmüller, C., Weber, I., Mendling, J., Leopold, H., Ludwig, A.: Increasing Recall of Process Model Matching by Improved Activity Label Matching. In: Daniel, F., Wang, J., Weber, B. (eds.) BPM 2013. LNCS, vol. 8094, pp. 211–218. Springer, Heidelberg (2013)
18. Leopold, H., Niepert, M., Weidlich, M., Mendling, J., Dijkman, R., Stuckenschmidt, H.: Probabilistic Optimization of Semantic Process Model Matching. In: Barros, A., Gal, A., Kindler, E. (eds.) BPM 2012. LNCS, vol. 7481, pp. 319–334. Springer, Heidelberg (2012)
19. Li, J., Bose, R.P.J.C., van der Aalst, W.M.P.: Mining Context-Dependent and Interactive Business Process Maps Using Execution Patterns. In: Muehlen, M., Su, J. (eds.) BPM 2010 Workshops. LNBIP, vol. 66, pp. 109–121. Springer, Heidelberg (2011)
20. Lohmann, N., Verbeek, E., Dijkman, R.M.: Petri net transformations for business processes - a survey. Petri Nets and Other Models of Concurrency **2**, 46–63 (2009)
21. Maggi, F.M., Bose, R.P.J.C., van der Aalst, W.M.P.: Efficient Discovery of Understandable Declarative Process Models from Event Logs. In: Ralyté, J., Franch, X., Brinkkemper, S., Wrycza, S. (eds.) CAiSE 2012. LNCS, vol. 7328, pp. 270–285. Springer, Heidelberg (2012)
22. Pérez-Castillo, R., Weber, B., de Guzmán, I.G.R., Piattini, M., Pinggera, J.: Assessing event correlation in non-process-aware information systems. Software and System Modeling **13**(3), 1117–1139 (2014)
23. Rogge-Solti, A., Weske, M.: Prediction of Remaining Service Execution Time Using Stochastic Petri Nets with Arbitrary Firing Delays. In: Basu, S., Pautasso, C., Zhang, L., Fu, X. (eds.) ICSOC 2013. LNCS, vol. 8274, pp. 389–403. Springer, Heidelberg (2013)

24. La Rosa, M., Lux, J., Seidel, S., Dumas, M., ter Hofstede, A.H.M.: Questionnaire-driven Configuration of Reference Process Models. In: Krogstie, J., Opdahl, A.L., Sindre, G. (eds.) CAiSE 2007 and WES 2007. LNCS, vol. 4495, pp. 424–438. Springer, Heidelberg (2007)
25. Weidlich, M., Dijkman, R., Mendling, J.: The ICoP Framework: Identification of Correspondences between Process Models. In: Pernici, B. (ed.) CAiSE 2010. LNCS, vol. 6051, pp. 483–498. Springer, Heidelberg (2010)
26. Weidlich, M., Dijkman, R., Weske, M.: Behaviour Equivalence and Compatibility of Business Process Models with Complex Correspondences. ComJnl (2012)

PQL - A Descriptive Language for Querying, Abstracting and Changing Process Models

Klaus Kammerer[✉], Jens Kolb, and Manfred Reichert

Institute of Databases and Information Systems, Ulm University, Ulm, Germany
{klaus.kammerer,jens.kolb,manfred.reichert}@uni-ulm.de
http://www.uni-ulm.de/dbis

Abstract. The increasing adoption of process-aware information systems (PAISs) has resulted in large process repositories comprising large and complex process models. To enable context-specific perspectives on these process models and related data, a PAIS should provide techniques for the flexible creation and change of process model abstractions. However, existing approaches focus on the formal model transformations required in this context rather than on techniques for querying, abstracting and changing the process models in process repositories. This paper presents a domain-specific language for querying process models, describing abstractions on them, and defining process model changes in a generic way. Due to the generic approach taken, the definition of process model abstractions and changes on any graph-based process notation becomes possible. Overall, the presented language provides a key component for process model repositories.

1 Introduction

Process-aware information systems (PAISs) provide support for business processes at the operational level. In particular, a PAIS separates process logic from application code relying on explicit *process models*. This enables a separation of concerns, which is a well-established principle in computer science to increase maintainability and to reduce costs of change [1]. The increasing adoption of PAISs has resulted in large process model collections. Thereby, a process model may comprise dozens or hundreds of activities [2]. Furthermore, process models may refer to business objects, organizational units, user roles and other resources. Due to this high complexity, the various user groups need customized views on the processes [3]. For example, managers rather prefer an abstract process overview, whereas process participants need a detailed view of the process parts they are involved in.

Several approaches for creating process model abstractions have been proposed in literature [4–6]. However, current proposals focus on fundamental abstraction concepts for aggregating or reducing process model elements to derive a context-specific process view. Existing approaches neither provide concepts to specify process model abstractions independent from a particular process model (e.g., a particular user may be involved in several processes) nor to define them in a more descriptive way. Accordingly, for each relevant process model, users must create respective abstractions manually. In particular, the operations for abstracting a process model need to be specifically defined referring to the elements of this model; i.e., abstractions must be specified separately for each individual process model, which causes high efforts when being confronted with large process model collections (cf. Figure 1a). A possibility to lower efforts is to reduce the number of operations required to abstract process models, i.e., elementary operations may be composed to high-level ones [5]. However, the application of respective operations is still specific to a particular process model.

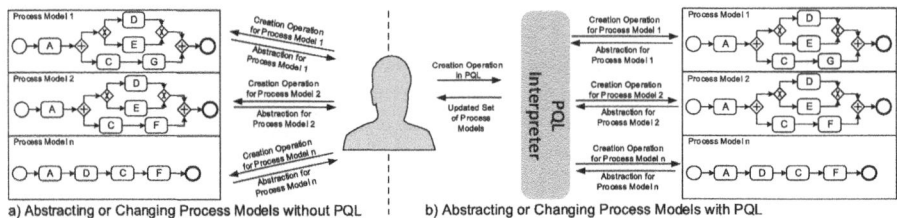

Fig. 1. Using the Process Query Language to Abstract or Change Process Models

In existing approaches, process changes refer to specific elements (e.g., nodes) of a process model rather than on generic process properties (e.g., process attributes). For example, it is usually not possible to replace a specific user role by another one in all process models stored in the repository [2], i.e., the change must be manually applied to each process model.

In other domains (e.g., database management), the use of domain-specific languages is common when facing large data sets. For example, *SQL* has been used to create, access and update data in relational databases [7]. However, to the best of our knowledge, no comparable approach exists for large process model repositories. To remedy this drawback, this paper introduces the *Process Query Language (PQL)*. PQL is a domain-specific language that allows defining process model abstractions in a declarative way as well as specifying changes on collections of (abstracted) process models from a process repository (cf. Figure 1b). In particular, such a declarative definition of a process can be automatically applied to multiple process models if required. Furthermore, users may use PQL to define personalized process views on their process, e.g., by abstracting process information not relevant for them. Additionally, process model collections can be easily changed based on such declarative descriptions. For example, same or similar process elements used in multiple models may be changed concurrently based on one PQL change description, e.g., if activities related to quality assurance shall be changed in all variants of a business process.

Section 2 introduces fundamentals on abstracting process models. Section 3 presents PQL and its syntax. Section 4 presents a proof-of-concept implementation. Section 5 discusses related work and Section 6 summarizes the paper.

2 Fundamentals on Process Model Abstractions

Section 2.1 defines basic notions. Section 2.2 then discusses how *process model abstractions* can be created and formally represented. It further describes how elementary operations can be composed to define high-level operations for abstracting process models.

2.1 Process Model

Basically, a process model comprises *process elements*, i.e., *process nodes* as well as the *control flow* between them (cf. Figure 2). The modeling of the latter is based on *gateways* and *control flow edges* (cf. Definition 1). Note that the data perspective is excluded in this paper to set a focus.

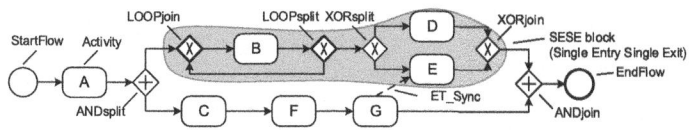

Fig. 2. Example of a Process Model

Definition 1 (Process Model). *A process model is defined as a tuple $P = (N, NT, CE, EC, ET, attr, val)$ where:*

- *N is a set of process nodes (i.e., activities, gateways, and start/end nodes).*
- *$NT : N \to NodeType$ with $NodeType = \{StartFlow, EndFlow, Activity, ANDsplit, ANDjoin, XORsplit, XORjoin, LOOPsplit, LOOPjoin\}$ is a function with $NT(n)$ returning the type of node $n \in N$.*
- *$CE \subset N \times N$ is a set of precedence relations (i.e., control edges): $e = (n_{src}, n_{dest}) \in CE$ with $n_{src}, n_{dest} \in N \land n_{src} \neq n_{dest}$.*
- *$EC : CE \to Conds \cup \{\text{TRUE}\}$ assigns to each control edge either a branching condition or $TRUE$ (i.e., the branching condition of the respective control edge always evaluates to true).*
- *$ET : CE \to EdgeType$ with $EdgeType = \{ET\_Control, ET\_Sync, ET\_Loop\}$. $ET(e)$ assigns a type to each control edge $e \in CE$.*
- *$attr : N \cup CE \to \mathcal{AS}$ assigns to each process element a corresponding attribute set $AS \subseteq \mathcal{AS}$.*
- *$val : (N \cup CE) \times \mathcal{AS} \to valueDomain(\mathcal{AS})$ assigns to any attribute $x \in \mathcal{AS}$ of a process element $pe \in N \cup CE$ its value:*

$$val(pe, x) = \begin{cases} value(x)^1, & x \in attr(pe) \\ null^2, & x \notin attr(pe) \end{cases}$$

---

[1] $value(x)$ denotes the value of process attribute x.
[2] Attribute is undefined for the respective process element.

Definition 1 can be used for representing the schemes of both process models and related process model abstractions. In particular, Definition 1 is not restricted to a specific activity-oriented modeling languages, but may be applied in the context of arbitrary graph-based process modeling languages. This paper uses a subset of BPMN elements as modeling notation. We further assume that a process model is *well-structured*, i.e., sequences, branchings (of different semantics) and loops are specified as blocks with well-defined start and end nodes having the same gateway type. These blocks—also known as SESE blocks (cf. Definition 2)—may be arbitrarily nested, but must not overlap (like blocks in BPEL). To increase expressiveness, *synchronization edges* allow for a *cross-block* synchronization of parallel activities (like BPEL links). In Figure 2, for example, activity E must not be enabled before G is completed. Additionally, process elements have associated attributes. For example, an activity has attributes like *ID*, *name* or *assignedUser* (cf. Figure 3).

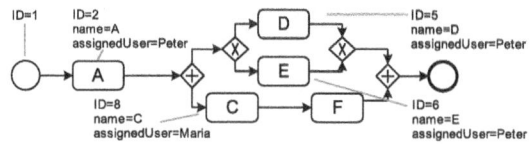

Fig. 3. Example of a Process Model with Attributes

Definition 2 (SESE). *Let $P = (N, NT, CE, EC, ET, attr, val)$ be a process model and $X \subseteq N$ be a subset of activities (i.e., $NT(n) = Activity \; \forall n \in X$). Then: Subgraph P' induced by X is called SESE (Single Entry Single Exit) block iff P' is connected and has exactly one incoming and one outgoing edge connecting it with P. Further, let $(n_s, n_e) \equiv MinimalSESE(P, X)$ denote the start and end node of the minimum SESE comprising all activities from $X \subseteq N$.*

How to determine SESE blocks is described in [8]. Since we presume a well-structured process model, a minimum SESE can be always determined.

2.2 Process Model Abstractions

In order to abstract a given process model, the schema of the latter needs to be simplified. For this purpose, *elementary operations* are provided (cf. Table 1) that may be further combined to realize *high-level abstraction operations* (e.g., show all activities a particular actor is involved in and their precedence relations) [5,9]. At the elementary level, two categories of operations are provided: *reduction* and *aggregation*. An elementary *reduction* operation hides an activity of a process model. For example, $RedActivity(P, n)$ removes activity n and its incoming and outgoing edges and re-inserts a new edge linking the predecessor of n with its successor in process model P (cf. Figure 4a). An *aggregation* operation, in turn, takes a set of activities as input and combines them to an abstracted node.

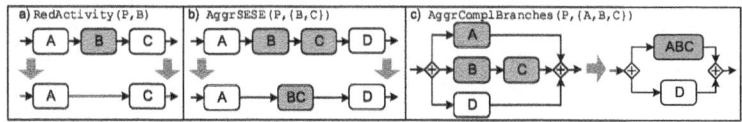

Fig. 4. Examples of Elementary Abstraction Operations

For example, $AggrSESE(P, N')$ removes all nodes of the SESE block induced by node set N' and re-inserts an abstract activity instead (cf. Figure 4b). Furthermore, operation $AggrComplBranches(P, N')$ aggregates multiple branches of an XOR/AND branching to a single branch with one abstracted node (cf. Figure 4c). An abstraction of a process model can be created through the consecutive application of elementary operations on a given process model (cf. Definition 3) [5,10]. Note that there exist other elementary operations, which address process perspectives other than control flow as well (e.g., data flow); we omit details here.

Definition 3 (Process Model Abstraction). *Let P be a process model. A process model abstraction V(P) is described through a creation set $CS_{V(P)}$ = (P, Op) with*

- *$P = (N, NT, CE, EC, ET, attr, val)$ being the original process model,*
- *$Op = \langle Op_1, \ldots, Op_n \rangle$ being a sequence of elementary operations applied to P: $Op_i \subseteq \mathcal{O} \equiv \{RedActivity, AggrSESE, AggrComplBranches\}$*

A node n of the abstracted process model $V(P)$ either directly corresponds to a node $n \in N$ of the original process model P or it abstracts a set of nodes from P. $PMNode(V(P), n)$ reflects this correspondence by returning either n or node set N_n aggregated in V(P), depending on creation set $CS_{V(P)}$.

$$PMNode(V(P), n) = \begin{cases} n & n \in N \\ N_n & \exists Op_i \in Op : N_n \xrightarrow{Op_i} n \end{cases}$$

Table 1. Examples of Elementary Abstraction Operations

Operation	Description
$RedActivity(P, n)$	Activity n and its incoming and outgoing edges are removed in P, and a new edge linking the predecessor of n with its successor is inserted (cf. Figure 4a).
$AggrSESE(P, N')$	All nodes of the SESE block defined by N' are removed in P and an abstract activity is re-inserted instead (cf. Figure 4b).
$AggrComplBranches(P, N')$	Complete branches of an XOR/AND branching are aggregated to a branch with one abstracted node in P. N' must contain the activities of the branches (i.e., activities between split and corresponding join gateway) that shall be replaced by a single branch consisting of one aggregated node (cf. Figure 4c).

When abstracting a process model, unnecessarily complex control flow structures could result due to the generic nature of the operations applied. For example, single branches of a parallel branching might become "empty" or a parallel branching might only have one branch left after applying reductions. In such cases, unnecessary gateways should be removed to obtain a more comprehensible schema of the abstracted model. Therefore, refactoring operations are provided. In particular, this does not affect the control flow dependencies of activities and, hence, does not change behavioral semantics of the refactored model [2].

To abstract multiple aspects of a process model several elementary operations may be applied in combination [5]. For example, *AggrSESE* and *AggrComplBranch* may be combined to high-level operation *AggregateControlFlow* (cf. Figure 5). Obviously, abstracting large process models becomes easier, when providing high-level operations in addition to elementary ones. In particular, the selection of the nodes to be abstracted should be more convenient. Current abstraction approaches, however, require the explicit specification of these nodes in relation to a particular process model. A declarative definition of these nodes (i.e., select all activities, a user is involved in), therefore, would enable users to abstract nodes in a more convenient and flexible way.

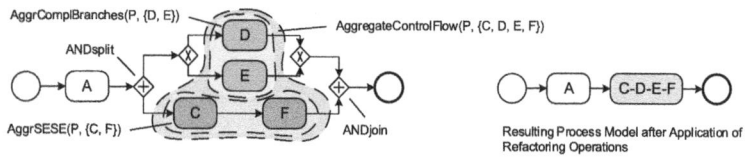

Fig. 5. Composition of Elementary Operations

3 The PQL Language

The presented operations for abstracting process models (i.e., process views) always refer to a process model they shall be applied to. Thus, their effects cannot be described independently from a particular process model, which causes high efforts in case an abstraction shall be introduced to multiple process models. To remedy this drawback, we introduce *Process Query Language (PQL)* that allows specifying process abstractions independent from a specific process model. PQL allows defining changes on a selected collection of process models as well.

3.1 Overview

PQL allows describing process model abstractions as well as process model changes in a declarative way. More precisely, respective descriptions may not only be applied to a single process model, but to a collection of selected process models as well. In the following, we denote such a declarative description of abstractions or changes on a collection of process models as *PQL request*. In general, a PQL request consists of two parts: First, *selection section* defines the process models concerned by the PQL request; Second, *modification section* defines the abstractions and changes respectively to be applied to the selected process models.

Figure 6 illustrates how a PQL request is processed: First, a user sends a *PQL request* to the *PQL interpreter* (Step ①). Then, those process models are selected from the process repository that match the predicates specified in selection section of the PQL request (Step ②). If applicable, changes of the modification section of the PQL request are applied to the selected process models (Step ③). Following this, the abstractions defined in the modification section are applied to the selected process models (Step ④). Finally, the selected, changed and abstracted process models are presented to the user (Step ⑤). Note that Steps 3+4 are optional depending on the modification section of the PQL request.

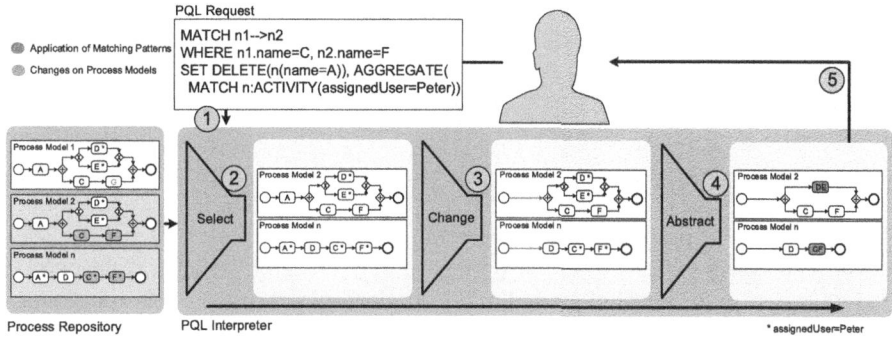

Fig. 6. Processing a PQL Request

The specification of PQL requests relies on the *Cypher Query Language*, which we *adopt* to meet the specific requirements of process modeling [11]. Cypher is a declarative graph query language known from the Neo4J graph database. In particular, it allows querying and changing the graphs from a graph database [12]. Furthermore, Cypher has been designed with the goal to be efficient, expressive and human-readable. Thus, it is well suited as basis for PQL. An example of a PQL request expressed with Cypher is shown in Listing 1.

```
1 MATCH    a1:ACTIVITY-[:ET_Control]->a2:ACTIVITY-[:ET_Control]->a3:ACTIVITY
2 WHERE    not (a1-[:ET_Control]->a3)
3 RETURN   a3
```

Listing 1. Example of a PQL Request

In the PQL request from Listing 1, Line 1 refers to process model that contain a path (i.e., a sequence of edges with type *ET_Control*) linking activities $a1$, $a2$ and $a3$. Note that $a1, a2$ and $a3$ constitute variables. To be more precise, the PQL request searches for process models comprising any sequence consisting of three activities. As a constraint (cf. Line 2), only directly adjacent nodes of $a2$ are returned. Additionally, the nodes must not be directly adjacent to $a1$. An

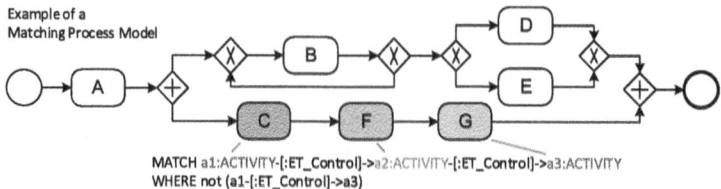

Fig. 7. PQL Request Determining a Sequence of Three Activities

application on the process models depicted in Figure 2 returns process activities C, F and G as the only possible match (cf. Figure 7).

Listing 2 presents the general syntax of a PQL request in BNF grammar notation [13]. Other relevant PQL syntax elements will be introduced step-by-step.

```
PQLrequest ::= match where? set?
```
Listing 2. BNF for a PQL Request

3.2 Selecting Process Models and Process Elements

PQL allows for a predicate-based selection of process models and process elements respectively. First, a search predicate may describe structural properties of the process models to be queried, e.g., to select all process models comprising two sequential nodes $n1$ and $n2$ (cf. Step ② in Figure 6). Second, process models and process elements can be selected by a predicate-based search on specific process element attributes. The latter are usually defined for each process model in a process repository, e.g., a user role designated to execute certain activities [5]. A *predicate* serves to assign properties (i.e., attributes) to process elements. For example, for predicate $PR \equiv \{x | x < 4, x \in \mathbb{N}\}$, we obtain $x \in \{1, 2, 3\}$. In general, a predicate PR may be described as boolean-valued function:

Definition 4 (Predicate). *Let* $P = (N, NT, CE, EC, ET, attr, val)$ *be a process model,* $x \in N \cup CE$ *be a process element, and* PR *be a predicate. Then:*
$$PR(x) : x \longrightarrow \{true, false\}$$

A predicate is used to compare attributes of process models and process elements respectively. In this context, *ordering functions* for numerical values (i.e., $\neq, <, \leq, =, \geq, >$) may be used. For example, string values may be compared on either equality against fixed values or based on edit distance to determine their similarity [14]. Two or more predicates may be concatenated using Boolean operations (i.e., AND, OR, NOT).

As aforementioned, PQL offers *structural* as well as *attributional matching patterns* to determine whether a specific process fragment is present in a particular process model. More precisely, *structural matching patterns* consider the control flow of a process model; i.e., they define the process fragments that need to be present in selected process models. In turn, *attributional matching patterns* allow selecting process models and process elements, respectively, based on process element attributes.

Structural matching patterns define constraints on process fragments to be matched against existing process models in a process repository. In PQL, structural matching patterns are initiated by a MATCH keyword (cf. Line 1, Listing 3) followed by a respective matching pattern, which describes the respective process fragment (cf. Line 2).

```
1  match        ::= "MATCH" match_pat (("," match_pat)+)?
2  match_pat    ::= (PQL_PATHID "=")? (MATCH_FUNCTION "(" path ")" | path)
3
4  path         ::= node ((edge) node)+)?
5
6  node         ::= PQL_NODEID (":" NODETYPE)? ("(" NODEID ")")?
7
8  edge         ::= cond_edge | uncond_edge
9  uncond_edge  ::= ("--" | "-->")
10 cond_edge    ::= (("-" edge_attrib "-") | ("-" edge_attrib "->"))
11 edge_attrib  ::= "[" PQL_EDGEID? (":"
12                    ((EDGETYPE ("|" EDGETYPE)* )? | edge_quant)?
13                  "]"
14 edge_quant   ::= "*" (EXACT_QUANTIFIER |
15                    (MIN_QUANTIFIER ".." MAX_QUANTIFIER)?)?
```

Listing 3. BNF for Structural Matching in a PQL Request

Structural matching patterns are further categorized into *dedicated* and *abstract* patterns. While *dedicated* patterns (cf. Lines 8-11 in Listing 3) are able to describe SESE blocks of a process model, an *abstract* pattern (cf. Lines 12-15) offers an additional edge attribute. The latter defines control flow adjacencies between nodes, i.e., the proximity of a pair of nodes. For example, to specify the selection of all succeeding nodes of activity A in Figure 3 requires abstract structural patterns. Table 2 summarizes basic PQL structural matching patterns.

Table 2. Examples of Structural Matching Patterns

Pattern	Description	Type
MATCH a-->b	Pattern describing the existence of an edge of any type between nodes a and b.	dedicated
MATCH a(2)-[:EDGE_TYPE]->b	Pattern describing a process fragment whose nodes a and b are connected by an edge with type EDGE_TYPE; furthermore, a has attribute ID with value 2	dedicated
MATCH a-[*1..5]->b	Pattern describing a process fragment with nodes a and b that do not directly succeed, but are separated by at least one and at most five nodes.	abstract
MATCH a-[*]->b	Pattern describing an arbitrary number of nodes between nodes a and b.	abstract
MATCH p = shortestPath(a-[:ET_Control*3]->c)	Pattern describing a minimum SESE block with a maximum of three control edges between nodes a and c.	abstract

Attributional matching patterns allow for an additional filtering of process fragments selected through a structural matching. For this purpose, predicates referring to process element attributes may be defined (cf. Listing 4). Attributional matching is initiated by a WHERE keyword, which may follow a

MATCH keyword (cf. Table 3). Note that attributional matching patterns refer to process elements pre-selected through a structural matching pattern. For example, nodes a and b selected by pattern MATCH a-->b can be further filtered with attributional matching patterns. If the attributional matching $a.ID = 5$ shall be applied to all activities of a process model, the MATCH keyword needs to be defined as follows: MATCH a:ACTIVITY(*) WHERE a.ID=5.

```
1 where             ::= "WHERE" predicate ((BOOL_OPERATOR predicate)+)?
2 predicate         ::= comparison_pred | regex_pred
3
4 comparison_pred   ::= PROPERTY_ID COMPARISON_OPERATOR any_val
5 regex_pred        ::= PROPERTY_ID "=~" REGEX_EXPRESSION
```

Listing 4. BNF for Attributional Matching in a PQL Request

Attributional matching patterns may be combined with structural ones. For example, PQL request MATCH a:ACTIVITY-->b matches with node attribute $NodeType = ACTIVITY$ for node a and any sequence of nodes a and b.

Table 3. Examples of Attributional Matching Patterns

Pattern	Description
MATCH (a) WHERE (a.NAME="Sell Item")	Select all nodes with name "Sell Item".
MATCH (a) WHERE HAS (a.attrib)	Select all nodes for which an attribute with name $attrib$ is present.
MATCH (a) WHERE a:ACTIVITY	Select all nodes with node type $ACTIVITY$.
MATCH (a-[*]->b) WHERE a:ACTIVITY(1), b(2)	Select (1) activity a with ID=1 and node type $ACTIVITY$ and (2) node b with ID=2.

Figure 8 illustrates the application of a matching pattern to the process model from Figure 3. Figure 8a matches a sequence of nodes a, b and c, with node a having attribute $ID = 2$, node b being an arbitrary node, and node c being of type $ACTIVITY$. Figure 8b matches for a node a with $a.ID = 8$ and arbitrary nodes succeeding a, having a maximum distance of 2 to a (i.e., nodes G and $ANDjoin$ match in the process model). In turn, Figure 8c matches all nodes having assigned attribute $assignedUser$. Finally, Figure 8d matches nodes whose ID either is 2 or 5 (i.e., nodes A and D match in the process model).

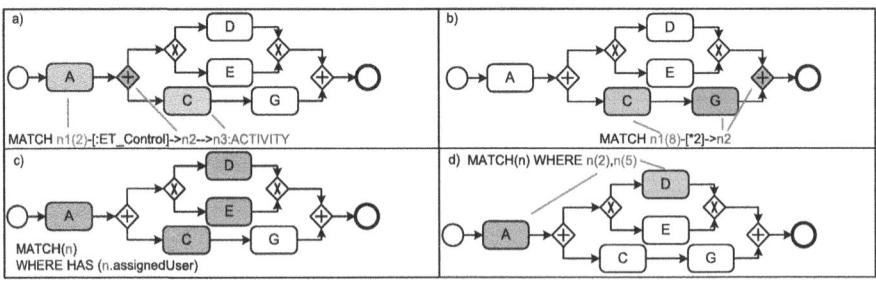

Fig. 8. Overview on PQL Matching Patterns

3.3 Abstracting Process Models

This section shows how to define abstractions on the process models referenced by the selection section of a PQL request.

Based on the matching patterns, PQL allows defining abstractions independent of a particular process model. As opposed to the elementary abstraction operations introduced in Section 2, two high-level abstraction operations AGGREGATE and REDUCE are introduced. The latter allow abstracting a set of arbitrary process elements (including data elements [15]). Thereby, the process elements to be abstracted are categorized into process element sets based on their type, e.g., node type. If an aggregation shall be applied to a set of process nodes, a minimum SESE block is determined to aggregate adjacent process nodes to an abstracted node. Hence, both well-structuredness and behavioral semantics of the respective process model are preserved.

In PQL, abstractions of process models are initiated by keyword SET. In turn, keywords AGGREGATE and REDUCE indicate the elements to be aggregated or reduced (cf. Listing 5 and Figure 8).

```
1 set           ::= "SET" operation (("," operation)+)?
2 operation     ::= abstraction | change_operation
3 abstraction   ::= ("AGGREGATE" | "REDUCE")+ "(" PQLrequest ")"
```

Listing 5. BNF for Structural Matching in a PQL Request

The PQL request depicted in Listing 6 selects all process models that contain any process fragment consisting of a sequence of two activities, i.e., variables a and b. Process elements selected by the first MATCH are then aggregated if their type is $ACTIVITY$ and $val(pe, assignedUser) = Peter$ holds (cf. Figure 8).

```
1 MATCH  a:ACTIVITY(NAME=C)-->b:ACTIVITY(NAME=F)
2 SET    AGGREGATE(
3            MATCH n:ACTIVITY(*)
4            WHERE n.assignedUser=Peter)
```

Listing 6. PQL Request to Aggregate Nodes

Listing 7 shows a PQL request reducing neighboring nodes C and F as described by PQL variables a and b. Note that a second PQL request is nested (cf. Line 3) utilizing the same variables as the parent PQL request does.

```
1 MATCH  a:ACTIVITY(NAME=C)-->b:ACTIVITY(NAME=F)
2 SET    REDUCE(
3            MATCH a-->b)
```

Listing 7. PQL Request to Reduce Nodes

3.4 Changing Process Models

In contemporary process repositories, changes related to multiple process models usually need to be performed on each process model separately. This not only causes high efforts for process designers, but also constitutes an error-prone task. To remedy this drawback, PQL allows changing all process models defined by the selection section of a PQL request at once, i.e., by one and the same change transaction. For example, structural matching patterns can be applied to select the process models to be changed.

Table 4 shows elementary change operations supported by PQL. These may be encapsulated as *high-level change operations*, e.g., inserting a complete process fragment through the application of a set of *InsertNote* operations.

Table 4. Examples of Change Operations Supported by PQL

Operation	Description
$InsertNode(P, n_{pred}, n_{succ}, n)$	Node n is inserted between preceding node n_{pred} and succeeding node n_{succ} in process model P. Control edges between n_{pred} and n as well as between n and n_{succ} are inserted to ensure compoundness of the nodes.
$DeleteNode(P, N')$	A set of nodes N' is removed from process model P.
$MoveNode(P, n, n_{pred}, n_{succ})$	Node n is moved from its current position to the one between n_{pred} and n_{succ}, control edges are adjusted accordingly.

Change operation $InsertNode(P, n_{pred}, n_{succ}, n)$, for example, inserts node n between nodes n_{pred} and n_{succ} in process model P. Thereby, the control edge between n_{pred} and n_{succ} is adjusted and another control edge is inserted to prevent unconnected nodes. Due to lack of space, we omit a discussion of other change operations here and refer interested readers to [16] instead.

Listing 8 shows how to insert a node with type *ACTIVITY* and name 'New Node' between nodes C and F (cf. Figure 3). Note that the insertion will be applied to any process model in which nodes C and F (cf. Line 2) are present; i.e., the insertion may be applied to a set of process models. Finally, abstractions and changes on process models may be defined in a single PQL request; in this case, changes on process models are applied first.

```
1 MATCH a-->b
2 WHERE a.NAME=C, b.NAME=F
3 SET   INSERTNODE(a, b, ACTIVITY, 'New Node')
```

Listing 8. PQL Request to Insert a Node

4 Proof-of-Concept Prototype

In order to demonstrate the applicability of PQL we developed a web-based PAIS called *Clavii BPM Platform*[1]. This platform implements a software architecture supporting the predicate-based definition and creation of process abstractions as well as predicate-based process model changes utilizing PQL [17]. Figure 9 illustrates the creation of a process model abstraction. Drop-down menu ① shows a selection of pre-specified PQL requests directly applicable to a process model. In turn, Figure 9b depicts a screenshot of Clavii's configuration window, where a stored PQL request may be altered. In this case, PQL request 'Technical Tasks'

---
[1] http://www.clavii.com/

is outlined in ②. The latter aggregates all nodes neither being service tasks nor script tasks (cf. ③). Future research will address the applicability of the prototype and PQL, respectively, in practical settings.

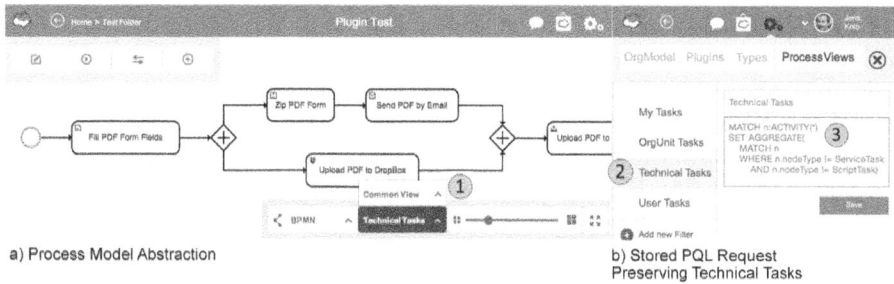

Fig. 9. Process Model Abstractions in the Clavii BPM Platform

5 Related Work

BPMN-Q allows querying process models in [18]. Queries are defined visually using a BPMN-like notation extended with additional attributes. As opposed to PQL requests, BPMN-Q queries are first defined visually and then converted into semantically expanded queries. Model selection is based on the comparison of process element attributes with predicate values specified in the BPMN-Q query, i.e., by measuring the edit-distance. However, BPMN-Q does not support abstractions and changes of the process models selected.

BP-QL is another language for querying process models, which is based on *statecharts* [19]. In particular, a query can be defined in terms of state chart patterns. BP-QL uses pattern matching in respect to node attributes as well as control flow structures when selecting process models. Like BPMN-Q, BP-QL does not allow for changes or abstractions of the selected process models.

A technique for searching and retrieving process variants based on similarity metrics is presented in [20]. More precisely, process variants may be compared by queries comprising structural, behavioral and contextual constraints.

None of the discussed approaches deals with abstractions or changes of the process models selected by a query. Furthermore, the various approaches are based on rather rigid constraints not taking practical issues into account, e.g., regarding the evolution of process models over time.

Several approaches for defining and managing process model abstractions exist. The approach presented in [21] measures semantic similarity between process activities by analyzing the structure of the respective process model. The discovered similarity is then used to abstract the process model. However, this approach neither distinguishes between different user perspectives nor does it provide concepts for flexibly describing process model abstractions. In turn, [22] applies graph reduction techniques to verify structural properties of process models similar to high-level abstraction operations AGGREGATE and REDUCE. However, no support for querying process models exists.

For defining and changing process models, various approaches exist. In [16], an overview of evidenced change patterns is presented. Furthermore, [23] summarizes approaches enabling flexibility in PAISs. In particular, [24] presents adaptions of well-structured process models, while preserving their correctness. In turn, [25] presents concepts for evolving process models over time. Finally, changes based on abstracted process models are described in [26–28].

The *Cypher* query language allows querying and modifying graphs stored in a graph database [11]. By contrast, the *Gremlin* graph query language is a Turing complete programming language realizing queries as chain of operations or functions. Hence, it is suited to construct complex queries for programming languages [29]. The *Structured Query Language (SQL)* offers techniques to manage sets of data in relational databases [7].

6 Summary

We introduced the *Process Query Language (PQL)*, which enables users to automatically select, abstract and change process models in large process model collections. Due to its generic approach, the definition of process model abstractions and changes on any graph-based process notation becomes possible. For this purpose, *structural* and *attributional matching pattern*s are used, which declaratively select process elements either based on the control flow structure of a process model or on element attributes. *PQL* has been implemented in a proof-of-concept prototype demonstrating its applicability. Altogether, process querying languages will be a key part of process repositories to allow for convenient management of process abstractions and changes on large sets of process models.

References

1. Weber, B., Sadiq, S., Reichert, M.: Beyond Rigidity - Dynamic Process Lifecycle Support: A Survey on Dynamic Changes in Process-Aware Information Systems. Computer Science - Research and Development **23**(2), 47–65 (2009)
2. Weber, B., Reichert, M., Mendling, J., Reijers, H.A.: Refactoring Large Process Model Repositories. Computers in Industry **62**(5), 467–486 (2011)
3. Streit, A., Pham, B., Brown, R.: Visualization support for managing large business process specifications. In: van der Aalst, W.M.P., Benatallah, B., Casati, F., Curbera, F. (eds.) BPM 2005. LNCS, vol. 3649, pp. 205–219. Springer, Heidelberg (2005)
4. Tran, H.: View-Based and Model-Driven Approach for Process-Driven, Service-Oriented Architectures. TU Wien, PhD thesis (2009)
5. Reichert, M., Kolb, J., Bobrik, R., Bauer, T.: Enabling personalized visualization of large business processes through parameterizable views. In: Proc 27th ACM Symposium On Applied Computing (SAC 2012), Riva del Garda, Italy (2012)

6. Chiu, D.K., Cheung, S., Till, S., Karlapalem, K., Li, Q., Kafeza, E.: Workflow View Driven Cross-Organizational Interoperability in a Web Service Environment. Information Technology and Management **5**(3/4), 221–250 (2004)
7. Information Technology - Database Languages - SQL - Part 11: Information and Definition Schemas (SQL/Schemata). Norm ISO 9075:2011 (2011)
8. Johnson, R., Pearson, D., Pingali, K.: Finding regions fast: single entry single exit and control regions in linear time. In: Proc ACM SIGPLAN 1993 (1993)
9. Kolb, J., Reichert, M.: A Flexible Approach for Abstracting and Personalizing Large Business Process Models. ACM Applied Comp. Review **13**(1), 6–17 (2013)
10. Bobrik, R., Reichert, M., Bauer, T.: View-based process visualization. In: Alonso, G., Dadam, P., Rosemann, M. (eds.) BPM 2007. LNCS, vol. 4714, pp. 88–95. Springer, Heidelberg (2007)
11. Panzarino, O.: Learning Cypher. Packt Publishing (2014)
12. Robinson, I., Webber, J., Eifrem, E.: Graph Databases. O'Reilly (2013)
13. Backus, J.: Can Programming Be Liberated from the Von Neumann Style?: A Functional Style and Its Algebra of Programs. Comm ACM **21**(8), 613–641 (1978)
14. Wagner, R.A., Fischer, M.J.: The String-to-String Correction Problem. Journal of the ACM **21**(1), 168–173 (1974)
15. Kolb, J., Reichert, M.: Data flow abstractions and adaptations through updatable process views. In: Proc 27th Symposium on Applied Computing (SAC 2013), Coimbra, Portugal, pp. 1447–1453. ACM (2013)
16. Weber, B., Reichert, M., Rinderle, S.: Change Patterns and Change Support Features - Enhancing Flexibility in Process-Aware Information Systems. Data & Knowledge Engineering **66**(3), 438–466 (2008)
17. Kammerer, K.: Enabling Personalized Business Process Modeling: The Clavii BPM Platform. Master's thesis, Ulm University (2014)
18. Sakr, S., Awad, A.: A framework for querying graph-based business process models. In: Proc ACM WWW 2010, pp. 1297–1300 (2010)
19. Beeri, C., Eyal, A., Kamenkovich, S., Milo, T.: Querying business processes. In: Proc VLDB 2006, pp. 343–354 (2006)
20. Lu, R., Sadiq, S., Governatori, G.: On managing business processes variants. Data & Knowledge Engineering **68**(7), 642–664 (2009)
21. Smirnov, S., Reijers, H.A., Weske, M.: A semantic approach for business process model abstraction. In: Mouratidis, H., Rolland, C. (eds.) CAiSE 2011. LNCS, vol. 6741, pp. 497–511. Springer, Heidelberg (2011)
22. Sadiq, W., Orlowska, M.E.: Analyzing Process Models Using Graph Reduction Techniques. Information Systems **25**(2), 117–134 (2000)
23. Reichert, M., Weber, B.: Enabling Flexibility in Process-aware Information Systems - Challenges, Methods. Springer, Technologies (2012)
24. Reichert, M., Dadam, P.: ADEPTflex - Supporting Dynamic Changes of Workflows Without Losing Control. Journal of Intelligent Inf. Sys. **10**(2), 93–129 (1998)
25. Rinderle, S., Reichert, M., Dadam, P.: Flexible Support of Team Processes by Adaptive Workflow Systems. Distributed and Par. Databases **16**(1), 91–116 (2004)

26. Kolb, J., Kammerer, K., Reichert, M.: Updatable process views for user-centered adaption of large process models. In: Liu, C., Ludwig, H., Toumani, F., Yu, Q. (eds.) Service Oriented Computing. LNCS, vol. 7636, pp. 484–498. Springer, Heidelberg (2012)
27. Kolb, J., Kammerer, K., Reichert, M.: Updatable process views for adapting large process models: the proview demonstrator. In: Demo Track of the 10th Int'l Conf on Business Process Management (BPM 2012), pp. 6–11 (2012)
28. Kolb, J., Reichert, M.: Supporting business and it through updatable process views: the proview demonstrator. In: Proc 10th Int'l Conf. on Service Oriented Computing (ICSOC 2012), Demonstration Track, Shanghai, 460–464 (2013)
29. TinkerPop: Gremlin. http://gremlin.tinkerpop.com, (last visited November 14, 2014)

Enhancing Declarative Process Models with DMN Decision Logic

Steven Mertens[(✉)], Frederik Gailly, and Geert Poels

Department of Business Informatics and Operations Management,
Faculty of Economics and Business Administration, Ghent University,
Tweekerkenstraat 2 9000, Ghent, Belgium
{steven.mertens,frederik.gailly,geert.poels}@ugent.be

Abstract. Modeling dynamic, human-centric, non-standardized and knowledge-intensive business processes with imperative process modeling approaches is very challenging. Declarative process modeling approaches are more appropriate for these processes, as they offer the run-time flexibility typically required in these cases. However, by means of a realistic healthcare process that falls in the aforementioned category, we demonstrate in this paper that current declarative approaches do not incorporate all the details needed. More specifically, they lack a way to model decision logic, which is important when attempting to fully capture these processes. We propose a new declarative language, Declare-R-DMN, which combines the declarative process modeling language Declare-R with the newly adopted OMG standard Decision Model and Notation. Aside from supporting the functionality of both languages, Declare-R-DMN also creates bridges between them. We will show that using this language results in process models that encapsulate much more knowledge, while still offering the same flexibility.

Keywords: Business process modeling · Declarative process models · Decision logic · Decision management · Healthcare processes

1 Introduction

BPMN takes an imperative approach to business process modeling as it provides a precise graph-based definition of the process control-flow [1]. While BPMN is suitable for modeling static and standardized business processes [2], specifying the complete control-flow for each variation of processes that require a high degree of run-time flexibility[1] is time consuming and results in overly complex models.

Goedertier et al. [1] state that dynamic, human-centric, non-standardized and knowledge-intensive business processes (KiP) are most likely to require the run-time flexibility offered by declarative process modeling. While imperative approaches focus on explicitly defining the exact path of activities to reach the process goals, declarative approaches determine only the activities that may be performed as well as constraints prohibiting undesired behavior [3]. Applying a declarative modeling

---

[1] Run-time flexibility: the flexibility allowed by a process after being deployed [22].

approach results in the specification of a collection of rules, constraints and assumptions that leaves enough freedom for various execution paths towards the process goals to exist. Additionally, explicitly specifying rules that remain tacit with imperative modeling, can enhance the knowledge management capabilities of the organization, allow for reuse of the rules in other process models, improve maintainability by way of high design-time flexibility[2], increase process compliance and improve traceability [4].

One of the most popular declarative process modeling languages is Declare [5, 6] (previously known as ConDec[3]). This language is based on Linear Temporal Logic (LTL), which is a formal language to express statements in modal temporal logic. It is very well suited to represent the control-flow of a process in a declarative manner as it does not offer a precise specification of the control-flow, but rather marks the rules to which a valid control-flow must oblige. Declare offers visual constructs to hide some of the complexity of LTL rules. To further improve its expressibility an extension has been proposed, Declare-R [7], that adds a resource perspective to the language.

The healthcare sector is one of the sectors where process modeling has had a hard time manifesting itself, as it remains mostly data-driven due to its knowledge-intensive nature. However, some of the main concerns trending in eHealth are very similar to other sectors, namely cost reduction and efficiency [8]. So a traditional focus on these two process goals could still create considerable value for both the patients as the healthcare personnel. The fact that process-orientation is nearly absent can be attributed mainly to the need to deliver a flexible and dynamic service. Medical professionals need to be prepared to handle a vast array of cases, where doctors are empowered to use their knowledge and judgment as a guide through the critical data-intensive situations. This makes it hard for traditional process modeling techniques to create added value and calls for a different approach. Languages like GLIF [9], Asbru [10] and PROforma [11] have been proposed to model the guidelines, used throughout the healthcare sector, visually as what are called Computer-Interpretable Guidelines (CIG). Mulyar et al. [12] state that these languages have problems with the dynamic and flexible nature of the healthcare processes (i.e., because they are hybrids of imperative modeling languages with decision modeling languages) and advise the use of a declarative language, CIGDec, which has since been integrated into Declare.

In this paper we will put Declare-R to the test by using it to model a realistic flexible process using a case example from the emergency department of a hospital. This will demonstrate that languages like Declare-R and CIGDec can model these types of processes well, but that they are missing essential information. This information can be seen as the intelligence of the process: the decision logic. The decision logic determines when a certain activity should be executed and this goes further than just specifying sequence constraints between activities. In imperative modeling languages like BPMN decision rules are implicitly modeled as part of the process control-flow (e.g., split gateways). Recently, OMG adopted (currently in finalization phase) a modeling method that allows for the separation of process logic and decision logic:

---

[2] Build-time or design-time flexibility: the intrinsic flexibility of a created model [22].
[3] http://www.win.tue.nl/declare/2011/11/declare-renaming/

the Decision Model and Notation (DMN) [13]. By separating decisions from the process control-flow, the decision rules can be explicitly specified. This means that the decision logic can be reused, be adjusted to evolve within an ever-changing environment and be used as justification for the choices being made. The decision logic is also specified in a declarative way [14], which makes it very well suited, if not more, to complement declarative process languages too. Consequently, it allows us to improve the way we model KiPs and the way we manage these processes. This results in value creation for all the stakeholders. This in turn can be an important incentive to establish a more process-minded way of thinking in KiPs.

The goal of the paper is to demonstrate how a combined Declare-R and DMN approach, we will call it Declare-R-DMN, can model flexible processes more completely with respect to their control-flow and decision logic, while still allowing for run-time flexibility. We will do this by first showing what can and cannot be modeled with Declare-R starting from a realistic healthcare process. Next, we will see what tools DMN can provide us with to model decisions, deontic rules and preferences. Finally, we discuss how the combined approach creates additional value, when compared to the use of only declarative process modeling, as it adds the decision logic which is an essential part of KiPs.

This paper is structured as follows. Section 2 gives a quick overview of the modeling languages used in this paper. In section 3, a case is presented, that demonstrates a realistic, dynamic and flexible process. This case will be modeled using Declare-R and Declare-R-DMN in section 4. In section 5 we provide a brief analysis of Declare-R-DMN. Finally, we conclude the paper and describe the future work in section 6.

2 Background

2.1 Declare-R

Declare is a graphical representation language proposed for declarative modeling of business processes based on LTL-logic [1, 5, 6]. A Declare model contains a set of activities and a set of constraints that can span multiple activities. It specifies the process environment in terms of what is necessary and what is not allowed (i.e., rules expressing the modal verb 'must'), restricting the possible process executions. Contrary to other declarative languages, Declare supports optional constraints (i.e., using a dotted line instead of a solid line). Such constraints offer guidance (i.e., rules expressing the modal verbs 'should' and 'ought to') through knowledge-intensive activities [1], while their soft character ensures flexibility is maintained (i.e., it is not necessary to enforce them).

There are four groups of Declare constraints (see **Table 1**):

- Existence constraints: unary cardinality constraints predicating the number of possible executions of an activity.
- Choice constraints: n-ary constraints expressing a choice between activities.

Table 1. Constraint templates of Declare

Type	Name	Graphical	Meaning
Existence	absence(n+1, a)	[0..n / a]	Activity a can be executed at most n times
	existence(n, a)	[n..* / a]	Activity a must be executed at least n times
	exactly(n, a)	[n / a]	Activity a must be executed exactly n times
	init(a)	[init / a]	Activity a must be the first executed activity
	absence(n+1, a)	[0..n / a]	Activity a can be executed at most n times
Choice	choice(n of m, [a₁,...,aₘ])	a₁ ◇ a₂ ... aₘ	At least n distinct activities among $a_1,...,a_m$ must be executed
	ex_choice(n of m, [a₁,...,aₘ])	a₁ ◆ a₂ ... aₘ	Exactly n distinct activities among $a_1,...,a_m$ must be executed
Relation	resp_existence(a, b)	a ●— b	If a is executed, then b must be executed at any time before or after a
	coexistence(a, b)	a ●—● b	Neither a nor b is executed, or they are both executed
	response(a, b)	a ●—→ b	If a is executed, then b has to be executed at any time after a
	precedence(a, b)	a —→● b	b can be executed only if a has been at some time previously executed
	succession(a, b)	a ●—→● b	a and b must be executed in succession (= response + precedence)
	...		
Negation	resp_absence(a, b)	a —‖— b	If a is executed, then b can never be executed
	...		

- Relation constraints: binary constraints enforcing the presence of an activity in combination with another activity. Table 1 presents the five most important constraint templates. There are six additional templates based on two variations (i.e., alternate and chain) of the response, precedence and succession templates.

The relation templates can also be extended to involve more than two activities. See [6] for more details.
- Negation constraints: negative version of the relation constraints. This is graphically represented with two parallel lines perpendicularly crossing the representation of the relation constraint in question.

To improve the expressibility and practical usability, an extension was proposed, called Declare-R [7], which allows for a textual specification of the information needed to reason about resources:

- Estimates of the duration of each activity
- The available resources
- For each activity, the resource(s) required for its execution

2.2 DMN

The primary goal of Decision Model and Notation (DMN) [13] is to provide a common notation for decision logic that is understandable for business users, business analysts and technical developers. DMN provides the constructs to model decision rules and the decision-making process itself. A DMN decision model consists of two levels: the decision requirements graph (DRG) and the decision logic. The former describes where the required information is coming from and can be depicted in one or more decision requirements diagrams (DRDs). The latter describes the logic behind the decision, which is depicted in Decision Tables [15]. The upper half of a decision table specifies the possible combinations of conditions that lead to certain actions, while the bottom half contains the actions to be taken (i.e., outcomes). A minimal scope is specified for the standardization by OMG, but the goal is to offer support for other decision logic notations (e.g., decision trees) and allow for references to other types of models (e.g., SBVR).

3 Case Example

We elaborate further on a case from [5] with additional information provided by a practicing surgeon[4]. The process of treating arm-related fractures takes place in the emergency department of a hospital. The process entails the registration, diagnosis and treatment phases for patients with one or more fractures of a finger, hand, wrist, forearm, upper arm, shoulder, and/or collarbone.

The process starts when a patient is registered at the reception of the emergency department. Alternatively, in acute emergency situations, the registration can be done at a later time. The next step will usually be to examine the patient. During this examination, the doctor will make a list of the symptoms (e.g., excessive pain and deformation) of the patient. Based on these symptoms he will make a preliminary diagnosis. Normally, this diagnosis is checked by making X-rays, which in turn always results in

---
[4] Dr. Kjell Fierens, AZ Sint-Lucas in Ghent

a new examination to evaluate the X-rays and make the actual diagnosis. In some situations the doctor can make the final diagnosis without X-rays (e.g., clearly no fracture) or there is just no time for this due to emergency conditions.

The next phase involves the treatment the patient will receive. Of course, a treatment is only possible after at least a preliminary examination by a doctor. There are five types of treatment: bandaging, providing support with a sling, fixating the fracture, applying a cast or performing surgery. Each patient will receive at least one of these treatments, even when no fracture is present the patient will be bandaged or receive a sling. Choosing one does not eliminate other treatments, as some strategies combine two or more treatments and patients can be treated for multiple fractures simultaneously. While some treatments do not necessarily require follow-up activities, others might require physiotherapy. For example, muscle atrophy will quickly take its toll after applying a cast or after surgery (due to the usual postoperative period of rest). With the latter the additional damage done to the muscles should also be considered. The other treatments might require physiotherapy, but this is more case dependent. It is also possible that the patient receives a sling (of course no more than one for each arm) or some bandages at any time during the process in order to make he/she as comfortable as possible, no matter the diagnosis.

When we look at the case on a more detailed level we can identify several different variations. These represent the classes of fractures that can occur. Each has a specific flow and different characteristics to be taken into account. One common characteristic is that all fractures require surgery if the fracture is open or complex or when there is extensive damage to the arteries or nerves. Also, if there is no emergency situation, the diagnosis will need to be confirmed by an X-ray. If the patient has multiple fractures, one process instance can combine more than one of these variations.

- A fractured finger or a fractured bone in his hand: in most cases a simple fixation is enough to let it heal. The patient will receive a sling before being sent home.
- A fractured wrist: a cast will be applied, possibly after performing surgery. Surgery is required if the patient is a child (under 16 years old) and also has a damaged periosteum. For adults, surgery is only performed when dealing with open or complex fractures. Afterwards a follow-up X-ray will be taken to confirm that the bone is positioned correctly to start the healing process. The patient will receive a sling before being sent home.
- A fractured forearm: usually this requires no more than a cast. Only when the bone parts are too far apart, surgery is required. To support the cast, the patient receives a sling before being sent home.
- A fractured upper arm: is commonly treated by applying a fixation. If the fracture is an open, surgery is performed. This surgery is also performed when the patient has broken both arms, there is extensive artery damage, there is extensive nerve damage or when there is no improvement over a period of 3 months. The patient will receive a sling before being sent home.
- A fractured shoulder: usually the conservative treatment is enough, letting the shoulder heal while wearing a sling. In the other cases, surgery is required. Physiotherapy is also needed, because the shoulder joint will be inactive for an extended period during each of the two treatments.

- A fractured collarbone: is treated in most cases by resting it while wearing a figure of eight bandage. Surgery is only required when dealing with open or complex fractures or extensive damage to the arteries or nerves.

Additionally, if surgery is required for a broken wrist or forearm, but the OR is unavailable, a temporary cast will be applied to bridge the time until surgery.

Another aspect is the prescription of medication. There is a general policy that states that no medication can be prescribed without being proceeded by an actual doctor's examination. For pain medication it also requires the doctor or surgeon to agree that the patient is in pain or could be in pain in the nearby future. Furthermore, the policy makes a distinction between patients between 0-16 years old (we will refer to them as children) and the older patients (we will refer to them as adults). For instance, if a child had surgery or is in excessive pain as determined by a doctor's examination, he/she will always receive a prescription for a weak painkiller at first. Only if the doctor finds this to be insufficiently effective, a stronger painkiller can be prescribed. This also holds for adults, except after surgery, when stronger painkillers will be prescribed immediately. For both children and adults, after surgery they will be prescribed anticoagulants and anti-inflammatory drugs as precaution. Likewise, patients that received a cast could be prescribed anticoagulants. Because there exist strong painkillers that do not mix well with anticoagulants and anti-inflammatory drugs, a distinction is made between classes of strong painkillers:

- Strong painkillers A: should not be taken while on anticoagulants or anti-inflammatory drugs, but are preferred in other cases.
- Strong painkillers B: can be taken while on anticoagulants or anti-inflammatory drugs.

Furthermore, we also need to consider that some activities require the availability of certain resources, which in turn are limited in number. Also, they are not only used for this process, but rather represent a pool of resources shared among multiple independent processes of the hospital. The inventory of the resources and the activities that require then are as follows:

- 3 reception desks are used to register patients
- 15 exam rooms are used for the examination of the patient as well as for applying a cast, an external fixation, a sling and bandages
- 1 X-ray room with 1 X ray machine used to make X-rays of the patients
- 4 operating rooms where surgery is performed on the patients
- 60 beds where the patients can rest after surgery
- 2 physiotherapy rooms used to provide in-house physiotherapy

Furthermore, there is also a list of human resources available:

- 3 receptionists to work at the reception
- 3 doctors to examine and treat patients (except for surgery)
- 10 nurses to apply casts/fixations/bandages or man the X-ray machine
- 2 surgeons to perform surgery
- 1 physiotherapist to provide physiotherapy sessions

Finally, estimates of the duration of the activities are provided. Of course, this is just an indication as this is dependent on the circumstances.

- Registration: 10 minutes
- Examination: 10 minutes
- Take an X-ray: 30 minutes
- Applying a cast: 15 minutes
- Applying a fixation: 10 minutes
- Applying a bandage: 5 minutes
- Applying a sling: 3 minutes
- Performing surgery: 120 minutes
- Resting after surgery: 180 minutes
- Physiotherapy: 60 minutes
- Prescribing painkillers, anticoagulants or anti-inflammatory drugs: 1 minute

4 Case Models

4.1 With Declare-R

If we model the process using Declare, we obtain the model in **Fig. 1** (the wavy line constraints are explained in section 4.2). By using the Declare-R extension of Declare we can also incorporate the resource constraints of the case as described in black in **Table 2**.

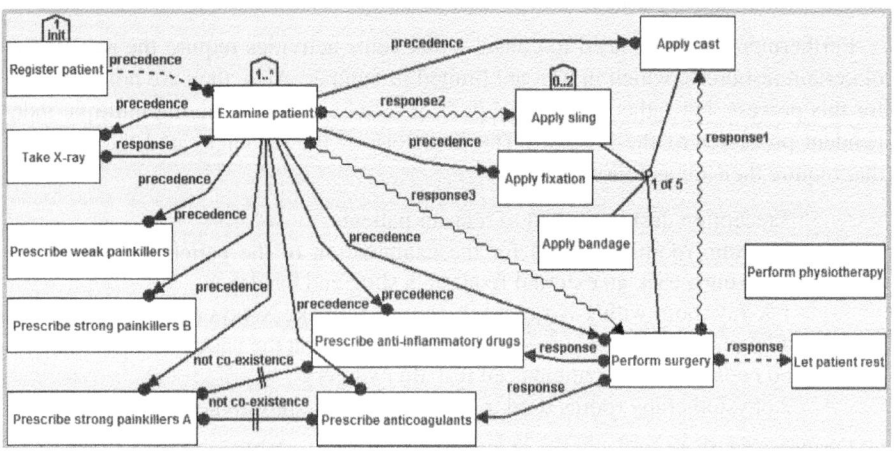

Fig. 1. Declare model of the arm fracture case

An important aspect of the case is the treatment of different types of fractures, each with a unique course of action. The Declare-R model describes a set of sequencing, timing and resource constraints, but does not consider this aspect. Creating a hierarchical process model is not supported according the original definition of Declare [6], however Zugal et al. [16] have described and evaluated a way to introduce it. But even hierarchy would not be enough to model these variations. This is because instead of specifying variations on one general activity, each variation adds different existence and sequential constraints between the activities that are already present on the general level. Each variation could be modeled in a separate Declare model, but how do we know which model is applicable in which situation (i.e., need for run-time flexibility)?

Each variation is in essence a different diagnosis, which can occur simultaneously (e.g., fracture in wrist and forearm). Since this diagnosis is not available at the start of the process, the type of treatment is chosen at runtime, and thus it is a decision made by the doctor conducting the examination. This decision does not only require information of the previously executed steps, but also information about the patient, the symptoms, the test results and perhaps the resource availability. Declare(-R) lacks expressibility to model these decisions and the data on which they are based.

Another shortcoming of this model is the absence of role responsibilities, which could lead to misuse of the model. Not every process actor can initiate each activity (e.g., a nurse cannot perform surgery). Support is however added to the official Declare tool, so this is a minor issue.

4.2 With Declare-R-DMN

The Declare-R-DMN language that we propose incorporates the Declare-R model presented in Fig. 1 and Table 2, while including the decision logic that was missing. The role responsibilities have also been explicitly added (in gray) in Table 2, specifying the process actors that can execute a certain activity.

By using DMN, as part of the proposed Declare-R-DMN language, to model the decision concerning the appropriate treatment, we get the Decision Requirements Graph (DRG) from Fig. 2 and the decision logic (i.e., decision table) from Table 3.

The decision requirements graph in Fig. 2 visualizes where we get the information needed to make the decision concerning the treatment to be applied. The patient record, examination notes and X-ray are documents containing explicit knowledge. This is combined with the implicit knowledge and experience of the responsible doctor to reach a decision on what treatment is appropriate.

The decision logic in Table 3 is presented in a simple syntactic and standardized structure [15], modeling the cause-and-effect relationships between the conditions and actions. The activity corresponding to the action of a decision table has to be executed at any time in the future. This means that other activities could be executed before the action activity, but the action activity has to be executed eventually. This is similar to how the response-constraint template of Declare works (see Table 1). For example, if the patient has an open (visually determined) or complex (determined with X-ray) fracture of the finger, surgery will be performed, but another X-ray might be taken first. The alternatives are visualized side by side to facilitate the analysis of combinations. The completeness property guarantees that every combination of condition values is considered [14]. Because of this structure the decision conditions are easy to understand and manipulate by analysts, programmers and non-technical users [15]. This makes it a great medium for documenting decisions and to allow for backwards

Table 2. Declare-R resource constraints of the arm fracture case

Estimates	Resource requirements and role responsibilities
Duration(Register patient) = 10	Register patient requires RECEPTIONDESK and is executed by RECEPTIONIST
Duration(Examine patient) = 20	Examine patient requires EXAMROOM and is executed by DOCTOR
Duration(Take X-ray) = 30	Take X-ray requires XRAYROOM and is executed by NURSE
Duration(Prescribe strong painkillers A) = 1	Apply cast requires EXAMROOM and is executed by NURSE or DOCTOR
Duration(Prescribe strong painkillers B) = 1	Apply fixation requires EXAMROOM and is executed by NURSE or DOCTOR
Duration(Prescribe weak painkillers) = 1	Apply bandage requires EXAMROOM and is executed by NURSE or DOCTOR
Duration(Prescribe anticoagulants) = 1	Apply sling requires EXAMROOM and is executed by NURSE or DOCTOR
Duration(Prescribe anti-inflammatory drugs) = 1	Perform surgery requires OPERATINGROOM and is executed by SURGEON
Duration(Apply cast) = 15	Let patient rest requires PATIENTBED
Duration(Apply fixation) = 10	Perform physiotherapy requires PHYSIOTHERAPYROOM and is executed by PHYSIOTHERAPIST
Duration(Apply bandage) = 5	Prescribe weak painkillers is executed by DOCTOR
Duration(Apply sling) = 3	Prescribe strong painkillers B is executed by DOCTOR
Duration(Perform surgery) = 120	Prescribe strong painkillers A is executed by DOCTOR
Duration(Let patient rest) = 180	Prescribe anti-inflammatory drugs is executed by DOCTOR
Duration(Perform physiotherapy) = 60	Prescribe anticoagulants is executed by DOCTOR
Resource and role availabilities	
#RECEPTIONDESK = 3	
#EXAMROOM = 15	
#XRAYROOM = 1	
#OPERATINGROOM = 4	
#PATIENTBED = 60	
#PHYSIOTHERAPYROOM = 2	
#RECEPTIONIST = 3	
#DOCTOR = 3	
#NURSE = 10	
#SURGEON = 2	
#PHYSIOTHERAPIST = 1	

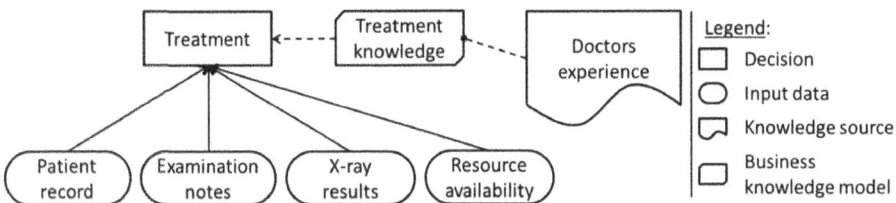

Fig. 2. Decision Requirements Graph for the arm fracture treatments

traceability as justification of decisions taken in actual cases [15]. Additionally, decision tables are also declarative [14], just like Declare, as the columns are rules with no particular order for conditions and actions to occur (i.e., stating the boundaries of the environment instead of a precise path through it). Lastly, decision tables can easily be annotated with statistical information to specify the likeliness of certain sets conditions, and thus of the chosen action [14].

Table 3. The decision table for the arm fracture treatments

	Context: *Examine patient*						Role: *Doctor*								
Open or complex fracture	Y			N											
Extensive damage to arteries or nerves	-			Y			N								
Fractured bone	Wrist or Forearm	Finger, Hand, Upper arm or Shoulder	Collarbone	Wrist or Forearm	Finger, Hand, Upper arm or Shoulder	Collarbone	Finger or Hand	Collarbone	Forearm or Shoulder	Upper arm	Upper arm	Upper arm	Wrist	Wrist	Wrist
Both arms broken	-	-	-	-	-	-	-	-	-	Y	N	-	-	-	-
No improvement in 3 months	-	-	-	-	-	-	-	-	-	-	Y	N	-	-	-
Age	-	-	-	-	-	-	-	-	-	-	-	-	-	0-16	>16
Periosteum torn	-	-	-	-	-	-	-	-	-	-	-	-	Y	N	-
OR available	Y	N	-	-	Y	N	-	-	-	-	-	-	-	-	-
Apply fixation	-	-	-	-	-	-	-	X	-	-	-	X	-	-	-
Perform surgery	X	-	X	X	X	-	X	X	-	-	X	X	-	X	-
Apply cast	-	X	-	-	-	X	-	-	-	X	-	-	-	X	X
Apply bandage	-	-	-	-	-	-	-	-	X	-	-	-	-	-	-
response1	X	X	-	-	X	X	-	-	-	-	-	-	-	-	-
response2	X	X	X	-	X	X	X	-	X	-	X	X	X	X	X
response3	-	X	-	-	-	X	-	-	-	-	-	-	-	-	-

Note that we added a context and role definition to each decision table. This specifies, respectively, the activities connected with the decision and the process actors that are responsible for taking it. The former provides us with a clear overview of what decisions are applicable during which activities, while the latter is a safeguard against unauthorized usage.

The decision table representation also has some drawbacks. When the decisions themselves are based on a very large amount of conditions and actions, the readability of a table gets lost. In such cases, the decision table will need to be split in multiple smaller tables to allow them to stay manageable. However, this comes at the price of cluttering the overall overview and understandability. Another problem arises when multiple actions are activated at the same time. Consider for example if the patient has a complex fracture of his forearm. The treatment for this diagnosis is surgery, followed by applying a cast. Decision tables do not allow a sequencing of these actions, but this is rather important in this situation. As a solution, we propose to introduce a new type of constraint in the Declare model: the *decision-dependent constraint*. It can embody all the templates from Table 1 and is visualized as a wavy line in the model. The decision-dependent constraint represents a constraint that can only be activated as outcome of a decision table. If we retake the previous example, we can model this by adding a decision-dependent response constraint between 'Perform surgery' and 'Apply cast' (i.e., response1) and adding the additional outcome activating this constraint in Table 3. In a similar way, we add a decision-dependent response constraint between 'Examine patient' and 'Apply sling' (i.e., response2) and another decision table outcome activating this constraint to model the fact that in all treatments, except of collarbone fractures, the patient will receive a sling eventually. Lastly, another one of these constraints (i.e., response3) is used to ensure that surgery will still be performed later on when a temporary cast is applied because the OR is not available at that time.

Table 4. The decision table for the prescription of medication

Context: *Examine patient or Perform surgery*	Role: *Doctor or Surgeon*									
In pain (now or in foreseeable future)	Y	Y	Y	Y	Y	Y	Y	Y	Y	N
Age	0-16	0-16	0-16	0-16	>16	>16	>16	>16	-	-
Previous step is surgery	Y	Y	N	N	Y	Y	N	N	-	-
On weak painkillers	Y	N	Y	N	-	-	Y	N	-	-
On anticoagulants or anti-inflammatory drugs	-	-	Y	N	-	-	Y	N	-	-
Prescribe strong painkillers A	-	-	-	X	-	-	-	X	-	-
Prescribe strong painkillers B	X	-	X	-	-	X	X	-	-	-
Prescribe weak painkillers	-	X	-	-	X	-	-	-	X	-
Prescribe anticoagulants	X	X	-	-	-	X	-	-	-	-
Prescribe anti-inflammatory drugs	X	X	-	-	-	X	-	-	-	-

Multiple outcomes in a decision table do not always lead to new sequencing constraints. Consider Table 4, which represents the decision to prescribe medication to patients. If the patient for example is in excessive pain, older than 16 years old and just had surgery, three actions are activated. The sequencing of these actions does not matter here, as they are just prescriptions (of course sequencing rules could apply when administering these drugs, but that is beyond the scope of the case).

Besides mandatory constraints, Declare allows for optional constraints to be modeled. We propose to extend this principle to the decision tables in Declare-R-DMN, as this allows them to represent 'wanted'-behavior (i.e., "should" and "ought to"). For example, patients that have had surgery, a cast applied and/or a fractured shoulder should do physiotherapy afterwards. But this is case dependent, so it should be possible to deviate from this structure. We modeled this in an optional decision table in Table 5. Following the representation in Declare, we visualized the 'optional'-property with a dotted line around the decision table. Table 6 does the same for the decision whether or not to take an X-ray.

Table 5. The optional decision table for the prescription of physiotherapy

Context: *Examine patient*	\multicolumn{4}{c}{Role: *Doctor*}			
Surgery was performed	Y	N		
Cast has been applied	-	Y	N	
Shoulder fracture	-	-	Y	N
Perform physiotherapy	X	X	X	-

Table 6. The decision table for an X-ray

Context: *Examine patient or Apply cast*	\multicolumn{4}{c}{Role: *Doctor*}			
Verifying fracture diagnosis	Y	N		
Fracture	-	Y	N	
Verify if bone is correctly positioned under cast	-	Y	N	-
Take X-ray	X	X	-	-

5   Analysis of the Declare-R-DMN Language

Declare-R-DMN has the expressive power needed to create a more complete model, compared to Declare-R, of the process of the arm fracture case. This is achieved by adding support to Declare to model decision logic, without losing sight of the need for flexibility of KiPs. Making the decision logic explicit also facilitates the justification of taken decisions, better conformance checking and reuse [18] across different processes or even organizations. Value is created for the users and the organization as declarative process modeling languages are better suited for these types of processes compared to traditional process modeling languages. Additionally, more value is created by adding support for decision logic to the declarative modeling language as this is essential information that would otherwise be omitted.

Recently, other approaches have been proposed [17, 18] that do similar work. They add a way to model constraints that deal with the data aspect. However, these approaches focus primarily on the expressibility of the language and therefore much less on understandability and the modeling aspect. It is our opinion that these aspects are crucial for the adoption of the technique, and thus should be more of a priority. The use of decision tables (the DMN standard is committed to also offer support for other techniques in the near future) for this purpose is pretty straightforward, because they are a known and proven way of representing decisions and they are understandable for both technical as business people. By aligning the interpretation of the decision tables with Declare (i.e., decision-dependent constraints), adding context definitions and role responsibilities and extending the optionality-concept of Declare, Declare-R-DMN becomes a comprehensive and coherent modeling language: the temporal logic is modeled using Declare, the resource perspective using the Declare-R extension and the decision logic using DMN.

However, a general problem still persists. Compared to imperative modeling, the increased support for flexibility by declarative modeling comes at a price of understandability (of Declare in particular) and maintainability issues arise [3, 6, 19]. Recently, a couple of hybrid approaches have been proposed [20, 21] that offer some improvements for semi-structured processes, but not for unstructured processes.

6 Conclusion and Future Work

This paper presents an idea that is similar to what DMN attempts to do for BPMN, but also differs in a fundamental way. First, BPMN already somewhat supported decision logic with its fundamental concepts. Second, in the context of dynamic, knowledge-intensive and flexible processes, where Declare finds its niche, the decision logic is of much greater importance than in case of static and standardized processes. Decision logic is essential when modeling these processes as it offers valuable insight and encapsulates the knowledge of the domain experts executing the process.

The proposed language, Declare-R-DMN, combines Declare-R and DMN in a way that both original languages are supported as well as some new concepts that bridge them together (i.e., decision-dependent constraints, context definitions, role responsibilities and extending the optionality-concept). The usefulness of Declare-R-DMN was demonstrated by modeling a case example, representing a realistic example of a dynamic, knowledge-intensive and flexible healthcare process, as is exhibited by the large variety of possible execution paths (i.e., theoretically infinite). Where Declare-R did not offer enough tools to model the process of the arm fracture case, Declare-R-DMN thrived by incorporating the knowledge that is essential to the case.

The scope of this paper was limited to a general elaboration of the idea. In the next phase we will formalize the semantics and metamodel of this new language. For this purpose, we need to analyze and propose solutions for all of its possible ambiguities, overlaps and shortcomings to obtain a clear and coherent language. Inspiration can come from the proposals for a data-aware Declare [17, 18]. The language will then be further evaluated by using it to model similar real-life cases from different areas.

References

1. Goedertier, S., Vanthienen, J., Caron, F.: Declarative business process modelling: principles and modelling languages. Enterp. Inf. Syst., 1–25 (2013)
2. Lu, R., Sadiq, W.: A Survey of Comparative Business Process Modeling Approaches. In: Abramowicz, W. (ed.) BIS 2007. LNCS, vol. 4439, pp. 82–94. Springer, Heidelberg (2007)
3. Haisjackl, C., Barba, I., Zugal, S., Soffer, P., Hadar, I., Reichert, M., Pinggera, J., Weber, B.: Understanding Declare models: strategies, pitfalls, empirical results. Softw. Syst. Model. (2014)
4. Krogstie, J.: Perspectives to Process Modeling – A Historical Overview. In: Bider, I., Halpin, T., Krogstie, J., Nurcan, S., Proper, E., Schmidt, R., Soffer, P., Wrycza, S. (eds.) EMMSAD 2012 and BPMDS 2012. LNBIP, vol. 113, pp. 315–330. Springer, Heidelberg (2012)

5. Van der Aalst, W.M.P., Pesic, M., Schonenberg, H.: Declarative workflows: Balancing between flexibility and support. Comput. Sci. - Res. Dev. **23**, 99–113 (2009)
6. Pesic, M.: Constraint-based workflow management systems: shifting control to users (2008)
7. Barba, I., Del Valle, C.: Filtering rules for ConDec templates - Pseudocode and complexity. http://www.lsi.us.es/quivir/irene/FilteringRulesforConDecTemplates.pdf (accessed on October 9, 2014)
8. Payton, F.C., Paré, G., LeRouge, C., Reddy, M.: Health care IT: Process, people, patients and interdisciplinary considerations. J. Assoc. Inf. Syst. 12 (2011)
9. Boxwala, A.A, Peleg, M., Tu, S., Ogunyemi, O., Zeng, Q.T., Wang, D., Patel, V.L., Greenes, R.A, Shortliffe, E.H.: GLIF3: a representation format for sharable computer-interpretable clinical practice guidelines. J. Biomed. Inform. 37, 147–61 (2004)
10. Seyfang, A., Kosara, R., Miksch, S.: Asbru's Reference Manual v7.3 (2002)
11. Fox, J., Johns, N., Rahmanzadeh, A.: Disseminating medical knowledge: the PROforma approach. Artif. Intell. Med. **14**, 157–181 (1998)
12. Mulyar, N., Pesic, M., van der Aalst, W.M., Peleg, M.: Declarative and Procedural Approaches for Modelling Clinical Guidelines: Addressing Flexibility Issues. In: ter Hofstede, A.H., Benatallah, B., Paik, H.-Y. (eds.) BPM Workshops 2007. LNCS, vol. 4928, pp. 335–346. Springer, Heidelberg (2008)
13. OMG: Decision Model and Notation (DMN). www.omg.org/spec/DMN/Current/ (accessed on November 10, 2014)
14. Software Testing Genius: Decision Table Based Testing-Black Box Software Testing Technique. http://www.softwaretestinggenius.com/decision-table-based-testing-black-box-software-testing-technique
15. Decision Table Task Group: A Modern Appraisal Of Decision Tables (1982)
16. Zugal, S., Soffer, P., Haisjackl, C., Pinggera, J., Reichert, M., Weber, B.: Investigating expressiveness and understandability of hierarchy in declarative business process models. Softw. Syst. Model. (2013)
17. Montali, M., Chesani, F., Mello, P., Maggi, F.M.: Towards data-aware constraints in declare. In: Proceedings of the 28th Annual ACM Symposium on Applied Computing, SAC, pp. 1391–1396. ACM Press, New York (2013)
18. Borrego, D., Barba, I.: Conformance checking and diagnosis for declarative business process models in data-aware scenarios. Expert Syst. Appl. **41**, 5340–5352 (2014)
19. Zugal, S., Pinggera, J., Weber, B.: Toward enhanced life-cycle support for declarative processes. J. Softw. Evol. Process., 285–302 (2012)
20. De Smedt, J., De Weerdt, J., Vanthienen, J.: Multi-Paradigm Process Mining: Retrieving Better Models by Combining Rules and Sequences. SSRN Electron. J. (2014)
21. Maggi, F.M., Slaats, T., Reijers, H.A.: The Automated Discovery of Hybrid Processes. In: Sadiq, S., Soffer, P., Völzer, H. (eds.) BPM 2014. LNCS, vol. 8659, pp. 392–399. Springer, Heidelberg (2014)
22. Reichert, M., Weber, B.: Enabling flexibility in process-aware information systems. Springer (2012)

Understanding and Sharing

Using the Process-Assets Framework for Creating a Holistic View over Process Documentation

Magnus Josefsson, Kim Widman, and Ilia Bider[✉]

Department of Computer and Systems Sciences (DSV),
Stockholm University, Stockholm, Sweden
{josefsson.magnus,kim.widman}@gmail.com,
ilia@dsv.su.se

Abstract. When an organization has not adopted a uniform and standardized way of producing and storing process documentation, keeping track of and maintaining process documents can be a real challenge. In this paper we suggest a framework for organizing process documentation which is created in different notations, for different purposes, and stored in different formats. We show how this framework has been applied in a real case in an organization where such problems are present. We discuss advantages and disadvantages of the framework and suggest further development and testing of the framework to improve its usability.

Keywords: BPM · Business process · Process documentation · Process map

1 Introduction

This experience report is aimed at attracting attention to an insufficiently researched practical problem that prevents reuse of existing process documentation. Besides stating the problem, we also present a solution tested in the course of our project. We do not insists that the solution (how) is optimal, but testing it in practice has helped us to further investigate the problem (what), and requirements on a solution. The paper presents the business case in which the problem has been encountered and the solution tested, as well as lessons learned about the problem and solution that could be of interest for both researchers and practitioners.

Ideally, all work revolving around business processes should be carried out in a systematic and carefully managed way, and be aligned with overall business goals and strategies [1]. However, as we discovered in the course of the project reported in this paper, this policy is not followed in practice in all organizations. Some organizations lack a structure to lead their work in Business Process Management (BPM). Still they perform business process related projects. Such projects are performed sporadically, i.e. on-demand. They can be completed locally by some department, or with a wider scope, e.g. analysis of the processes that run through the whole organization, or even several organizations. The on-demand BPM projects can be completed in connection to reorganization, acquiring or developing new IT-support, etc. They produce process maps and other types of process documents specifically aimed at the purpose of the project.

After the given BPM project ends, the documentation produced is filed in an unsystematic way. Reuse of the documents is mainly based on the people engaged in the project remembering its details. They can retrieve the documentation through search based on the project name, date, participants, etc. Possibilities of reuse under these circumstances are limited, especially considering that people may leave the organization. Lack of systematization can result in loss, or underuse of valuable knowledge contained in process maps and other process documents. It can also result in the same process being investigated several times as people who decide on new projects may not be aware of the previous projects.

The problem of unsystematic BPM encountered in the project reported in this paper, is not unique for the organization we investigated. The same phenomenon has also been encountered at least by some other researchers; see, for example, [2][1]. BPM consultants whom we interviewed also pointed out that many of their clients experienced the problems with unsystematically stored process documentation.

This paper reports on the project aimed at developing and testing a solution for systematizing and organizing process documentation created in an ad hoc, on demand fashion. The solution was worked out for a specific business case of an organization that has produced around 100 process maps in an unsystematic way. As the problem we dealt with was of general nature, we were looking not for a specific solution for this organization, but for a generic solution that could be used in other organizations independent of their business domain. The implementation of this solution in a specific organization presented in this paper thus could be considered as proof of concept of the suggested generic solution.

As a basis for systematizing process documentation, we used the process-assets framework [3]. This framework has been developed for finding all processes within an organization, starting from the main processes. As the framework was created with another purpose in view, it, in its original form, did not directly fit our goal. Therefore, we adjusted the framework for a new purpose while testing it on the material from the organization we investigated. Consequently, the goal of the project became twofold:

1. Solve a concrete problem in a specific organization while devising a generic solution that could be used in other organizations. This goal included creating a solution independent of business domain.
2. Test the usefulness of the assets-process framework from [3] for a new purpose, namely, systematization of BPM documentation produced in ad hoc, on-demand manner[2].

The rest of the paper is structured as follows. Section 2 presents the project context: the case organization, problem, and project participants. Section 3 is devoted to analyzing requirements on a solution. Section 4 gives an overview of our efforts to find a standard solution, and a suitable basis for a new solution. Section 5 explains the

---

[1] Actually, [2] is the only paper, we were able to find that reports on this problem.
[2] This sub-goal is in-line with the special theme of BPMDS 2014 as it is directed for increasing the value of the previous business process modeling projects.

process-assets framework chosen as a basis for our solution. Section 6 describes how this framework has been transformed to fit the new purpose. Section 7 deals with creating a solution for the case organization. Section 8 is devoted to evaluating the solution for our business case. Section 9 discusses lessons learned and limitations of the work completed so far.

2 The Project Settings

2.1 The Investigated Organization

The project took place in a Swedish organization operating in the area of betting, in the text below referred to as the *betting company*. The betting company offers betting both through physical shops and online. The main office currently has 250 employees, whereof nearly half are employed by the IT department. The turnover of the company is around thirteen billions Swedish Crowns (SEK) annually.

2.2 The Practical Goal of the Project

Since year 2000 the IT department of the betting company has conducted 11 process mapping projects that have rendered a large amount of process documentation. The reasons for these projects have been of varying kind, to improve efficiency, to develop computer systems, to oblige to new legislative demands, only to mention a few. There is no formal repository in place, and the different projects have made their own decisions about what modeling languages, methods, and tools to use. The documentation has therefore not been uniformly designed and stored, and has over the years become very difficult to retrieve and reuse. By the beginning of 2013 this problem had reached such dimensions that there was uncertainty of the exact number of existing process maps as well as where to find them. As a result, already gained knowledge on business processes had sometimes been lost, and redundant work had been done in investigating the same processes several times. A need to create a *holistic and comprehensible overview* of the existing documentation had arisen, and a project for this task was initiated.

2.3 The Project Participants

The project was carried out in 10 weeks by two business analysts (BAs) who were entrusted the task of finding a solution for the problem identified, a scientific adviser, and two employees of the betting company's IT-department. The roles of BAs were held by the first two authors, while the third author served as a scientific adviser. As far as participants from the betting company are concerned, one person held a position of chief systems architect; the other one was responsible for all internal support systems. The BAs were given full access to company's all known process documentation, and unlimited access to the participating members of the IT department.

3 Establishing the Requirements

It was a strong and explicit wish from the betting company that the holistic overview of the process documents would be presented in a graphical form. The holistic overview should also be easily understood by any person with knowledge of the company's terminology, not only by IT managers and developers. Furthermore, the holistic overview was not to change the existing BPM documents in order for them to better fit the overview. The latter requirement concerned both content of the documents, and the format in which they were stored.

Further requirements were elicited by studying the process documents. By examining the documentation we found that a total of 137 processes had been identified for the company's regular operations. 85 of these processes had been thoroughly described with start, end and activity flow, i.e. documented in detailed process maps. Some of the 85 process maps overlapped, since the processes had been investigated within different projects with little or no cooperation between them. In some maps inconsistencies were found, implicating that these documents were not checked for logical or formal correctness. There were also maps that described only parts of processes. A variety of mapping notations was found, as well as different methods of document storage. The remaining 52 documents described processes in a somewhat loosely manner as functional business areas with no clear starts, ends and activity flows.

Based on the investigation completed, the following list of requirements was compiled on holistic overview of process documentation, which we below refer to as the *holistic process model*:

1. The holistic process model should be presented in a graphical form.
2. The holistic process model should be easily understood by any person with knowledge of the organization's terminology.
3. The holistic process model must not in any way introduce changes into the existing process documentation.
4. The holistic process model must not rely on that all process documents are produced using the same notation or technique.
5. The holistic process model must not rely on that all process documents are stored in a specific way.
6. The holistic process model must not rely on that all process documents are produced having the same purpose in mind.
7. The holistic process model must not rely on that all process maps included in the process documentation are correct.
8. The holistic process model must not rely on all processes within the organization are mapped.
9. The holistic process model must not assume that a process document or map depicts exactly one process. It can depict more than one process or "half" of a process.
10. The holistic process model must not assume that only one map exists for a particular process.

4 In Search for a Solution

4.1 Searching for a Standard Solution

Before inventing a new solution, we checked whether there is a known solution for the problem identified in the betting company. For this, we, first, investigated whether this problem is unique for the betting company, or if it had been discovered in other organizations. The investigation has been done in two directions: (a) interviewing business management consultants with experience in BPM projects (b) literature search.

Two BPM consultants were contacted to check for real-life experience of the problem and possible solutions. Consultant No 1 belonged to a consultancy that the betting company had engaged for one of their projects. This consultant estimated that around 70% of their clients had experienced problems with finding documents on a given process. Among the other 30% of their clients, this was a lesser problem, due to the fact that they had implemented a process management structure with committed process owners and centralized standards for documentation.

Consultant No 2 was not bound to the betting company. He stated that most of their clients had experienced problems with finding process documents, and also with understanding their contents. Some clients had solved this problem by adopting formal repositories.

None of the BPM consultants interviewed knew a standard solution for the problem or had encountered the problem as a subject at any conference or congress. The search through the research literature has not resulted in a solution found for the problem of creating a holistic overview of already existing process documentation. The search was done on different variation of the phrase "organizing process documentation". A number of works has been found that proposed solutions for related problem, for example, works on automatic process clustering [4], and business process models repositories [5,6]. However, we found no works explicitly devoted to the problem at hand. The only paper we found that refers to this problem explicitly was [2] that investigated BPM experts' points of view on the problems in BPM. This paper includes extracts of the interviews of the following kind:

— "When one looks at the way that an organization gets its work done, you see that part of this is an important strategic level and part of this is an important operational level"
— "Often 4,5,6, different places in the organization run BPM projects and then you have the problem how to bring these local projects together, in an overall process architecture. I see a lot of bottom up projects but no way to tie that all into an overall business strategy or process strategy of the organization."

Though [2] confirms that the problem is spread, it does not suggest any solution for it.

4.2 Choosing a Suitable Framework

Though, to the best of our knowledge, there is no standard solution for the problem discovered, there exist several ways of classifying business processes that could be used for building a solution for the problem. To such classifications, for example, belong:

1. The classification based on the Porter's value chain [7] that differentiates primary and supporting processes, and inside each of categories differentiate sub-categories, like in-bound logistics (main process), or procurement (supporting process).
2. Process Classification Framework from the American Productivity & Quality Center (APQC) [8] which classifies processes dependent on the industry.

Other classification schemes are in details analyzed in [6] devoted to developing semantic annotation for business process models repositories. Besides classification schemes, we also considered a fractal process-assets framework [3,9] that allows to arrange all processes in an organization in a tree-like structure connecting them through the assets used in each processes.

Neither existing classification schemes, nor the process-assets framework have been tested for the practical task of organizing ad-hoc created process documentation, at least, to the best our knowledge. All these frameworks concern business processes not documents that can depict only part of the process or several processes at the same time. In addition, some of the classifications use quite abstract taxonomies and it has not been clear whether such taxonomy would be easily understood by practitioners.

As we did not have any special reasons to prefer one framework over all others, we decided to start with the process-assets framework [3,9] because of having more intrinsic knowledge of it and limitation on time for completing the project. As the process-assets framework is not widely known, in Section 5, we give a short overview of it, before describing how it was transformed to suit the task at hand.

5 The Process-Assets Framework

The process-assets framework consists of the process-assets archetype (Fig. 1) and the asset-processes archetype (Fig. 2), building on the idea that any business process in a company rely on the company's assets in order to be executed properly, and that the assets themselves need supporting processes to keep them in shape.

The process-assets archetype shows a process and various assets that are needed for this process to function friction free on a repetitive basis. The first process in the framework is the main process of the organization. The assets are divided into six different categories.

1. *Paying stakeholders,* e.g. customers of a company or paying members of a club.
2. *Business Process Templates (BPT).* For a manufacturing company this is both the design of a product and a scheme of technological process of its production.
3. *Workforce.* People that are qualified to work in the main process.

4. *Partners,* e.g. suppliers for the manufacturing process of a company.
5. *Technical and Informational Infrastructure.* Equipment that is needed to run the main process. For a manufacturing company this can be production lines, computers, etc.
6. *Organizational Infrastructure,* e.g., departments, teams and management within the organization.

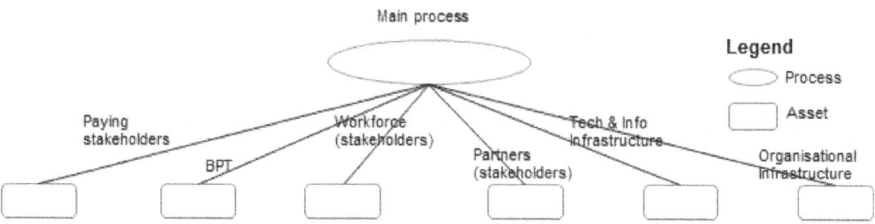

Fig. 1. The process-assets archetype

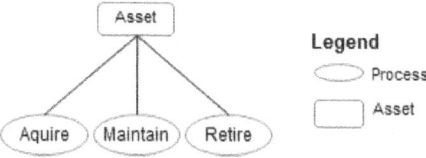

Fig. 2. The asset-processes archetype

For the above mentioned assets to be able to support a process the assets themselves need to be supported by a number of processes. This is where the asset-processes archetype comes in. This archetype shows what processes are needed to acquire, maintain, and retire an asset, see Fig. 2:

1. *Acquire.* Processes that are used to get hold of a certain asset, e.g. a recruiting process to add new employees to the workforce.
2. *Maintain.* Processes for keeping assets. This can be customer-relationship processes to keep customers, or training for employees.
3. *Retire.* Processes used to remove assets that are no longer needed. For the BPT category this can be to phase out a product that is not in demand anymore.

Having the above archetypes can help to unveil the dynamic process structure of an enterprise starting from the main process and going downwards via repeating the pattern "a main process->its assets->processes for each asset->assets for each process->...". In this way it is possible to discover all processes within an organization. First one defines the main processes and what assets are needed to run the processes. After that focus goes to the assets and what processes are needed to acquire, maintain and retire them. Then one starts to look at different assets that are needed to support the acquiring, maintaining, and retiring processes. This procedure can be repeated until all processes within the organization have been found and arranged in a kind of a tree with a repeating pattern of nodes. Such kinds of structures are known in the scientific literature under the name of fractal structures.

Note that in practice, the process structure will not form an indefinite tree as the same process or assets can have multiple usages, i.e. serve different assets or processes. This will, in the end stop the tree expansion. For details on the process-assets framework we refer the reader to [3,9].

6 Transforming the Process-Assets Model for Our Purpose

For the purpose of our task to get a holistic overview of all process documents that exist within an organization, there is no need to go too deep into the fractal tree described in the previous section. The full fractal view would make the holistic process model too large and difficult to overview. Therefore, the unveiling of the model stops after two iterations as it is represented in Fig. 3. In addition, the ovals that represent the processes in the original model from [3] (see Fig 1, and 2) are substituted by cylinders that are called "buckets" where the related process documentation can be placed. Parallelograms are added to symbolize process documents.

As a result, we get a relatively simple structure where all existing process documentation can be mapped. The disadvantage of stopping at the second level of iteration is that the documentation related to processes on the third or deeper levels will be placed one or more levels up in the hierarchy. However, if the number of documents in each bucket is small, it will not be difficult to look through all documents in it manually to find out which process documents related to the given branch of the tree exist. Extending the tree too deep can make it difficult for a non-technical person to orient in the structure, thus there is a risk of breaking requirement #2 from Section 3 (understandability). Staying with two process levels and using the terminology accepted in the organization makes it easier to navigate through the tree for the members of the staff of the organization.

Naturally, the process documents themselves are not placed in the buckets. Instead, the model works with references to the documents, e.g. their serial numbers. Therefore, the holistic process model of Fig. 3 does not require making any changes to the existing documentation. Working with references also supports the solution to the problem of documents that describe more or less than one process, see requirement #9 in Section 3. If the document describes only part of a process, the reference to it is put in the bucket where it belongs, as if it were a fully documented process. In a situation where a process supports multiple assets, a reference to the process document will be placed in all buckets connected to these assets (see process 17 in Fig. 3 that is placed in three different buckets). In cases where multiple documents describe the same process, the references to them are placed on the same reference holder.

Using the Process-Assets Framework for Creating a Holistic View 177

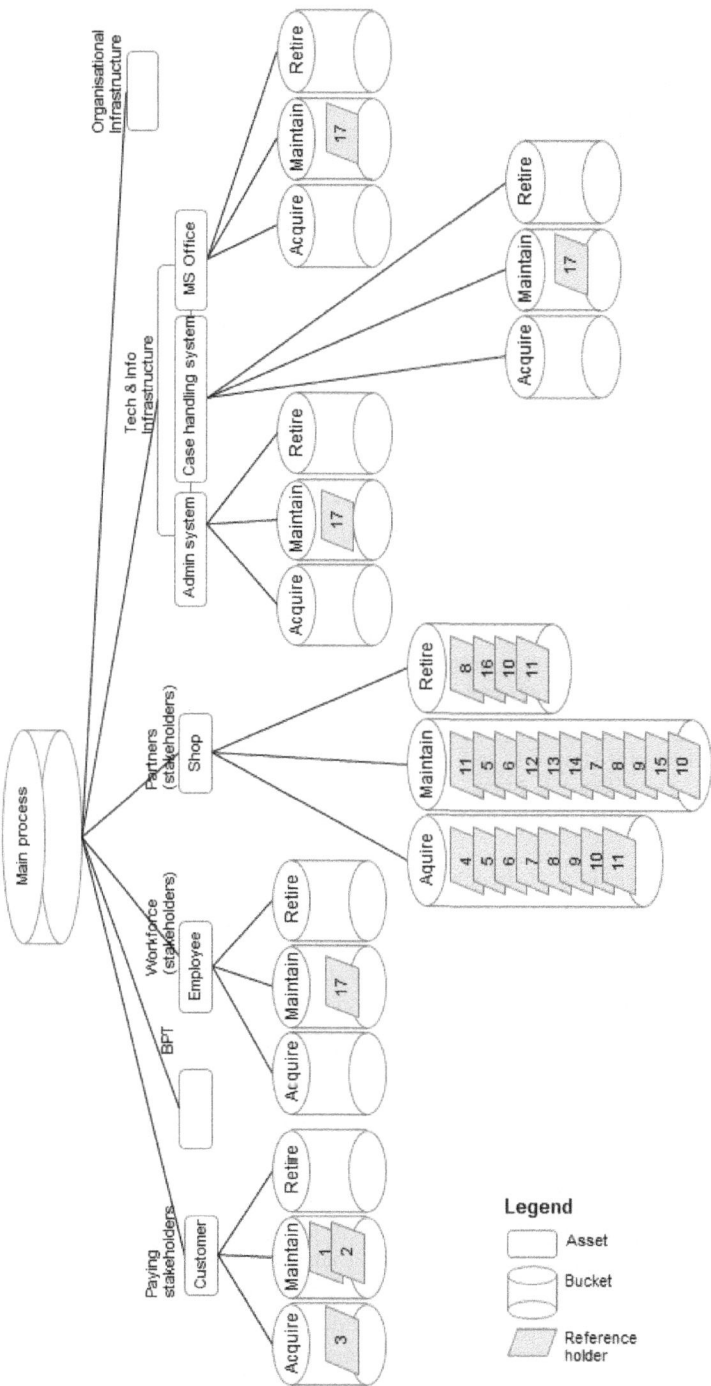

Fig. 3. The holistic process model after referencing 17 process maps

Using references in the holistic process model automatically satisfies requirements #3-7 from Section 3. For example, the documentation of the processes could be made with any methods or techniques, and the holistic model of Fig. 3 does not require that the existing documentation is correct. Neither does the model require that all processes of the organization have been documented. If there are no process documents related to a particular place in the tree, the bucket will be empty (see Fig. 3). This can be considered as an additional feature that shows areas in which there are no documented processes within the organization.

7   Building the Holistic Process Model for the Betting Company

Building a holistic model for the betting company based on the augmented process-assets framework was done in three steps.

1. Make a graphical representation of the upper level of the model. Go through the existing process documentation in order to find all assets mentioned in it. Put them under appropriate categories of assets on the upper level. In the case of the betting company, customers are to be placed under *Paying stakeholders*, employees under *Workforce stakeholders*, shops under *Partner stakeholders*, and computer systems under *Tech and info infrastructure*, see Fig. 3.
2. Extend the model by attaching buckets for *acquire*, *maintain* and *retire* processes for all assets identified in Step 1.
3. Investigate each process document in order to determine which assets the process described in it supports, and which of the asset process buckets acquire, maintain, or retire it belongs to. Place a reference to the process documentation into all buckets where it belongs.

To facilitate step three of the above, we compiled a document that contained metadata for each process map at our disposal. This document helped in determining in which bucket(s) to place references to any given map. The metadata represents a kind of semantic annotation of the process at hand; it includes the name of the process map as found in the documentation, and gives a short description of start and end conditions for the process. The latter helped us to understand the business goal(s) of the process. An example of process metadata is presented in Table1. This table refers to the process related to applying for membership in Club XP. Club XP is a club that unties players interested in a game called Game XP. As a Club XP member, a player gets access to additional information about Game XP competitors on the weekly basis.

The resulting holistic process model after references to 17 process maps have been placed in the buckets is presented in Fig. 3. As can be seen from Fig. 3, most of the processes analyzed are related to the asset labeled *Shop*. A shop, e.g. a news agency, is a partner who takes bets on behalf of the betting company, and pays prizes to the winners. The processes in the *acquire bucket* deals with finding and setting formal agreements with new partners. The bucket *maintain* contains processes related to existing partners, the bucket *retire* contains processes related to canceling the agreement with a partner.

Table 1. An example of process map metadata

Name of process	Register new member in Club XP
Start conditions	A person/organization has applied for membership in Club XP
End conditions	The applicant is registered as a member in the Club XP membership database, or rejected. The applicant has been informed about the outcome of the application.
Assets supported	Member of Club XP [acquire], Club XP [maintain], Club XP membership database [Maintain][1]
Name of document	Handling Club XP membership.doc
Placement in storage	Directory //G:/Club XP/Process descriptions/ [2]
Serial #	36

[1] Note that assets not explicitly named in the start and end conditions may also be supported by the process, their presence being discovered in the detailed descriptions of the process activities.
[2] It can also be something of the form "bookshelf in printer room 504" if the document exists only in the printed form.

To give better understanding of the holistic process model built for the betting company, below, we present some details on process documents 11 and 17 from Fig. 3.

Document 11 in the *acquire*, *maintain* and *retire* buckets of asset *Shop* deals with situations when a shop owner is retiring and sells his or her business to a new owner. In essence, this process serves multiple purposes, i.e., retiring the current shop owner, acquiring a new shop owner and keeping the shop open for betting services during the change of ownership.

Document 17 in the maintain bucket of employees and maintain buckets of IT-software systems describes the process for internal IT support. When an employee reports a problem encountered with any of the company's software, it can result in a bug report filed in the company's case handling system. The latter will lead to fixing the software, which means maintenance of the software asset. But the problem report can also come from an employee who misunderstands the functionality of the software. In this case, the problem report does not lead to a bug report filed in the case handling system, but to a training session that extends the employee's knowledge of the software, which is maintenance of employees.

8 Evaluating the Results

The evaluation of the holistic process model built for the betting company was completed in two phases. *Phase one* of the evaluation was conducted with the chief systems architect of the betting company, hereafter called SA (Systems Architect). The evaluation started with a description of the method and the model, to be continued by a semi-structured interview where SA was given possibilities to comment on fulfillment of requirements, applicability of the results, and their suitability for the purpose.

SA did not convey any negative opinion that would indicate that the requirements from Section 3 were not fulfilled. SA made positive comments regarding the graphical model. To SA, it was refreshing to see the information about processes presented in a new way, since it gave a new perspective, different from the one that is given by process flow diagrams. SA found that the model was suitable to be presented to the upper management since they are more used to look at organizational charts than business process diagrams, and that the holistic process model was somewhat similar to an organizational chart. The model also made it easier to find the places where there was lack of process documentation. SA did however point out that it was not possible to see how different processes from the same bucket are connected to each other, and that this could be an issue for further investigation aimed at extending the model.

Phase two of the evaluation was designed as an test consisting of two steps. It was carried out with a person responsible for all internal support systems at the betting company, hereafter called IS (Internal Systems). IS was first given a short tutorial on the structure of the model, and semantics of the graphical symbols in it. Thereafter, the test began.

Step one of the test was aimed at examining whether IS could find the information in a graphical model. IS was given a model that was pre-populated with assets and references to process maps, and a table containing metadata of the referenced process documents in the form as in Table 1. Thereafter, IS should complete assignments of retrieving information from the model and the table. IS navigated easily through the model and found all information that was asked for in the assignments, even in cases where the name of a process did not clearly depict its contents.

Step two of the test examined whether IS could populate a half-filled model. IS was given a model that was populated with some assets, but with their buckets empty. IS was also provided with a set of cards that held metadata about various process maps (in the form of Table 1). IS's task was to place references to the process maps described on the cards in the right buckets. In case an asset that was not already present in the model appeared on one of the cards, IS had also to place this new asset under the right asset category. This step of the test required more concentration from IS, and each placement was preceded by a monologue in which IS considered the right place or places for each process map. IS finally finished the task and had no problem with either process maps that concerned more than one place in the model, or process maps that dealt with new assets.

Interesting in step two of the test was that IS at one point decided not to place a reference in one of the buckets that we considered appropriate. When asked about this after all references had been placed, IS explained the reasons for the decision. We had apparently misunderstood the meaning of some terms used in the process documentation when building our holistic process model. This underlines the fact that the model is easy to understand and operate only for those who have good knowledge about the organization and terminology used in it. The incident did not reveal a fault built in the modeling principles, it merely showed that IS knew the organization better than we did.

In total, the model, and the method of building and using it showed to be operational and useful for arranging and finding process documentation. There were no substantial difficulties for IS to perform different tasks in the test, but without the introductory tutorial before the test, IS would probably not have understood how to complete the assignments. This is due to the processes and process documentation were visualized in the model in an unusual manner.

9 Lessons Learned and Limitations

We summarize our experience from the project reported in the previous sections in the form of a list of lessons learned and limitations of the work completed so far. While discussing the lessons, we also point out areas where our work can produce *impact on research and practice*.

1. Not all organizations work with BPM in a systematic way, but rather adopt an ad hoc, on-demand approach to their BPM projects. This may lead to *unintended and undesirable* consequences reported earlier in this paper. This problem is known to BPM consultants, and but is not sufficiently covered by research literature.
2. To the best of our knowledge there is no solution for how to create a holistic view of large amounts of process documentation created in an ad hoc manner. The approach to a solution proposed and tested in this paper, even if not ideal, presents a starting point for discussing and comparing new solutions to the problem.
3. During our work, we found that known ways of classifying processes are not particularly suitable for the task of creating a holistic process view on the existing documentation. Either they are domain dependent, or do not take into account the diversity of process documents created in the ad hoc manner. The diversity concerns multiple aspects, e.g. many-to-many relationships between the documents and processes, diversity of notations and quality of the documents, etc.
4. The process-assets framework, originally developed for unveiling the dynamic process structure of an enterprise, showed to be a suitable foundation for creation of a holistic process model, but needed modification to fit this purpose.
5. The initial tests have shown that the model is understandable and have value for people who belong to the business depicted in the model, but who have not been directly engaged in its building. Though the results of these tests are promising, more validation is required.
6. So far, our approach to building a holistic process model has been tested only in one organization. Further validation of the approach requires its testing in other organizations. We hope that publishing of this work and spreading it among BPM researchers and consultants may help in promoting the adoption of our approach by the BPM industry.
7. So far, the model was built using Power Point as a drawing tool. Its usage did not create any major problems while the model was small (see Fig. 3). However,

using Power Point became cumbersome when the model grew to 85 documents, see Fig. 4, and the tree structure became impossible to keep on one page. There is a need to find a better tool, or create a new one specifically designed for building the holistic process models. In the latter case, the tool can even be integrated with the storage where all process documents are stored. In this case, clicking on the document reference can result in opening it in some document viewer. In a specialized tool, it would also be possible to expand or collapse parts of the tree dynamically to concentrate on parts of interest to the viewer. It would also be possible to trace maps that are referenced in more than one bucket, for example, by clicking and highlighting.

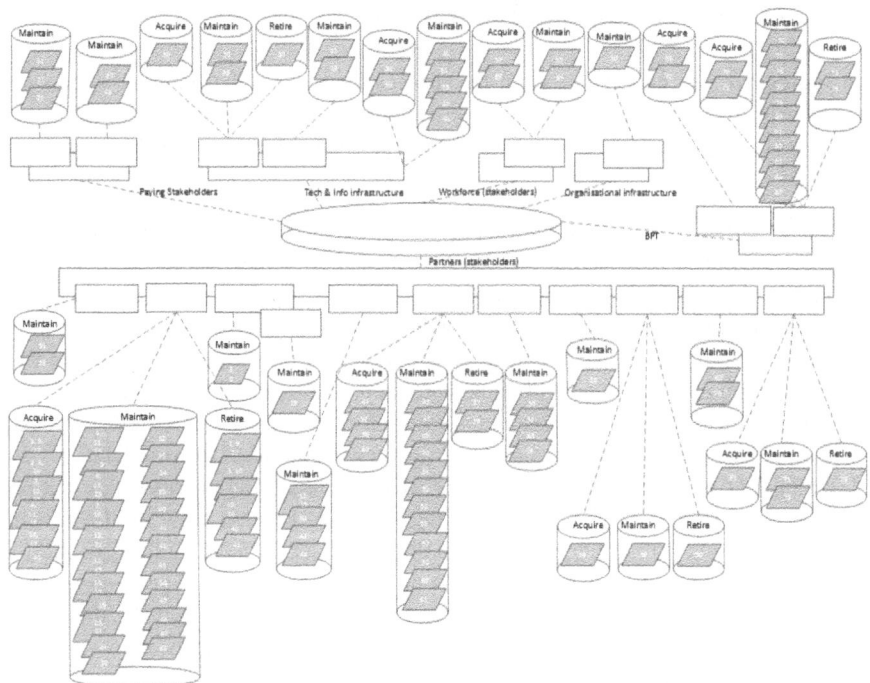

Fig. 4. The holistic process model after referencing 85 process maps

8. Right now, the holistic process model represents only process-assets-processes relationships. As was mentioned by the betting company's system architect, it would be advantageous to complement the model with means to show how processes are connected to each other in other ways, for example, how the output from one process is being used as an input to another.
9. The process-assets model, though proved to be useful for our task, requires further development. For example, it is not clear where to place and how to handle economic assets. Other organizations may require adding additional type of assets as well.

10. The biggest challenge in building the holistic process model for the betting company was to understand the nature of the processes depicted in each document. None of the notations used in these documents has been good at explaining semantic content in a simple and understandable way. They all rely on the reader's skill of how to interpret process maps. Variety of notations used in the documents made our task even more difficult. Though we created a special form for semantic annotation of the processes depicted in the documentation, see Table 1, we do not have a strict method for how to extract information from a document to fill this form.
11. The holistic model seems to work well for a mid-range size of process documentation (about 100 processes). It is not clear whether it will scale up to thousands documents. Most probably, a sophisticated software tool will be required to allow the scalability upwards.

References

1. Jeston, J., Nelis, N.: Business Process Management Practical Guidelines to successful Implementations, 2nd edn. Elsevier Ltd., Oxford (2008)
2. Bandara, W., Indulska, M., Chong, S., Sadiq, S.: Major Issues in Business Process Management: An Expert Perspective. In: Proceedings ECIS 2007 – The 15th European Conference on Information Systems, pp. 1240–1251. St Gallen, Switzerland (2007)
3. Bider, I., Perjons, E., Elias, M.: Untangling the Dynamic Structure of an Enterprise by Applying a Fractal Approach to Business Processes. In: Sandkuhl, K., Seigerroth, U., Stirna, J. (eds.) PoEM 2012. LNBIP, vol. 134, pp. 61–75. Springer, Heidelberg (2012)
4. Srivastava, B., Mukherjee, D.: Organizing Documented Processes. In: Proceedings of the 2009 IEEE International Conference on Services Computing, September 21-25, pp. 25–32 (2009)
5. Reijers, H., La Rosa, M., Dijkman, R. (eds.) Managing Large Collections of Business Process Models. Special Issue of Computers in Industry 63(2), pp. 91–180 (2012)
6. Mturi, E., Johannesson, P.: A context-based process semantic annotation model for a process model repository. BPMJ **19**(3), 404–430 (2013)
7. Porter, M.: On Competition, Updated and Expanded Edition. Harvard Business School Press (2008)
8. APQC, Retail Process Classification Framework (PCF) (Electronic) http://www.apqc.org/knowledge-base/download/275010/K04040%20Retail%20PCF%20version%206_0_0.pdf (used December 8, 2013)
9. Elias, M., Bider, I., Johannesson, P.: Using Fractal Process-Asset Model to Design the Process Architecture of an Enterprise. Experience Report. In: Bider, I., Gaaloul, K., Krogstie, J., Nurcan, S., Proper, H.A., Schmidt, R., Soffer, P. (eds.) BPMDS 2014 and EMMSAD 2014. LNBIP, vol. 175, pp. 287–301. Springer, Heidelberg (2014)

Process Fragmentation: An Ontological Perspective

Asef Pourmasoumi[1,2(✉)], Mohsen Kahani[2], Ebrahim Bagheri[1], and Mohsen Asadi[3]

[1] Department of Electrical and Computer Engineering, Ryerson University, Toronto, Canada
{a.pourmasoumi,bagheri}@ryerson.ca
[2] Web Technology Lab, Ferdowsi University of Mashhad, Mashhad, Iran
kahani@um.ac.ir
[3] School of Interactive Arts and Technology, Simon Fraser University, Surrey, Canada
masadi@sfu.ca

Abstract. Process fragmentation provides the basis for re-usability and process improvement. Various re-searchers have already introduced different definitions for what constitutes a reasonable process fragment, and have offered algorithmic support for identifying such fragments. As we will show in this paper, some of these definitions suffer from ambiguity or imprecision. Therefore, the objectives of this paper are twofold: first, we provide an ontological assessment of the various process fragment definitions based on the well-known Bunge's Ontology and its process representational model, GPM. On this basis, we then extract the most important features of these definitions in order to formalize a precise definition for process fragments and propose a precise and non-ambiguous definition: *morphological fragmentation*. We present our work through a case study and report on our observations.

Keywords: Process model fragmentation · Ontological theory · Generic process model (GPM)

1 Introduction

Organizational mining focuses on discovering organizational structures, social networks, and resource allocation patterns [1]. Organizational mining was traditionally introduced within a single organization. The growing increase of IT infrastructure needs has led many organizations to reuse or share resources and processes leading to the introduction of *cross-organizational mining* [2]. Cross-organizational mining considers organization's IT infrastructures from two perspectives. In the first case, different organizations work with each other to perform the same process instances [3]. In the second case, different organizations are separately handling the same process while sharing experiences, knowledge, or a common infrastructure [2]. In this case, each organization could be executing a variant of the same process family, such as the sale process offered by Salesforce.com. In other words, these organizations use the common infrastructures of Saleforce.com for handling their processes. However, they do not execute the exact same process and often customize and build a variation of the sale process. These customized processes share many *commonalities* and some degree of *variability*. The analysis and mining of these commonalities and variabilities can lead to valuable insight for the organizations.

In [1], a good review of process mining from the organizational perspective and its existing challenges has been reported. One of the challenges that is of interest to our work in this paper is the identification and analysis of *common process fragments* from among multiple variants of the same business process within different peer organizations. To the best of our knowledge, most of the existing process fragment definitions have only considered the practical implications of their work and little, if any, theoretical analysis has been done [4]. In this paper, we investigate the use of ontological theories for the theoretical analysis of process fragmentation models. An ontological theory defines necessary constructs for describing the processes and structure of the world in general [5]. Ontological theories have been used to evaluate modeling languages in terms of the correspondence of ontological concepts to modeling constructs. Bunge's ontology ([5]) is a widely used ontological theory that has been used to evaluate several conceptual modeling languages [5] [6]. For example in [7], the authors used Bunge's ontology for evaluating BPMN and workflow nets. This ontology includes a set of high level constructs for representing real world phenomena. The evaluation of modeling languages is based on the assumption that an information system is an artifact that represents a real-world domain.

Bunge's ontology has also been used for rep-resenting process models. In [8], a generic model, called GPM, is derived from Bunge's ontology for semantically describing process models. GPM gives a formal abstract view of a process model in terms of state transitions that occur rather than using common notions such as control flows and activities [8]. In the other words, GPM can be viewed as a mapping between process models and the real world.

Since GPM is an ontological representation of real world process models, we presume that it is appropriate for analyzing the fragments of processes models that represent real-world domains. Therefore, our work is systematically grounded in concepts from Bunge's Ontology and GPM.

In this paper, we provide the following concrete contributions:

— We gather and classify some of the main process fragment definitions and systematically discuss their pros and cons. We present a comparative analysis of these definitions.
— We present a formal representation of process fragments derived from existing definitions. This formal representation can be used as a theoretical basis for comparing and designing new process fragmentation techniques.
— Using Bunge's ontological representation of process models (GPM) we theoretically present a process fragment definition, which covers the weaknesses of previous process fragment definitions.

The paper proceeds as follows. In Section 2, we briefly introduce Bunge's ontology and the Generic Process Model (GPM). We then introduce a running example, which will be used throughout the paper. In Section 3, we review different definitions of process fragments and discuss the pros and cons of each. Here, we are not focused on the algorithmic support for process fragmentation and are only analyzing the theoretical basis for the process fragment definitions. We theoretically define the notion of process fragments using Bunge's ontology and GPM. Section 4 applies GPM for presenting a process fragment definition. In Section 5 we analyze the various fragment definitions through a case study. Finally, in Section 6 we conclude the paper and suggest directions for future work.

Table 1. Fundamental ontological constructs in the Bunge-Wand-Weber representational model [5]

Fundamental Ontological Construct	Description
Thing	The basic unit in the Bunge's ontological model is thing and can be in two types: simple and compound. A compound thing is made up of other things.
Property	Each property can be described using a function (called attribute function) that maps the thing into some values. Things can have several properties.
State	The state of a thing is a vector of values for each property functions of that thing.
Transformation	A mapping from a domain containing states into a co-domain containing states is called transformation.

2 Preliminaries

2.1 Bunge's Ontology

Bunge's ontology has been widely used for evaluating several conceptual modeling languages [5] [6]. In this paper, we are interested in the theoretical analysis of process fragments through Bunge's ontology. Bunge models the world as a world of systems [9]. In Bunge's ontological model, the "*world is made up of substantial things which possess properties*" [9]. Since Bunge's ontology provides concepts for representing real world phenomena, it seems appropriate to be used for analyzing process models in real software systems, which represents real-world domains [7].

Bunge's ontological model contains four essential concepts: thing, property, state and event. Table 1 provides an overview of the fundamental ontological constructs [5]. Things are elementary units and can be specific instances of person, building, car, book and etc. A "*property*" of a thing can be any intrinsic or mutual (meaningful only in the context of two or more things) feature of it like height, color, weight, and shape. A "*state*" is the vector which contains the values of all property functions of a thing [10]. A "*state law*" makes a restriction on the values of the thing's properties to a subset that is considered lawful. The set of thing's states that conform to the state laws of the thing are called the "*lawful state space*". An "*event*" occurs when a change in the state of a thing can be seen. The set of all possible events of a thing is called its "*event space*" [10]. A set of things that possess a common property is termed a "*class*". Wand et al. have extended Bunge's ontology with 28 (real-world) constructs [10] [5].

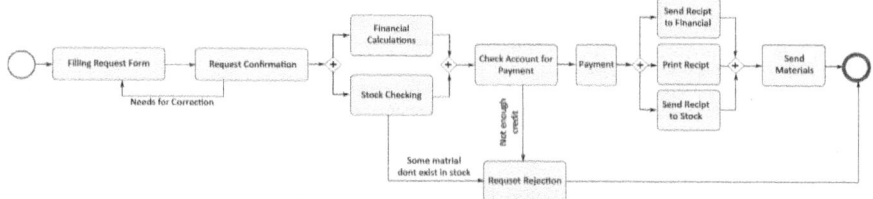

(a) An example of purchasing process model.

Things	Properties (attributes)	State Space	Lawful Event Space
Specific instances of Request, Stock, Material, Employee, Applicant, Account	Request(ID, Date, Materials) Stock(ID, Materials, No), Material(ID, Name, Price), Employee(ID, Name, Grade), Applicant(Code, Requests), Account(ID, Credit, ApplicantID)	Request-Status(Filled, Confirmed, Rejected, Audited, Manager Confirmed) Material-Status (Available, Not Exist) Account-Status(A, B, C, D)	Request { (Filled→Confirmed), (Filled→Filled), (Confirmed →Audited), (Confirmed → Stock Checked), (Stock Checked → Rejected), (Stock Checked → Manager Confirmation), (Audited →Manager Confirmation)}, Account-Status{(A→B),(B→C), (C→B), (B→A)}

(b) Ontological representation of mentioned purchasing process.

Fig. 1. An example of purchasing process

2.2 Generic Process Model (GPM)

The Generic Process Model (GPM) provides a process specification semantics based on ontological constructs [8]. GPM is considered as a framework for reasoning about process models according to their real-world meaning based on Bunge's ontology. GPM was used for various purposes such as analyzing the validity of process models, describing the conception of goals in business processes, and to interpret of control flow elements [8].

GPM focuses on the notion of *domain* by the use of Bunge's ontology concepts. A domain is represented by a set of state variables. The state of the domain can change for two reasons: first, due to the internal events which occur within the domain and second by external events which stimulate from outside the domain [8]. A state that changes due to actions in the domain is called *"unstable state"* and a state that only changes due to the actions of the domain's environment is called *"stable state"* [8]. Unstable states can be manifested as internal events and stable states as external events. A complete definition of GPM can be found in [8].

In [11], a process model is defined based on GPM as following:

Definition 1. [11]: *a process model is a tuple $< I, G, L, E >$ where,*

I: *is a subset of unstable states of the domain (initial states),*

G: *is a subset of stable state (goal set),*

L: *is the set of state transitions,*

E: *is a set of relevant external events.* ∎

One of the advantages of this definition is that GPM explicitly addresses the goal of a process and checks the validity of a process design against its defined goal [11]. In this paper, we will use GPM for theoretically analyzing process fragments.

Table 2. A review of process fragment definitions

Group	References	Fragment Definition
G1	[16] [17] [18] [19]	Components with single input and single output control flow arc (SESE).
G2	[13]	SESE components which have connected nodes and their nodes have high label similarity.
G3	[20]	A portion of design process adequately created and structured for being reused during the composition and enactment of new design processes.
G4	[21] [22] [23] [24]	A connected portion of a process intended for reuse and contains no cycles and a single control flow linking up two distinct activities. It is made of at least one activity and several controls.
G5	[12]	A partition of a workflow model, and consists of a source transition, all the transitions are reachable from the source transition, and all the linking places of these transitions.
G6	[25] [26]	A logically different, smaller model part of input process model which extracted with the intention to distributing over different execution and controlling partners.
G7	[4]	A part of process models that contain process's elements such as activities, data flows, and controls.
G8	[27]	A connected graph with significantly relaxed completeness and consistency criteria compared to an executable process graph which contains a process start or end node and at least one activity and is not necessarily directly executable.

2.3 A Running Example

In this section, we present a running example of a "purchasing process" for better clarifying Bunge's ontological constructs.

Figure 1-a shows the flow of this process model. It starts by "filling request form" and ends with "send material" or "request rejection" activities. After the "filling request form" activity, the request will be checked and if it is confirmed then two actions can be done simultaneously: the stock will be checked and the financial calculation will be done. If the requested material exists in stock, then the user account will be checked and payment will be processed. Afterwards, the receipt will be printed and sent along with the material to the customer. This process is a simplified purchasing process. The fundamental Bunge's ontological constructs for this process are shown in Figure 1-b. We will use this example throughout the paper.

3 Process Fragmentation

Process fragmentation is referred to the act of categorizing process model elements such as activities, data flows, and control flows into groups [4]. The created groups are known as *process fragments*. *Process fragmentation* is the basis for techniques supporting reusability, parallel execution, management and analysis of process models [12]. It is also known as *process decomposition* or *process modularization* [13].

In this section, we systematically review some of the existing process fragment definitions from the literature and discuss their pros and cons. As mentioned earlier, we are only interested in the process fragment definitions in this paper and not in the algorithms that facilitate the identification of such fragments. By identifying the strengths and weaknesses of different process fragment definitions, one can propose a definition that would cover the pros of existing definitions and cover their cons.

3.1 A Review of Process Fragment Definitions

In order to identify the main work in process fragment definition, we first started by focusing on the main survey papers in this domain [4] [14]. We then gathered additional papers by searching for the keywords mentioned in these survey papers on reliable databases including *CiteSeerX*, *ScienceDirect*, ACM Portal, *SpringerLink* and *IEEEXplore*. We then extracted and classified these papers based on their fragment definition and then selected only those papers that had been cited more than 10 times ac-cording to Google Scholar. The result is shown in Table 2.

In order to analyze these definitions and identify their pros and cons, we required some comparative criteria. In [4], some classification criteria for process fragmentation techniques is provided. These criteria provide a foundation for the classification of process fragmentation algorithms and can be useful for the evaluation of these techniques. For example, some of the criteria state *why is the process model fragmented, how is the fragmentation performed, who performs the fragmentation, when is the fragmentation performed in the process model lifecycle*, among others. Most of these criteria are independent of the fragment *definition* and are with respect to different aspects of the fragmentation *algorithms*. Therefore the criteria introduced in [4] cannot be directly applied for our work.

We suggest three criteria, *structural restrictions for input/output process models*, *ambiguity* and *determinism*. We don't claim that these criteria are complete, but they can highlight main weaknesses regarding the fragment definitions. These criteria are presented as research questions in the following:

Q1: Does the definition impose any structural restriction on the input process model or the output fragments?

Some definitions consider limitations on the input or output processes. For example, some definitions require the process fragment not to have any cycle or all transitions should be reachable from the start node.

Q2: Does the definitions have any elements of ambiguity or leave room for different interpretation?

Given some of the definitions do not have a theoretical representation and are written in natural language, there might be room for different interpretations of the definitions. For example, some definitions just state the portions of process models that are suitable for reusability are fragments. Such definitions have ambiguity and there could be different interpretations.

Q3: Is the definition precise and deterministic?

This criterion specifies whether the definition will always guarantee the extraction of the exact same process fragments for the same input process model or not. This criterion is different from ambiguity. In the case of determinism, an implementation of a non-deterministic definition can be viewed as being not a function and for the same input, could produce different outputs. So, a definition might not be ambiguous but be non-deterministic.

Table 3. Analyzing various definitions of process fragments based on three criteria: structural restriction, ambiguity, and determinism

Group	Structural Restriction	Ambiguity	Non-Determinism
G1	—	—	■
G2	■	—	—
G3	—	■	■
G4	■	—	■
G5	—	—	■
G6	—	■	■
G7	—	■	■
G8	■	—	■

Table 4. All features extracted from various definitions of process fragments

F#	Features
F_1	Process fragment has single input and single output control flow arc (SESE)
F_2	Process fragment must be connected.
F_3	Fragment's node must have label similarity.
F_4	Process fragment should contain no cycles.
F_5	It is made of at least one activity and, of several control (dangling or not) and data flows.
F_6	All the transitions must be reachable from the source transition and all the linking places of these transitions.

Table 5. Extracted features against the various definitions

Definition Group ID	F1	F2	F3	F4	F5	F6
G1	■	—	—	—	—	—
G2	■	■	■	—	—	—
G3	—	—	—	—	—	—
G4	—	■	—	■	■	—
G5	■	—	—	—	—	■
G6	—	—	—	—	—	—
G7	—	—	—	—	—	—
G8	■	■	—	—	■	—

In Table 3, the results of the evaluation of existing process fragment definitions based on the three criteria are shown[1]. As can be seen, most of the definitions are non-deterministic and have no structural restrictions. For example, the definition in Group 1 has no ambiguity and is clear, but is non-deterministic, because in this definition any part of process can be considered as a fragment and for the same input process, different fragments can be produced. Group 2 has a more precise definition and has structural restriction on process inputs (must be connected). This definition is clear and has no ambiguity. Since it uses precise label similarity, it is deterministic. The definition in group 3 do not have clear structural restriction on process input and the definition is not clear which parts of input can be taken as fragments. The definition in group 4 states that fragments should be connected, have no cycles, and have at least one activity and a single control flow linking up two distinct activities. It is clear but is non-deterministic because it does not work as a function. The definition in group 5 is explicit and has no structural restriction but again it is non-deterministic, since for the same input process different process fragments can be generated. The definitions in Groups 6 and 7 are descriptive

---

[1] The evaluation was conducted with the first author by investigating the coverage of the criteria for each definition. In the cases that there were uncertainties for the coverage, they have been discussed between the authors to reach conclusions about the coverage.

and contain ambiguity. Finally, in the definition in group 8, a fragment is a connected graph (structural restriction) that has a start and end node and at least one activity and is not necessarily directly executable. So, this definition is clear and has no ambiguity. It is non-deterministic because with the same process we can have lots of fragments that have start and end nodes and be also connected.

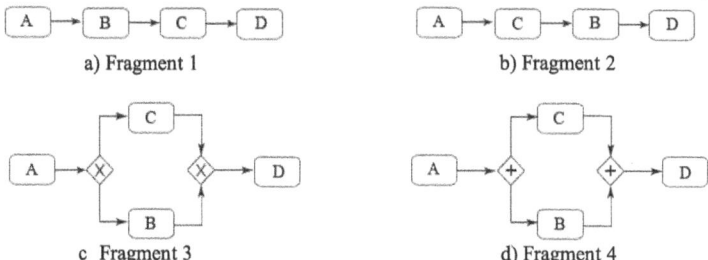

Fig. 2. Four fragments of *purchasing process* with different structures which operationally are identical. Activity A is "filling request form", B is "financial calculation", C is "stock checking" and D is about department management confirmation.

We further analyzed the various fragment definitions and identified the top six features that they had in common as shown in Table 4. These features represent the characteristics of the process fragments that are described by the definitions. Feature F_1 is the most widely seen feature for fragments. This feature implies that each fragment must have a single input and single output. Features F_2 and F_6 are concerned with the concept of connectivity. A node belongs to a fragment if it has a connection with at least one node within the fragment. Feature F_3 has a semantics point of view. It implies that the nodes that have more similar labels have a higher probability of belonging to the same fragments [13]. Natural language processing (NLP) tools can be used for detecting the similarity of the labels of the nodes in the fragments. Feature F_4 creates a structural limitation on the fragments that they should not contain cycles. Feature F_5 implies that in each fragment at least one action needs to be done.

In Table 5, we show each feature against each definition group. In other words, Table 5 shows for each definition group, what the features of their fragments are. In this table, the rows show each definition group's id (from Table 3) and the columns show the features (from Table 4). From this table, it can be understood that Features F_1 and F_2 are the most frequent features for process fragments within the literature, respectively.

Now, we exploit the features of Table 4 in order to provide a formal definition for a process fragment as follow:

Definition 2. *A process fragment F is a directed graph described as* $F(A, G, R, s, e)$ *where:*

1) $R \subseteq (A \times G) \cup (G \times A) \cup (G \times G)$,
2) $|A| \geq 1$,
3) $\forall t \in (A \cup G) \; \exists \; v_0 = s, v_1, v_2, \ldots, v_k = t \; ((v_{i-1}, v_k) \in R, 1 \leq i \leq k)$,
4) $if \; \exists \; n_1, n_2 \; (n_1 = s, n_2 = s) \; then \; n_1 = n_2$,
5) $if \; \exists \; n_1, n_2 \; (n_1 = e, n_2 = e) \; then \; n_1 = n_2$,
6) $\nexists t \in \; (A \cup G) \left((v_t, v_i) \, and \, (v_k, v_t) \in R \right) and \; ((v_j, v_{j+1}) \in R, \; i \leq j < k)$ ■

A is the set of activities, G represents gateways and R is the set of control flow relations and s is single input and e is single output nodes. In this formalism, $|A| \geq 1$ (line 2) ensures feature F_5 and the expression in line 3 is enforces F_2 and F_6. Two if-clauses in this formalism (lines 4 and 5) together represent F_1 and the later expression is equal to F_4. In this definition we did not include feature F_3. The reason is that label similarity is based on the assumption that process names are always selected meaningfully and consistently [13]. For small processes or cross-organizational processes, this assumption is not necessarily always true.

4 Analyzing Process Fragments Using Bunge's Ontology

In Definition 2, we included the most prominent non-ambiguous features identified in the definition groups for process fragments. This definition is non-ambiguous and does not contain any ambiguous feature, has no structural limitation on inputs and outputs and is deterministic. Nevertheless, it would not be useful for identifying process similarity or multi-process analysis. This is due to the following reason: in Definition 2, two fragments within two process variants are shared, if and only if they are identical [15]. For example, in Figure 2, fragments 1, 2, 3 and 4 are not identical (Based on all of the definitions in Table2). Based on *Definition2*, since the R set of these four fragments are not identical, so they cannot be captured as common fragments between two process models. Therefore, with these definitions, common fragments are either complete matches or not considered at all.

In the example shown in Figure2, All fragments try to check a purchasing request; however, each achieves this in a different way. Activity A is filling a request by the customer for purchasing some goods. All fragments start with this activity. Activity B is the computation of tax and total price of goods and activity C checks the stock for requested goods. Finally, activity D is about department management confirmation. As seen in Figure 1, activities B and C can be placed in different relations to each other. This difference can be probably due to various branch managers' choices. In fact, these fragments are performing a similar task and would be considered to be very similar fragments that only have structural variances.

In the next section, we will theoretically define the notion of process fragments using GPM and derive a new conceptual definition for a process fragment that would address the above issue among others.

4.1 Process Fragment Based on GPM

Using Bunge's ontology, it can be inferred that there is a mapping between information systems and concepts in the real world. The concepts (domain) in the real world are made of sub-concepts (sub-domain). The structures and processes of the real world can be represented by constructs in the ontological models. So, we can define sub-processes or fragments of a process model based on its ontological mapping in the real world. In [8], the sub-domain is defined based on GPM as following:

Definition3. (*sub-domain*) [8]: *A sub-domain is part of the domain described by a subset of the set of domain state variables.* ■

It must be noted that using this definition, there might be many ways to divide a domain into sub-domains and not all of them will be meaningful. One of the main applications of process fragmentation is reusability. It means that process fragments can be used in the design of different processes models. In this case, process fragments should be run independently. In this respect, in the real world, partitioning of a domain into independent sub-domains is possible. Partitioning of a domain into independently-behaving sub-domains is the result of different actors existing in the domain. In [8], an independent sub-domain is defined as follows:

Definition 4. *(independent sub-domain)* [8]: *A sub-domain will be called independently behaving (or independent) in a given state (of the sub-domain) if the law projection on the sub-domain is a function for this state.* ∎

In this case, the law projection is a function that depends only on the sub-domain's state variables. In the other words, the meaning of *Definition 4* is that each sub-domain behaves independently and ends on a stable state of the sub-domain. This stability can lead to the stability of the whole domain in the goal states or other states [8]. So, in this definition the stable states (initial and goal states) of the sub-domain are important.

We use *Definition 4* for creating a mapping between sub-domains in the real world and process fragments in information systems. So we can define process fragments based on GPM as follows:

Definition 5. *(process fragments): A subset of a process model is called a process fragment if it starts and ends with stable states and there exists at least one transformation inside it.* ∎

Now, using GPM-based definition of process fragment (*Definition 5*), we can develop a model for process fragments that does not necessarily require a complete and exact match for finding similar fragments. As we will show, this allows us to assess the similarity of two process fragments beyond a binary match or no-match. In the next Section, the new notion for process fragment, building on Definitions 2 and 5, is introduced.

4.2 Morphological Fragments

In *Definition 5,* a process fragment is defined based on GPM. The main focus of this definition is on the stable states of fragments as a point of separability. Unstable states inside the sub-domain are responsible for the behavior of the sub-domain. It can be understood that two fragments f_1 and f_2 have equal behavior if they have equal stable states ($I_1=I_2$ and $G_1=G_2$) and also have equal *transformation* sets (In [8] transformation set of a process is equal to its Activities set). The order of transformation sets is equal to the way which processes execute. In the other words, with equal stable states and transformation sets, different transformation orders (law) for two fragments show that the fragments do similar tasks in different ways. In other words, they represent the same behavior but not necessarily the same structure. Now, we can define some measures for extracting common fragments from a family of process variants based on this observation.

Definition 6: Two fragments $f_1(A_{f_1}, G_{f_1}, R_{f_1}, s_{f_1}, e_{f_1})$ and $f_2(A_{f_2}, G_{f_2}, R_{f_2}, s_{f_2}, e_{f_2})$ are behaviorally similar, called morphological fragments, iff:

$$s_{f_1} = s_{f_2} \ \& \ e_{f_1} = e_{f_2} \ \& \ A_{f_1} = A_{f_2}$$ ∎

In this definition, the start and end points are equal to the stable states in the *Definition 5* and the relation between internal nodes is ignored given the above explanation. The reason is that two fragments that have equal start/end points and have equal activities sets, and performing similar tasks, would be considered to be very similar fragments that only have structural variances hence, the term *morphological fragment*. For example in Figure 2, there are four fragments that check the purchasing request of a customer in different ways. These differences can affect the efficiency and effectiveness of the whole process. So detecting these fragments as common fragments among a family of process variants can lead to added value for organizations. All of these four fragments have equal start/end point and their internal activities are identical. So they are morphologically identical. The different relationships among internal activities show the different ways of doing same the task.

Table 6. Similarity patterns between two sub-processes

Behavioral Similarity		
Stable State	**Transformation Space**	**Class Name**
Full Similarity-Double Side	Full Similarity-Double Side	Full Similarity-Double side among Process
	Full Similarity-One Side	Partial Similarity among Process
	Partial Similarity	
	Dissimilarity	Complete Dis-similarity among process
Full Similarity-One Side	Full Similarity-Double Side	Complete Dis-similarity among process
	Full Similarity-One Side	
	Partial Similarity	
	Dissimilarity	
Partial Similarity	Full Similarity-Double Side	Complete Dis-similarity among process
	Full Similarity-One Side	
	Partial Similarity	
	Dissimilarity	
Dissimilarity	Full Similarity-Double Side	Complete Dissimilarity among process
	Full Similarity-One Side	
	Partial Similarity	
	Dissimilarity	

Definition 6 allows the identification of process fragments that are behaviorally similar but not structurally identical. It is now possible to define the degree of morphological similarity based on the degree of the transformation spaces similarity of the fragments. But before that, at first we use the definitions mentioned in [28] for a set of general similarity patterns between two sets of phenomena (By phenomena we refer to any possible observation that can be made about the domain or part of it).

Assume $A = \{a_1, a_2, ..., a_n\}$ is a set of phenomena belonging to domain D_1 and $B = \{b_1, b_2, ..., b_m\}$ is a set of phenomena belonging to domain D_2. We can see one of the following situations with respect to similarity between these two sets:

Definition 7 (Equivalent Sets of Phenomena) [28]. *Phenomenon A_1 is equivalent to A_2 (denoted as $A_1 \equiv A_2$), if and only if there is a unique mapping between elements in A_1 and elements in A_2.* ∎

Definition 8 (Similar Sets of Phenomena) [28]. *Phenomenon A_1 is similar to A_2 with respect to p (denoted as $A_1 \equiv A_2$) if and only if there is a subset of A_1 (i.e., $A'_1 \subset A_1$) and of A_2 (i.e., $A'_2 \subset A_2$) which are equivalent $A'_1 \equiv A'_2$. p is the equivalent subset i.e. $p = A'_1 = A'_2$.* ∎

Definition 9 (Completely Dissimilar Set of Phenomena) [28]. *Phenomenon A_1 is completely dissimilar to A_2 (denoted as $A_1 \neq A_2$) if and only if there are no subsets of A_1 (i.e., $A'_1 \subset A_1$) and of A_2 (i.e., $A'_2 \subset A_2$) that are equivalent.* ∎

Based on Definitions 7-9, we can define the following similarity patterns between two different sets of phenomena:

- *Full similarity double side*: when the sets A_1 and A_2 are *equivalent* (i.e., $A_1 \equiv A_2$).
- *Full similarity one side*: when the sets A_1 and A_2 are *similar* (i.e., $A_1 \cong_p A_2$) and when we have either $A'_1 \subset A_1$ and $A'_1 \equiv A_2$ or $A'_2 \subset A_2$ and $A'_2 \equiv A_1$.
- *Partial similarity*: when the sets A_1 and A_2 are *similar* (i.e., $A_1 \cong_p A_2$) and there is no subset of one set that is equivalent to the other set.
- *Complete Dissimilarity*: when two sets are *completely disjoint*.

All of the above similarity patterns can occur between any two sets in the real world. In the case of *partial similarity* we can define the amount of similarity as:

$$S_p(A_1, A_2) = \frac{A_1 \cap A_2}{A_1 \cup A_2} \quad (1)$$

The value S_p is a positive value between 0 and 1 where values of S_p closer to 1 represent higher similarity between A_1 and A_2.

In Table 6, we show all of the similarity patterns that can happen between the two subset of process based on fundamental ontological constructs (we just show *stable state space* and *transformation space* constructs based on GPM, because we want to analyze behavioral similarities between process fragments and structural similarity is not our concern in this paper).

In Table 6, the first row is equivalent to our definition for morphological fragments. It means that two fragments are behaviorally identical if their *stable state space* and *transformation sets* are equivalent (have *full similarity double side* pattern). In the other words, the order of events does not matter. The only important point is that their internal event space (*transformation set*) is equivalent and they have equal start and end nodes (*stable state space*). So, the first row of highlighted part of Table 6 describes full similarity. Also, we can define some degree of similarity between morphological process fragments. If T_1 and T_2 be *transformation* sets of process fragments F_1 and F_2 respectively, then we can define the degree of the similarity between these two process fragments using α:

$$\alpha = S_p(T_1, T_2) = \frac{T_1 \cap T_2}{T_1 \cup T_2} \quad (2)$$

The two fragments F_1 and F_2 have fully- similarity if their degree of the similarity is equal to 1 ($S_p(p_1, p_2) = 1$). The reasons that why we do not consider the other parts of Table 6 as morphological fragments is discussed in Section 5.2.

5 Discussion

In this section, we analyze the running example of Section 2.3 based on the proposed morphological fragment definition. For this purpose, we compare this running example to three other purchasing systems that are similar to the system in our example. In this section, we analyze the running example of Section 2.3 based on the proposed morphological fragment definition. For this purpose, we compare this running example to three other purchasing systems that are similar to the system in our example.

Table 7 shows the ontological representation of all these four systems. Now, using the morphological fragment definition (Definition 6), we show how some structurally different sub-processes of these purchasing processes can be considered to be similar. These fragments are shown in Figure 3. All fragments start and end with the same activities "*filling request form*" and "*department manager confirmation*", respectively. Regardless of their order or composition, the set of internal activities of all these fragments are identical. However, the order and relationships of them are different. All of them check the validity of the purchase request and check the existence of the goods in stock, each in their own way. These differences can be due to various reasons and can effect efficiency and performance of the process. For example, one manager may decide that checking the stock and doing the financial calculations should be done in parallel and another might decide that it should be done sequentially. Undoubtedly this decision will affect various execution characteristics of the process, e.g. time to completion. By identifying similar morphological fragments, one can

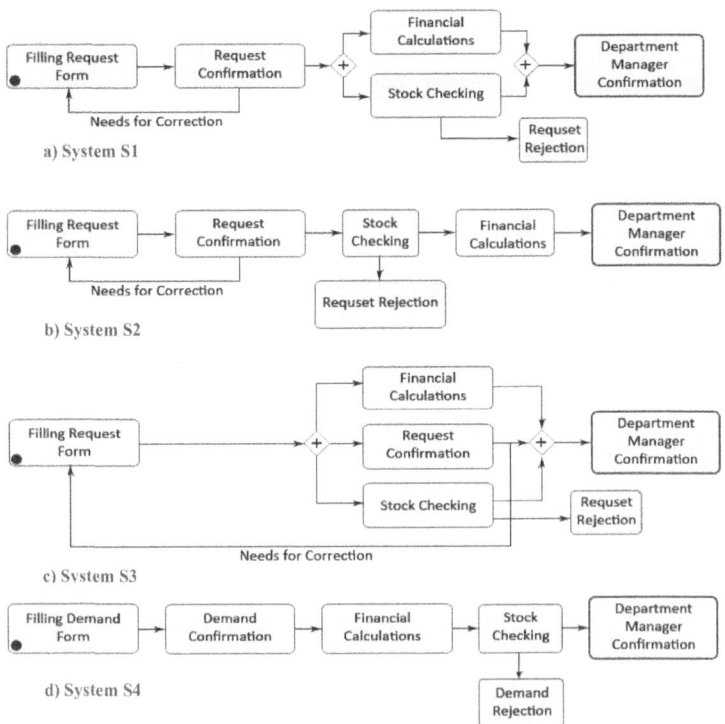

Fig. 3. Four similar fragments corresponding to systems that are shown in Table 3

determine various quantifiable measures for these fragments and use them for the purpose of process improvement. For instance, given the fragments start and end with the same activities, one can decide to replace one morphological fragment with another similar morphological fragment in the hopes to reduce the time to completion. As well, other criteria such as complexity, and cost can be used for optimizing processes through morphological fragments. It should be noted that existing definitions of process fragments as given in Table 2 do not support for this important point.

5.1 Analysis of Degree of Morphological Similarity

The use of the morphological fragment definition can lead to insights for organizations that cannot be otherwise obtained if a strict fragment definition is employed. Under real world scenarios, the number of *exact* fragments that can be mined may not be numerous; therefore, one can use the degree of morphological similarity (α in Equation 2) in order to relax the requirements to some extent. Two fragments are morphologically *identical* if their α similarity is equal to 1. Likewise, two fragments are completely dissimilar if $\alpha = 0$. Otherwise, there is a degree of morphological similarity where degrees closer to 1 represent higher similarity between the identified fragments.

With regards to the criteria shown in Table 3, we can note the following points:

- The morphological fragment definition imposes structural restriction on the process input, i.e., the input process should be connected.
- The definition does not have any ambiguity and is clearly defined; therefore, does not leave room for different interpretations.

The definition is deterministic and would lead to the same set of morphological fragments given a similar value for α.

5.2 Stable States of Morphological Fragments

There are two reasons why we only consider the first tow of Table 6 and not the other three other rows of Table 6 in our morphological fragment definition:

- First, the primary goal of eliciting morphological fragments is business process improvement and/or reusability. For this purpose, the most important property of morphological fragments is *composability*. Therefore, having the same input and output activities for morphological fragments can significantly facilitate replacement and composition of process fragments.
- Morphological fragments do not necessarily guarantee full goal compatibility but rather they point to sub-processes that are likely to be related to similar goals/objectives but this might not be necessarily the case. It should be noted that we would relax attention to accuracy and full compatibility for the sake of finding more potential matches.
- In many cases, if there is partial similarity between two fragments, it is possible by taking a window of smaller size on the main process, to reach to *full similarity-double side*. For example, imagine that we have two fragments: $F_1 =< A,B,C,F,E,D,G,H >$ and $F_2 =< M,H,C,D,E,F,G,A >$. These fragments are not morphological fragments because their stable states are not equals. However, if take a window of smaller size on these fragments, these morphological fragments can be achieved: $F_1' =< C,F,E,D,G >$ and $F_2' =< C,D,E,F,G >$.

6 Conclusion

In this paper, we have systematically analyzed various definitions of process fragments and compared them based on three criteria: structural restriction on input, ambiguity and determinism. We further formally analyzed the definitions based on Bunge's ontological model and its process representational model, GPM. We extracted the most important features of these definitions and formalized a precise definition for process fragments, called *morphological fragments*. Morphological fragments can be valuable for cross-organizational mining and especially useful for process improvement in peer-organizations when the processes are similar but not completely identical.

As future work, we are interested in pursuing the following three areas:

- Designing algorithms that can automatically detect morphological fragments from existing families of process models.
- Extracting common morphological fragments directly from collections of event logs as opposed to formal business process models.
- Providing a more operational definition for morphological fragments. In this case, the goal is to find fragments, which are operationally identical even if the set of activities are not the same.

References

1. Zhao, W., Zhao, X.: Process Mining from the Organizational Perspective. Advances in Intelligent Systems and Computing **277**, 701–708 (2014)
2. van der Aalst, W., et al.: Process mining manifesto. In: Daniel, F., Barkaoui, K., Dustdar, S. (eds.) BPM Workshops 2011, Part I. LNBIP, vol. 99, pp. 169–194. Springer, Heidelberg (2012)
3. Buijs, J.C.A.M., van Dongen, B.F., van der Aalst, W.M.P.: Mining configurable process models from collections of event logs. In: Daniel, F., Wang, J., Weber, B. (eds.) BPM 2013. LNCS, vol. 8094, pp. 33–48. Springer, Heidelberg (2013)
4. Mancioppi, M., Danylevych, O., Karastoyanova, D., Leymann, F.: Towards classification criteria for process fragmentation techniques. In: Daniel, F., Barkaoui, K., Dustdar, S. (eds.) BPM Workshops 2011, Part I. LNBIP, vol. 99, pp. 1–12. Springer, Heidelberg (2012)
5. Wand, Y., Weber, R.: On the Deep Structure of Information Systems. Information Systems Journal **5**, 203–223 (1995)
6. Evermann, J., Wand, Y.: Ontology based object-oriented domain modelling: fundamental concepts. Requirements Engineering **10**(2), 146–160 (2005)
7. Jan, R., Marta, I., Michael, R., Peter, G.: How good is BPMN really? Insights from theory and practice. In: Ljungberg, J., Andersson, M. (eds.) Proceedings 14th European Conference on Information Systems. Goeteborg, Sweden (2006)
8. Soffer, P., Kaner, M., Wand, Y.: Assigning Ontological Meaning to Workflow Nets. Journal of Database Management **21**(i3), 35 (2010)
9. Bunge, M.: Treatise on basic Philosophy. In: Ontology I: The Furniture of the World, vol. 3. Reidel, Boston (1977)

10. Wand, Y., Weber, R.: On the Ontological Expressiveness of Information Systems Analysis and Design Grammars. Journal of Information Systems **3**, 217–237 (1993)
11. Soffer, P., Yehezkel, T.: A state-based context-aware declarative process model. In: Halpin, T., Nurcan, S., Krogstie, J., Soffer, P., Proper, E., Schmidt, R., Bider, I. (eds.) BPMDS 2011 and EMMSAD 2011. LNBIP, vol. 81, pp. 148–162. Springer, Heidelberg (2011)
12. Tan, W., Fan, Y.: Dynamic workflow model fragmentation for distributed execution. Computers in Industry **58**(5), 381–391 (2007)
13. Reijers, H.A., Mendling, J., Dijkman, R.M.: Human and automatic modularizations of process models to enhance their comprehension. Inf. Syst. **36**(5), 881–897 (2011)
14. Reijers, H.A., Mendling, J.: Modularity in process models: review and effects. In: Dumas, M., Reichert, M., Shan, M.-C. (eds.) BPM 2008. LNCS, vol. 5240, pp. 20–35. Springer, Heidelberg (2008)
15. Pourmaoumi, A., Kahani, M., Bagheri, E., Asadi, M.: Mining common morphological fragments from process event logs. In: CASCON, (2014)
16. Leymann, F.: Workflows make objects really useful. EMISA Forum **6**(1), 90–99 (1996)
17. Basu, A., Blanning, R.: Synthesis and decomposition of processes in organizations. Information Systems Research **14**(4), 337–355 (2003)
18. Vanhatalo, J., Volzer, H., Leymann, F.: Faster and more focused control-flow analysis for business process models through sese decomposition. In: Proceedings of the 5th International Conference on Service-Oriented Computing, Vienna, Austria (2007)
19. Vanhatalo, J., Völzer, H., Koehler, J.: The Refined Process Structure Tree. In: Dumas, M., Reichert, M., Shan, M.-C. (eds.) BPM 2008. LNCS, vol. 5240, pp. 100–115. Springer, Heidelberg (2008)
20. Seidita, V., Cossentino, M., Hilaire, V., Gaud, N., Galland, S., Koukam, A., Gaglio, S.: The meta model: a starting point for design processes construction. International Journal of Software Engineering and Knowledge Engineering **20**(4), 575–608 (2010)
21. Eberle, H., Unger, T., Leymann, F.: Process fragments. In: Meersman, R., Dillon, T., Herrero, P. (eds.) OTM 2009, Part I. LNCS, vol. 5870, pp. 398–405. Springer, Heidelberg (2009)
22. Seidita, V., Cossentino, M., Chella, A.: A proposal of process fragment definition and Documentation. In: Cossentino, M., Kaisers, M., Tuyls, K., Weiss, G. (eds.) EUMAS 2011. LNCS, vol. 7541, pp. 221–237. Springer, Heidelberg (2012)
23. Zemni, M.-A., Hadj-Anouane, N.B., Yeddes, M.: An approach for producing privacy-aware reusable business process fragments. In: Proceedings of the 2012 IEEE 19th International Conference on Web Services, p. 659-661, June 24-29, 2012
24. Assy, N., Chan, N.N., Gaaloul, W.: Assisting business process design with configurable process fragments. In: IEEE SCC, pp. 535–542, (2013)
25. Hens, P., Snoeck, M., De Backer, M., Poels, G.: Process Fragmentation, Distribution and Execution Using an Event-Based Interaction Scheme. J. Syst. Softw. **89**, 170–192 (2014)
26. Khalaf, R., Kopp, O., Leymann, F.: Maintaining data dependencies across BPEL process fragments. International Journal of Cooperative Information Systems **17**, 259–282 (2008)
27. Schumm, D., Leymann, F., Ma, Z., Scheibler, T., Strauch, S.: Integrating compliance into business processes: process fragments as reusable compliance controls. In: Proc. of the MKWI 2010 - Integriertes ERM in automatisierten Geschäftsprozessen (2010)
28. Asadi, M., Gasevic, D., Wand, Y., Hatala, M.: Deriving variability patterns in software product lines by ontological considerations. In: Atzeni, P., Cheung, D., Ram, S. (eds.) ER 2012 Main Conference 2012. LNCS, vol. 7532, pp. 397–408. Springer, Heidelberg (2012)

Identifying and Quantifying Visual Layout Features of Business Process Models

Vered Bernstein[✉] and Pnina Soffer

University of Haifa, Mount Carmel, Haifa 3498838, Israel
vbernste@campus.haifa.ac.il, spnina@is.haifa.ac.il

Abstract. Business process models abstract complex business processes by representing them as graphical models. Their layout, solely determined by the modeler, affects their understandability. It would be beneficial to systematically study this effect and support the construction of understandable models. However, this requires a basic set of measureable key visual features, depicting the layout properties that are meaningful to the human user. The aim of this research is thus twofold. First, to empirically identify key visual features of business process models which are perceived as meaningful to the user. Second, to quantify the features into metrics, which are applicable to business process models. The paper reports an exploratory study which resulted in a set of key visual layout features. The metrics derived from these features are presented and demonstrated by applying them to example models.

Keywords: Business process modeling · Metrics · Visual layout · Qualitative empirical study

1 Introduction

Business process modeling is a broad and important area for applied research. Process modeling refers to the representation of organizational or business processes in a graphical manner, usually as a flow of activities [6]. It is common across industries – important for designing and improving business processes, analyzing industry goals and outcomes – including organizational efficiency, revenues, and social impact [6]. For these purposes, the quality of the process model is of importance. Model quality has been classified to syntactic ("correctness" of a model), semantic (to the extent to which the modeled behavior is captured) and pragmatic (usefulness) quality [9]. An important aspect of pragmatic quality is the understandability of the model by a human user. Indeed, attempts were made to study and to improve user comprehension of process models [14], [24].

A large body of research has addressed influencing factors of model understandability (both business process models and other kinds of conceptual models). Much attention in this respect has been given to semantic clarity of the modeling language [15], [22] and to the graphical elements of the modeling language [17], [23]. Additional factors identified relate to properties of an individual model (e.g., complexity metrics [3], [12], [29], [30]). Visual features of elements in a model [25], [27] have been studied, and specifically the effect of what is sometimes called "secondary notation" on model understandability [10], [26]. In contrast, the specific layout of a model has received little attention. In fact, we are not aware of studies investigating how layout features of a process model affect its understandability. Model layout has mainly been addressed as design research, where new tools and algorithms were developed for rearranging the elements in the model in order to improve its readability (e.g., [5]).

However, cognitive psychology research has showed that the appearance of a model has a significant effect on user comprehension of the model [11], [17]. Thus, the visual layout of a process model is central to achieving its aims – effectively communicating the intended process, ensuring comprehension by its users, and enabling revision and improvement of the process model. Yet, currently there is no agreed upon set of concepts with which layout properties can be described and consistently characterized so they can be reasoned and communicated about. Such set of concepts should be precisely defined and allow quantification and measurement of layout properties, serving several purposes.

First, it will permit a more focused study of process model layout and its effect on model understanding. Second, it can become a basis for process models creation guidance. Currently, although there have been some broad efforts to guide modeling from a visual perspective – such as 7PMG [13] – process modelers individually decide how to design the process model layout. Third, systematically developed layouting guidelines may support the need for training modelers. Finally, they can serve for the development of automatic layout features of modeling tools.

While some layout properties are already well defined and underlie automatic layout features of modeling tools (e.g., [4], [5], [7]), there is no indication regarding how comprehensively they correspond to layout properties which are meaningful for the human perception of the model. In fact, there is no cognitive anchoring of these features.

The primary aim of this paper is to develop a set of measurable layout features of process models, including features which are meaningful to users. We start by qualitatively identifying key visual layout features in an exploratory study based on human perceptions. We then quantify the features into metrics that may permit further study of model layout.

Accordingly, the remainder of the paper is structured as follows: Section 2 presents the methodology used in the exploratory study; Section 3 reports the findings of this study as a list of identified layout features; Section 4 presents the metrics of layout features derived from the key concepts in section 3; Section 5 reviews related work; Section 6 provides the conclusions and highlights implications of this study for future research.

2   Methodology

Our aim in this paper was twofold. First, we aimed to identify a set of concepts describing features of process model layout, which are perceived as meaningful by users. Second, we aimed to develop metrics that quantify the identified features. The first step, an exploratory study for identifying key layout features, is discussed in this section. The second step, the metrics development, will be discussed in section 4.

The first research step was guided by the research question of what are the layout features of process models that are perceived as meaningful by users. Due to the nature of the question, which sought to discover features rather than to corroborate hypotheses, the study was exploratory in nature. To identify candidate visual features of a process model, an empirical qualitative study was conducted. Qualitative data gathering was needed in order to get an understanding of a user's point of view [1]. Acquiring knowledge from participants was essential to understand how they perceive the visual layout of business process models.

2.1 Setting

The empirical study took place at the Department of Information Systems at the University of Haifa. Participants in the study were 15 undergraduate and 7 graduate students. Participants all had some knowledge of business processes modeling. All participants came with similar educational background – all took a variety of information systems analysis courses and studied modeling languages. Participation in the study was voluntary.

The study included questionnaires and interviews. First, 15 undergraduate students filled the questionnaires. Following an initial screening of the answers that were obtained, additional 7 graduate students were interviewed to gain a deeper understanding. The interviews were based on the questionnaires, but allowed for interaction and prompting deeper explanations. Thus, interviews were used in order to get a better understanding of the user perspective on the visual layout of business process models [18].

2.2 Data Collection Process

Questionnaires. The goal of the questionnaire was to understand participants' beliefs and perceptions of the visual layout features of process models. A pilot questionnaire was given ahead of time to three participants in order to simplify and improve the questions in the questionnaire. The questionnaire had 5 pairs of BPMN business process models. The models were made small to fit a single page, yet their structure was clear visually. Any label in the model elements or on the edges was blurred in order to have participants address the visual aspect of the model exclusively and not "read into it". Some of the pairs included two models which appeared to be visually different according to the judgment of one of the researchers, while others appeared visually similar. The models were all presented in black and white in order to have participants focus on layout features of the models. Color might have drawn much attention and blur the effect of less dominant features. An example pair from the questionnaire can be found below in Fig. 1. The questionnaire consisted of the same set of questions referring to each of the pairs. This set included one question asking the participants to rate the visual similarity of the models on a 7-point Likert scale. The goal of using a Likert scale was to prompt participants to actually look at the figures, compare their layout, and evaluate their similarity. Following the Likert scale evaluation, two open-ended questions were presented to the participants, asking them to indicate differences and similarities between the models (at least three of each). The information which was sought and addressed in the data analysis was the answers of the open-ended questions. The Likert-scale evaluation was only intended to prompt the comparison of the models and was not analyzed afterwards.

Interviews. The interviews were semi-structured, based on the questionnaires. First, participants were asked to complete the questionnaire. Next, the participants were asked specific questions about their answers, prompting additional explanations about specific differences or similarities between the models of each pair. In particular, the questions related to features that support or hamper the understandability of the models in order to encourage the participants to engage in the specific appearance features. Participants were asked questions such as: At a first glance, which model seems easily readable and why? Which model is less understandable? What visual features would you change here to make it easier to understand? The participants were also asked to point out differences between models and indicate their preferences in regards to comprehension of the models.

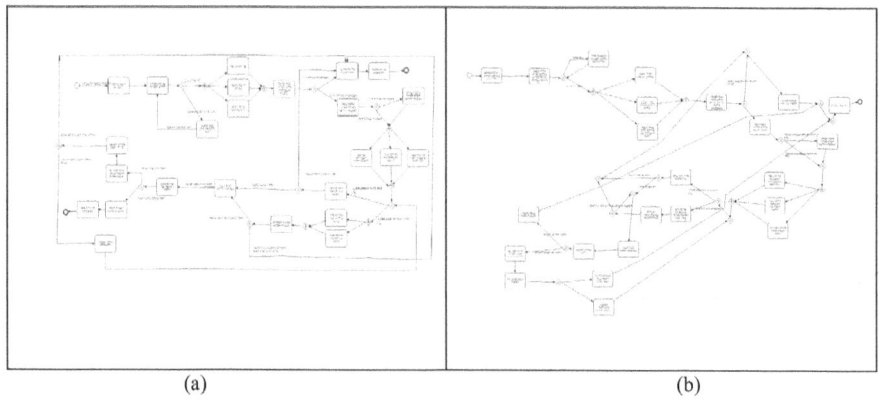

Fig. 1. Example from the questionnaire

3 Analysis and Findings

The data collected from the questionnaires and interviews was qualitatively coded and classified into categories based on data collected from participants.

3.1 Analysis

The analysis considered the text of the written questionnaires and the interviews text. Using qualitative methods, [28] textual segments were coded by the model(s) they referred to and classified to categories of features, which were later on aggregated to higher-level categories. Table 1 describes the steps taken during the data analysis phase.

Table 1. Data Analysis Process

Stage I	Getting familiar with the data – reading and interpreting the data collected using the questionnaires; collecting additional data through interviews
Stage II	Identifying repeating elements – color categorizing concepts which repeat in the different questionnaires and interviews. Saturation of categories was reached by the 4th interview.
Stage III	Recognizing super categories – finding higher level categories grouping lower-level ones.

3.2 Categories

The next stage was to identify and define categories of visual features in the models. Table 2 summarizes the categories found by qualitative analysis. It provides a definition for each category, and examples of supporting statements taken from the questionnaires and interviews. The lower-level categories are grouped under higher-level ones (where applicable).

Table 2. Categories of features found in the study

Feature	Description	Example Reference from Data
Edges		
Length of Edges	The length of each edge in the model	"The model on the right doesn't seem right since there are many long edges throughout the model"
'Broken' Edges	Edges that are divided into two segments or more	"Need to straighten all the broken edges"
Crossing Edges	Edges that cross each other – intersect with other edges	"…there are edges here that just go one on top of the other" "This looks like a 'spider web'"
Text on Edges	Existing text annotations on edges. The text can either be descriptive or conditional	"When something is written on the edge, it is difficult to understand which edge it refers to"
Total Ending Points	The total amount of ending points in the model	"There are many ending points" "One ending point connected to many edges, appears like a loop"
Angles	The angles in the model: 90° angles, 180° angles, angles larger than 45°, angles smaller than 45°	"I would improve the angles in this model to be 90° angles" "Change the edges to be straight lines"
Model's Structure		
Model's Shape	The shape of the model is referred to the way the model is spread on the canvas	"The structure in both models is horizontal"
Model's Size	The area the model takes on the canvas	"The size of the models is different"
Model's Direction		
General Direction	The general direction of the	"This model goes in a clear direction"

Table 2. (*continued*)

		model. Left-right, right-left, top-bottom, bottom-top	"Both models are vertical"
Placement of Ending Event		The location of ending points in the model in relation to the starting point of the model	"Location of the ending point makes it clear where the process ends"
Branching off		Branching off of the model from one main branch to more than one in different directions	"I don't like to wonder where an arch leads to"
Change in Direction		Change in the direction of the model	"There is a change in the direction of the model" "Both models are built stepwise"
Symmetry in blocks		Referring to structured blocks in the model – symmetry of element arrangement in the block	"This block in the model is very symmetrical and therefore very understandable"
Alignment in the Model		Alignment of the elements in the model with each other	"This model is clearer because of the alignment of the whole model. It is very aesthetic"

4 Quantification

Following the identification of key layout features in the model, we have analyzed them and derived measurable metrics, applicable to the layout of business process models. The quantification was guided by the following general guidelines:

(1) The metrics should relate to well defined elements in the model (e.g., nodes, node types, and edges) marked by coordinates on the canvas
(2) The metrics should not be dependent on the model's size
(3) The metrics should represent the entire model and not just a specific part of it. Specifically, we realized that the Symmetry feature was indicated with respect to specific parts of the observed models, and hence decided to leave this feature out of the set of metrics.

The metrics, presented in Table 3, relate to the following model elements:

Start / End events: each process model has a start event and at least one end event.

Model nodes (elements): Events, Activities, or Gateways. A node has the following attributes: Coordinates (X, Y), Label, Type (start/end event, activity, gateway)

Edges: An edge is a directed line linking two nodes in the model. It is denoted as Edge (source, end)

Segments: An edge can be broken into two or more segments, each has a start and an end point.

Single Entry Single Exit (SESE) fragment: A SESE fragment has exactly one incoming and exactly one outgoing edge, regardless of the internal structure of the fragment [4].

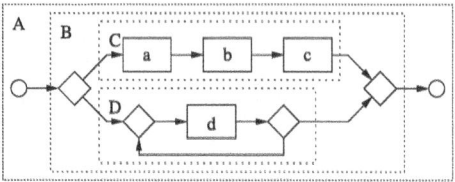

Fig. 2. Example of business process model and its SESE fragments [4]

Fig 2 [4], illustrates nested SESE fragments (A, B, C, and D) in a process model. The lowest level fragments are fragments C and D, contained by B, contained by A. There exist known algorithms for breaking a model into SESE fragments where possible.

Using the above terms, Table 3 presents the derived metrics, indicating the key concepts they were derived from, and specifying their formulas.

Table 3. Metrics of key layout features

Name	*Edges' Length: Minimal, Maximal, Average, Standard deviation*
Description	The length of an edge in the model: For straight edges, the Euclidian distance from the source to the end. When measuring a 'broken' edge, aggregating its segments.
Concept	Length of Edges
Example	For the model in Fig. 1(a), avgEdgeLength = 3.22; for the model in Fig. 1(b), avgEdgeLength = 2.8
Formula	SegmentLength = $\sqrt{(SegEnd.X - SegStart.X)^2 + (SegEnd.Y - SegStart.Y)^2}$ EdgeLength = Sum (SegmentLength) minEdgeLength = Min (EdgeLength) maxEdgeLength = Max (EdgeLength) avgEdgeLength = AVG (EdgeLength) STDevEdgeLength = STDev (EdgeLength)
Name	*Edges' Style*
Description	The style of the edges in the model: ratio of curved edges, broken edges (edges with breaking points), and simple edges (straight lines – this is the default style)
Concept	'Broken' Edges
Example	For the model in Fig. 1(a), %brokenEdges = 15%; for the model in Fig. 1(b), %brokenEdges = 9%

Table 3. (*continued*)

Formula	totalEdges = the total number of edges in the model countBroken = The number of 'broken' edges in the model countCurved = The number of curved edges in the model countSimple = The number of simple edges in the model $\%\text{brokenEdges} = \frac{countBroken}{totalEdges}$ $\%\text{curvedEdges} = \frac{countCurved}{totalEdges}$ $\%\text{simpleEdges} = \frac{countSimple}{totalEdges}$
Name	**Crossing Edges**
Description	Edges that cross each other – intersect – in the model. Creates a look of more edges than there actually exist in the model. Recognize an intersection of edges in the model and counting how many crossing edges exist in a model
Concept	Crossing Edges
Example	For the model in Fig. 1(a), %totalCrosses = 3.7%; for the model in Fig. 1(b), %totalCrosses = 12%
Formula	totalEdges = the total number of edges in the model countCross = The number of intersections in the model $\%\text{totalCrosses} = \frac{countCross}{totalEdges}$
Name	**Text on Edges**
Description	Existing text box on the edges in the model
Concept	Text on Edges
Example	For the model in Fig. 1(a), %textEdge = 55%; for the model in Fig. 1(b), %textEdge = 21%
Formula	totalEdges = the total number of edges in the model countTextEdge = The number of edges with a text annotation in the model $\%\text{textEdge} = \frac{countTextEdge}{totalEdges}$
Name	**Number of ending points**
Description	The number of ending points in the model. An ending point is an end event or an element with no outgoing edges; it is not used as a source element for any flow, just as a destination element.
Concept	Total Ending Points
Example	For the model in Fig. 1(a), countEndPoint = 2; for the model in Fig. 1(b), countEndPoint =1
Formula	countEndPoint = the number of ending points in the model
Name	**Orthogonal Segments (segments aligned to a grid layout)**
Description	Orthogonal segments are parts of edges which are aligned with a grid layout

Table 3. (*continued*)

	of the model, X axis or Y axis. For an orthogonal segment ΔX = 0 or ΔY = 0.
Concept	Angles
Example	For the model in Fig. 1(a), %orthogonalSegments = 55%; for the model in Fig. 1(b), %orthogonalSegments = 11%
Formula	ΔX= Start.X – End.X over all the model segments ΔY= Start.Y – End.Y over all the model segments totalSegments = the total number of segments in the model countOrthSegment = The number of orthogonal segments in the model % orthogonalSegments = $\frac{countOrthSegment}{totalSegments}$
Name	***Model's General Layout***
Description	The model's general layout describes how the model is spread/ laid out on the canvas. It includes: Area: the area taken by the model on the canvas Shape: numerical value indicating the general shape of the model as the ratio between width and height (1 denotes a square, <1 a vertically rectangular model, >1 a horizontally rectangular model)
Concept	Model's Structure: Model's Shape, Model's Size
Example	For the model in Fig. 1(a), areaModel = 468, shapeModel = 1.92; for the model in Fig. 1(b), areaModel = 477, shapeModel = 1.47
Formula	ΔX= Max(X) – Min(X) over all the model elements ΔY= Max(Y) – Min(Y) over all the model elements areaModel = ΔX × ΔY shapeModel = ΔX/ ΔY

Direction of a model

The direction of a model refers to the general flow of the model. The direction of the model can be a definite direction – from the beginning of the model all edges are directed at the same direction till the end of the model. On the other hand, the direction of the model can be unclear and changing throughout the process to different directions.
The direction of each edge is recorded. Segments of 'broken' edges are considered as individual edges. The direction of the model is measured by its horizontal direction and by its vertical direction. Each segment is measured for the horizontal and the vertical metrics.

Name	***Horizontal Direction of a model***
Description	Horizontal direction of a model. This feature checks the horizontal flow of the model (from left to right, from right to left). Each segment is measured in relation to the X axis. Segment is considered as horizontal right if ΔX>0, it is considered as horizontal left if ΔX<0.

Table 3. (*continued*)

	horizontalDirection is a number between -1 and 1. 1 indicating a stronger horizontal direction to the right. -1 indicating a stronger horizontal direction to the left.
Concept	Model's Direction: General Direction, Placement of Ending Event
Example	For the model in Fig. 1(a), %rightDirection = 40%, %leftDirection = 36%, horizontalDirection = 0.04; for the model in Fig. 1(b), %rightDirection = 59%, %leftDirection = 38%, horizontalDirection = 0.21
Formula	SegDirection=Right if $\Delta X > 0$, Left otherwise. totalSegments = the total number of segments in the model countRight = The number of segments directed right in the model countLeft = The number of segments directed left in the model $\%\text{rightDirection} = \frac{countRight}{totalSegments}$ $\%\text{leftDirection} = \frac{countLeft}{totalSegments}$ $\text{horizontalDirection} = \frac{countRight - countLeft}{totalSegments}$
Name	***Vertical Direction of a model***
Description	Vertical direction of a model. This feature checks the vertical flow of the model (from top to bottom, and from bottom to top). Each segment is measured in relation to the Y axis. Segment is considered as vertical up if $\Delta Y > 0$, it is considered as vertical down if $\Delta Y < 0$. verticalDirection is a number between -1 and 1. 1 indicating a stronger vertical direction upwards. -1 indicating a stronger vertical direction downward.
Concept	Model's Direction: General Direction, Placement of Ending Event
Example	For the model in Fig. 1(a), %downDirection = 33%, %upDirection = 24%, verticalDirection = -0.09; for the model in Fig. 1(b), %downDirection = 38%, %upDirection = 44%, verticalDirection = 0.06
Formula	SegDirection = Up if $\Delta Y > 0$, Down otherwise. totalSegments = the total number of segments in the model countDown = The number of segments directed down in the model countUp = The number of segments directed up in the model $\%\text{downDirection} = \frac{countDown}{totalSegments}$ $\%\text{upDirection} = \frac{countUp}{totalSegments}$ $\text{verticalDirection} = \frac{countUp - countDown}{totalSegments}$
Name	***Change in model's direction***
Description	Change of direction between SESE fragments (assuming SESE fragments direction is definite). Comparing the direction of SESE fragments to determine whether or not they have different direction, indicating a change in the

Table 3. (*continued*)

	general direction of the model. This metric first calculates a single horizontal and vertical direction value for each SESE fragment. Each can have a value between -1 and 1. When comparing a pair of SESE fragments, we compare their horizontal direction value and vertical direction value looking for a change in direction. V represents the vertical direction; H represents the horizontal direction.
Concept	Model's Direction: Branching off, Change in Direction
Example	For the model in Fig. 1(a), changeInDirection = True; for the model in Fig. 1(b), changeInDirection = True.
Formula	Given SESE1 and SESE 2, H1 = (End1.x − Start1.x)/((End1.x − Start1.x)+(End1.y − Start1.y)) horizontal direction of SESE 1 H2 = (End2.x − Start2.x)/((End2.x − Start2.x)+(End2.y − Start2.y)) horizontal direction of SESE 2 V1 = (End1.y − Start1.y)/((End1.x − Start1.x)+(End1.y − Start1.y)) vertical direction of SESE 1 V2 = (End2.y − Start2.y)/((End2.x − Start2.x)+(End2.y − Start2.y)) vertical direction of SESE 2 changeInDirection = True if |H2 − H1| > threshold or |V2 − V1| > threshold
Name	***Alignment of elements***
Description	Alignment of the events in the model. Calculating the alignment of elements in the model that share a coordinate (x or y). This metric builds a matrix which counts all the commonality of X's and Y's of the different elements in the model.
Concept	Alignment in the Model
Example	For the model in Fig. 1(a), %totalAlignment = 2.1%; for the model in Fig. 1(b), %totalAlignment = 6.3%
Formula	For elements (X_1, Y_1), (X_2, Y_2), Alignment=True if $(X_1=X_2)$ OR $(Y_1=Y_2)$, False otherwise. countPairAlignment = the total number of pairs for which Alignment =True in the model (N(N-1))/2 = total number of possible pairs in the model $\%\text{totalAlignment} = \dfrac{countPairAlignment}{(N(N-1))/2}$

5 Related Work

Research on the visual layout of business process models has largely relied on studies done in the field of graph drawing. A considerable body of knowledge exists on how to automatically set the layout of graphical models in order to improve their readability. Studies done in the graph drawing field mainly explored the following visual layout features: edge crossing, edge bends, the minimum angle between edges leaving a node, orthogonality, symmetry, flow direction, edge length variation, and width of layout. [19, 20, 21]. The direct relation between these metrics and understandability was also investigated. Research on aesthetics of graph layout in general [20] found that an important feature to users is minimizing line crosses; less important are: minimizing bends, maximizing symmetry; other features were not found to have a significant effect. Research on users' preferences of UML layout / appearance [21] indicates that users rated features as follows: arc crossings, orthogonality, direction of flow, arc bends, text direction, width of layout and font type. Considering process models, [26] explored understanding of process models by experts and novices in regards to the following layout features: line crossings, edge bends, symmetry, and vicinity of related elements. [2] investigated user preferences of layout aesthetics for BPMN models, considering heterogeneous user groups with the goal of designing a modeling tool for BPMN. They used line crossings, orthogonal lines, drawing area, line bends, and flow. Findings showed that the aforementioned layout criteria were most relevant for users with average or greater experience and at least basic education in business process modeling. The layout features described above were all identified as part of the findings in our exploratory study.

Other work has developed or evaluated algorithms that will change an existing layout of a business process model manually or automatically to match a desirable aesthetical pattern for effective visual layouting of a model. In [4, 5] both studies present algorithms which are based on a set of constraints targeting a readable layout of a process model (unified flow direction, minimal edge crossing, minimal bend-points, usage of Manhattan layout). Automatic layout of BPMN is presented in [8] and is focused on edge positioning. The study in [23] presents a comprehensive framework which allows for a personalized process model visualizations, meaning that the model's visual appearance can be tailored to the specific needs of different user groups. In addition, in the field of graph drawing, applied research has developed algorithms and related tools to automatically or manually improve visual layout of graphs and thus improve their understandability. GraphEd system in [7] compared and evaluated different algorithms of graph drawing while considering the following layout criteria: edge length, edge distribution, area, density, bends, crossings, and orthogonal edges. The work presented in [16] suggests an algorithm which reorders a diagram using orthogonal ordering while preserving the 'mental map' of the diagram.

The conclusion from the reviewed works is that the visual layout of business process models is important for understandability. Yet, as far as we know, all existing work addresses a conveniently selected set of features – ones that are immediate to think of and possible to automatically address. This paper is the first to ask the question of what features should be selected to reflect the human perception, and to use this as a basis for a collection of layout features.

6 Conclusions and Future Work

The visual layout of the model, the way elements of the model are laid out on the canvas, is an important factor for the user's understanding of the model. Since layout properties are mostly not addressed by modeling languages, and in the absence of enforced layout conventions, the modeler is the sole decider on how a model will be laid out. A common terminology in which layout properties can be specified is an essential basis needed for developing an understanding, appropriate conventions, and tools that enforce them. The aim of this study was to identify key visual features which are important to users, and to quantify those features into measurable layout features – metrics.

The contribution of this paper is twofold. First, the human-centered approach, according to which key visual layout features in business process models were elicited based on what users perceive as important. Second, developing the identified layout features into measurable metrics, paves the way to clear and effective communication about model layout. These metrics can be used for a body of research that would establish the effect of model layout on its understandability, and for the development of automatic layout support in process modeling tools.

As limitations of the exploratory study we can mention the use of students as participants. This is sometimes considered as a threat to external validity and may compromise generalization of the findings. Nevertheless, since the study addressed the human perceptions, not relying on any professional knowledge or experience, the effect should be minimal. In addition, although saturation was reached, it might be that additional models in the questionnaire would result in additional features. This will be addressed in future.

Currently, the metrics are implemented in a process model collection, allowing to precisely "measure" the layout properties of every model. Yet, validation is still needed, to establish the correlation between the metrics and the human perception of models' layout. In addition, future work may include quantitative studies to experimentally test to what degree these layout features indeed affect user comprehension.

References

1. Creswell, J.W.: Research design: qualitative, quantitative, and mixed methods approaches. Sage (2013)
2. Effinger, P., Jogsch, N., Seiz, S.: On a study of layout aesthetics for business process models using BPMN. In: Mendling, J., Weidlich, M., Weske, M. (eds.) BPMN 2010. LNBIP, vol. 67, pp. 31–45. Springer, Heidelberg (2010)
3. Gruhn, V., Laue, R.: Complexity metrics for business process models. In: 9th international conference on business information systems (BIS 2006), vol. 85 (2006)
4. Gschwind, T., et al.: Edges, structures, and constraints: the layout of business process models. Technical Report RZ 3825, IBM Research, Zurich (2011)
5. Gschwind, T., et al.: A linear time layout algorithm for business process models. Journal of Visual Languages & Computing **25**(2), 117–132 (2014)
6. Havey, M.: Essential Business Process Modeling. O'Reilly Media, Inc. (2005)
7. Himsolt, M.: Comparing and evaluating layout algorithms within GraphEd. Journal of Visual Languages and Computing **6**(3), 255–273 (1995)
8. Kitzmann, I., et al.: A simple algorithm for automatic layout of bpmn processes. In: IEEE Conference on Commerce and Enterprise Computing, 2009. CEC 2009. IEEE (2009)
9. Krogstie, J., Sindre, G., Jørgensen, H.: Process models representing knowledge for action: a revised quality framework. European Journal of Information Systems **15**(1), 91–102 (2006)

10. La Rosa, M., et al.: Managing process model complexity via concrete syntax modifications. IEEE Transactions on Industrial Informatics **7**(2), 255–265 (2011)
11. Mayer, R.E.: Multimedia Learning. Cambridge University Press, New York (2001)
12. Mendling, J.: Metrics for Process Models: Empirical Foundations of Verification, Error Prediction, and Guidelines for Correctness, vol. 6. Springer (2008)
13. Mendling, J., Reijers, H.A., van der Aalst, W.M.P.: Seven process modeling guidelines (7PMG). Information and Software Technology **52**(2), 127–136 (2010)
14. Mendling, J., Reijers, H.A., Cardoso, J.: What makes process models understandable? In: Alonso, G., Dadam, P., Rosemann, M. (eds.) BPM 2007. LNCS, vol. 4714, pp. 48–63. Springer, Heidelberg (2007)
15. Milton, S.K., Rajapakse, J., Weber, R.: Ontological Clarity, Cognitive Engagement, and Conceptual Model Quality Evaluation: An Experimental Investigation. Journal of the Association for Information Systems, **13**(9) (2012)
16. Misue, K., et al.: Layout adjustment and the mental map. Journal of visual languages and computing **6**(2), 183–210 (1995)
17. Moody, D.L.: The "physics" of notations: toward a scientific basis for constructing visual notations in software engineering. Software Engineering, IEEE Transactions on **35**(6), 756–779 (2009)
18. Myers, M.D., Newman, M.: The qualitative interview in IS research: Examining the craft. Information and organization **17**(1), 2–26 (2007)
19. Purchase, H.C.: Metrics for graph drawing aesthetics. Journal of Visual Languages & Computing **13**(5), 501–516 (2002)
20. Purchase, H.: Which aesthetic has the greatest effect on human understanding?. Graph Drawing. Springer, Berlin Heidelberg (1997)
21. Purchase, H.C., Allder, J.-A., Carrington, D.: User Preference of Graph Layout Aesthetics: A UML Study. In: Marks, J. (ed.) GD 2000. LNCS, vol. 1984, pp. 5–18. Springer, Heidelberg (2001)
22. Recker, J., et al.: Business process modeling-a comparative analysis. Journal of the Association for Information Systems **10**(4), 1 (2009)
23. Reichert, M.: Visualizing Large Business Process Models: Challenges, Techniques, Applications. In: La Rosa, M., Soffer, P. (eds.) BPM Workshops 2012. LNBIP, vol. 132, pp. 725–736. Springer, Heidelberg (2013)
24. Reijers, H.A., Mendling, J.: A study into the factors that influence the understandability of business process models. IEEE Transactions on Systems, Man and Cybernetics, Part A: Systems and Humans **41**(3), 449–462 (2011)
25. Rinderle, S.B., et al.: Business process visualization-use cases, challenges, solutions (2006)
26. Schrepfer, M., Wolf, J., Mendling, J., Reijers, H.A.: The Impact of Secondary Notation on Process Model Understanding. In: Persson, A., Stirna, J. (eds.) PoEM 2009. LNBIP, vol. 39, pp. 161–175. Springer, Heidelberg (2009)
27. Streit, A., Pham, B., Brown, R.: Visualization Support for Managing Large Business Process Specifications. In: van der Aalst, W.M., Benatallah, B., Casati, F., Curbera, F. (eds.) BPM 2005. LNCS, vol. 3649, pp. 205–219. Springer, Heidelberg (2005)
28. Taylor-Powell, E., Renner, M.: Analyzing qualitative data. University of Wisconsin-Extension Program Development & Evaluation (2003)
29. Vanderfeesten, I. et al.: Quality metrics for business process models. BPM and Workflow handbook **144** (2007)
30. Vanderfeesten, I.T., Reijers, H.A., Mendling, J., van der Aalst, W.M., Cardoso, J.: On a quest for good process models: the cross-connectivity metric. In: Bellahsène, Z., Léonard, M. (eds.) CAiSE 2008. LNCS, vol. 5074, pp. 480–494. Springer, Heidelberg (2008)

Quality and Security Issues

Identifying Quality Issues in BPMN Models: An Exploratory Study

Cornelia Haisjackl[1], Jakob Pinggera[1], Pnina Soffer[2], Stefan Zugal[1(✉)], Shao Yi Lim[1], and Barbara Weber[1]

[1] University of Innsbruck, Innsbruck, Austria
{cornelia.haisjackl,jakob.pinggera,stefan.zugal,
barbara.weber}@uibk.ac.at, shao.lim@student.uibk.ac.at
[2] University of Haifa, Haifa, Israel
spnina@is.haifa.ac.il

Abstract. Even though considerable progress regarding the technical perspective on modeling and supporting business processes has been achieved, it appears that the human perspective is still often left aside. In particular, we do not have an in-depth understanding of how process models are inspected by humans, what strategies are taken, and what cognitive processes are involved. This paper takes a first step towards such an understanding and reports an exploratory study investigating how humans identify quality issues in BPMN process models. Providing preliminary answers to initial research questions, we also indicate other research questions that can be investigated using this approach. Our qualitative analysis shows that humans adapt different strategies on how to identify quality issues. Finally, we observed for different quality dimensions quality issues that were spotted by a large number of subjects (e.g., deadlocks), but also quality issues that did not seem to bother the participants of this study (e.g., line crossings).

Keywords: Process Model Quality · Empirical Research · Human–Centered Support

1 Introduction

Much conceptual, analytical and empirical research has been conducted during the last decades to advance our understanding of conceptual modeling. Especially process models have gained significant relevance in recent years due to their critical role for the management of business processes [1]. Business process models help to obtain a common understanding of a company's business processes [2], serve as drivers for the implementation and enactment of business processes and enable the discovery of improvement opportunities [3]. Even though considerable progress regarding process modeling languages and –methods has been achieved,

This research is supported by Austrian Science Fund (FWF): P23699-N23, P26140-N15.

the question how *humans* can be efficiently supported in creating, understanding and maintaining business process models is still a lingering problem. As a consequence, for instance, process models still display a wide range of quality problems impeding their comprehensibility and maintainability [4]. Similarly, literature reports, for example, on error rates between 10% and 20% in industrial process model collections [5]. Moreover, *process model smells* like non-intention-revealing or inconsistent labeling [6] are typical quality issues, which can be observed in existing process model collections. In addition, layout conventions are often missing [7], resulting in models that lack a consistent graphical appearance, thereby introducing an additional burden for humans when building an understanding of process models.

While some quality issues, mostly syntactic, can be detected automatically by verification algorithms (e.g., [8]), many others cannot. Hence, human inspection of process models is still essential [9]. This manual inspection is currently not supported. Furthermore, we do not have an in-depth understanding of how it is conducted, what strategies are taken, and what cognitive processes are involved.

Taking a first step towards such an understanding, this paper reports a study which explores the model inspection process. When exploring a question which has not been addressed so far, a main issue is to identify an appropriate research approach and show that it can be applied to the current question. In the reported study we investigate the strategies taken by humans when inspecting a process model and the kinds of quality problems they notice.

In order to tackle these research questions, an exploratory study utilizing *think-aloud* is conducted, asking humans to find different types of quality issues, i.e., syntactic, semantic and pragmatic, in process models of varying sizes. By videotaping the exploratory study, we were able to observe different strategies on how to accomplish this objective. Further, we observed that the importance of quality issues does not correlate with the quality dimensions of the SEQUAL framework, i.e., we did not observe that one dimension is "more important" than the other. For each quality dimension, we observed quality issues that were spotted by a large number of subjects, but also quality issues that did not seem to bother the participants of this study. For example, we were surprised to find out that line crossing did not seem to bother our subjects. The sample size in the study is small; yet we show how these questions can be addressed. Providing preliminary answers to initial research questions, we also indicate other research questions that can be investigated using this approach. More specifically, the question arises which quality issues should be supported with appropriate tool support and which parts of a modeling notation cause difficulties while creating process models.

The remainder of the paper is structured as follows. Sect. 2 gives background information. Sect. 3 describes the setup of the study, whereas Sect. 4 deals with its execution. Sect. 5 presents the findings of the study and Sect. 6 a corresponding discussion. Related work is presented in Sect. 7. Finally, Sect. 8 concludes the paper.

2 Background

This section describes backgrounds regarding quality issues in business process modeling. In particular, we discuss different types of quality issues along the dimension of the SEQUAL [10] framework, i.e., syntactic, semantic, and pragmatic quality. This classification of quality issues builds the basis for the exploratory study described in Section 3.

Syntactic Quality. Syntactic quality refers to the correspondence between the model and the language the model is described in. Typical examples for syntactic quality issues are deadlocks or the lack of synchronization in process models. Note that most of these issues can be detected automatically. Yet, existing works have analyzed typical syntactical errors at IBM [11], in the SAP reference model [12] and other large industrial model collections [13].

Semantic Quality. Semantic quality refers to both the validity (i.e., statements in the model are correct and related to the problem) and completeness (i.e., the model contains all relevant and correct statements to solve this problem) of the model. Typical errors at this level include missing or superfluous activities. Semantic quality issues have been addressed only to a limited extent so far [14], and can only be identified by human.

Pragmatic Quality. Finally, pragmatic quality can be described as the correspondence between the model and people's interpretation of it, which is typically measured as model comprehension [10]. Significant research has been conducted in recent years on factors that impact *process model comprehensibility*, such as the influence of model complexity [15], modularity [16–18], grammatical styles of labeling activities [6] and secondary notation [19]. Respective insights led to the development of empirically grounded guidelines for the modeling of business processes [20], describing *process model smells* [4,20]. Examples of process model smells are non–intention revealing names of activities [4], redundant process fragments [21], unnecessarily complex process fragments [20] or edge crossings [22]. While some of these issues can be handled automatically [4], a comprehensive quality inspection by human is still essential.

3 Defining and Planning the Exploratory Study

Despite the importance of human inspection of process models, there has been no considerable research on how subjects identify quality issues. Hence, since no theories exist we can base our investigation on, we address the topic in an exploratory manner using a qualitative research approach [23]. In particular, we use the think-aloud method, i.e., we ask participating subjects to voice their thoughts, allowing for a detailed analysis of their reasoning process [24]. Then, we turn to grounded theory [25], where theory emerges when analyzing data, identifying recurring aspects and grouping them to categories. These categories are validated and refined throughout the analysis process. First of all, we describe setup and planning of the exploratory study.

Research Questions. The goal of this study is to investigate how humans identify quality issues in BPMN process models. In particular, we are interested in common strategies that humans apply for identifying quality issues. The research question RQ_1 can be stated as follows:

Research Question RQ_1 *What are common strategies that humans take for inspecting and identifying quality issues in BPMN process models?*

Further, it might be argued that not all quality issue are of equal importance to humans. Therefore, in research question RQ_2 we investigate which types of quality issues were typically mentioned by humans when inspecting a process model and which quality issues did not seem to bother humans when understanding the process models:

Research Question RQ_2 *Which types of quality issues are typically spotted by humans when inspecting a BPMN process model? Which quality issues gain less attention?*

Subjects. In order to ensure that obtained results are not influenced by unfamiliarity with BPMN process modeling, subjects need to be sufficiently trained. Even though we do not require experts, subjects should have at least a moderate understanding of imperative processes' principles. For information on the actual subjects, see Sect. 4.

Objects. Since we were interested in how subjects identify quality issues, we created two models (P_1 and P_2) containing several quality issues. The models vary regarding different aspects, e.g., the number of activities (between 13 and 45) and number of message flows (between 8 and 4). The process models are based on two different domains describing a company selling self-mixed muesli and a new product development process. Both models contain quality issues of all three dimensions, i.e., syntactic, semantic, and pragmatic. P_1 and P_2 comprise between 5 and 9 syntactic, between 3 and 4 semantic, and between 6 and 12 pragmatic quality issues.[1]

Design. Fig. 1 shows the overall design of the study: First, subjects obtain introductory assignments and demographical data is collected. Afterwards, subjects are confronted with the actual tasks. Each subject works on two process

---
[1] The study's material can be downloaded from: http://bpm.q-e.at/QualityIssues BPMN

models. For each model, the subjects are asked to find as many quality issues as they can. After each model, we asked for assessment of mental effort as well as for feedback on domain knowledge, understandability of the model, difficulties with think-aloud, and difficulties understanding the model's language (English). Finally, subjects are shown the same process models with marked quality issues and are asked to comment on the quality issues they were not able to find.

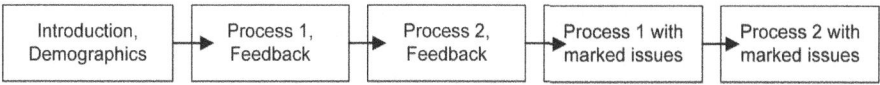

Fig. 1. Design of the exploratory study

Instrumentation. For each model, subjects received separate paper sheets showing the process models, allowing them to use a pen for highlighting quality issues. No written answers were required, only free talking. Audio and video recording are used as it has proven being useful for resolving unclear situations in think-aloud protocols [26].

4 Performing the Exploratory Study

Execution. The study was conducted in October and December 2014 at the University of Innsbruck, i.e., a total of twelve subjects participated. Each session was organized as follows: First, the subject was welcomed and instructed to speak thoughts out loudly. One supervisor handed sheets containing the study's material out and collected them as soon as the subject finished the task. Meanwhile, the subject's actions were audio- and video-recorded to gather any uttered thoughts.

Data Validation. In each session, only a single subject participated, allowing us to ensure that the study setup was obeyed. In addition, we screened whether subjects fitted the targeted profile, i.e., were familiar with process modeling and BPMN. We asked questions regarding familiarity on process modeling, BPMN, and domain knowledge; note that the latter may significantly influence performance [27]. For this, we utilize a 7-point Likert Scale, ranging from *"Strongly agree" (7)* over *"Neutral" (4)* to *"Strongly disagree" (1)*. Results are summarized in Table 1. Finally, we assessed the subjects' professional background: six subjects were students, four subjects had an academic background (i.e., were either PhD students or postdocs), and two subjects indicated a professional background. We conclude that all subjects had an adequate background in process modeling (the least experienced subject had 2 years of modeling experience) and were moderately familiar with BPMN.

Table 1. Demographics (5–9 based on 7-point Likert Scale)

	Minimum	Maximum	Median
1) Years of modeling experience	2	6	4
2) Models read last year	2	250	5
3) Models created last year	0	50	2.75
4) Average number of activities	0	30	11
5) Familiarity BPMN	2	6	5
6) Confidence understanding BPMN	3	6	6
7) Confidence creating BPMN	3	6	5.5
8) Familiarity mixing muesli company	5	7	6
9) Familiarity new product development company	2	7	5

Data Analysis. Our research focused on identifying quality issues of BPMN process models. On one hand, we investigated strategies applied by subjects in identifying quality issues, on the other, we explored which types of quality issues were typically mentioned by humans when inspecting a process model and which quality issues gain less attention. For this purpose, data analysis comprised the following stages. As a starting point, we transcribed the subjects' verbal utterances. Afterwards, we applied grounded theory to the transcripts in order to answer our research questions. For example, to investigate what kind of strategies the subjects applied to identify quality issues, we inspected the transcripts, marking aspects that indicated such strategy. In a second iteration, we revisited the marked areas and searched for new aspects. This process of open coding analysis was repeated until no new aspects could be found, so saturation has been reached. Afterwards, we performed axial coding, i.e., we repeatedly grouped aspects to form high level categories. We counted the number of identified markings per category.

5 Findings

In the previous section, we have discussed the design and execution of the study. In the following, we use the gathered data to investigate research questions RQ_1 and RQ_2.

5.1 RQ_1: What are Common Strategies that Humans Take for Inspecting and Identifying Quality Issues in BPMN Process Models?

When analyzing the transcripts, we observed that subjects consistently adopted similar strategies when identifying quality issues in BPMN process models. Fig. 2 shows the process model P_1[2]. The model consists of two pools. The first one represents the customer, the second one comprises the muesli mixing company. We could identify three different strategies:

---

[2] A high resolution of this figure can be downloaded from: http://bpm.q-e.at/QualityIssuesBPMN

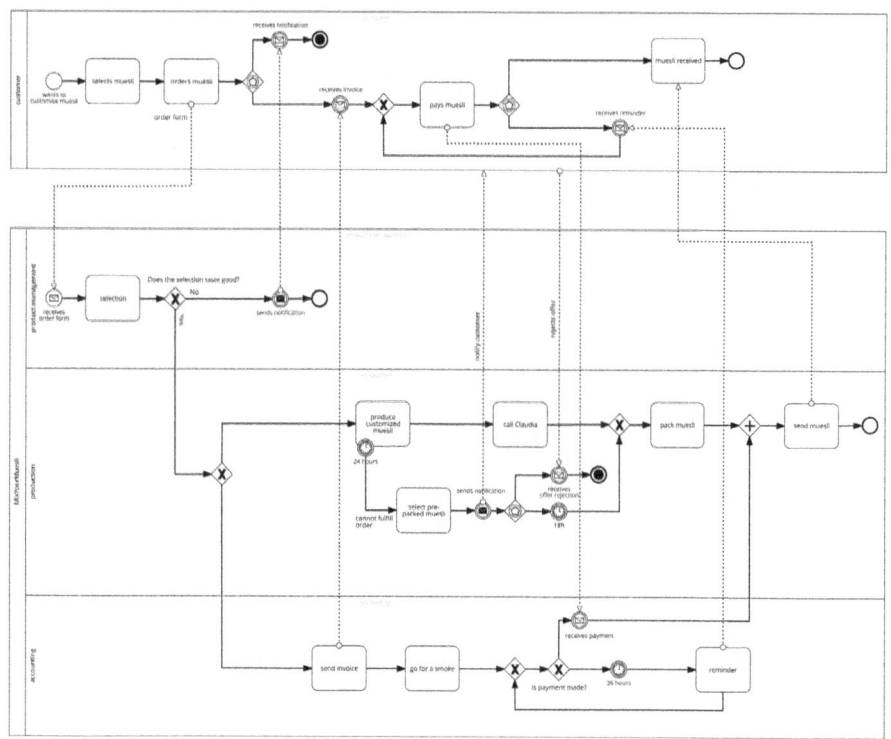

Fig. 2. Process model P_1

Strategy S_1. The majority of subjects identified quality issues in the model by first getting an overview of the model and then checking the whole model for quality issues. For P_1 7 out of 12 subjects (58.33%) and for P_2 additional 2 subjects (9 out of 12 subjects, 75.00%) used strategy S_1. In detail, there are three ways how subjects overviewed the model. First, they started by reading the pool and lane descriptions. Second, they read through the whole model before they checked it for any quality issue. Third, they looked at the structure of the process model. Relating to P_1 depicted in Fig. 2, 4 subjects listed the pools (e.g., *"First I'm looking at the different kinds of lanes and pools... So apparently it's the company 'Mix Your Muesli' and the customer."*), 2 subjects read silently the whole model, and one subject glanced over the structure of the model (*"Ok, so... first of all I look at the structure."*). Afterwards, they started reading the model while looking for quality issues: *"Let's start at the start point. The customer wants to customize muesli... Ok, this is the start event."*

Strategy S_2. Subjects adopting this strategy renounced getting an overview of the process model and directly started with reading the model while checking for quality issues. Regarding our example (cf. Fig. 2), subjects began to read the model: *"So let's start here. So we're starting with muesli to select, then we* *order the muesli.")*. For P_1 3 out of 12 subjects (25%) and for P_2 only 1 out of 12 subjects (8.33%) used strategy S_2. Note that P_2 (45 activities) was much larger than P_1 (13 activities). The subjects that used this strategy for P_1 but not for P_2 abandoned it for strategy S_1, i.e., they got an overview of P_2 before they started identifying quality issues.

Strategy S_3. 2 out of 12 subjects (16.67%) preferred to identify quality issues in P_1 and P_2 by checking off a mental list of possible quality problems, i.e., they were looking at specific parts of the process model one after another. For example, one subject started analyzing P_1 (cf. Fig. 2) by checking properties of message flows: *"Well first I check if there are messages in between of the lanes because that's not allowed."* Afterwards, she got an overview of the process model by reading the pool descriptions: *"Now I check what kind of pools there are..."* Then, she validated if all labels were conform to the verb-object style: *"Now I check for the labels [...] and there should always be a verb and a subject."* She moved on to checking the logic of the whole process: *"Alright now I go over the logic... Alright, first the customer selects the muesli..."* She continued to look for sequence flow line crossings and controlling the correctness of loops: *"I check for crossings which are not necessary... but there aren't. And I check for loops..."* Afterwards, she checked if the placement of activities were correct regarding their lanes: *"The payment is in the accounting... that's all production... product management... yes, that's customer activities."* After searching for message flow crossings, she finished her analysis of P_1 by a quick scan of the whole model.

5.2 RQ_2: Which Types of Quality Issues are Typically Spotted by Humans when Inspecting a BPMN Process Model? Which Quality Issues Gain Less Attention?

When analyzing what kind of quality issues were marked by subjects, we observed that subjects typically mentioned specific issues more often than others. In the following, we explain our findings with respect to BPMN notational elements.

Gateways. All syntactical errors with respect to parallel (AND) and data-based exclusive (OR) gateways (1 in P_1 and 9 in P_2) were identified by at least 8 out of 12 subjects (66.67%). Overall, 87.50% of these issues were marked. Also, pragmatic issues relating to gateways, i.e., implicit gateways (2 in P_2), were identified by at least 6 out of 12 subjects (50.00%, overall 75.00% issues marked). However, all syntactical errors relating to event-based exclusive gateways (2 in P_1) were found by at most 3 out of 12 subjects (25.00%). In total, only 25.00% of these errors were marked.

Message flows. In P_1, we deleted from 5 sending tasks the filled envelope marker that distinguished these tasks from normal activities. Only 2 out of 12 subjects (16.67%) marked each one of those 5 activities. In addition, 1 out of 12 subjects mentioned that not all message flows have a line description.

Activity out of context. We inserted in P_1 2 and P_2 3 superfluous activities. From these 5 activities, 3 were completely out of context, i.e., we named them 'go for a smoke', 'complain to boss' and 'discuss where to go for lunch'. These activities where identified by almost all subjects (at least 10 out of 12, 75.00%). Overall, these activities were marked to 91.67%. The other 2 superfluous activities were labeled 'call Claudia' and 'check stock market news', which were discovered by at least 6 out of 12 subjects (50.00%, overall 62.50% identified).

One subject had serious understandability issues because she did not recognize that the activity 'check stock market news' in P_2 is out of context. She got influenced by this activity to the extent that she thought that the whole new product development process is about stocks: *"Stock market news. It seems like that's not a product, it's more like, stocks... That would make sense, because then you wouldn't have a development of the product itself."* Therefore, she concluded that the parts of the model about the materials for the product development are incorrect, because there is no need for materials if the process is about stocks: *"Are materials available... it's not are stocks available. That's really bad! That's awful."*

Semantic issues. In addition to 5 superfluous activities, we added 1 semantic issue to each model. In P_1, it can happen that the customer has to pay more than once for his muesli. In P_2, we switched the labels of the lanes. 29.17% of these issues were marked by at most 4 out of 12 subjects (33.33%).

Labels. P_1 and P_2 contain each 2 activities with non-intention revealing labels, P_2 additionally 1 activity that is not according to verb-object style. Overall, 63.33% of these quality issues were identified by at least 6 out of 12 subjects (50.00%).

We could observe twice (once in P_1 and once in P_2) that a subject claimed that an activity with a non-intention revealing label should be skipped, because they could not derive a purpose of these activities without a correct label.

Line crossings. We introduced 1 unnecessary line crossing to P_1 and 2 to P_2. However, only 3 out of 12 subjects (25.00%) ever mentioned a line crossing. One of them identified all line crossing because she used a strategy to identify quality issues where she explicitly looked out for crossings of the sequence and message flows (cf. Sect. 5.1, S_3). Therefore, line crossings in P_1 and P_2 were only mentioned to 16.37%.

Reverse sequence flow direction. In P_2, at 3 parts of the model the sequence flow is from right to left instead of the other way round. These issues were identified to 38.89% by at most 7 out of 12 subjects (58.33%). Knowing that persons with western culture background prefer layout from left to right [28], this finding corroborates the assumption that an unexpected layout direction may be indeed perceived as modeling flaw.

6 Discussion

Strategies. We could identify three different strategies that our subjects adopted for inspecting and identifying quality issues in BPMN process models. However, it is not clear if there are any other strategies how humans spot quality issues. Our findings indicate that humans that adopt either Strategy S_1 (first getting an overview of the process model before checking it for quality issues) or Strategy S_3 (having a mental list of possible quality problems to inspect a process model) use these strategies irrespective of the model's size or complexity. Otherwise, it seems that humans that renounced looking over the process model before starting to read and check it (Strategy S_2) mostly preferred to switch to Strategy S_1. The question remains unsettled which strategy should be used to spot a specific kind of quality issue, or, on the contrary, which strategy should not be used to identify a specific kind of quality issue. In addition, we do not know if humans with less BPMN knowledge prefer other strategies than professionals. In our study, subjects adopted strategies irrespective of their demographic background, i.e., no strategy was only used by students, academics or professionals. This was rather surprising, as problem-solving relates mental lists (Strategy S_3) to experts as opposed to novices [29]. Moreover, besides accuracy of identifying quality issues, the time aspect stays an open question. These issues can be addresses with a bigger sample size and remain for future work.

Notational Elements. We observed that the importance of quality issues does not correlate with the quality dimensions of the SEQUAL framework, i.e., we did not observe that one dimension is "more important" than the other. For each quality dimension, we observed quality issues that were spotted by a large number of subjects, but also quality issues that did not seem to bother the participants of this study.

Regarding gateways, our subjects spotted quality issues relating to parallel (AND) and data-based exclusive (OR) gateways more often than quality issues with respect to event-based exclusive gateways. Also, quality issues regarding message flows were nearly never spotted by our participants. Relating to semantic issues, our findings indicate that obvious superfluous activities are identified easily. However, it seems that one unapparent superfluous activity is enough to hamper the understandability of a whole model. Apart from superfluous activities, it seems likely that subjects miss semantic issues. For example, we were surprised to find out that only 3 out of 12 subjects noticed that in P_1 it can happen that the customer has to pay more than once—an issue that should never occur in practice. Also, even though 10 out of 12 subjects read the labels of the lanes in P_2, only 4 of them noticed that these were switched. Besides superfluous activities, activities with non-intention revealing labels can pose a hindrance in understanding a process model. Regarding quality issues with respect to line crossing or reverse sequence flow direction, we were surprised that these issues did not seem to bother our subjects.

Limitations. This study has to be viewed in the light of several generalization limitations. In particular, the detection of quality issues presumably also depends on the support that is offered. First, in the sense of notational support, models may be specifically designed for a group of stakeholders, making the models particularly suitable for understanding and thus detecting quality issues. In this work, we focused on BPMN—a notation that is typically taught to a large group of stakeholders [30]. In this sense, our findings are difficult to generalize to more specifically tailored notations. Second, in the sense of computer support, such as scrolling or syntax highlighting, we specifically decided to conduct our study without computer support to establish a base–line further studies can be compared against. Apparently, this makes it difficult to transfer our insights to process models presented with computer support. We would like to stress that these decisions were made deliberately and this study should be seen as exploratory, highlighting future research directions. Quantitative and more focused studies are still to follow. Regarding further limitations, it should be noted that the number of subjects in the study is relatively low (12 subjects). Nevertheless, it is noteworthy that the sample size is not unusual for this kind of empirical investigation due to the substantial effort to be invested per subject [31,32]. Finally, half of the participating subjects were students. However, all subjects indicated profound background in business process management.

7 Related Work

The goal of this study is to investigate how humans identify quality issues in BPMN process models. Errors at the syntactic level can be automatically detected for a large class of process models using verification techniques, which also support the incremental validation of process models [33]. The identification of semantic errors can only be partially automated [14]. For instance, [34] describes a two-step procedure for measuring process model quality at the semantic level by comparing a process model with a reference model. First, activities present in the process model must be mapped to the activities of the reference process model, e.g., using measures like the Levenshtein distance for activity labels [35] or combining edit distances measures with the detection of synonyms [36]. Similarly, the ICoP framework [37] provides means to automatically detect potential activity matches between process models. After establishing an activity mapping, the similarity to the reference model can be assessed [34] by measuring edit distances between graphs, e.g., [36], focusing on causal dependencies of activities, e.g., [38]. In turn, [21] suggests a technique for detecting redundant process fragment and for automatically extracting them to sub–processes. In turn, techniques for modularizing large process models and for automatically labeling the extracted sub–process fragments are proposed in [39].

Nonetheless, the question how humans can be efficiently supported in creating, understanding and maintaining business process models is still a lingering problem. Therefore, this paper takes a first step toward an in-depth understanding of how BPMN process models are inspected, what strategies are taken, and what cognitive processes are involved. This study is also an indication what other research questions can be investigated using this approach (cf. Sect. 6).

8 Summary and Outlook

While some quality issues can be detected automatically, many others cannot. Even though human inspection of process models is still essential [9], this manual inspection is currently not supported. This paper takes a first step toward an in-depth understanding of how BPMN process models are inspected, what strategies are taken, and what cognitive processes are involved. The presented exploratory study investigates the strategies taken by humans when inspecting BPMN process models and the kinds of quality problems they notice. Even though the sample size is small, we show how these questions can be addressed. Our qualitative analysis shows that humans adapt different strategies on how to identify quality issues. Also, we observed for each quality dimension quality issues that were spotted by a large number of subjects (e.g., deadlocks), but also quality issues that did not seem to bother the participants of this study (e.g., line crossings). Further, we also indicate other research questions that can be investigated using this approach.

Future research can build upon these initial findings by performing more comprehensive studies. More specifically, the question arises which quality issues should be supported with appropriate tool support and which parts of a modeling notation cause difficulties while creating process models. Likewise, we plan to extend our research focus by additionally asking practitioners and business managers to inspect process models for quality issues.

References

1. Becker, J., Rosemann, M., von Uthmann, C.: Guidelines of business process modeling. In: van der Aalst, W.M.P., Desel, J., Oberweis, A. (eds.) Business Process Management. LNCS, vol. 1806, pp. 30–49. Springer, Heidelberg (2000)
2. Rittgen, P.: Quality and perceived usefulness of process models. Proc. SAC 2010, pp. 65–72 (2010)
3. Scheer, A.W.: ARIS–Business Process Modeling, 3rd ed. Springer (2000)
4. Weber, B., Reichert, M., Mendling, J., Reijers, H.A.: Refactoring large process model repositories. Computers in Industry **62**, 467–486 (2011)
5. Mendling, J.: Empirical studies in process model verification. In: Jensen, K., van der Aalst, W.M.P. (eds.) Transactions on Petri Nets and Other Models of Concurrency II. LNCS, vol. 5460, pp. 208–224. Springer, Heidelberg (2009)
6. Mendling, J., Reijers, H., Recker, J.: Activity Labeling in Process Modeling: Empirical Insights and Recommendations. Information Systems **35**, 467–482 (2010)
7. Rosa, M.L., ter Hofstede, A., Wohed, P., Reijers, H., Mendling, J., van der Aalst, W.P.: Managing Process Model Complexity via Concrete Syntax Modifications. IEEE Trans. Industrial Informatics **7**, 255–265 (2011)
8. Wynn, M.T., Verbeek, H.M.W., van der Aalst, W.M.P., ter Hofstede, A.H.M., Edmond, D.: Business process verification - finally a reality!. Business Proc. Manag. Journal **15**, 74–92 (2009)

9. Soffer, P., Kaner, M.: Complementing business process verification by validity analysis: A theoretical and empirical evaluation. J. Database Manag. **22**, 1–23 (2011)
10. Krogstie, J.: Model-Based Development and Evolution of Information Systems: A Quality Approach. Springer (2012)
11. Koehler, J., Vanhatalo, J.: Process Anti-Patterns: How to Avoid the Common Traps of Business Process Modeling. Technical report, IBM ZRL Research Report 3678 (2007)
12. Mendling, J.: Metrics for Process Models: Empirical Foundations of Verification, Error Prediction and Guidelines for Correctness. Springer (2008)
13. Roy, S., Sajeev, A., Bihary, S., Ranjan, A.: An Empirical Study of Error Patterns in Industrial Business Process Models. IEEE Transactions on Services Computing **99**(2013). doi:10.1109/TSC.2013.10
14. Soffer, P., Kaner, M., Wand, Y.: Towards understanding the process of process modeling: theoretical and empirical considerations. In: Proc. ER-BPM 2011. (2011) 357–369
15. Reijers, H.A., Mendling, J.: A Study into the Factors that Influence the Understandability of Business Process Models. IEEE Transactions on Systems, Man and Cybernetics, Part A **41**, 449–462 (2011)
16. Zugal, S., Soffer, P., Haisjackl, C., Pinggera, J., Reichert, M., Weber, B.: Investigating expressiveness and understandability of hierarchy in declarative business process models. Software & Systems Modeling 1–23 (2013)
17. Zugal, S., Soffer, P., Pinggera, J., Weber, B.: Expressiveness and understandability considerations of hierarchy in declarative business process models. In: Bider, I., Halpin, T., Krogstie, J., Nurcan, S., Proper, E., Schmidt, R., Soffer, P., Wrycza, S. (eds.) EMMSAD 2012 and BPMDS 2012. LNBIP, vol. 113, pp. 167–181. Springer, Heidelberg (2012)
18. Zugal, S., Pinggera, J., Mendling, J., Reijers, H., Weber, B.: Assessing the impact of hierarchy on model understandability-a cognitive perspective. In: Proc. EESSMod 2011, pp. 123–133 (2011)
19. Schrepfer, M., Wolf, J., Mendling, J., Reijers, H.A.: The impact of secondary notation on process model understanding. In: Persson, A., Stirna, J. (eds.) PoEM 2009. LNBIP, vol. 39, pp. 161–175. Springer, Heidelberg (2009)
20. Mendling, J., Reijers, H.A., van der Aalst, W.M.P.: Seven process modeling guidelines (7pmg). Information & Software Technology **52**, 127–136 (2010)
21. Dumas, M., García-Bañuelos, L., La Rosa, M., Uba, R.: Fast Detection of Exact Clones in Repositories of Business Process Models. Information Systems **38**, 619–633 (2013)
22. Purchase, H.: Which aesthetic has the greatest effect on human understanding? In: Proc. GD 1997. (1997) 248–261
23. Bassey, M.: Case study research in educational settings. Doing qualitative research in educational settings. Open University Press (1999)
24. Ericsson, K.A., Simon, H.A.: Protocol Analysis: Verbal Reports as Data. MIT Press (1993)
25. Corbin, J., Strauss, A.: Basics of Qualitative Research: Techniques and Procedures for Developing Grounded Theory. SAGE Publications (2007)
26. Zugal, S., Haisjackl, C., Pinggera, J., Weber, B.: Empirical Evaluation of Test Driven Modeling. International Journal of Information System Modeling and Design **4**, 23–43 (2013)
27. Khatri, V., Vessey, I., Ramesh, P.C.V., Park, S.J.: Understanding Conceptual Schemas: Exploring the Role of Application and IS Domain Knowledge. Information Systems Research **17**, 81–99 (2006)

28. Figl, K., Strembeck, M.: On the importance of flow direction in business process models (2014) Poster presented at ICSOFT-EA
29. Schoenfeld, A.H., Herrmann, D.: Problem perception and knowledge structure in expert and novice mathematical problem solvers. Journal of Experimental Psychology: Learning, Memory, Cognition **8**, 484–494 (1982)
30. Recker, J.: Opportunities and constraints: the current struggle with bpmn. Business Process Management Journal **16**, 181–201 (2010)
31. Costain, G.F.: Cognitive Support During Object oriented Software Development: The Case of UML Diagrams, Ph.D thesis, University of Auckland (2007)
32. Nielsen, J.: Estimating the number of subjects needed for a thinking aloud test. Int. J. Hum.-Comput. Stud. **41**, 385–397 (1994)
33. Kühne, S., Kern, H., Gruhn, V., Laue, R.: Business process modeling with continuous validation. JSEP **22**, 547–566 (2010)
34. Becker, M., Laue, R.: A comparative survey of business process similarity measures. Computers in Industry **63**, 148–167 (2012)
35. Levenshtein, W.: Binary codes capable of correcting deletions, insertions and reversals. Soviet Physics Doklady, 707–710 (1966)
36. Dijkman, R.M., Dumas, M., van Dongen, B.F., Käärik, R., Mendling, J.: Similarity of business process models: Metrics and evaluation. Inf. Syst. **36**, 498–516 (2011)
37. Weidlich, M., Dijkman, R., Mendling, J.: The ICoP framework: identification of correspondences between process models. In: Pernici, B. (ed.) CAiSE 2010. LNCS, vol. 6051, pp. 483–498. Springer, Heidelberg (2010)
38. Weidlich, M., Mendling, J., Weske, M.: Efficient Consistency Measurement Based on Behavioral Profiles of Process Models. IEEE Trans. Software Eng. **37**, 410–429 (2011)
39. Smirnov, S., Reijers, H., Weske, M.: From fine-grained to abstract process models: A semantic approach. Inf. Syst. **37**, 784–797 (2012)

Modeling and Reasoning about Information Quality Requirements in Business Processes

Mohamad Gharib[✉] and Paolo Giorgini[✉]

DISI, University of Trento, 38123 Povo, Trento, Italy
{gharib,paolo.giorgini}@disi.unitn.it

Abstract. Information Quality (IQ) is particularly important for the successful and efficient execution of any Business Process (BP). Despite this, most existing BP approaches either ignore IQ needs, or they deal with them as mere technical issues, without considering the social and organizational aspects that underlie such needs. In this paper, we propose a goal-oriented approach to capture IQ requirements (needs) and map these requirements into workflow net (WFA-net) that is a formal language for modeling and analyzing IQ requirements in BP. We illustrate our approach with an example concerning a stock market system.

Keywords: Information Quality · Business process · Work flow · Requirements engineering

1 Introduction

A Business Process (BP) can be defined as a set of activities that has a clear structure describing their sequencing order and dependencies [1]. Traditionally, the BP literature has focused on control-flow (activity flow) perspective of the process with less emphasis on information perspective. However, information related problems can be a main reason of different kinds of errors in BP [2]. Yet, in recent years some efforts have been devoted to information-aware process design (e.g., Sadiq et al. [3]; Sidorova et al. [4]; Trcka et al. [2]). However, the focus of attention in these works is combining information flow with activity flow, i.e., they are able to detect when an activity in BP relay on information that does not exist, but they say nothing about Information Quality (IQ) concerns.

IQ is a key success factor for most BPs, since depending on low-quality information may result in undesirable outcome [5], or it might even prevent the BP from achieving its goals. In the literature, we can find several techniques for dealing with IQ (e.g., preventing, detecting and correcting IQ related issues [6]). Yet most of these techniques propose solutions that are able to address the technical aspects of IQ, and seem to be limited in addressing the social and organizational IQ related aspects. Such aspects are particularly important for BPs, since they are mainly executed by social actors and not only machines [7]. More specifically, most BPs occur in social context (e.g., socio-technical systems [8]), where humans and technical components are considered as an integral part of the BP.

Thus, understanding the social and organizational context where the BPs are executed is essential to detect different kinds of vulnerabilities.

For example, Fisher and Kingma [9] showed how existing IQ techniques are not able to capture IQ needs in their social and organizational context, where different kinds of vulnerabilities might manifest themselves in the actors' interactions and dependencies. The Flash Crash (a main U.S market crash) is an example where the problem was not caused by a mere technical failure, but it was also due to several socio-technical IQ related vulnerabilities [10]. This introduce the need of analyzing the social and organizational environment where the BP operates [11]. In this paper, we propose a goal-oriented approach to capture IQ requirements of the social and organizational context where the BP is executed, and then introduce mechanisms for mapping these requirements into workflow net with actors (WFA-net). The paper is organized as follows; Section (§2) describes the research baseline, a motivating example concerning a stock market system is presented in section (§3). We propose our approach in section (§4). The prototype is summarized in (§5). Related work is presented in Section (§6), we conclude and discuss the future work in Section (§7).

2 Research Baseline

Our research baseline is based on three main areas; we briefly discuss each of them as follows:

(i) Goal-Oriented Requirements Engineering (GORE): several approaches that adopts GORE paradigm have been proposed in the literature (e.g., KAOS [12], i* [13], secure Tropos [14], etc.). Among the existing ones, we adopt an extended version of secure Tropos [15] as a baseline for our approach, which support the basic modeling concepts offered by secure Tropos, and provides concepts for capturing IQ requirements. In particular, it introduces primitives for modeling actors of the system, which covers two concepts, a role and agent. Goals are used to represent actors' strategic interests, and they can be refined through and/or decomposition into finer sub-goals. While information is used to represent any informational entities, and it has volatility attribute that can be used to determine its timeliness (validity). An actor can be a legal owner of information item, which gives it a full control over its use. While goals may produce, read and send information. Finally, it adopts the notion of delegation to model the transfer of entitlements and authorities among actors, and it adopts the notion of trust and distrust to capture the actors' expectations of one another concerning their delegated entitlements and authorities.

(ii) Information Quality: IQ is a hierarchal multi-dimensional concept [16,17], that can be characterized by several dimensions [17,18], including: accuracy, completeness, timeliness, accessibility, trustworthiness, etc., where each of these dimensions can be used to represent a certain aspect of IQ. We focus on 3 main IQ dimensions that enable us to address the IQ related issues that we consider in this paper, namely: *Accessibility*: the extent to which information is available, or easily and quickly retrieved [16], we limit accessibility definition to information availability and having the required permission to perform a task at hand; *Accuracy*: means that information should be true or error free with respect to some known values [17]; and *Timeliness*: can be defined as to which extent information is valid in term of time[16], i.e., sufficiently up-to-date.

(iii) Petri nets/WF-nets/WFD-nets: several workflow modeling languages have been proposed, yet we focus on Petri-nets-based languages (e.g., Petri nets, WF-nets, and WFD-nets). In particular, a petri net [19] is a graphical and formal language that can be used to model different kinds of BPs. Formally: a Petri net N = $\langle P, T, F \rangle$, where P is a finite set of places, T is a finite set of transitions, and $F \subseteq (PXT) \cup (TXP)$ is a set of arcs (flow relation). At any time, a place contains zero, one, or more tokens, while a transition $t \in T$ is said to be *enabled*, **iff**, each input place p of t contains at least one token. An *enabled* transition t may *fire*, **iff**, a transition t consumes one token from each input place p of t, and it produces one token in each output place p of t. Furthermore, a marking of a Petri net is a multi-set of its places $M : P \longrightarrow \mathbb{N}$. Transitions are the active components in a Petri net, i.e., they change the state of the net. For example, given a Petri net N and a marking M_1, we say that $M_1 \xrightarrow{t} M_2$: if transition t is *enabled* at marking M_1, and firing t at M_1 results in M_2. While $M_1 \xrightarrow{\sigma} M_n$: $\sigma = t_1, t_2, \ldots, t_{n-1}$ is a firing sequence leading from M_1 to M_n. Finally, we say that a marking M_n is *reachable* from M_1, **iff**, there is a firing sequence $\sigma = t_1, t_2, \ldots, t_{n-1}$ such that $M_1 \xrightarrow{\sigma} M_n$. At the other hand, a workflow-net (WF-net) [20] is a Petri net with well-defined starting point (**start**) and a well-defined ending point (**end**), and every place or transition is on a path from **start** to **end**, where transitions in a WF-net are called *tasks*. While a workflow net with data (WFD-net) [4] is a workflow net with data elements in which tasks can read, write, or delete data elements. Moreover, a task can also have data dependent guards that block its execution when it is evaluated to false.

3 US Stock Market System

A stock market (equity market) system is the aggregation of investors, traders, trading markets, along with several firms that provide different kinds of financial services. Based on [21,22], we can identify several main stakeholders of a stock market system, including: *stock investors* are individuals or companies, who have a main goal of making profit from trading securities. *Stock traders* are persons or companies involved in trading securities in *stock markets* either for their own sake or on behalf of their *investors*. *Stock markets* are places where *traders* gather and trade securities (e.g., NYSE, CME, NASDAQ, etc.). In particular, markets make profit by facilitate security trading among traders, i.e., they receive, match, and perform trades among different *traders*. Moreover, they should guarantee a fair and stable trading environment for their traders.

Accounting firms can be defined as firms that provide accounting services to companies for a fee. While *auditing firms* are responsible of auditing the financial statements of legal entities (e.g., persons, companies), where a financial statement is a formal record of the financial activities of such entities. *Consulting firms* provide professional advices concerning financial securities for a fee to both *traders* and *investors*. *Credit assessment ratings firms* are specialized for providing assessments of the credit worthiness of companies' securities, i.e., they help traders in deciding how risky it is to invest money in a certain security.

4 Approach for Modeling and Reasoning About IQ Requirements in Business Process

In this section, we propose our approach for modeling and analyzing IQ requirements in BP. An overview of the methodological process that underlies our approach is shown in Figure 1, the process is composed of 3 main phases[1]:

(1) Modeling phase: aims to model the social, organizational and IQ requirements of the overall system where the BP occurs; this phase is composed of 5 main steps: (1.1) *Actors modeling*: aims to model the actors of the system in terms of agents and the role(s) they play; (1.2) *Goals modeling*: identify and model actors' top-level goals and refine them, if needed, through and/or-decomposition into leaf goals; (1.3) *Information modeling*: identify and model the legal owners of information items, which is essential to identify who has full control concerning information permissions. Moreover, we model the different relations between goals and information they use (e.g., produces, reads, modifies and sends); (1.4) *Social dependency modeling*: aims to model actors' dependencies for information provision, and the delegation of both authorities and entitlements, i.e., based on actors' capabilities some goals might be delegated to other actors, who have the capabilities to achieve them; and based on actors' needs, information and permissions are provided/ delegated to them respectively; and (1.5) *Trust modeling*: model trust/ distrusts among actors concerning goals/ permissions delegation, based on their expectations in one another. When modeling phase is complete, and if the model does not require any refinements, we proceed to the mapping phase.

(2) Mapping phase: in which, we map the requirements model, that has been produced in the previous phase, into workflow net with actors (WFA-net) that is a formal language we propose for modeling and analyzing BP control-flow, information flow, and IQ requirements.

(3) Analysis phase: aims to verify the correctness and consistency of the BP model. In particular, we define a set of properties to check the correctness and consistency of the BP control-flow, information flow along with IQ requirements, i.e., BP is correct and consistent, if all of these properties hold.

4.1 Modeling Phase

In order to model IQ requirements of the BP in their social and organizational context, we rely on an extended version of secure Tropos [15], which provides

---

[1] We discuss each of these phases in details in sections 4.1, 4.2, and 4.3 respectively.

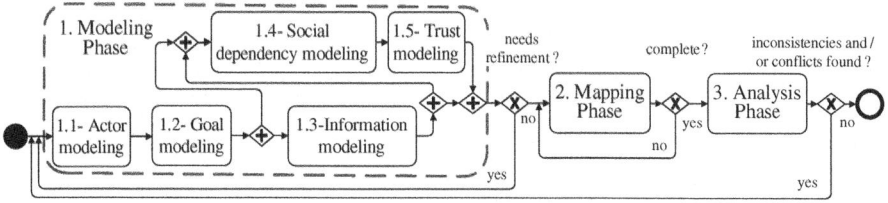

Fig. 1. The process for modeling and reasoning about IQ requirements in BP

concepts for modeling IQ requirements. Figure 2 shows a portion of a goal model concerning the stock market system represented in the extended secure Tropos modeling language. For instance, John is an agent that plays stock investor role, which has a main goal of G1. Make profit from trading securities that is or-decomposed into G1.1 Trade securities by itself and G1.2 Delegate trading activities to a trader. In the case of or-decomposition, the parent goal is achieved, if any of its sub-goals is achieved, i.e., the goal G1. is achieved, if G1.1 or G1.2 is achieved. Moreover, G1.1.1 Trade securities by itself is and-decomposed into G1.1.1.1 Produce and send orders and G1.1.1.2 Finalize the trade. In the case of and-decomposition, the parent goal is achieved only if all its sub-goals are achieved.

Moreover, the goal G1.1.1.1 (P)roduces and (S)ends investor's orders, and the goal G1.1.1.2 (R)eads trade settlement. Each of the previously mentioned relations between goals and information are required for the achievement of the goals, i.e., if a goal could not use (e.g., reads, sends, or produces) information as intend, it will not be achieved (it will be prevented). Moreover, the stock investor provides (Integrity Provision (IP)[2]) investor's orders to a trader, and it delegates the goal G1.2 Delegate trading to a trader to *stock trader*, and trusts it for its achievement. Extended secure Tropos does not support modeling of permission, but secure Tropos does. Thus, we refine the modeling language by proposing 4 different types of permissions concerning the 4 types of information usage (e.g., (P)roduces, (R)eads, (M)odifies and (S)ends). Moreover, we extend the language to model permission delegation among actors, and to model trust/ distrusts concerning the delegated permissions. For example, the stock investor is the owner of investor's orders, and it delegates (R)eads, (M)odifies and (S)ends permissions concerning it to the trader.

4.2 Mapping Phase

In this section, we propose a workflow net with actors (WFA-net) that is a formal language, we propose, for modeling and analyzing IQ requirements for BPs. In particular, WFA-net is able to model and analyze the control-flow, information flow, and IQ requirements of BPs. Moreover, we discuss the mechanisms that are used for mapping IQ requirements model into WFA-net.

---

[2] IP provision preserves the integrity of the provided (transferred) information [23].

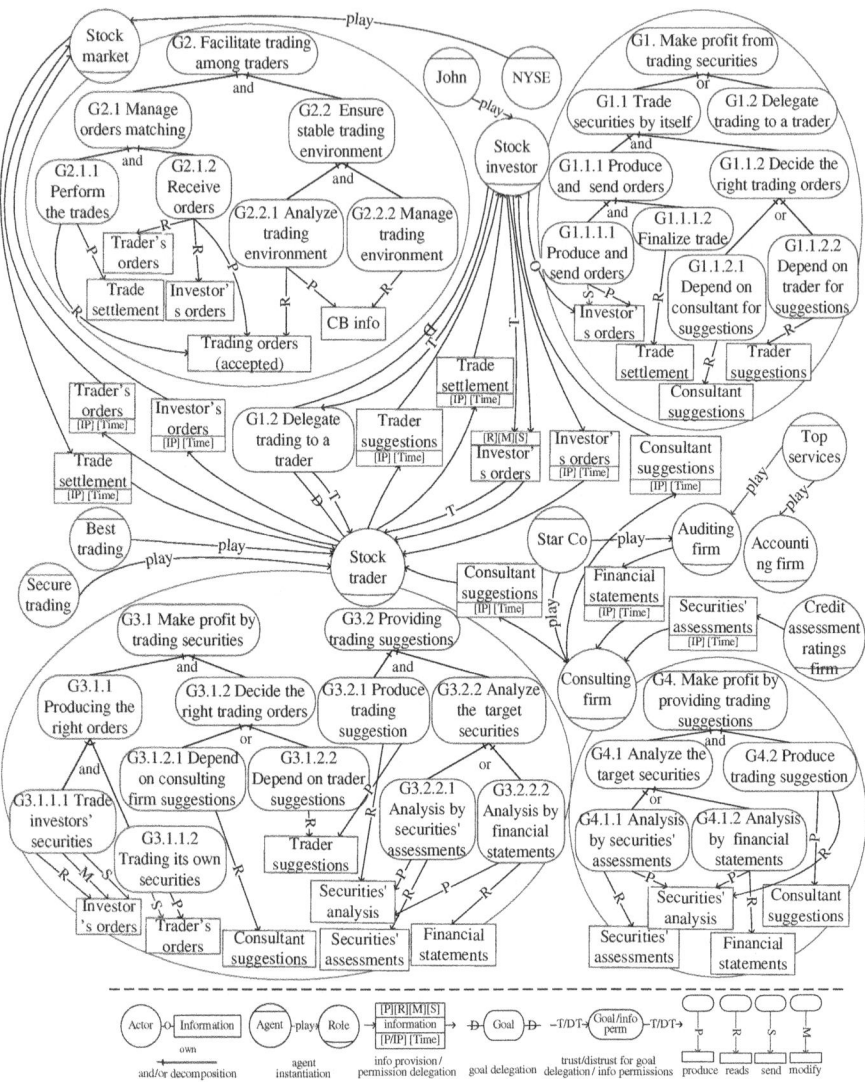

Fig. 2. A partial goal model concerning the stock market structure

Work Flow Net with Actors (WFA-Net) a workflow net with actors (WFA-net) adopts workflow net (WF-net) and extends it with the notion of social actor, and IQ related concerns. In WFA-net, activities (tasks) are assigned to social actors, and they may produce, read, modify, and send information items. In what follows, we define the semantics of WFA-nets. Let us consider a finite set of social actors $A = \{a_1, a_2, ..., a_n\}$, a finite set of information elements $I = \{i_1, i_2, ..., i_m\}$, and a finite set of time intervals $T = \{t_1, t_2, ..., t_m\}$, and we define $I_v \subseteq \{I \times T\}$ to describe information items along with their volatility values. Moreover, to capture information send relations, we define $S \subseteq \{A \times I \times T\}$ that describe the target (actor) of send information, information to be send, and the required send time. Further, we define a set of responsibility predicates $\Pi_A = \{\pi_{a_1}, \pi_{a_2}, \ldots, \pi_{a_k}\}$ to capture the relation between actors and activities they are responsible of (e.g., responsible(actor); a set of produce predicates $\Pi_P = \{\pi_{p_1}, \pi_{p_2}, \ldots, \pi_{p_j}\}$ to capture the relation between activities and information they produce (e.g., produce(info)); a set of read predicates $\Pi_R = \{\pi_{r_1}, \pi_{r_2}, \ldots, \pi_{r_k}\}$ to capture the relation between activities and information they read (e.g., read(info)); a set of modify predicates $\Pi_M = \{\pi_{m_1}, \pi_{m_2}, \ldots, \pi_{m_l}\}$ to capture the relation between activities and information they modify (e.g., modify(info)); a set of send predicates $\Pi_S = \{\pi_{s_1}, \pi_{s_2}, \ldots, \pi_{s_l}\}$ to capture the relation between activities and information they send (e.g., send(actor, info, send_time)).

Furthermore, we define the following functions: $f_{\pi_a} = \Pi_A \longrightarrow \{A\}$, responsibility function that assign responsibility predicates with actors responsible of achieving the related activities; function $f_{\pi_p} = \Pi_P \longrightarrow 2^{I_v}$, production function that assigns produce predicates with information items that activities produce; function $f_{\pi_r} = \Pi_R \longrightarrow 2^{I_v}$, reading function that assigns read predicates with information items that activities read; function $f_{\pi_m} = \Pi_M \longrightarrow 2^{I_v}$, modify function that assigns modify predicates with information items that activities modify; and function $f_{\pi_s} = \Pi_S \longrightarrow 2^S$, send function that assigns send predicates with information items that activities send.

Now, we define a WFA-net as a WF-net where every transition t is described with: an actor being assigned to perform the activity (res); a set of information items being produced (pd) by the activity when t fires; a set of information items being modified (md) by the activity when t fires; a set of information items being read (rd) by the activity when t fires; and a set of information items being send (sd) by the activity when t fires.

Definition 1 (WFA-net). *A workflow net with actors (WFA-net) $N = \langle$ P, T, F, res, pd, rd, md, sd $\rangle$ consist of WF-net $N= \langle$ P, T, F $\rangle$, an actor assigning function res: $T \longrightarrow A$, information producing function pd : $T \longrightarrow 2^{I_v}$, information reading function rd : $T \longrightarrow 2^{I_v}$, information modifying function md : $T \longrightarrow 2^{I_v}$, and information sending function sd: $T \longrightarrow 2^S$.*

Example 1. A WFA-net of a stock investor for trading securities is shown in Figure 3. Its actor set is A = {credit_firm, audit_firm, trader, investor, consulting_firm, stock market}, and its information set is I= {securities_assessment, financial_statement, trader_suggestion, trading_orders, consultant _suggestion , investors_orders, trading_settlement}. Considering the transition T12: Produce and send orders, the responsibility function res(Produce and send orders) = {investor}, the production function pd (Produce and send orders) = {investor's order}, the sending function sd(Produce and send orders) = {(stock market, investor's order, time)}, modify and read functions md(Produce and send orders) = rd(Produce and send orders) = {∅}.

To capture the work flow in WFA-net, we should be able to evaluate the activities related predicates either to true ($\top$) or to false ($\bot$) based on an already

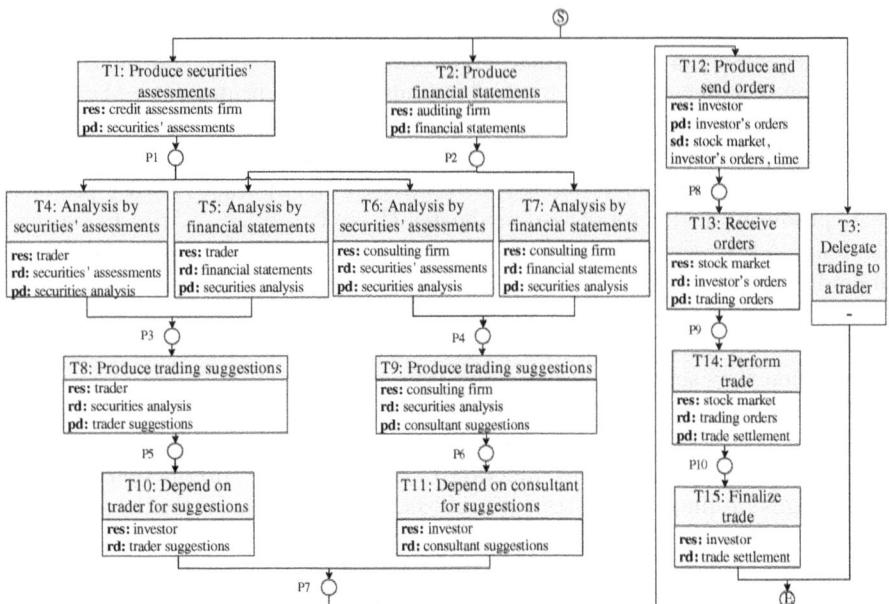

Fig. 3. A WFA-net of a stock investor for trading securities

defined criteria. Thus, we define the following functions, $\sigma_{\pi_a}: \Pi_A \longrightarrow \{\top, \bot\}$ assigns to each responsibility predicate either $\top$, when the responsible actor can achieve the activity, or it assigns $\bot$, when the responsible actor cannot achieve the activity. Similarly, we define $\sigma_{\pi_p}: \Pi_P \longrightarrow \{\top, \bot\}$, $\sigma_{\pi_r}: \Pi_R \longrightarrow \{\top, \bot\}$, $\sigma_{\pi_m}: \Pi_M \longrightarrow \{\top, \bot\}$, and $\sigma_{\pi_s}: \Pi_S \longrightarrow \{\top, \bot\}$ that assign to each produce/ read/ modify/ send predicate either $\top$, when information item i can be produce/ read/ modify/ send by the activity, or it assigns $\bot$ otherwise.

Finally, we define σ_Π function that sums the values of the previously mentioned functions over their related predicates. $\sigma_\Pi: T \to (\top, \bot)$, where $\sigma_\Pi = \sigma_{\pi_a} \wedge \sigma_{\pi_p} \wedge \sigma_{\pi_r} \wedge \sigma_{\pi_m} \wedge \sigma_{\pi_s}$. Following WFD-net, we refer to a **state**[3] of WFA-net as a configuration, where a WFA-net configuration is a state that includes responsible actors along with the produce, read, modify, and send information that the activity perform. Moreover, a **state** can be represented as σ_Π, and the set of all **states** is denoted by Σ.

Definition 2 (Configuration). *Let $N = \langle$ P, T, F, res, pd, rd, md, sd $\rangle$ be a WFA-net, let m be a **marking** of N, and let σ_Π be as defined above. Then, $c = \langle m, \sigma_\Pi \rangle$ is a configuration of N. With Ξ we denote the set of all configurations, and the start configuration of N is defined by $\langle [start]$, $\sigma_\Pi \rangle$, $(I_v = \emptyset) \rangle$. While $C_e = \{\langle [end]$, $\sigma_\Pi \rangle \mid \sigma_\Pi \in \Sigma \}$ defines the set of final configurations.*

In the initial configuration, only one place is marked [start], and I_v set is initialized to empty set. While a configuration is a final configuration, if it contains

---
[3] A state (also called marking) of a Petri net is a distribution of tokens over its places.

a marking [end]. A transition t of a WFA-net N can be enabled at a configuration $c = \langle m, \sigma_\Pi \rangle$, IFF: (1) the transition t is enabled at marking m (activity flow), and (2) the predicates related to configuration c enabling (σ_Π) must be true, which guarantees that IQ requirements are met. When a transition t is enabled, it may fires, where firing of transition t changes the marking as well as the related predicates and information set I_v, i.e., the firing of t enable a set of successor configurations $\langle m^{'}, \sigma_\Pi^{'} \rangle$, and changes predicates and information.

Definition 3 (Firing a transition for WFA-nets). Let $N = \langle$ P, T, F, do, pd, md, nd, up $\rangle$ be a WFA-net. A transition $t \in T$ of N is enabled at a configuration $c = \langle m, \sigma_\Pi \rangle$ of N, if $m \xrightarrow{t}$, σ_Π is assigned true ($\top$). Firing t enables a set of configurations $C \subseteq \Xi$, with $C = \{ \langle m^{'}, \sigma_\Pi^{'} \rangle \mid m \xrightarrow{t} m^{'} \wedge (\forall\ i \in pd(t) = \top: I = i \cap I) \}$ and it is denoted by $c \xrightarrow{t} C$.

Example 2. Consider transition T13: Receive orders in Figure 3, and suppose there is a token in place p8. The transition is enabled if *stock market* (responsible actor) has the capability to achieve such activity, information investors_orders fits for read from the perspective of the *stock market*, i.e., information has been already produced, *stock market* has it, and it does not suffer from any IQ related issue (e.g., information is valid (timeliness), accurate, *stock market* has the read permissions over it, etc.). Firing this transition means that the token in p8 is removed, and a token is produced in p9. Moreover, information trading_orders is produced by the *stock market*. Note that we only consider information producing when a transition fires, since it affects all the other information related operations (e.g., sends, modifies). While other IQ aspects can be captured with the help of the automated reasoning support (discussed in section 4.3).

Mapping IQ Requirements from the Goal Model into WFA-Nets in this section, we describe how IQ requirements model can be mapped into WFA-net. In particular, we define rules for identifying complete building blocks that are used to represent the extended secure Tropos constructs, which can be mapped into WFA-net activities. Moreover, we define several sets of constraints that should be followed during the mapping process to ensure the correctness of both the mapping and the resulting WFA-net:

Building blocks: we define 3 rules for identifying complete building blocks: (a) a goal that is not and/ or-decomposed of any other goal, and it is not composed into sub-goals as well, can be considered as a complete building block, and it can be mapped into a WFA-net activity (task) taking into consideration the actor, who is responsible of its achievement, and information it relies on (if any); (b) goals that are and-decomposed from a parent goal are considered as a complete building block, and they can be mapped into a sequence of WFA-net activities, where each of these activities represents a sub-goal; (c) goals that are or-decomposed from a parent goal are considered as a complete building block, and they can be mapped into parallel (alternatives) WFA-net activities, where each of these activities represents a sub-goal.

Consistency constraints: we define 3 consistency constrains that can be used to ensure a correct mapping between the identified building blocks and WFA-net activities: (i) mapping is allowed for complete building blocks only, i.e., no goal is allowed to be mapped unless it can be considered as a complete building block; (ii) if the WFA-net is used to model a plan to achieve a top-level goal, the full plan to achieve the top-level goal should be considered in the WFA-net; and (iii) no information is allowed in the WFA-net unless its source (the goal that produces it) exists in the WFA-net.

Sequencing constraints: we define 3 sequencing constraints that can be used to ensure the proper ordering of the WFA-net activities: (i) activities in WFA-net should be consistent with their sequencing order in their own building blocks; (ii) if an activity depend on another activity, it should appear after the activity it depends on in the WFA-net; (iii) if an activity depend on the outcome of another activity (e.g., information), it is desirable to appear after the activity it depends on its outcome.

Refinement constraints are constrains derived from the WFA-net semantics, and used to refine the WFA-net that results from the sequencing phase. We define two simple *refinement constraints*: (i) no two places (p_1, p_2) can appear in sequence without a transition (t) separating them; and (ii) no two transitions (t_1, t_2) can appear in sequence without a place (p) separating them.

Example 3. In this example, we show how investor process for trading securities shown in Figure 2 can be mapped into WFA-net shown in Figure 3. The investor aims to achieve the top-level goal G1, but G1 cannot be considered as a building block, since it is or-decomposed into G1.1 and G1.2. Thus, instead of G1 we have G1.1 and G1.2 that can be mapped as parallel activities into WFA-net. G1.2 can be mapped into T3 activity, while G1.1 is and-decomposed into G1.1.1 and G1.1.2, which can be mapped into two sequential transactions. Moreover, G1.1.1 is also and-decomposed into G1.1.1.1 and G1.1.1.2, which can be mapped into two sequential transactions T12 and T15 respectively. While G1.1.2 is or-decomposed into G1.1.2.1 and G1.1.2.2, and they can be mapped into two parallel transactions T11 and T10 respectively.

Furthermore, transaction T10 needs to read trader suggestion information, and transaction T11 needs to read consultant suggestion information. Since no information is allowed to exist without its source, we add T8 and T9 transactions that produce trader suggestion and consultant suggestion respectively. However, T8 requires to read securities analysis that is produced by trader, which can be produced either by T4 or T5. Similarly, T9 requires to read securities analysis that is produced by consultant, which can be produced either by T6 or T7.

Moreover, T4 and T6 need to read securities assessment. Thus, T1 is added since it is responsible of producing such information. Similarly, T5 and T7 need to read financial statement information. Thus, T2 is added since it is responsible of producing such information. At the other hand, T15 needs to read trade settlement information, thus, T13-14 are also added. Finally, following the refinement constraints, we add some position between transactions when required.

Table 1. Properties of the design

Pro1	:- position(end), not reached(end)
Pro2	:- read(A, I), information(I, T), not has(A, I, T)
Pro3	:- need_perm(P, A, I), not has_perm(P, A, I)
Pro4	:- dele_perm(P, A, B, I), not has_perm(P, A, I)
Pro5	:- has_perm(P, B, I), owner(A, I), not trust_perm_chain(P, A, B, I)
Pro6	:- read(G, I), not fits_read(G, I)
Pro7	:- send(T, G, B, I), not fits_send(T, G, B, I)

4.3 Analysis Phase

In this section, we describe the automated reasoning support that our approach proposes to guarantee the correctness and consistency of information-flow, IQ requirements, and control-flow of BPs. In order to verify the correctness and consistency of BP model, we provide a Datalog [24] formalization of all the concepts that have been introduced in the paper, along with the reasoning axioms[4]. Moreover, we define a set of properties of the design (shown in Table 1) that can be used to verify the correctness and consistency of the BP model; in what follows we discuss each of these properties:

Pro1 states that the end position in BP should be reached. If this property holds, the BP reaches its end configuration, which verifies the correctness of the BP information-flow, IQ requirements, and control-flow. We can rely on this property to quickly verify the correctness and consistency of the BP.

Pro2 states that actors should have all information that is required for the transitions they are responsible of, i.e., it is used to verify information-flow related issues. For instance, consider transition T4 in Figure 3, if trader did not have security analysis information (e.g., it is not created), the analysis will detect such situation and notify the analyst about it.

Pro3-5 are used to verify information permissions related properties. For instance, **Pro3** states that actors should have all permissions they require to achieve the transition they are responsible of. **Pro4** states that actors cannot delegate permissions they do not have. While **Pro5** states that BP should not include any actor who have permissions, and there is no trust/ trust chain between the actor and information owner concerning such permissions. This property enables information owners to guarantee that actors will not misuse permissions concerning information they own. For instance, consider transition T14 in Figure 3, if stock market does not have (R)ead permission concerning trading orders, or (P)roduce permission concerning trading settlement, the analysis will detect and notify the analyst about such situation. While if stock market has (M)odify permission concerning investor's orders, and no trust relation hold between the market and the investor (information owner), the analysis will notify the analyst to solve such situation.

---

[4] The formalization of the concepts and axioms is omitted due to space limitation, yet they can be found at https://mohamadgharib.wordpress.com/iqbp/

Pro6-7 are used to verify IQ related properties in the BP[5]. For instance, **Pro6** states that the model should not include any information that does not fits for the purpose of read (appropriate for read), where information should be accessible, accurate, and valid to be considered appropriate for read. In particular, *Pro6* is able to detect: (1) information accessibility: information is inaccessible to an actor, if the actor does not has read permissions; (2) information accuracy: information is inaccurate, if it is produced with no permissions, or modified intentionally/ intentionally during its transfer; (3) information timeliness (validity), where timeliness can be analyzed depending on information *currency* that is the time interval between information creation and its usage time [16], and information *volatility* that is the change rate of information value [16], i.e., information is not valid, if its currency is bigger than its volatility interval, otherwise it is valid. While **Pro7** states that the model should not include any information that does not fits for send, where information should be accurate and valid at its intended destination to be considered appropriate for send, where information is accurate at its destination, if it was transferred through IP provision that guarantee its integrity, and it is valid if its transmission time is less than the required send time.

5 Prototype Implementation

Our proposed approach belongs to the design area. Thus, it can be evaluated by simulation method (experimental) [25], i.e., developing a prototype and test its applicability with artificial data. To this end, we developed a prototype implementation of our approach[6] to test its applicability, i.e., test its ability for modeling and analyzing IQ requirements in BPs. In what follows, we briefly describe the prototype, and then discuss its applicability over scenarios abstracted from the Flash Crash case study[7].

Prototype implementation: our prototype has been developed depending on Eclipse integrated development environment (IDE), and it consist of 3 parts: (1) A graphical user interface (GUI)[8]: that support designers while designing BPs. In particular, it enable designer to model the overall system where the BP occur, and then map the requirements model into WFA-nets by drag-and-drop modeling elements from palettes; (2) Model-to-text transformation: supports the transformation of the graphical BP model into Datalog formal specifications depending on Acceleo[9]; and (3) automated reasoning support (DLV system[10]) that takes the Datalog specifications as an input, and then perform the required analysis that help to verifies the correctness and completeness of the BP against the properties of the design.

---

[5] Produce/ Modify related issues can be addressed by permissions properties (Pro3-5)
[6] The prototype tool is available at https://mohamadgharib.wordpress.com/iqbp/
[7] For more information about the case study refer to [26]
[8] Developed by Sirius https://projects.eclipse.org/projects/modeling.sirius
[9] https://projects.eclipse.org/projects/modeling.m2t.acceleo
[10] http://www.dlvsystem.com/dlv/

Applicability: we tested the applicability of our approach by applying it to several scenarios abstracted from the Flash Crash case study. In particular, we modeled several BPs, each of them violates one or more of the properties of the design, and we transformed these BP models into Datalog specification, and then we run the automated analysis to test the analysis ability in discovering these violations to the properties of the design. The analysis was able to detect and notify the analyst about all of these violations.

6 Related Work

Traditionally, BP literature has focused on the control-flow perspective of the processes with less emphasis on information perspective. However, in recent years some efforts have been devoted to data-aware process design. For example, Trcka et al. [27] introduced data anti-patterns that represent undesirable data-flow behaviors in BPs, while [4] proposed WFD-nets that is able to address data-flow issues along with the control-flow of a BP. Moreover, data-related process analysis methods have also been proposed by [3]. Furthermore, Deutsch et al. [28] propose *TNest* that is a data-centric workflow modeling language, which allows for expressing data dependencies along with time constraints. However, all these approaches do not specifically consider IQ related issues.

At the other hand, combining Goal models and BPs is not new, for example, Cysneiros and Yu [29] discuss agents autonomy in modeling and supporting business processes (BPMN). While Koliadis et al. [11] propose a preliminary work for mapping i^* to BPMN. Lapouchnian et al. [30] propose requirements-driven approach for BP design that uses requirements goal models to capture alternatives in process configuration. Still, to the best of our knowledge, there is no previous work in GORE that considers and map IQ requirements to BPs.

Several approaches for improving IQ by design have been proposed. For instance, Wang [31] proposes the Total Data Quality Management (TDQM) methodology for delivering high quality information products (IP) to information consumers. Furthermore, Ballou et al. [32] presented an information manufacturing system that can be used to determine the data quality in terms of timeliness, quality and cost. Moreover, Shankaranarayanan et al. [33] extend Ballou's work, and propose a formal modeling method for creating an IP-MAP.

7 Conclusions and Future Work

In this paper, we discussed the importance of capturing IQ requirements in BPs from the early design phase. Moreover, we introduced a goal-oriented approach to model and analyze IQ requirements in BPs from a socio-technical perspective. In particular, our approach is based on an extended version of secure Tropos that is able to model and analyze IQ requirements in their social and organizational context, and then map these requirements into workflow net with actors (WFA-net). Moreover, we provide detailed execution semantics for the WFA-nets, which enable to capture design flaws related to control-flow, information-flow, and IQ requirements. We illustrated the applicability of our framework by an example concerning a U.S stock market crash. For the future work, we intend to extend the IQ dimensions we considered, and we believe that the different interrelations among IQ dimensions need to be studied in more details. . Finally, we aim to better validate our approach by applying it to more complex case studies that belong to different domains.

Acknowledgments. This research was partially supported by the ERC advanced grant 267856, "Lucretius: Foundations for Software Evolution", http://www.lucretius.eu/.

References

1. Aguilar-Saven, R.S.: Business process modelling: Review and framework. International Journal of Production Economics **90**(2), 129–149 (2004)
2. Trčka, N., van der Aalst, W.M.P., Sidorova, N.: Data-flow anti-patterns: discovering data-flow errors in workflows. In: van Eck, P., Gordijn, J., Wieringa, R. (eds.) CAiSE 2009. LNCS, vol. 5565, pp. 425–439. Springer, Heidelberg (2009)
3. Sadiq, S., Orlowska, M., Sadiq, W., Foulger, C.: Data flow and validation in workflow modelling. In: Proceedings of the 15th Australasian database conference, vol. 27, pp. 207–214. Australian Computer Society, Inc. (2004)
4. Sidorova, N., Stahl, C., Trčka, N.: Workflow soundness revisited: checking correctness in the presence of data while staying conceptual. In: Pernici, B. (ed.) CAiSE 2010. LNCS, vol. 6051, pp. 530–544. Springer, Heidelberg (2010)
5. Redman, T.: Improve data quality for competitive advantage. Sloan Management Review **36**, 99–99 (1995)
6. Hamming, R.: Error detecting and error correcting codes. Bell System Technical Journal **29**(2), 147–160 (1950)
7. Yu, E., Mylopoulos, J.: From er to armodelling strategic actor relationships for business process reengineering. In: Entity-Relationship Approach ER 1994 Business Modelling and Re-Engineering, pp. 548–565 (1994)
8. Emery, F., Trist, E.: Socio-technical systems. management sciences, models and techniques. churchman cw et al. (1960)
9. Fisher, C., Kingma, B.: Criticality of data quality as exemplified in two disasters. Information & Management **39**(2), 109–116 (2001)
10. Sommerville, I., Cliff, D., Calinescu, R., Keen, J., Kelly, T., Kwiatkowska, M., Mcdermid, J., Paige, R.: Large-scale complex it systems. Communications of the ACM **55**(7), 71–77 (2012)
11. Koliadis, G., Vranesevic, A., Bhuiyan, M., Krishna, A., Ghose, A.: Combined approach for supporting the business process model lifecycle. In: PACIS, p. 76. Citeseer (2006)
12. Dardenne, A., Van Lamsweerde, A., Fickas, S.: Goal-directed requirements acquisition. Science of Computer Programming **20**(1-2), 3–50 (1993)
13. Yu, E.S.K.: Modelling strategic relationships for process reengineering. PhD thesis, University of Toronto (1995)
14. Mouratidis, H., Giorgini, P.: Secure tropos: A security-oriented extension of the tropos methodology. International Journal of Software Engineering and Knowledge Engineering **17**(2), 285–309 (2007)

15. Gharib, M., Giorgini, P.: Modeling and reasoning about information quality requirements. In: Fricker, S.A., Schneider, K. (eds.) REFSQ 2015. LNCS, vol. 9013, pp. 49–64. Springer, Heidelberg (2015)
16. Pipino, L.L., Lee, Y.W., Wang, R.Y.: Data quality assessment. Communications of the ACM **45**(4), 211–218 (2002)
17. Bovee, M., Srivastava, R.P., Mak, B.: A conceptual framework and belief-function approach to assessing overall information quality. International Journal of Intelligent Systems **18**(1), 51–74 (2003)
18. Wand, Y., Wang, R.: Anchoring data quality dimensions in ontological foundations. Communications of the ACM **39**(11), 86–95 (1996)
19. Murata, T.: Petri nets: Properties, analysis and applications. Proceedings of the IEEE **77**(4), 541–580 (1989)
20. van der Aalst, W.M.: The application of petri nets to workflow management. Journal of Circuits, Systems, and Computers **8**(01), 21–66 (1998)
21. Kirilenko, A., Kyle, A.S., Samadi, M., Tuzun, T.: The flash crash: The impact of high frequency trading on an electronic market. Manuscript, U of Maryland (2011)
22. Mishkin, F.S.: Policy remedies for conflicts of interest in the financial system. In: Macroeconomics, Monetary Policy and Financial Stability Conference. Citeseer (2003)
23. Gharib, M., Giorgini, P.: Analysing information integrity requirements in safety critical systems. In: The 3rd International Workshop on Information Systems Security Engineering WISSE 2013 (2013)
24. Abiteboul, S., Hull, R., Vianu, V.: Foundations of databases. Citeseer (1995)
25. Hevner, A.R., March, S.T., Park, J., Ram, S.: Design science in information systems research. MIS Quarterly **28**(1), 75–105 (2004)
26. Gharib, M., Giorgini, P.: Detecting conflicts in information quality requirements: the may 6, 2010 flash crash. Technical report, Università degli studi di Trento (2014)
27. Trcka, N., van der Aalst, W., Sidorova, N.: Analyzing control-flow and data-flow in workflow processes in a unified way. Computer Science Report (08–31) (2008)
28. Deutsch, A., Hull, R., Patrizi, F., Vianu, V.: Automatic verification of data-centric business processes. In: Proceedings of the 12th International Conference on Database Theory, pp. 252–267. ACM (2009)
29. Cysneiros, L.M., Yu, E.: Addressing agent autonomy in business process management-with case studies on the patient discharge process. In: Proc. 2004 IRMA Conference (2004)
30. Lapouchnian, A., Yu, Y., Mylopoulos, J.: Requirements-driven design and configuration management of business processes. In: Alonso, G., Dadam, P., Rosemann, M. (eds.) BPM 2007. LNCS, vol. 4714, pp. 246–261. Springer, Heidelberg (2007)
31. Wang, R.: A product perspective on total data quality management. Communications of the ACM **41**(2), 58–65 (1998)
32. Ballou, D., Wang, R., Pazer, H., Tayi, G.K.: Modeling information manufacturing systems to determine information product quality. Management Science **44**(4), 462–484 (1998)
33. Shankaranarayanan, G., Wang, R., Ziad, M.: Ip-map: representing the manufacture of an information product. In: Proceedings of the 2000 Conference on Information Quality, pp. 1–16 (2000)

From Secure Business Process Models to Secure Artifact-Centric Specifications

Mattia Salnitri[1](✉), Achim D. Brucker[2], and Paolo Giorgini[1]

[1] University of Trento, Trento, Italy
{mattia.salnitri,paolo.giorgini}@unitn.it
[2] SAP SE, Karlsruhe, Germany
achim.brucker@sap.com

Abstract. Making today's systems secure is an extremely difficult and challenging problem. Socio and technical issues interplay and contribute in creating vulnerabilities that cannot be easily prevented without a comprehensive engineering method. This paper presents a novel approach to support process-aware secure systems modeling and automated generation of secure artifact-centric implementations. It combines social and technical perspectives in developing secure systems. This work is the result of an academic and industrial collaboration, where SecBPMN2, a research prototype, has been integrated with SAP River, an industrial artifact-centric language.

1 Introduction

Today's systems are more and more similar to complex organizations, where autonomous and independent components interact one another to achieve common and local objectives. An air traffic management system is, for instance, composed of several autonomous elements, such as the communication service provider network, the tower control, the meteorological services provider, the Very High Frequency (VHF) network, the ground management system, and so forth. Some of them can be considered as pure technical components (e.g., satellite communication network or the aircraft router) while others are human/social elements (e.g., the controllers in the control tower, or airport rescue team). In other words, socio and technical elements are components of the same Socio-Technical System (STS) where they interact as autonomous elements.

STSs can easily become complex and hard to control systems, where human factors may introduce an high level of unpredictability. To regulate the system's interactions, process modeling languages are commonly used to design the flow of activities and to prescribe roles and responsibilities. Business Process Management and Notation (BPMN) 2.0 [1] and Business Process Execution Language (BPEL) [2] are well known examples of process-centric modeling languages. The design of a STS cannot leave out, however, the artifacts (entities, data and documents) that are used, consumed and shared within the system. SAP River [3] and Oracle PeopleCode [4] are largely used artifact-centric approaches to model business artifacts and their business logic.

In STS, security is not exclusively a technical problem, very often it is the combination of socio and technical factors that gives origin to the most critical vulnerabilities of a system [5]. To guarantee desirable levels of security, artifact-centric approaches offer, for example, access control and authentication security controls to constraint the access to the data and related executable functions (business logic) [6]. However, the security strategies that are beyond the usage of such security controls, should be consistent with the security choices adopted in the business process model; namely, any security strategy adopted for the STS should be first implemented into the business process and then, as consequence, coded at level of artifacts. For example, in a payment engine (e.g., SAP Payment Engine [7]), security choices of creating a process that maintains the integrity of the invoice or ensuring the confidentiality of the credit card information, should be enforced on related business artifacts (e.g., implementing authentication controls for accessing the credit card data).

The literature offers a number of process-centric languages for modeling security concerns along the activities' flow of a system. SecBPMN [8] and SecureBPMN [9] are two examples of modeling languages where specific annotations are introduced to extend BPMN with security concepts. However, no approach has been proposed so far to handle with security as a global concern across process-centric and artifact-centric dimensions. For example, SAP proposes SAP River [10] and ABAP [11] as artifact-centric languages without any related support for modeling security at process level.

In this paper, we present an approach to deal with security that combines the advantages of business process modeling with the advantages of artifact development. We implemented our approach using SecBPMN2 [12] as a process-centric modeling language to define the business processes and the security choices and SAP River platform for the artifact-based implementation. The overall approach guarantees that the artifact-based implementation complies to the high-level security-aware process specification.

In more detail, our contributions are three-fold: first, we present an integrated approach for modeling and implementing secure process-aware socio-technical systems. Second, we present a mapping from control-flow-centric business process models to artifact-centric implementations that include the translation of security and compliance properties. And, third, we implemented our approach on an industrial platform.

2 Baseline

In this section, we introduce the foundations of our work: the security aware, process-centric modeling language SecBPMN2 and the artifact-centric framework SAP River.

2.1 SecBPMN2

Among various process-centric modeling languages, SecBPMN2 stands out for its expressiveness, and the possibility to model both business processes and

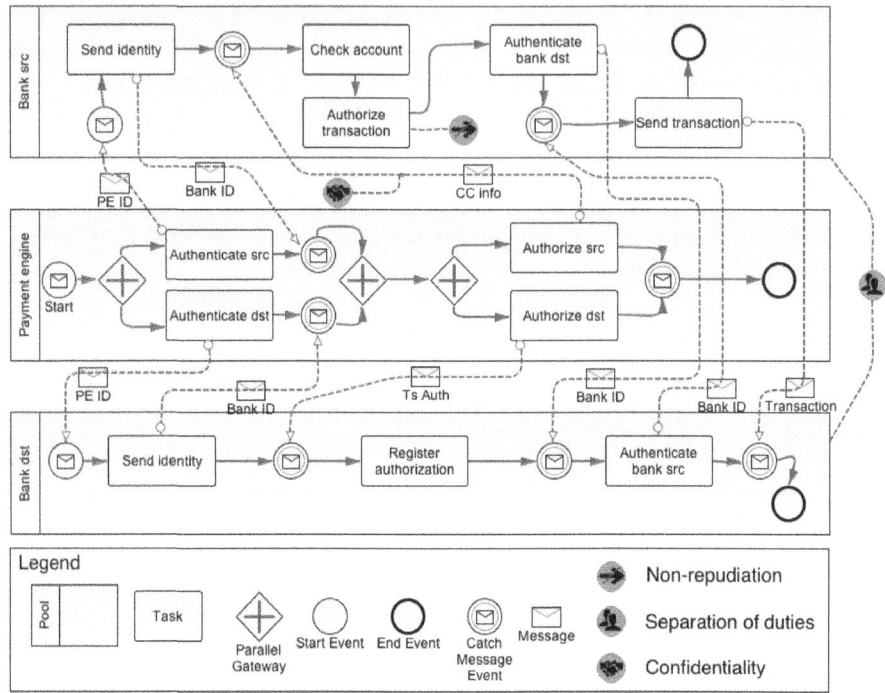

Fig. 1. Example of SecBPMN2-ml model

security policies. It is composed of SecBPMN2 - modelling language (SecBPMN2-ml), a modeling language that extends BPMN 2.0with security concepts, and SecBPMN2-Query (SecBPMN2-Q), a modeling language for security policies. Figure 1 shows an example of SecBPMN2 concerning the SAP Payment Engine (PE). SAP Payment Engine [7] is a flexible single-payment platform aimed for processing payments into one central hub. It can be used as a single entry point where the company orders/receives payments. It interacts with any bank it is required to be connected using the proper interface and security level required by the bank. The process in Figure 1 starts when a money transfer is executed between two banks.

SecBPMN2 extends BPMN 2.0 adding security choices, represented as eleven security annotations: accountability, auditability, authenticity, availability, integrity, privacy, binding of duties, non-delegation, non-repudiation, separation of duties, and confidentiality. In Figure 1, we have: *non-repudiation*, linked with "Authorize transaction", specifies that "Bank src" cannot be able to deny the execution of that task; *separation of duties* specifies that "Bank src" and "Bank dst" cannot be the same bank; *confidentiality*, linked to the message flow that transmits the "CC info", specifies that only the authorized receivers can read the message. More details can be found in [12].

```
1   @OData
2   type LocalDate { date : UTCTimeStamp; state: String; }
3   application PizzaCloud.SalesApp {
4     role Approver;
5     export entity SalesOrder accessible by Approver {
6       key element ID : String;
7       element transactionDate : LocalDate;
8       element items : association[0..*] to SalesOrderItems via backlink order;
9       action approveOrder() { [..] } } }
```

Listing 1. A Simple River Example: Modeling a Sales Order

2.2 Artifact-Centric Business Process Modeling

Well known business-process or workflow modeling languages such as BPMN 2.0 or BPEL are based on activity flows: data that is processed within the processes is often an afterthought. In contrast, artifact-centric business process modeling [10, 13] puts the business artifacts (e.g., data, documents) into the center of the process modeling.

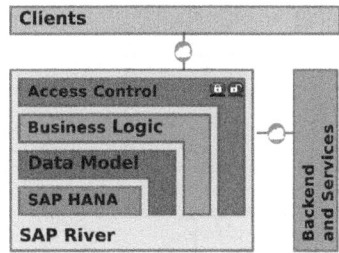

Fig. 2. The SAP River Platform

For our prototype, we use the SAP River, a framework for developing business applications on top of the SAP HANA. Figure 2 shows an high-level overview of SAP River: SAP HANA provides the persistency layer as well as the container in which the enterprise applications are executed. Clients as well as back end systems or external services can communicate with the River platform using standard protocols. The business artifacts (i.e., the data model) and their behavior (i.e., the business logic) as well as the access control are specified in the River Definition Language (RDL) [3].

RDL is an executable specification language that allows to specify declaratively the artifacts (e.g., entities), the relations between them (e.g., associations) as well as the business logic (e.g., actions) on the artifacts. Listing 1 illustrates an excerpt of a SAP River application. Most importantly, RDL allows for specifying entities (e.g., SalesOrder, their attributes (e.g., transactionDate), custom types of the attributes (e.g., LocalDate), and the relation (associations) to other entities. For example, the entity SalesOrder as a bidirectional association (similar to associations in UML [14]) to the entity SaledOrderItems. Besides this pure data modeling, RDL also supports to specify the actions (e.g., approveOrder in a declarative style. Finally, RDL supports to specify role-based access control restrictions: the actions of the entity SalesOrder are only accessible by members of the role Approver.

By default, artifacts in RDL are private. To enable access outside of their scope, they need to explicitly marked with the **export** keyword. Moreover, the annotation @OData enables remote access using the OData protocol (www.odata.org). Such a remote access is controlled by the same access control restrictions of internal access.

3 From Procedural Specification to Business Artifact Specification

Figure 3 provides an overview of our approach: Step 1 consists in defining the procedural specification, where a team composed of domain experts (i.e., the customers and consultants) and security engineers work together to express security needs over the activities' flow of the system. The procedural model is then used to automatically generate a set of River application skeletons (step 2) that will be used by developers to implement the business logic (step 3). Whenever the system changes (e.g., to adapt to new organizational processes or new legislations), a revised version of the SecBPMN2 processes or new specifications for the artifacts can be introduced. Compliance between SecBPMN2 models and artifact specifications has to be checked again and new changes on the process models or artifact specifications are introduced (step 4).

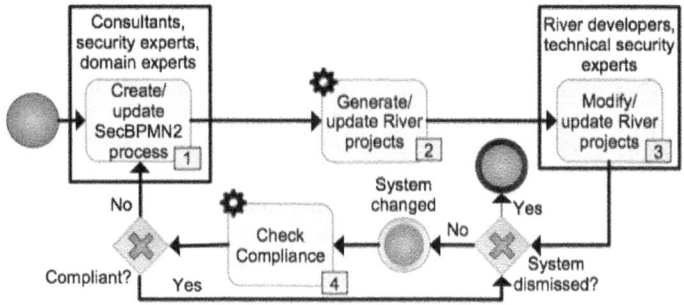

Fig. 3. The process proposed with the approach

The generation of River applications skeleton from SecBPMN2 models requires extra information that are specific of business artifacts and that are not represented in the process models. Particularly, SecBPMN2 data objects are complex data structure with heterogeneous data types in River; for instance, in the procedural specification in Figure 1, the credit card data object is specified only by its name, while in River it will be specified as complex business artifact with information such as the number of the credit card, Card Verification Value (CVV), name of the owner, or the issue date. Companies such as SAP or Oracle have created repositories of templates, that can be reused whenever the same business artifacts are requested. Our approach includes such repositories, so to support the generation of River applications that already incorporate a complete definition of business artifacts and therefore to reduce the amount of work required to River developers to customize the River application once generated. Developers will, then, complete the specification of the business artifacts and their business logic with domain dependent details. For example, in PE, the business logic related to the credit card strictly depends on the context in which the system applies.

Moreover, since River applications can enforce only SecBPMN2 security choices regarding information handled in the process, they cannot be the only enforcement point for all security choices defined in the procedural specification. We include, as an output of the approach, a "security specification" document which contains a list of the security controls that cannot be implemented in River applications.

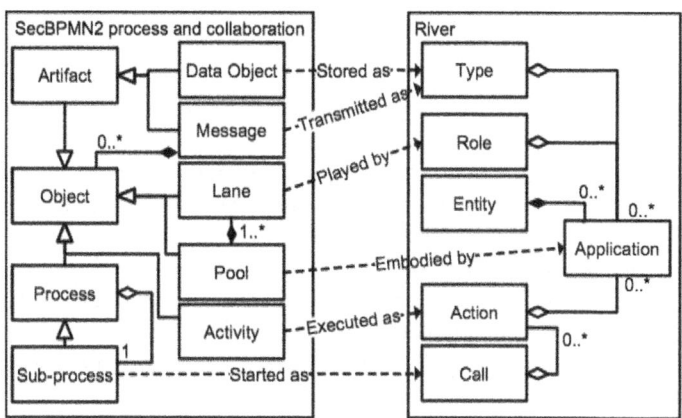

Fig. 4. Part of meta-models of SecBPMN2 and River and the mapping relations

Figure 4 shows the mapping relations between part of SecBPMN2 and River meta-models. The six mapping relations are described below.

Stored as "Data object", which represents a set of information, is "stored as" a River "Type", which represents the structure of the information of an element in entity.

Transmitted as "Message", that represents a set of information sent between pools, is "transmitted as" a River "Type", which represents the structure of a message sent.

Embodied by "Pool", which defines a company or an actor such as a buyer or a manufacturer, is "embodied by" an "Application", that represents a set of business artifacts, which can be accessed only using the APIs, and their business logic. From an artifact-centric perspective, a pool and a River application are use to identify organizations or a well defined parts of them.

Executed as A "Task" represents an operation performed by a participant. "Action" represents the business logic linked to a data structure, i.e., they are the operations executed to set/maintain some properties of the business objects. The operations represented by a task are implemented and executed in a River action.

```
1   application paymentEngine
2   {
3           role controller;
4           role validator;
5           type VCNUM {
6                   element ClientIDs   : Integer;
7                   element PaymentCardType : String;
8                   element CardNumber  : String;
9                   [..]}
10          export entity VCNUMEntity {
11                  key element id: Integer;
12                  element VCNUMData: VCNUMType; }
13          export namespace VCNUMnamespace accessible by sap.hana.All {
14                  export action ValidateCreditCard() {}
15                  export action FilterSensitiveInformation() {} } }
```

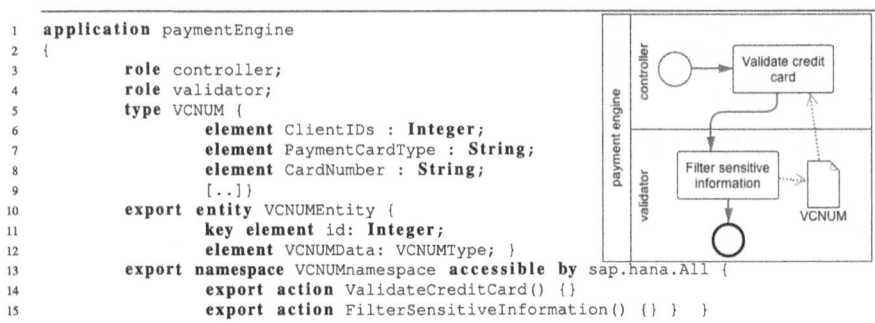

Fig. 5. A SecBPMN2 model representing part of a PE business process

Played by A "Lane" represents a participant, i.e., a person, a service or a set of them. A "Role" represents a set of authorizations assigned to a (set of) physical entity(ies), i.e., the it represents any entity that can receive an authorization. A lane is "played by" a River role because any entity represented by a role can perform all the actions required by the tasks in the lane.

Started as A "Sub-process" is a task that encapsulates a business process, which contains a whole new set of SecBPMN2 elements. Similarly a "Call" is a reference to another River application, which, in turn, contains a whole set of business artifacts. A sub-process is "started as" a call because a call starts a new River application.

The creation of River skeletons is based on generation rules that follow the mapping relations defined in Figure 4. Events and gateways elements are, however, not part of the model transformation since they are used to define the control flow. The main generation rule specifies that a River entity and a River type are generated for each data object and for each message in the SecBPMN2 model. Each entity contains one element of the type generated together with the entity. Each task linked to the data object is transformed in an action, that is placed in an ad-hoc namespace, created for the data object (see Section 5).

Figure 5 shows an example of generation. The name of the application reflects the name of the pool and two roles, which correspond to the lanes in the business process, are specified. From the data object "VCNUM" are generated: (i) a type "VCNUM" that contains the structure of the data that makes tangible the information; (ii) an entity "VCNUMEntity" that contains the actual information; (iii) a namescape "VCNUMnamespace" that contains all the actions derived from the SecBPMN2 tasks linked to the "VCNUM" data object. The structure of "VCNUM" is retrieved from the SAP repository, indeed "VCNUM" is a template for the information related to credit cards.

The fourth step of process described in Figure 3 consists in checking if the security requirements are satisfied in SecBPMN2 models and River applications. The former control can be performed using SecBPMN2 verification engine, while the latter may be perform using software verification techniques.

Table 1. SecBPMN2 security annotations

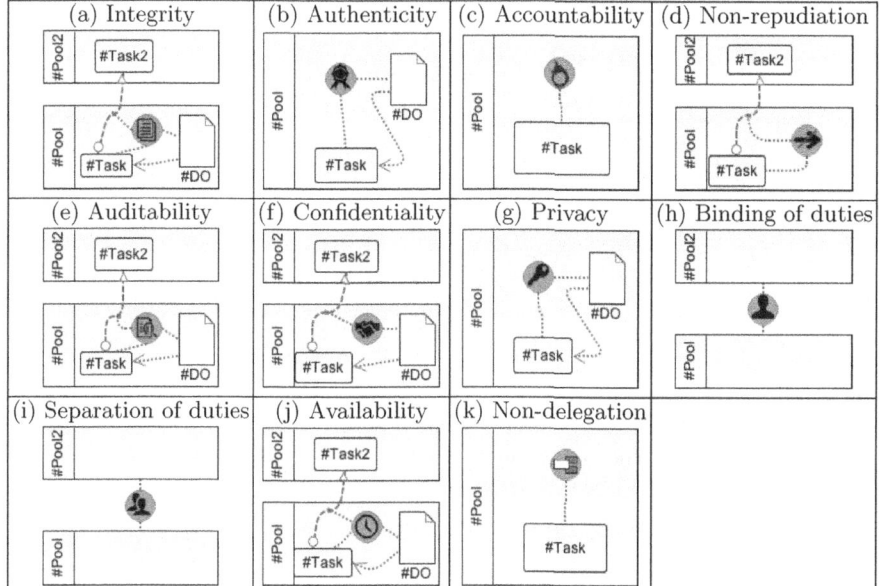

4 Security Enforcement Rules

In this section, we present a set of rules that are used to enforce the security choices of SecBPMN2 into River applications. Table 1 shows SecBPMN2 security annotations. To simplify, security annotations shown in the table are linked to all elements they can be linked to; however, SecBPMN2 allows for only one link (except for separation of duties and binding of duties, that can be linked to two pools). In this paper we report only the enforcement rules for process and collaboration models, while details about choreographies can be found in [12]. For each SecBPMN2 security annotation, we briefly describe in the following its meaning and the corresponding Enforcement Rule (ER).

ER1: Integrity. It requires a system to ensure completeness, accuracy and absence of unauthorized modifications in all its components [8]. It can be linked to one task, data object or message flow (Table 1-a). Although, it can be partially enforced by filtering the users who can access the River entities (i.e. using authentication and access control), backup mechanisms should be used to avoid loosing potentially precious information (when linked to a data object), or loosing functionalities offered by the system (when linked to the message flow or to an activity). Since, such configurations cannot be specified in a River application, they are enlisted in the security specification document.

ER2: Authenticity. It is defined as the ability of a system to verify identity and to establish trust in a third party and in information it provides [8]. It can be connected to one task or data object (Table 1-b). When it is linked to #Task, it

```
1   application #Pool {
2     entity repositorySignatures {
3       signature : String;
4       user : String; }
5     export entity SignatureLogs {
6       element signature: Association to repositorySignatures;
7       element date: UTCTimestamp;
8       element #DO : Association to #DO; }
9     export namespace #DOnamespace accessible by sap.hana.All {
10      export action #Task() {
11        let newSignatureLogs : SignatureLogs = SignatureLogs{
12          date : sap.hana.utils.dateTime.currentUTCTimestamp(),
13          signature : SELECT ONE repositorySignatures FROM repositorySignatures
14                     WHERE user = sap.hana.services.session.getUserName(), #DO : this };
15        Add newSignatureLogs to SignatureLogs; } } }
```

Listing 2. Enforcement of accountability, implementing signature security control

can be enforced with an authenticity security control that verifies the identities of users who execute the action generated from #Task. When it is linked to #DO, an element which contains the hash of the type generated from the #DO will be included in the type itself.

ER3: Accountability. It is defined as the ability of a system to hold users responsible for their actions (e.g., misuse of information) [8]. It can be linked to a task, as shown in Table 1-c. It is enforced using the signature security control, which stores the private key of the user who performs the action generated from to #Task. We used private key to unequivocally identify users that performed the action.

Listing 2 shows part of the River template used to generate the River code for the signature security control. A River template is a piece of River application with placeholders, marked with a "#", that are substituted with an appropriate string. For example, in Listing 2, #Pool will be substituted with the name of the pool in SecBPMN2. The entity in lines 2–4 contains the private keys associated to the users, while the entity defined in lines 5–8 stores the link to the signature of the user who executed the action, the date in which the action is performed and the link to the entity that contains the action performed. Lines 10–15 show how the signature security control is implemented in each action generated from #Task: a new entry is inserted in the entity "SignatureLogs".

ER4: Non-repudiation. It is defined as the ability of a system to prove (with legal validity) occurrence/non-occurrence of an event or participation/non-participation of a party in an event [8]. It can be connected to one activity or one message flow (Table 1-d). We use the signature security control to enforce the non-repudiation security choice. If the security annotation is linked to a message flow, every time send and receive actions are executed, the information about the execution is inserted in the signature entity. If the security annotation is linked to #Task, the information is inserted whenever the action, generated from #Task, is executed.

```
1   application #Pool {
2       type ActionType : enum { READ; WRITTEN; SENT; RECEIVED; }
3       export entity ActionLogs#DO {
4           element date: UTCTimestamp;
5           element actionType : ActionType;
6           element user : String; }              [..]
7       export namespace #DOnamespace accessible by sap.hana.All {
8           export action get#DO(idEntity: Integer) : #Pool.#DOType { [..]
9               let log:ActionLogs#DO = ActionLogs#DO{
10                  user : sap.hana.services.session.getUserName(),
11                  actionType : ActionType.READ,
12                  date : sap.hana.utils.dateTime.currentUTCTimestamp() };
13              log.save(); }
14      export action #Task() { [..] } } }
```

Listing 3. Enforcement of auditability, implementing logging security control

ER5: Auditability. It is defined as the ability of a system to conduct persistent, non-by-passable monitoring of all actions performed by humans or machines within the system [8]. It can be linked to one task, data object or message flow (Table 1-e). It is enforced with the logging security control, which stores information of the actions performed. If the security annotation is linked to #DO, all actions in the entity that contains the type generated from #DO are logged; if it is linked to #Task, only the calls to the action that is generated from #Task are stored; if it is linked to a message flow only the actions send/receive, generated from the message flow, are stored.

Listing 3 shows part of the River template for the logging security control. The type "ActionType" (line 2) defines the type of actions. The entity in lines 3–6 contains: type of the action, date of execution and user who performed the action. Lines 9–13 show how the information about the execution of an action is stored in the entity actionLog#DO. If the security annotation is linked to #DO, information about the execution is inserted in actionLog#DO every time an action, defined in an entity that contains the type generated from #DO, is performed; if the security annotation is linked to #Task then the information is stored every time the action, generated from #Task, is performed; if the security annotation is linked to the message flow, then the information is stored every time the send and receive actions, generated from the message flow, are executed.

ER6: Confidentiality. It requires a system to ensure that only authorized users access information [8]. It is a security annotation that is linked to one message flow or one data object (Table 1-f). We enforced it using authentication, access control and implementing encryption security control.

Listing 4 shows part of the River template for the encryption security control. In lines 5 and 6 the encryption and decryption functions are defined. For the sake of brevity, the algorithms used to encrypt and decrypt data are not shown. Lines 8–11 show how the functions are used to enforce confidentiality: the content of the entity/message is decrypted when retrieved/received and encrypted when is stored/sent. Therefore, the content of entity/message will be visible only in the River application. The encryption/decryption functions are inserted in the send/receive actions when the security annotation is linked to a message flow,

```
1    application #Pool {
2        type #DO {[..]}
3        export entity #DOEntity {
4            [..]
5            action encrypt(data :#DO): #DO { [ENCRYPTION ALGORITHM] }
6            action decrypt(data :#DO): #DO { [DECRYPT ALGORITHM] } }
7        export namespace #DOnamespace accessible by sap.hana.All {
8            export action get#DO(idEnt: Integer) : #DOEntity {
9                let #DOInst: #DOEntity = SELECT * FROM #DOEntity WHERE id == idEnt;
10               #DOInst.#DOData = #DOInst.decrypt(#DOInst.#DOData);
11               return #DOInst.#DOData; } } }
```

Listing 4. Enforcement of confidentiality, implementing encryption security control

while are inserted in getters and setters of the entity which contains the type generated from #DO.

ER7: Privacy. It requires a system to obey privacy legislation and it should enable individuals to control, where feasible, their personal information (user-involvement) [8]. It is linked to one message flow or one data object (Table 1-g). With authentication and access control, we restrict the access to authorized users. We further enforce it, encrypting the content of the entity that contains the type generated from #DO (when the security annotation is connected to #DO), or encrypting the entities sent and received (when the security annotation is linked to a message flow).

ER8: Binding of duties. It requires the same person to be responsible for the completion of a set of tasks [15]. It is linked to two pools (Table 1-h). It is enforced using authentication and access-control, ad-hoc security controls. Listing 5 shows the template for the enforcement of binding of duties. Element "BoDUser" in line 3 contains the first user who accesses the entity and, therefore, the only one that is authorized to access the entity(s) contained in the River applications generated from #Pool and #Pool2. In lines 4–14 the function "CheckBoD" is defined: it checks if the variable "BoDUser" is set locally and in the application generated from #Pool2. If the variable is not set, it sets the variable both local and remotely, otherwise it checks if the user who is executing the action in which the "CheckBoD" method is called, is the same as the one memorized in the variable. The "CheckBoD" method will be called in any action of the entities contained in the applications generated from #Pool and #Pool2 (lines 16–19).

ER9: Separation of duties. It requires two or more different people to be responsible for the completion of a set of tasks [16]. It is linked to two pools (Table 1-i). Static separation of duties [17] is enforced using authentication and access control, while dynamic separation of duties [17] is enforced with authentication, access control and ad-hoc security controls. The template for enforcing dynamic separation of duty is similar to the one for binding of duty (Listing 5).

ER10: Availability. It requires a system to ensure that all its components are available and operational when they are required by authorized users [8].

```
1   application #Pool{ [..]
2     export entity #DOEntity {
3       element BoDUser : String;
4       action checkBoD(): Boolean {
5         if (BodUser is null && getBodUser('#poolURL',id) is null) {
6           BodUser = sap.hana.services.session.getUserName();
7           setBodUser('#poolURL', id, sap.hana.services.session.getUserName());
8           return true; }
9         else
10          if (BodUser == sap.hana.services.session.getUserName() &&
11              getBodUser('#poolURL',id) == sap.hana.services.session.getUserName())
12            return false;
13          else
14            return true; } }
15    export namespace #DOnamespace accessible by  sap.hana.All {
16      export action get#DO(idEnt: Integer) : #DOEntity {
17        let #DOInst: #DOEntity = SELECT * FROM #DOEntity WHERE id == idEnt;
18        if (!#DOInst.checkBoD()) return null;
19        return #DOInst.#DOData; }
20    export action setBoDUser(urlPool: String, idEnt: String, BodUser: String){[..]}
21    export action getBoDUser(urlPool: String, idEnt: String): String{[..]}  } }
```

Listing 5. Implementation of dynamic binding of duties

Table 1-j shows its graphical representation. It cannot be enforced in a River application because it requires configuration of the system (e.g., the configuration of the data-base management system), so the specification of using backup mechanism for #DO will be added to the security specification document.

ER11: Non-delegation. It requires the system to ensure that the actions are performed only by indicated actor(s). It can be linked to one task (Table 1-k). It is enforced using access control: when #Task is transformed in an action in River, it is executed by the roles authorized to access the tenancy/server where the River application is deployed. Once the action is implemented, it will not be anymore delegated to a third party.

5 Implementation and Evaluation

5.1 Prototypical Implementation

Our prototype (available from [12]) tool takes as inputs an XML specification of the SecBPMN2 model, a repository of business artifact definitions and, optionally, a set of enforcement rules. Using the generation rules described earlier, the prototype generates the River skeletons form templates using Freemarker (http://freemarker.org): a Java template library.

Alg. 1 shows the generation of the River skeletons. It follows the generation and enforcement rules described in Section 3 and 4. It uses the function GENERATE that retrieves the Freemarker templates and instantiate them using the information contained in the SecBPMN2 model. For each pool of the SecBPMN2 model (line 1), the algorithm creates a new River application (line 2) and it adds, to the application generated, all roles generated from all lanes contained in the pool (lines 3–5). For each data object, it creates a River type, entity and namespace, and add them to the application (lines 6–10). After that, for each task in the pool, it generates the corresponding action and adds it to the entity(ies) that is(are) generated from the data object that is linked to the task (lines 11–16). The RETRIEVE function checks for this link. If no data object is linked the task, the generated action is added to the application. The last part of the algorithm is for the enforcement of the security annotations: for each security annotation in the pool, GENERATESC instantiates the Freemarker template for the corresponding security control(s) and after that GERNERATESP generates the security specifications that are added to the security specification document.

Algorithm 1.. Algorithm for generation of River applications

GENERATERIVERAPPLICATIONS(*SecBPMN2 model*)
1 **for each** *pool* ∈ *model*
2 **do** *riverApplication* ← GENERATE(*pool*)
3   **for each** *lane* ∈ *pool*
4   **do** *riverRole* ← GENERATE(*lane*)
5   *riverApplication*.ADD(*riverRole*)
6   **for each** *dataObject* ∈ *pool*
7   **do** *riverType, riverEntity, riverNamespace* ← GENERATE(*dataObject*)
8   *riverApplication*.ADD(*riverType*)
9   *riverApplication*.ADD(*riverEntity*)
10   *riverApplication*.ADD(*riverNamespace*)
11   **for each** *task* ∈ *pool*
12   **do** *riverAction* ← GENERATE(*task*)
13   **if** LINKEDDO(*task*)
14   **then** *riverEntity* ← RETRIEVE(*DataObject*)
15   *riverEntity*.ADD(*riverAction*)
16   **else** *riverApplication*.ADD(*riverAction*)
17   *securitySpecificationDoc* ← NEW()
18   **for each** *securityAnnotation* ∈ *pool*
19   **do** *securityMeachanisms* ← GENERATESC(*securityAnnotation*)
20   *riverApplication*.ADD(*securityMechanisms*)
21   *securitySpecifications* ← GENERATESP(*securityAnnotation*)
22   *securitySpecificationDoc* ← ADD(*securitySpecification*)

Due to lack of resources, our prototype is currently limited to public and private process models and collaboration models. We do not foresee any fundamental problem in extending the prototype to support SecBPMN2 choreography models.

5.2 Evaluation and Discussion

To evaluate our approach, we used the framework to generate River applications for the PE system. Following the process proposed in Figure 3, we modeled the business processes and then we generated the River applications, using the software tool we implemented. The choice of SecBPMN2 was appropriate since the modeling language was expressive enough to specify the business processes and the security choices. The generation required no effort, and its execution took only few seconds. We successfully deploy the generated River applications on a River server sandbox. The usage of our approach saved lot of effort and time in the first part of the implementation phase, where River skeletons are defined. While the overall evaluation of the approach was very positive, we also observed several limitations that need to be addressed before a commercialization is possible. In particular, we identified the following restrictions:

Manual written code: we generate skeletons of River applications. Although, we try to minimize the human intervention after the generation of the River applications, we believe that, with the current technologies, is hardly possible to completely remove the intervention of developers after the generation. The price to pay in order to automatically generate complete River applications is to collect all the information required, for example the actual implementation of the business logic, before the generation. This would only move the effort required to developers before the generation and, nevertheless, it would lead to a less intuitive, and less flexible framework.

Choice of security controls: Security constraints can often be fulfilled by different security controls. For example, a confidentiality requirement can be implemented by role-based access control or by encryption based access-control. In our prototype, we decided to limit the choice of security controls to, first, increase the usability, and, second, to be able to ensure the compositionality of the security controls. For applying our approach to further application domains, we would need to guide a security expert in selecting the most suitable security control as part of the generation.

Limitations of implementing "security-by-design": While our framework is designed with "security-by-design" in mind, due to technical and fundamental limitations, it cannot be fully achieved. First, there are security controls that require a run-time configuration (e.g., access control) and, second, security controls that are part of the generated implementation could be modified during the development process. With respect to the first item, we are generating requirements documents that need to be fulfilled while configuring the actual system. With respect to the second item, we would need to integrate consistency checks that ensure that the generated source is not modified during development. Moreover, the generated security controls require that manually developed parts to not violate the security requirements. To ensure this, we envision to implement static source code checks (see[18] for a first work in this area).

Cross-organizational security constraints: Currently, our approach has only very limited support for cross-organizational security requirements such as separation-of-duties across multiple organizations. This is a challenge which is out of scope of our work, as it requires collaborations between the organizations on the overall system level, e.g., by using a federated identify management systems. As soon as such federated security systems are used, our framework will support cross-organizational requirements.

Not all of those restrictions are limiting the wide-spread use of our approach similarly. For example, relying on a "honest developer" is not considered to be a roadblock as the current framework already advanced the state of the art with respect to building secure process-aware systems significantly.

5.3 Related Work

In the area of secure process-aware systems, a variety of BPMN-based approaches have been proposed for modeling security, privacy, and compliance aspects (e.g., [9,19,20]). The BPMN meta-model is extended with new attributes and properties, and different selections of security, privacy, or compliance properties are considered. However, none of these proposals provide support to map BPMN models into artifact centric implementations. The approaches to implement and enforce security properties mainly focus on integrating security control mechanisms (e.g., access control infrastructures) into business process execution engines [21]. Lohmann et al. [22,23] discuss also the integration of compliance aspects into artifact centric business processes.

In the area of mapping or transforming control-flow centric business process specifications to artifact based models, the number of existing proposals is surprisingly small. Estañol et al. [24] present a mapping of BPMN to UML models with OCL constraints. The data model of the target language, i.e., UML/OCL class models, is conceptual very close to the River language. The pure mapping of business process artifacts results very similar to our approach. In contrast to our work, Estañol et al. [24] do not discuss security at all. Moreover, their approach is not supported by an actual implementation.

6 Conclusions and Future Work

To our knowledge, this paper presents the first automated framework for translating security-aware control-flow-centric business-process-models to a secure artifact-centric implementation of process aware systems. While our prototype used SecBPMN2 and SAP River, the underlying approach is generic and can be applied as well to other security-aware business process languages as well as other artifact-centric frameworks and languages. Adapting the approach to a different security-aware business process languages, e.g., SecureBPMN [9], changes the set of supported security properties, which might require the development of new mappings. Adapting the mapping to different artifact-centric frameworks, e.g., ABAP (used by the SAP Business Suite) or PeopleCode (used by Oracle PeopleSoft) that already support access control is easy.

We plan to extend our approach along at three lines of research: (i) automated generation of validation checks to be executed after each update of security-related configurations; (ii) as preliminary discussed in [18], automated check for the implementation validation; (iii) integration with monitoring and process mining frameworks.

Acknowledgments. This research was partially supported by the ERC advanced grant 267856, 'Lucretius: Foundations for Software Evolution', www.lucretius.eu.

References

1. OMG: BPMN 2.0. OMG, January 2011. www.omg.org/spec/BPMN/2.0
2. OASIS: Web Services Business Process Execution Language. OASIS, April 2007. http://www.docs.oasis-open.org/wsbpel/2.0/wsbpel-v2.0.html
3. SAP SE: SAP River Developer Guide. Document Version 1.0, SAP HANA SPS 08 (2014)
4. Doolittle, J.: PeopleSoft Developer's Guide for PeopleTools and PeopleCode. McGraw-Hill Osborne Media (2008)
5. Paja, E., Dalpiaz, F., Giorgini, P.: Managing security requirements conflicts in socio-technical systems. In: Ng, W., Storey, V.C., Trujillo, J.C. (eds.) ER 2013. LNCS, vol. 8217, pp. 270–283. Springer, Heidelberg (2013)
6. Reichert, M., Weber, B.: Enabling Flexibility in Process-Aware Information Systems - Challenges, Methods, Technologies. Springer (2012)
7. SAP SE: SAP Payment Engine Website. www.sap.com/services-support/svc/custom-app-development/cnsltg/prebuilt/payment-engine/ (last visited March 28, 2015)
8. Salnitri, M., Dalpiaz, F., Giorgini, P.: Modeling and verifying security policies in business processes. In: Bider, I., Gaaloul, K., Krogstie, J., Nurcan, S., Proper, H.A., Schmidt, R., Soffer, P. (eds.) BPMDS 2014 and EMMSAD 2014. LNBIP, vol. 175, pp. 200–214. Springer, Heidelberg (2014)
9. Brucker, A.D.: Integrating security aspects into business process models. it - Information Technology **55**(6), 239–246 (2013)
10. Nigam, A., Caswell, N.S.: Business artifacts: an approach to operational specification. IBM Syst. J. **42**(3), 428–445 (2003)
11. Keller, H., Krüger, S.: ABAP Objects. SAP PRESS (2007)
12. SecBPMN Website. www.secbpmn.disi.unitn.it (last visited March 28, 2015)
13. Cohn, D., Hull, R.: Business artifacts: A data-centric approach to modeling business operations and processes. IEEE Data Eng. Bull. **32**(3), 3–9 (2009)
14. OMG: OMG Unified Modeling Language, Infrastructure, V2.1.2 (2007). www.omg.org/spec/UML/2.1.2/Infrastructure/PDF
15. Wainer, J., Barthelmess, P., Kumar, A.: W-RBAC - a workflow security model incorporating controlled overriding of constraints. Int. J. Cooperative Inf. Syst. **12**(4), 455–485 (2003)
16. Simon, R., Zurko, M.: Separation of duty in role-based environments. In: CSFW 1997, pp. 183–194. IEEE Computer Society (1997)
17. Ferraiolo, D., Kuhn, R.: Role-based access control. In: 15th NIST-NCSC National Computer Security Conference, pp. 554–563 (1992)
18. Brucker, A.D., Hang, I.: Secure and compliant implementation of business process-driven systems. In: La Rosa, M., Soffer, P. (eds.) BPM Workshops 2012. LNBIP, vol. 132, pp. 662–674. Springer, Heidelberg (2013)
19. Mülle, J., von Stackelberg, S., Böhm, K.: A security language for BPMN process models. Technical report, University Karlsruhe (KIT) (2011)
20. Rodríguez, A., Fernández-Medina, E., Piattini, M.: A BPMN extension for the modeling of security requirements in business processes. IEICE - Trans. Inf. Syst. **E90–D**, 745–752 (2007)

21. Brucker, A.D., Hang, I., Lückemeyer, G., Ruparel, R.: SecureBPMN: modeling and enforcing access control requirements in business processes. In: Atluri, V., Vaidya, J., Kern, A., Kantarcioglu, M., eds.: SACMAT 2012, pp. 123–126. ACM (2012)
22. Lohmann, N.: Compliance by design for artifact-centric business processes. Information Systems **38**(4), 606–618 (2013)
23. Lohmann, N., Nyolt, M.: Artifact-centric modeling using BPMN. In: Pallis, G., Jmaiel, M., Charfi, A., Graupner, S., Karabulut, Y., Guinea, S., Rosenberg, F., Sheng, Q.Z., Pautasso, C., Ben Mokhtar, S. (eds.) ICSOC 2011 Workshops. LNCS, vol. 7221, pp. 54–65. Springer, Heidelberg (2012)
24. Estañol, M., Queralt, A., Sancho, M.R., Teniente, E.: Artifact-centric business process models in UML. In: La Rosa, M., Soffer, P. (eds.) BPM Workshops 2012. LNBIP, vol. 132, pp. 292–303. Springer, Heidelberg (2013)

New Areas for BPMDS

# PROtEUS: An Integrated System for Process Execution in Cyber-Physical Systems

Ronny Seiger[✉], Steffen Huber, and Thomas Schlegel

Institute of Software and Multimedia Technology, Technische Universität Dresden,
01062 Dresden, Germany
{Ronny.Seiger,Steffen.Huber,Thomas.Schlegel}@tu-dresden.de

Abstract. Cyber-physical systems (CPS) are networks of a large number and variety of components. Sensors, actuators and software create a closed feedback loop between the physical and the cyber world. Current process execution systems mostly focus on web-based workflow applications and are not directly compatible with CPS environments. Especially the processing of low-level sensor events as well as the integration of heterogeneous components–including the users–as process resources are limited in current workflow systems. In this work, we propose PROtEUS: an integrated system for process execution in cyber-physical systems. PROtEUS is able to bridge the gap between hardware-related processes and human-centered workflows. It enables the integration of sensor data from various sources into model-based processes, the flexible access to heterogeneous resources, as well as user interactions in processes. We demonstrate the feasibility of our approach for process execution within data-centric and event-driven CPS in a Smart Home case study.

Keywords: Process execution · Cyber-physical systems · Workflow system · System architecture · Middleware · Event processing

1 Introduction

Cyber-physical systems (CPS) represent an emerging type of distributed system that integrates a multitude of sensors, actuators and software applications into large networks of interconnected components and things (Internet of Things). A closed feedback loop (MAPE-K) exists between local sensing (*Monitor*) and processing on embedded systems (*Analyze*), computing on local or remote cloud-based servers (*Plan*), and controlling local devices and applications (*Execute*), which are able to influence the physical world. That way, the virtual (cyber) and the real world (physical) are interweaved to a new degree [1]. The trend towards ever more intelligent environments (*Smart Spaces*) shows the increasing importance of CPS throughout all areas of life [2].

Processes have been widely used for formalizing sequences of activities in order to facilitate the automation of repetitive tasks. They can be regarded as high-level programs consisting of method calls and additional logic defining the flow of data and activations between the components of a system. The introduction of processes into CPS carries a great potential for automating tasks and creating intelligent environments assisting users with their everyday activities. However, current process execution systems have been designed with a strong focus on web-based workflows involving web services, high-level human activities and organizational processes [3]. From another perspective, processes can be considered as a more formal abstraction of hardware-related programs regarding only a closed set of homogeneous resources [4]. Several properties of CPS impose requirements that are not met by state-of-the-art process/workflow execution systems (e. g., the combination of sensors and actuators including the derivation of high-level workflow events, the integration of heterogeneous components, the varying availability of resources, and the necessity for user interactions).

In this paper, we present an integrated process execution system for cyber-physical systems (PROtEUS). PROtEUS introduces processes into cyber-physical environments by incorporating a model-based execution engine, a complex event processing engine, a service platform and a service caller. In addition, a WebSocket server enables bi-directional access to the execution environment for the purpose of user interaction and process management. With PROtEUS, it is possible to describe and integrate complex event streams and data flow as well as use a wide variety of resources–including the users–in interactive processes. This facilitates the autonomous execution of event-driven processes within dynamic, intelligent environments and enables the loose coupling of CPS components.

The paper is structured as follows: Section 2 presents a scenario process in the Smart Home domain. Section 3 describes the goals and challenges in process execution for CPS that we address with this work. Section 4 introduces our high-level concept for an integrated process execution system and its proof-of-concept implementation in the form of PROtEUS. Section 5 evaluates our approach and results by means of a case study. Section 6 discusses related research with respect to the requirements for process execution identified in section 3. Section 7 concludes the paper and shows starting points for future work.

2 Scenario

A typical application domain for CPS is the domestic environment [5]. *Smart Homes* are equipped with a multitude of sensors for measuring physical properties or more complex data, e. g., with the help of cameras, microphones or infrared sensors. Actuators controlling household applications enable the manipulation of the physical world by means of software. These applications can either be running on local computers and embedded systems (e. g., WiFi routers, Smart TVs and service robots) or they can be executed in the form of Cloud services. Stationary and mobile devices provide access to CPS components for interaction and control. Ambient Assisted Living (AAL) employs this Smart Home technology to support elderly people in living a more self-determined life. In the following, we present a typical process in the AAL domain: Alice is a 73 years old woman living alone in her AAL-enabled apartment. She is wearing a fitness tracker, which senses movement, heart rate and blood pressure. The fitness tracker's data can be accessed and processed by the Smart Home. When Alice gets up from the kitchen table, she suddenly faints due to postural hypotension (sudden drop in blood pressure). The Smart Home detects that Alice's health might be in a critical condition and asks Alice if something is wrong. After a defined time without response, an emergency call is placed automatically. Upon the paramedics' arrival at the apartment, the door is unlocked by the medics to provide Alice with medical assistance.

3 Goals and Challenges

From the Smart Home scenario in Section 2, several goals were identified for a process engine operating within a CPS:

High-Level Event Detection: CPS are highly data and event-driven. The detection of high-level events inside the Smart Home represents the reactive side of the *physical* part within the CPS loop. In the scenario, events allow for detecting Alice's health state and the paramedics' arrival at her door. The main challenge for high-level event detection is the processing of a multitude of data from sensors and other sources. This includes homogenizing sensor readings and other input data and defining patterns that lead to a high-level event.

Events and Data in Processes: The integration of events into process control-flow facilitates the connection between the *physical* and the *cyber* world. In our application scenario, a health warning event triggers a process that eventually leads to an automatic emergency call. Additionally, a door unlocking process is triggered when the paramedics' arrival is detected. Due to the mainly reactive nature of CPS, simple and complex events from a large number of event sources may occur [6]. In order to handle these events in a process, process models need a suitable representation of an event, which the execution system is able to listen and react to. Similarly, the flow of typed data inside a process needs to be formalized in order to process data in high-level process steps and pass data between them (e.g., to evaluate Alice's response).

Heterogeneous Components as Resources: Heterogeneous components (e.g., actuators, services and applications) are used as process resources for the active execution of tasks within CPS. In the scenario, an in-house service is called to ask for Alice's well-being on a tablet device. After the timeout, an external WebService placing an emergency call is invoked. The door lock actuator is controlled by the door unlocking process. The main challenge for integrating heterogeneous CPS components and services as process resources is to coherently define available functionality and execution capabilities. Reactive (event-driven) components have to be combined with active systems, whose availability and capabilities may vary during runtime depending on various context factors.

Human Interaction Support: Sensor readings may be error-prone, resulting in the detection of false or ambiguous high-level events. Due to the close coupling of events and process control flow, we also need to account for falsely triggered processes. User interactions are needed for performing manual tasks, reacting to errors and to provide data in case of uncertainty. In the scenario, a *Human Task* is used to ask for Alice's feedback after the detection of a health warning event. The context-adaptive selection of appropriate modalities for human interaction with and within processes as well as their process model integration represent major challenges in this regard. In addition, users should also be able to control and monitor process execution.

The execution of processes in cyber-physical systems poses new requirements for process-aware information systems as they have to be able to handle novel properties introduced by CPS [1,2]. The aim of this work is to develop a process execution system integrating various components for achieving the goals mentioned above and thereby introducing the concept of processes into CPS. The main challenges are posed by the combination of the concepts and properties of low-level sensors and actuators as key elements of CPS with formalized high-level workflows. On the other hand, users also need to be considered as part of the automated processes involving sensors and actuators. PROtEUS is designed to bridge this gap between hardware-related automatic processes and human-centered workflows, which is prevalent in current process/workflow systems.

4 System Architecture

The process execution system (PROtEUS) consists of several components designed to meet the challenges CPS pose for process execution (cf. Section 3). The core of PROtEUS consists of a meta-model describing the structure of process models and a process engine responsible for instantiating these models and executing actual process instances. Access to the engine's process control functionality is achieved by the process manager. A complex event processing (CEP) engine is able to process large amounts of external sensor data and trigger high-level events within process instances. PROtEUS also offers a service platform for deploying and calling services during the execution of processes as well as an external service invoker. Human interactions are enabled by a human task handler sending interaction requests to interactive clients. The bi-directional communication with the process execution system from remote clients for monitoring, interaction and control purposes is enabled by a WebSocket server. The overall systems architecture is depicted in Figure 1. The following sections explain the individual components and their interactions in more detail.

4.1 Process Engine and Process Manager

The core of the process execution environment is a Petri net-based *Process Engine*. We decided to use a self-developed process engine in order to achieve a better integration of all components necessary for executing CPS processes according to the challenges. In addition, the engine facilitates further developments of advanced features meeting future requirements. The engine instantiates a process model and walks through the process description based on the process

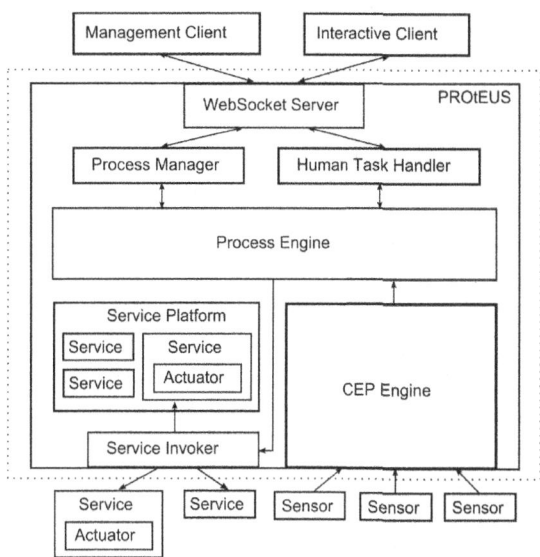

Fig. 1. Overview of the process execution system's (PROtEUS) architecture

meta-model described in [7]. The meta-model follows a component-based approach regarding process steps as atomic or hierarchically composed activities. Typed ports at the start and end of each process step describe its interface, i.e., activations (control flow) or data produced/consumed (data flow) by the process step. Transitions between ports define routes of control and data flow within a process. Through subtype polymorphism, the process engine is able to call a specific execution method according to the type process step to be executed. There exist special types for control flow logic (e.g., for splits, joins and loops), data flow logic (e.g., for mapping, replication and composition) and for the invocation of various types of services. Through specializations of an atomic or a composite process step, the meta-model can easily be extended to support a wide range of process elements. The engine has to be provided with specific attributes and an execution method for each type of process step. Figure 5 shows the graphical representation of the process presented in Section 2. The *Process Manager* component is responsible for managing process models and process instances. The manager provides means for uploading, parametrizing and deploying process models in the execution system as well for controlling process instances. The manager provides access to monitoring information about process models and the states of process instances, which can be queried by clients.

4.2 Complex Event Processing Engine

An important part of PROtEUS is the complex event processing engine. As CPS are characterized by being highly data and event-driven (cf. Section 3), the process execution environment has to be able to process data from sensors and other sources of events. Upon its start, the CEP engine subscribes and listens to a configurable set of event sources. The process meta-model includes a special type of process step allowing the definition of high-level events within a process. In order to define a specific pattern of low-level events that will lead to the triggering of a high-level event, an EPL statement (*Event Processing Language*) can be specified as part of the high-level event's attributes. When executing a process instance, the process engine will call the execution method specific to the *TriggeredEvent* process step, which registers a listener for the EPL pattern at the CEP engine. The CEP engine then analyzes the incoming stream of low-level events looking for registered patterns. Upon recognizing a pattern, the corresponding listeners are informed and the high-level events are activated within the process instance continuing with the execution. New sources and types of events can be added to the CEP engine at runtime.

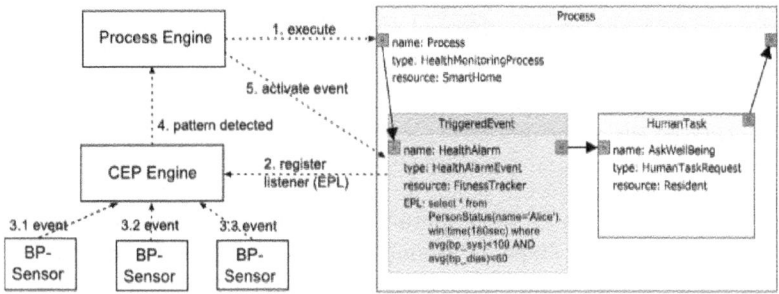

Fig. 2. Communication between process engine and CEP engine during execution

Figure 2 shows the communication between the process engine and the CEP engine during the execution of a process instance. After starting the process (1), a listener for the given EPL statement is registered with the CEP engine (2). The CEP engine triggers a high-level event if Alice's average blood pressure drops below 100 mmHg (systolic) and 60 mmHg (diastolic) over a period of 180 seconds (cf. EPL in Figure 2). Sensors built in the fitness tracker measure her vital signs and publish the respective events to the CEP Engine (3.1–3.3). Upon the detection of the EPL pattern, the CEP engine informs the process engine through the listener (4). The process engine then activates the high-level event *HealthAlarm* in the process and continues with the execution (5), i.e., it activates the subsequent *HumanTask* process step.

4.3 Service Invoker and Service Platform

With web servers and pervasive services also being integrated into embedded systems, it becomes possible to transparently network more and more devices and web-based applications with each other. Therefore, we follow a service-based approach for controlling actuators and invoking remote functionality from within a process. The *Service Invoker* component provides access to external web services via protocol specific adapters. The process meta-model contains specializations of atomic process steps for various service protocols and platforms (SOAP, REST, OSGi[1]) as well as for special purpose solutions. The meta-model enables the extension of these types in order to support further protocols and services. Adapters for communicating with the servers have to be implemented for each new type of service. Within PROtEUS, the functionality provided by actuators and other physical devices or virtual applications is currently encapsulated by one of the service types. Upon reaching a specific service invocation process step, the process engine passes necessary attributes from the process model (e. g., URI, method names, parameters) to the service invoker. The service invoker can be regarded as a middleware, which creates requests according to the specific type of service. The process engine continues after the service invoker received a response from the server and passes the result back to the engine.

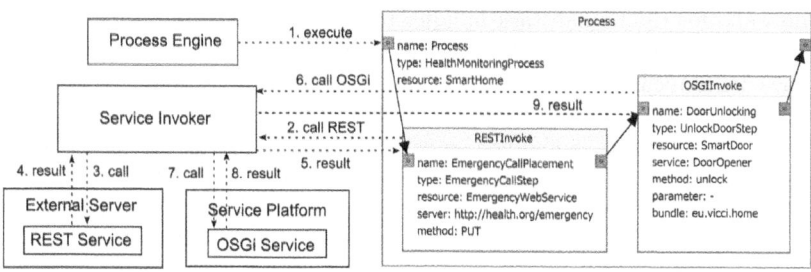

Fig. 3. Message flow during execution of local and remote service invokes

Thus far we are able to call a large variety of services on remote servers from within a process. However, there is a need for deploying local services only accessing in-house devices and applications (e. g., for security reasons). For this reason, we also integrate a *Service Platform* into the process execution system allowing the local operation of self-developed services. The OSGi-based platform allows for the deployment of bundles and services at runtime. In addition, the integrated service registry enables clients to search for services. The service's location and further attributes are specified in the process model. Figure 3 shows the flow of messages during the invocation of a remote REST service (2–5) and of a local OSGi service (6–9) via the service invoker.

4.4 Human Task Handler

The users are an important factor and component within CPS. Automated processes may require user interactions in order to, e. g., solve manual tasks, enter data or handle errors that have occurred. Analogous to the WS-HumanTask

---

[1] http://www.osgi.org/

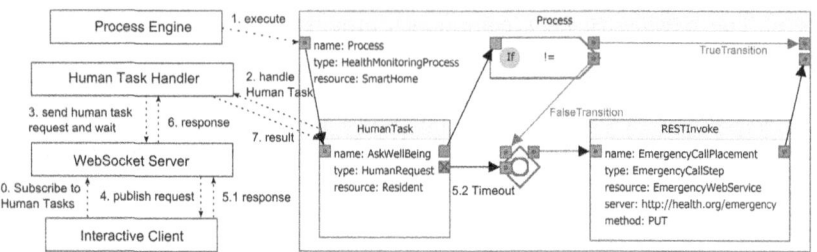

Fig. 4. Message flow during execution of a human task

and BPEL4People concepts known from WS-BPEL, we introduce a special process step called "HumanTask" into the process meta-model. When called by the process engine, the *Human Task Handler* is responsible for sending interaction requests to the interaction devices connected to the execution system. The engine waits for the human task handler to deliver a user response according to the specific human task before it continues execution afterwards.

The communication is done on a publish/subscribe basis using the WebSocket server. Interactive clients capable of handling human tasks subscribe to these requests. The human task handler sends out messages containing the description of the human task to all subscribers. On the client side, a user interface is displayed, which presents the human task step's attributes and parameters. After a response is sent back to the human task handler, the user data is incorporated into the corresponding process instance and execution continues. Figure 4 shows the message flow between the process engine and an interactive client subscribed to human tasks (0–7). With respect to the scenario process, either a user response is received (5.1) OR a timeout is triggered automatically after a predefined period of time (5.2). If Alice confirms her well being, the process finishes. In case she answers negatively or a timeout is triggered, the emergency call is placed.

5 Case Study

In order to provide a proof-of-concept evaluation of our approach, we conducted a case study employing a prototype of PROtEUS in a Smart Home setting. Performance and real-time aspects are important factors of CPS, too. However, the focus of this work is on presenting an initial reference architecture for process execution in cyber-physical systems. We will provide a more elaborate evaluation of the system's performance as part of future work.

5.1 Implementation

PROtEUS is implemented as a Java-based prototype and is currently employed within a Smart Home lab. As the Eclipse Modeling Framework (EMF)[2] provides

---

[2] http://www.eclipse.org/modeling/emf/

a large set of tools for creating meta-models, models and implementations in an automated way, we based the process meta-model and engine on Ecore. The event processing engine Esper[3] is used for realizing the CEP engine. A model-based semantic middleware able to collect and unify sensor data from various sensors serves as the main source of event data for the CEP engine. PROtEUS's service platform is based on the OSGi component platform. Using protocol specific adapters, the service invoker is able to call OSGi services running on the local platform or services on remote servers. We implemented adapters for calling SOAP, REST and XMLRPC services; local Java classes; to invoke robot services on ROS[4]; and to call services on the semantic middleware. To realize the WebSocket server, we use an implementation of the Web Application Messaging Protocol (WAMP)[5]. Clients supporting the processing of human tasks and monitoring of processes are implemented on Android and tabletop devices.

We modeled all steps from the health monitoring scenario (cf. Section 2) as a process according to the meta-model. This process model ready to be executed by PROtEUS is shown in Figure 5. An active instance of the HealthMonitoring-Process listens for a HealthAlarmEvent in the HealthAlarm process step. The EPL pattern modeled in this high-level event defines the activation of the event when a person's average blood pressure drops below 100/60 mmHg for at least 180 seconds, which indicates postural hypotension. Analyzing the sensor readings from the fitness tracker, the CEP engine triggers the event when the pattern appears within the sensor data. The subsequent process step is activated and a HumanTask is sent out to a tablet device asking for the resident's well being. The IF process step defines that if the response is positive, the process terminates; in case of a negative response OR the activation of a timeout (defined as a special type of port), a REST service placing an emergency call is being invoked. Afterwards, a subprocess of type SecureOpeningProcess is instantiated. An EPL statement for sensing the physical world by means of a fingerprint scanner is modeled within a high level event. The event is activated when a person

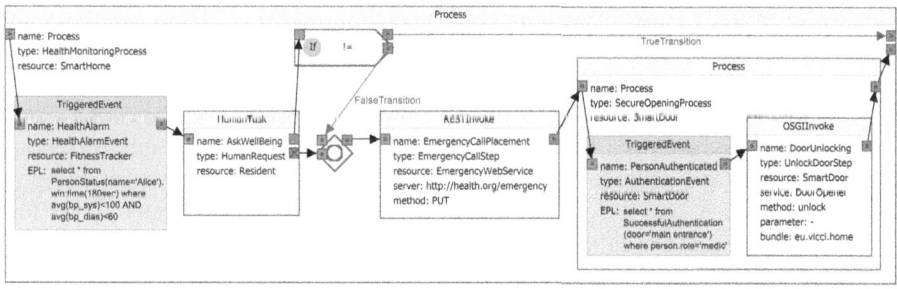

Fig. 5. Health monitoring process from scenario (cf. Section 2)

---

[3] http://esper.codehaus.org/
[4] http://ros.org/
[5] http://wamp.ws/

having the active role of a 'medic' is authenticated successfully at the main entrance door. This leads to the invocation of a local OSGi service within the smart home to unlock the door and the process instance terminates.

5.2 Discussion

With respect to the challenges identified in Section 3, we find that PROtEUS meets the requirements for executing processes in CPS:

High-Level Event Detection: The CEP engine integrated into PROtEUS allows for an efficient processing of large amounts of low-level sensor events (up to 500,000 Events\s) from various sources. EPL statements provide expressive semantics to define temporal and logical dependencies among low-level events that will lead to the activation of high-level events. However, sensor data have to be unified before being injected into the CEP, which employs its own data model. The semantic description of sensor data provides a remedy to this issue as it facilitates the modeling of EPL conditions. As long as PROtEUS is able to convert event data into its data format, arbitrary event sources can be used.

Events and Data in Processes: The introduction of high-level events into the process meta-model enables the combination of reactive behavior and active service calls within processes. As event-driven architectures are the predominant paradigm within CPS, the ability to process and react to simple and complex events is an important requirement for process-enabled systems controlling CPS. The meta-model supports the definition of events and their properties in the form of a specialized process step. Upon reaching an event defined in a process, PROtEUS creates a listener, which will be notified if the low-level event pattern is detected. In this way, hardware-related sensing and high-level processes can be connected. Additionally, the meta-model provides means for formalizing the flow of data between process steps via data ports. We achieve more sophisticated ways of expressing data flow with the introduction of specialized types of process steps for mapping and replication of complex data structures.

Heterogeneous Components as Resources: Integrating various sources and types of event data, the CEP engine provides a unified view on sensors and other event sources. As, with PROtEUS, we follow a service-based approach for invoking functionality, a wide range of devices and applications can be called from within a process instance. The service invoker supports several standardized as well as proprietary protocols and, therefore, acts as a middleware enabling a homogeneous view on CPS components. The underlying meta-model allows for an easy extension of the supported service types. Loading the service parameters from a standardized interface definition (e.g., a WSDL file for SOAP services) simplifies the modeling of a service call. A drawback of following a service-oriented approach for actuators is that their functionality has to be encapsulated in a service deployed on a server. This can introduce an overhead with respect to the effort for implementing and running the service as well as an increased response time. Therefore, we provide a service platform for deploying local services based on OSGi which yields a basic level of security as not all services have to be accessible on remote servers (e. g., the door opener). Further protocols may govern access to ensure security but this is out of scope of this work. When controlling physical objects and machines, a service-based approach may not be feasible for safety-critical CPS. We will therefore investigate the consideration of physical machines as first class citizens (i. e., not encapsulated by a service) within CPS processes as part of our future work. In order to cope with the fluctuation of CPS components, various approaches have been developed for the discovery of resources [8,9] and search for services at runtime [10–12]. The integrated service platform provides lookup functionality for registered services based on process parameters. That way, services can be found and called at runtime according to the properties of the current process step.

Human Interaction Support: The human task handler allows for publish/-subscribe access to interact with processes in the form of high-level human tasks. Clients capable of providing interactions receive requests for human tasks specifying the required data and type of interaction. Responses only have to include the data necessary for the process instance to continue. This leaves the processing of ingoing and outgoing task data completely to the client. Arbitrary interactive clients can receive and respond to human task requests and provide data in case of errors as well as access to monitoring information. Upon state changes within the process engine, monitoring messages are published to all subscribers, which allows for (near) real-time monitoring of process instances. In addition, the process manager and WebSocket server enable the control of the execution environment. Hence, users are provided with extensive means for interacting with CPS processes through various interactive clients. A timeout mechanism for process steps is integrated into PROtEUS to handle unpredictable and safety-critical behavior (e. g., errors, missing responses or missing data).

The process execution system integrates various components for handling the specific properties of processes within CPS beyond the state-of-the-art. PROtEUS discusses a model-driven architecture of a CPS control system based on processes and events, which consists of established concepts and components. Specialized solutions lack the capability of handling the specific CPS properties: the data and event-driven nature, the integration of heterogeneous resources–sensors, actuators, software and humans–and the dynamic availability of resources. The case study discussed in this section shows that PROtEUS and the underlying process meta-model meet these requirements and, therefore, provide a solution for the execution of interactive processes in CPS. PROtEUS is able to bridge the gap between hardware-related processes and high-level human-centered workflows by integrating low-level sensor events and data as well as interactions and human tasks into model-based processes.

6 Related Work

Cyber-physical systems research relates to a multitude of new and established research fields. The challenges for process execution in CPS identified in Section 3 have each been addressed to some extent by various research projects.

Several approaches for the processing and integration of large amounts of event data already exist from the complex event processing (CEP) field. In [13], a programming system providing means for analysis, querying and processing of large sensor systems is presented. Wombacher [14] discusses an approach for correlating sensor and workflow data for observing the physical effects of workflow activities. We primarily focus on using sensor data for situation and context recognition in order to generate high-level events within processes. However, the integration of Wombacher's approach will be a part of our future work to monitor the effects of workflow activities in the physical world and thereby increase the support for the *MAPE-K* loop within CPS. BPMN supports a large set of event types for business processes. As pointed out by Talcott [6], events and their semantics are important aspects of CPS. However, the number and payload of events that need to be processed in CPS is significantly higher than in business processes due to the large number of event sources (sensors). Current process execution environments focus on executing high-level service-based business processes and, therefore, are only able to handle certain amounts of abstract events. In [15], Schiefer et al. present a framework for defining event-triggered rules for sensing and responding to business situations. In [16], Tuysuz et al. discuss a framework for interactive mobile workflows also integrating sensor and user events as well as web-services. This approach enables event-based behavior in workflows on a basic level. As we focus on executing workflows in event-driven CPS, we need more elaborate methods for the processing of a large number of events. Therefore, we integrate the concept of events and CEP on different levels into the process execution environment. Closely related to the handling of events is the formalization and processing of data in processes. In contrast to the limited possibilities for formalizing the flow of data in current workflow languages (e.g., BPMN [17]), we see a strong need for expressing complex flows of typed data within high-level processes inside CPS. Advanced mechanisms enabling the composition of and operation on complex data for formalizing high-level data flow patterns within processes are described by Montagnat et al. [18]. We use these approaches for the handling of data flow between the tasks of high-level CPS processes in a more sophisticated way and to integrate data from sensors and others sources into processes. The workflow language YAWL [19] allows for the definition of typed input and output data within various scopes of tasks of a workflow. We extend this data modeling by mappings, compositions and replications of high-level process data. On the other hand, it is not feasible to define events and data flow for every low-level sensor and actuator of a CPS in a high-level process model. In contrast to workflow languages only enabling the modeling of single data instances, we use queries and pattern matching for the processing of low-level CPS data. YAWL's workflow engine also follows a Petri-net based approach for executing processes. An example of a modern cloud-based process engine is the CPEE [20]. In order to meet all the challenges identified in section 3, extensions to both engines and additional components are necessary. We therefore decided to use a self-developed engine connecting all components for executing CPS workflows as the core component of PROtEUS.

Various service-based approaches for integrating heterogeneous resources into workflows including the aspects of service description [10,11], discovery [21], composition [11,12] and dynamic assignment of resources [9,10] already exist. We also embrace a service-based approach for the integration of actuators and software components into our process execution environment. In [8], the authors propose a workflow system supporting long-running transactions for handling the fluctuation of available resources during process execution. As this work addresses an important issue for CPS–the dynamics of components–we incorporate concepts from it into our own execution system. In current process execution systems, the aspect of human interaction within processes and the access to suitable user interfaces is often realized through web-based client or desktop applications. BPEL4People and WS-HumanTask provide a framework for integrating user interactions into processes on a formal level not considering the practical realization of these interactions. An approach for the introduction of implicit interaction into pervasive workflows is presented in [22]. Regarding service-based business processes, web-based applications are often sufficient for providing necessary data and conducting high-level tasks. However, the increased focus on the users in CPS requires more sophisticated means for context-aware interactions within processes and with the process execution than it is possible with current BPM systems. Our aim is to provide user interfaces to enable realtime interactions and monitoring of processes from arbitrary interactive clients, independent of type, location or modality. As described in [16] and [8] ubiquitous access to workflows is of increasing importance for future human-centered information systems to increase the user experience and usability. Hence, the processes and process execution system have to be accessible from multi-modal user interfaces.

Looking at related work, we find that several aspects have already been solved in order to be applied within the context of process execution in cyber-physical systems. We integrate these existing solutions into our own process execution system. However, the combination of the identified CPS requirements (e. g., the processing of sensor and event data as part of the process execution, the handling of user interactions, and the integration of heterogeneous components) has not been addressed in a satisfying way by state-of-the-art process execution environments. There is a considerable gap between hardware-oriented processes and human-focused workflows. The previous sections presented how PROtEUS addresses these issues by enabling a closer integration of humans and things into a process-aware information system for cyber physical systems.

7 Conclusion

In this paper, we presented PROtEUS–an integrated system for process specification and execution in cyber-physical systems. The execution of processes in CPS poses new challenges that cannot be completely handled by current workflow engines. Especially the data-centric and event-driven nature of CPS resulting from the combination of various low-level sensors, actuators and software components requires process execution systems that are able to integrate a heterogeneous set of resources on an active and reactive basis. For CPS there is a need to handle the dynamics of resource-constraint, loosely coupled devices as well as to support user interactions. Therefore, we developed a comprehensive process execution environment consisting of a core engine for executing model-based processes, a complex event processing engine for the integration and processing of sensor data, and a service invoker for calling on-site or external services. Services can be deployed on a local platform and found at runtime by means of an integrated registry. Remote access for clients is provided through a WebSocket server enabling remote procedure calls and publish/subscribe access for process control, high-level process interaction and monitoring purposes. Processes are based on a component-based meta-model that allows for the composition of process steps and the definition of control and data flow. By combining these components into an integrated execution environment we are able to cope with the challenges of using processes for controlling cyber-physical systems beyond the state-of-the-art. The gap between hardware-related sensing and actuating in processes and human interactions in abstract workflows is reduced by PROtEUS. The system serves as a basis for more advanced mechanisms and future developments regarding loose coupling and late binding of process resources to further the handling of process-based CPS with respect to runtime adaptation.

Regarding future work, we will conduct comprehensive performance studies in order to evaluate the feasibility of our prototype within smart environments. To optimize the usage of available resources and reduce availability issues caused by a centralized approach, we will decentralize parts of the process execution system. The distribution of the process engine and management components across a hierarchical overlay network structure as described in [23] allows for creating scalable and resource-efficient process execution systems that can be applied to large scale systems of systems (e. g., Smart Factories and Smart Cities). We will also introduce semantics for describing properties and relations of processes and resources as well as role-based inference mechanisms for assigning resources [11], finding services and reconfiguring processes at runtime.

References

1. Lee, E.: Cyber physical systems: Design challenges. In: 2008 11th IEEE International Symposium on Object Oriented Real-Time Distributed Computing (ISORC), pp. 363–369 (2008)
2. Broy, M., Cengarle, M.V., Geisberger, E.: Cyber-Physical systems: imminent challenges. In: Calinescu, R., Garlan, D. (eds.) Monterey Workshop 2012. LNCS, vol. 7539, pp. 1–28. Springer, Heidelberg (2012)
3. Scheer, A.-W., Nüttgens, M.: ARIS architecture and reference models for business process management. In: van der Aalst, W.M.P., Desel, J., Oberweis, A. (eds.) Business Process Management. LNCS, vol. 1806, pp. 376–389. Springer, Heidelberg (2000)

4. Cao, J., Jarvis, S.A., Saini, S., Nudd, G.R.: Gridflow: Workflow management for grid computing. In: Proceedings of the 3st International Symposium on Cluster Computing and the Grid, CCGRID 2003, pp. 198–205. IEEE Computer Society, Washington, DC (2003)
5. Shi, J., Wan, J., Yan, H., Suo, H.: A survey of cyber-physical systems. In: 2011 International Conference on Wireless Communications and Signal Processing (WCSP), pp. 1–6, November 2011
6. Talcott, C.: Cyber-Physical systems and events. In: Wirsing, M., Banâtre, J.-P., Hölzl, M., Rauschmayer, A. (eds.) Soft-Ware Intensive Systems. LNCS, vol. 5380, pp. 101–115. Springer, Heidelberg (2008)
7. Seiger, R., Keller, C., Niebling, F., Schlegel, T.: Modelling complex and flexible processes for smart cyber-physical environments. Journal of Computational Science (2014)
8. Montagut, F., Molva, R., Golega, S.T.: The pervasive workflow: A decentralized workflow system supporting long-running transactions. IEEE Transactions on Systems, Man, and Cybernetics, Part C, 319–333 (2008)
9. Kalasapur, S., Kumar, M., Shirazi, B.: Dynamic service composition in pervasive computing. IEEE Transactions on Parallel and Distributed Systems 18(7), 907–918 (2007)
10. Bellur, U., Narendra, N.: Towards service orientation in pervasive computing systems. In: International Conference on Information Technology: Coding and Computing, ITCC 2005, vol. 2, pp. 289–295, April 2005
11. Shen, J., Yang, Y., Yan, J.: A p2p based service flow system with advanced ontology-based service profiles. Advanced Engineering Informatics, 221–229
12. Qian, Z., Wang, Z., Xu, T., Lu, S.: A dynamic service composition schema for pervasive computing. Journal of Intelligent Manufacturing 23(4), 1271–1280 (2012)
13. Jiang, N., Parashar, M.: A programming system for sensor-based scientific applications. Journal of Computational Science 1(4), 206–220 (2010)
14. Wombacher, A.: A-Posteriori detection of sensor infrastructure errors in correlated sensor data and business workflows. In: Rinderle-Ma, S., Toumani, F., Wolf, K. (eds.) BPM 2011. LNCS, vol. 6896, pp. 329–344. Springer, Heidelberg (2011)
15. Schiefer, J., Rozsnyai, S., Rauscher, C., Saurer, G.: Event-driven rules for sensing and responding to business situations. In: Proceedings of the 2007 Inaugural International Conference on Distributed Event-based Systems, DEBS 2007, pp. 198–205. ACM, New York (2007)
16. Tuysuz, G., Avenoglu, B., Eren, P.: A workflow-based mobile guidance framework for managing personal activities. In: 2013 Seventh International Conference on Next Generation Mobile Apps, Services and Technologies (NGMAST), pp. 13–18 (2013)
17. Wohed, P., van der Aalst, W.M.P., Dumas, M., ter Hofstede, A.H.M., Russell, N.: On the suitability of BPMN for business process modelling. In: Dustdar, S., Fiadeiro, J.L., Sheth, A.P. (eds.) BPM 2006. LNCS, vol. 4102, pp. 161–176. Springer, Heidelberg (2006)
18. Montagnat, J., Glatard, T., Lingrand, D.: Data composition patterns in service-based workflows. In: Workshop on Workflows in Support of Large-Scale Science, WORKS 2006, pp. 1–10, June 2006
19. van der Aalst, W., ter Hofstede, A.: Yawl: yet another workflow language. Information Systems 30(4), 245–275 (2005)
20. Mangler, J., Rinderle-Ma, S.: Cpee - cloud process execution engine. In: BPM (Demos) 2014, p. 51 (2014)

21. Montagut, F., Molva, R.: Enabling pervasive execution of workflows. In: 2005 International Conference on Collaborative Computing: Networking, Applications and Worksharing, p. 10 (2005)
22. Giner, P., Cetina, C., Fons, J., Pelechano, V.: Implicit interaction design for pervasive workflows. Personal Ubiquitous Comput. **15**(4), 399–408 (2011)
23. Seiger, R., Niebling, F., Schlegel, T.: A distributed execution environment enabling resilient processes for ubiquitous systems. In: 2014 IEEE International Conference on Pervasive Computing and Communications Workshops (PERCOM Workshops), pp. 220–223, March 2014

Fundamental Issues in Modeling

Applying Predicate Abstraction to Abstract State Machines

Alessandro Bianchi[✉], Sebastiano Pizzutilo, and Gennaro Vessio

Department of Informatics, University of Bari, 70125, Bari, Italy
{alessandro.bianchi,sebastiano.pizzutilo,
gennaro.vessio}@uniba.it

Abstract. Abstract State Machines (ASMs) represent a general model of computation which subsumes all other classic computational models. Since the notion of ASM state naturally captures the classic notion of program state, ASMs are suitable to be verified through a *predicate abstraction* approach. The aim of this paper is to discuss how predicates over ASM states can support the formal verification of ASM-based models. The proposal can overcome the main limitations that penalize traditional model checking techniques applied to ASMs.

1 Introduction

Abstract State Machines (ASMs) represent a general model of computation which subsumes all other classic computational models [27], [12]. Indeed, they suffice to capture the behavior of wide classes of sequential [20], parallel [8] and distributed algorithms [18]. The origin of this generality lies in the notion of ASM *state*. In classic formalisms, such as finite state machines and Turing machines, states are symbolically represented by (sequence of) symbols belonging to finite alphabets [21]. Conversely, ASM states are syntactically and semantically represented by algebraic structures defined over finite signatures. Therefore, ASM states can model any object of arbitrary complexity [20].

Thanks to their generality, ASMs have been successfully applied for modeling critical and complex systems in a wide range of application domains, and for analyzing their computationally interesting properties, both domain-independent (e.g. the *termination* of the execution, *deadlock-* and *starvation-freedom*, and so on) and domain-specific (e.g. security issues, the movement of robotic arms, synchronization issues of real-time controllers, and so on) [9]. However, despite the advantages they provide, the computational power and the arbitrary complexity of the formalism cause an unavoidable drawback: several computationally interesting properties are undecidable, so the formal verification of ASM-based models cannot be fully automated [29].

Traditional model checking approaches to the problem of verifying properties typically model systems with finite state machines (or variants) and express the properties to be verified using some temporal logic [5]. Analogously, when model checking techniques are applied to ASMs, the ASM-based model under study is transformed into the input required by the adopted model checker [14], [3], which is in general less expressive. Therefore, this approach suffers from two main limitations: the loss of expressive power due to the translation of the ASM specifications, and the difficulty in using the declarative notations of temporal logics, considered less comfortable for practitioners with respect to operational specifications of the properties (e.g. [4]).

Our long term research is aimed at providing a theoretical framework for treating properties analysis entirely within the ASM framework, i.e. without translating ASMs, and without using temporal logics. To this end, the present paper takes a step towards the application of a *predicate abstraction* approach for formally analyzing ASMs. Predicate abstraction is a popular and widely used technique for automatizing the verification of programs [19], [11]. It consists in the approximation of the program states into a finite number of predicates defined over these states. In this context, the goal of the present paper is to support the verification of ASM properties through predicates over the states. In this way, the goal of formally verifying ASM models entirely within the ASM framework can be achieved.

The rest of this paper is organized as follows. Section 2 is about related work. Section 3 provides background knowledge about the ASM formalism. Section 4 deals with predicate abstraction for ASMs. Section 5 depicts some illustrative examples. Finally, Section 6 concludes the paper and sketches future developments.

2 Related Work

The ASM formalism allows both manual and automated formal verification of systems. In [9] numerous proofs are provided to illustrate how a modeler can verify properties of a given ASM. These proofs range from simple to complex and, since ASMs are executable machines which lend themselves to traditional inductive proofs, they are often formulated in a mathematical way. For example, in [15] a manual verification calculus based on the Hoare logic is proposed. Conversely, in [14] and [3] the authors use an automatic model checking approach for verifying ASMs. However, the translation of the ASM into the input required by the model checkers may cause a loss of expressive power. Moreover, in all these cases, properties are expressed in some temporal logic [5]. But, the effectiveness of this hybrid approach is not unanimously recognized: several authors emphasize the need of an entirely operational specification of properties within the same formalism, e.g. [22], [4].

ASMs have been successfully used for modeling several systems, often concurrent and distributed, and for investigating their properties. For example, an ASM specification to model concurrency in a Web browser and to prove some consistency properties has been proposed in [17]; ASMs have also been used to model and validate vision-based robot control applications in [25]; and they have been applied for studying Grid systems in [6]. However, all these works are characterized by the lack of a theoretical framework for systematically treating the analyzed properties.

Concerning the application of predicate abstraction to formal methods, other formalisms, such as Petri nets [10], already employs predicates on states whenever properties are to be analyzed. However, Petri nets typically provide only few levels of abstraction, so they are not able to support refinements till to implementation details.

Conversely, the expressive power of ASMs provides a way to describe algorithmic issues in a simple abstract pseudo-code, which can be translated into a high level programming language source code in a quite simple manner [9]. Furthermore, predicate abstraction has been already used within the ASM formalism in [16]; however, the authors use what they call *test predicates* in a way which is different from ours. Indeed, they use predicates on states in order to generate test sequences.

3   Background on ASMs

Abstract State Machines are finite sets of so-called *rules* of the form **if** *condition* **then** *updates* (possibly with the **else** clause) which transform *abstract* states [9]. An ASM state is an algebraic structure, i.e. a domain of objects with functions and relations defined on them. Partial functions are turned into total functions by using the special value *undef*. Moreover, without loss of generality, relations are treated as particular functions that evaluate to *true* or *false*. On the other hand, the concept of rule reflects the notion of transition occurring in traditional transition systems: *condition* is a first-order formula whose interpretation can be *true* or *false*, whereas *updates* is a finite set of assignments of the form $f(t_1, ..., t_n) := t$, whose execution consists in changing in parallel the value of the specified functions to the indicated value.

Pairs of function names, fixed by a signature, together with values for their arguments are called *locations*: they abstract the notion of memory unit. Therefore, a state can be viewed as a function that maps locations to their values: the current configuration of locations together with their values determines the current state of the ASM. As usual in computational models, an ASM *step* is a pair (s, s') of states: in a given state, all conditions are checked, so that all updates in rules whose conditions evaluate to *true* are simultaneously executed, and the result is a transition of the machine from that state to another. Note that for the unambiguous determination of the next state, updates must be *consistent*, i.e. no pair of updates must refer to the same location.

A generalization of basic ASMs is represented by Distributed ASMs (DASMs) [9], [18], capable to capture the formalization of multiple agents acting in a distributed environment. Essentially, a DASM is intended as an arbitrary but finite number of independent agents, each executing its own underlying ASM. In a DASM the keyword **self** is used for supporting the relation between local and global states and for denoting the specific agent which is executing a rule.

4   Predicates Over ASM States

Classic computational models, such as finite state machines and Turing machines, represent the current state of the computation with (sequence of) symbols belonging to finite alphabets [21]. This poses a limitation: the representation of states is restricted to a specific data structure. Instead, as explained in Section 3, ASMs allow any algebraic structure to serve as representation of states. This results in a great amount of details specifying the states, so making the analysis of the properties of the whole system more difficult, mainly for what concerns the comprehension of the semantics of each state, with respect to the computational behavior of the modeled system.

For better explaining this issue, the next section will elaborate two examples, both concerning distributed systems: the analysis of *starvation-freedom* [2] and *deadlock-freedom* [28], and the analysis of the computation of an agent capable to play two or more roles at the same time. In the former case, the simple execution of one or more updates does not necessarily involve the change of the locations values in such a way that the process makes real computational progress, so driving to starvation. In fact, an ASM could starve even if the computation continues to evolve through different states. In other words, it is difficult to recognize effective progress. As an extreme case, this computational behavior can lead to deadlock. On the other hand, the second case concerns, for example, the case of a process that acts both as a client with respect to a service, and, simultaneously, as a server with respect to another service. In this case, ASMs easily capture in a same state different computational activities to be run in parallel. However, it is difficult for the modeler to distinguish, inside the same state, what computational branches have been entered or not.

In order to overcome these problems, the need of an abstraction framework capable to capture the semantics of the ASM states arises. More precisely, there is the need to partition the set of locations into subsets and extract from them the locations specifically interesting for the verification purposes. To this end, we apply a predicate abstraction approach. Predicate abstraction is a popular and widely used technique proposed to analyze programs [19], [11]. It aims at generating an abstract model from the concrete system to be verified, so checking the former instead of the latter. Briefly speaking, the system states are mapped to model states according to their evaluation with respect to a finite set of predicates defined over the system states. The model has the same control flow of the original program but it concerns only the predicates over the states. The model can then be used in the place of the original program when performing model checking, theorem proving, or other kinds of verification techniques.

Literature agrees that a program state coincides with the configuration of program variables and their current values, e.g. [24]. Analogously, an ASM state coincides with the configuration of ASM locations and values. So, since there exists a natural parallelism between classic program states and ASM states, predicate abstraction can be applied to ASMs as much as to programs of traditional programming languages. In this context we can apply predicate abstraction to ASMs through the following:

Definition. *A predicate ϕ over an ASM state s is a first-order formula defined over the locations in s, such that $s \vDash \phi$.*

Predicates over the states serve to represent the semantics of each state, i.e. the properties locally satisfied, and can be regarded as a non-injective labeling function that maps predicates to each state. An ASM model can then be equipped with a set of predicates $\Phi = \{\phi_1, ..., \phi_n\}$, such that, in the current state, each ϕ_i can be satisfied or not. In this way, the ASM control flow can be represented by the truth value of the predicates over the states, i.e. by composing the local properties of the various states. So, global properties to be verified can be analyzed by focusing on this composition.

Note that our use of predicate abstraction is quite different with respect to the traditional way: instead of extracting abstract models from the ASMs to be verified, our aim is to use predicates over the states in order to support the verification of ASM models. In particular, applying predicate abstraction to ASMs induces the partition of locations we need for expressing the semantics of the states.

5 Two Examples

In order to show the application of predicate abstraction over ASM states two examples are discussed: the Dining Philosophers problem, and the Ad-hoc On-demand Distance Vector (AODV) routing protocol for Mobile Ad-hoc NETworks (MANETs). The first example deals with starvation-freedom and deadlock-freedom analysis: here the same value for a predicate holds for several states. The second example discusses the case of several predicates holding over the same state in the context of an agent playing several different roles within the same system.

5.1 Dining Philosophers

The Dining Philosophers problem, due to Dijkstra [13], is a well-known metaphor for discussing concurrent processes. Five philosophers are sitting around a table with a bowl of spaghetti in the middle. For them, life consists only of two moments: thinking and eating, rigorously using two forks. Since each philosopher has a right fork and a left fork, (s)he thinks till both forks become available, eats for a certain amount of time, then stops eating (putting back both forks on the table), and starts thinking again. The problem is that in between two neighboring philosophers there is only one fork: each one shares a fork with a neighbor. The ASM-based model of Dining Philosophers is in [9]: it is a DASM with a set of *philosophers* = $\{p_1, ..., p_5\}$, i.e. the agents of the system, and a set of *forks* = $\{f_1, ..., f_5\}$, i.e. their shared resources. The computation evolves through the states characterized by the following predicates:

- thinking: $\neg(owner(rightFork(\textbf{self})) = \textbf{self} \vee owner(leftFork(\textbf{self})) = \textbf{self})$. The philosopher is thinking, so (s)he is waiting for both forks to become available;
- eating: $owner(rightFork(\textbf{self})) = \textbf{self} \wedge owner(leftFork(\textbf{self})) = \textbf{self}$. The philosopher is eating, so (s)he has obtained both forks,

  where:

- *rightFork*: *philosophers* → *forks* indicates a philosopher's right fork;
- *leftFork*: *philosophers* → *forks* indicates a philosopher's left fork;
- *owner*: *forks* → *philosophers* denotes the current user of a fork.

Initially, each philosopher p_i thinks, and has fork f_i on the right and fork f_{i-1} on the left, except for p_1 that has fork f_5 on the left. The ASM program for p_i is shown below:

PhilosopherProgram(p_i) =
    if $owner(rightFork(\textbf{self})) = undef \wedge ower(leftFork(\textbf{self})) = undef$ **then** {
        $owner(rightFork(\textbf{self})) := \textbf{self}$
        $owner(leftFork(\textbf{self})) := \textbf{self}$
    }
    if $owner(rightFork(\textbf{self})) = \textbf{self} \wedge owner(leftFork(\textbf{self})) = \textbf{self}$ **then** {
        $owner(rightFork(\textbf{self})) := undef$
        $owner(leftFork(\textbf{self})) := undef$
    }

Ideally, p_i, denoted by **self**, would like to execute alternatively the two rules above to get and later to release the desired forks. Indeed, even if ASM rules are executed in parallel by definition, the second rule (i.e. the second **if-then** statement in the program above) can be performed if and only if the first rule has been previously executed.

During the waiting for both forks, the computation of p_i can go through four states:

1. (*owner*(*rightFork*(**self**))=philosopher on the right) ∧ (*owner*(*leftFork*(**self**))=*undef*);
2. (*owner*(*rightFork*(**self**))=*undef*) ∧ (*owner*(*leftFork*(**self**))) = philosopher on the left);
3. (*owner*(*rightFork*(**self**)) = philosopher on the right) ∧ (*owner*(*leftFork*(**self**))) = philosopher on the left);
4. (*owner*(*rightFork*(**self**)) = *undef*) ∧ (*owner*(*leftFork*(**self**))) = *undef*).

In all four states, the thinking predicate holds. The state changes whenever an update is executed by the neighboring philosophers over the shared locations; however, the ASM could not make a real computational step towards the state characterized by the eating predicate. In fact, only state (4) allows the first rule to be executed, so the desired state can be reached. In this particular case, even if the ASM state changes, the computation could not make a real progress, i.e. the process risks to starve.

For verification purposes, predicate abstraction is very suitable for capturing starvation. In particular, starvation could arise if there are rules: (*i*) whose condition concerns functions which represent the dependency of the agent from external resources; and (*ii*) whose execution/non-execution could have effects that does not change the value of the predicate over the states which represents the "waiting for something" issue. In our case, the first rule of the *PhilosopherProgram* shows these issues.

Finally, it is worth noting that, if resource holding holds, the scenario above is affected by the risk of deadlock: each philosopher picks up his/her right fork and waits for the left fork to become available. Thanks to predicate abstraction, this issue can be captured by the following predicate: $owner(rightFork(p)) = p, \forall p \in Philosophers$. Therefore, the model is deadlock-free if, during the DASM computation, it is not possible that its global state fulfill the predicate above, i.e. at any moment there must be at least one ASM whose state fulfills ¬(*owner*(*rightFork*(**self**)) = **self**).

5.2 A MANET Routing Protocol

Mobile Ad-hoc NETworks (MANETs) [1] are wireless networks designed for communications among nomadic hosts, in absence of fixed physical infrastructure. Each node plays a twofold role: end-point of a communication session and router supporting other communications. Both activities evolve concurrently. A MANET that adopts the AODV routing protocol [26] can be modeled by a DASM including a homogeneous set of $hosts = \{h_1, ..., h_n\}$. Each ASM can be in one of different states, which are characterized by the following predicates over the states:

- idle: the host is inactive. Its formula is given by: *wishToInitiate*(**self**, *dest*) = *false* ∧ *receivedRREQ*(**self**, *dest*) = *false* ∧ *isEmpty*(*replies*(**self**)) = *true*, $\forall$ *dest* ∈ *hosts*;
- router: the host has received a control packet directed to it. It is characterized by *receivedRREQ*(**self**, *dest*) = *true*;

- `initiator`: the host has to start a new communication session. It is characterized by *wishToInitiate*(**self**, *dest*) = *true*;
- `forwarding`: the host is forwarding a control packet to another recipient. It is characterized by *isEmpty*(*replies*(**self**)) = *false*.

where:

- *wishToInitiate*: *hosts* × *hosts* → *boolean* indicates whether a new communication session to a destination is required;
- *receivedRREQ*: *hosts* × *hosts* → *boolean* indicates whether an RREQ packet has been received;
- *isEmpty*: *queues* → *boolean* states if a queue of messages is empty or not.

In fact, in order to model broadcasting and unicasting, each host is associated with two queues of messages: *requests* and *replies*, including: RREQ (Route REQuest) packets for requesting a route to a desired destination, and RREP (Route REPly) packets for replying this request, respectively. This allows us to model sending/receiving of packets by means of enqueuing/dequeuing abstract messages into the corresponding queue. In addition, each ASM includes the following functions:

- *routingTable*: *hosts* → PowerSet(*records*), which represents the information about the hosts recorded into the host's routing table;
- *hostsInRT*: PowerSet(*records*) → PowerSet(*hosts*), which returns the set of the hosts stored in a given routing table, for checking information about hosts.

The ASM pseudo-code of the i-th host is shown below.

HostProgram(h_i) =
   If ¬*isEmpty*(*requests*(**self**)) **then** {
      RREQ = *top*(*requests*(**self**))
      nextHop = sender of *top*(*requests*(**self**))
      updateRoutingTable(**self**, *RREQ*)
      receivedRREQ(**self**, *dest*):= *true*
      Router(*RREQ*, *nextHop*)
   }
   if *wishToInitiate*(**self**, *dest*) = true **then**
      Initiator(*dest*)
   if ¬*isEmpty*(*replies*(**self**)){
      RREP = *top*(*replies*(**self**))
   if *RREP.init* ≠ **self then**{
      nextHop = select *c.nextHop* ∈ *hostsInRT*(*routingTable*(**self**))
         with *RREP.init* = *c.dest*
      updateRoutingTable(**self**, *RREP*)
      UnicastRREP(*RREP*, *nextHop*)
      dequeue *RREP* from *replies*(**self**)
   }
}

It is worth noting that the activation of a host unfolds different computational branches: two of them lead to the execution of the *Router* or *Initiator* submachine, respectively; in the third case, the forwarding of RREPs is executed. In particular, the *Router* submachine models the behavior of the node when it supports communications between other end-points; instead, the *Initiator* submachine models the behavior of the node when it acts as the initiator of a new communication session. Note that if initiator does not know a route to reach destination, then it starts a *route discovery* process aimed at discovering this route. For clarity, an ASM *submachine* is a parameterized rule [9]: it allows the declaration of *local* functions, so that each call of a submachine works with its own instantiation of its local functions.

For verification purposes, predicate abstraction can help in expressing the node's behavior. In our case, the simultaneous fulfillment of different predicates over the same ASM state is very suitable for capturing the intrinsic concurrency of the nodes' computation. Indeed, when the MANET starts operating, each host is idle. But, during the normal execution of the MANET, a host can, for example, fulfill the `router` predicate with respect to a destination, but at the same time it can fulfill other predicates, e.g. `initiator`, for what concerns other destinations. The values of the arguments help in distinguishing the various cases.

Moreover, predicate abstraction can help in investigating some specific properties for MANETs: the correctness of the activities of sending/receiving packets, the starvation-freedom of initiator when it starts a route discovery process, and so forth. For example, concerning the starvation issue, the presence of a timeout in the *Initiator* submachine allows initiator to escape the waiting for RREPs if a route to destination cannot be found. So, for that specific destination, after the timeout expiration, the ASM state does not fulfill the `initiator` predicate but the `idle` predicate.

For the purposes of the present work, it is not necessary to further detail the *updateRoutingTable* and *UnicastRREP* rules, as well the *Router* and *Initiator* submachines. The interested reader can find the full specification of the model and the proof of its correctness in [7].

6 Conclusion

This paper proposes predicate abstraction over ASM states. The proposed approach can support the verification of ASM-based models by overcoming the main limitations that penalize classic model checking techniques applied to ASMs. In fact, applying predicate abstraction to ASMs enables the analysis of the global properties of the system to be verified through a representation of its local properties through predicates over the states. In this way, the analysis is executed entirely within the ASM framework, without the need of less expressive models and without the burden of temporal logics. Possible applications are represented by various kinds of critical and complex systems that can benefit from a formal approach: Internet-based services, security protocols, Cloud, Grid and mobile systems, and so on.

It is worth specifying that researchers usually distinguish between two classes of properties [23]. *Safety* properties specify that "something bad never happens", e.g. deadlock-freedom. Instead, liveness properties stipulate that "something good eventually happens", e.g. starvation-freedom. From this point of view, safety properties require that certain predicates over the states, which represent a "bad thing", must never be satisfied, or, alternatively, their negation must always hold during the computation. Conversely, liveness properties require that certain predicates over the states, which represent a "good thing", must eventually be satisfied during the computation. Future directions of this research should investigate specific features of predicate abstraction with respect to the specific kinds of properties to be analyzed.

Acknowledgements. This work has been partially funded by the Italian Ministry of Education, University and Research, within the "Piano Operativo Nazionale" PON02_00563_3489339.

References

1. Agrawal, D.P., Zeng, Q.A.: Introduction to Wireless and Mobile Systems. Thomson Brooks/Cole (2003)
2. Alpern, B., Schneider, F.B.: Defining Liveness. Information Processing Letters **21**(4), 181–185 (1985)
3. Arcaini, P., Gargantini, A., Riccobene, E.: AsmetaSMV: AWay to link high-level ASM models to low-level NuSMV specifications. In: 2th International Conference on Abstract State Machines, Alloy, B and Z, pp. 61–74 (2010)
4. Arcaini, P., Gargantini, A., Riccobene, E.: CoMA: conformance monitoring of java programs by abstract state machines. In: Khurshid, S., Sen, K. (eds.) RV 2011. LNCS, vol. 7186, pp. 223–238. Springer, Heidelberg (2012)
5. Baier, C., Katoen, J.P.: Principles of Model Chacking. The MIT Press (2008)
6. Bianchi, A., Manelli, L., Pizzutilo, S.: An ASM-based Model for Grid Job Management. Informatica (Slovenia) **37**(3), 295–306 (2013)
7. Bianchi, A., Pizzutilo, S., Vessio, G.: Suitability of Abstract State Machines for Discussing Mobile Ad-hoc Networks. Global Journal of Advanced Software Engineering **1**, 29–38 (2014)
8. Blass, A., Gurevich, Y.: Abstract State Machines Capture Parallel Algorithms. ACM Transactions on Computational Logic **4**(4), 578–651 (2003)
9. Börger, E., Stärk, R.: Abstract State Machines: A Method for High-Level System Design and Analysis. Springer (2003)
10. Chen, Z., Zhou, C., Ding, D.: Automatic abstraction refinement for petri nets verification. In: 10th Int. Workshop on High-Level Design, Validation and Test, pp. 168–174 (2005)
11. Das, S., Dill, D.L., Park, S.: Experience with predicate abstraction. In: Halbwachs, N., Peled, D.A. (eds.) CAV 1999. LNCS, vol. 1633, pp. 160–171. Springer, Heidelberg (1999)
12. Dershowitz, N.: The Generic Model of Computation. Electronic Proceedings in Theoretical Computer Science (2013)
13. Dijkstra, E.W.: Hierarchical Ordering of Sequential Processes. ACTA Informatica **1**(2), 115–138 (1971)
14. Farahbod, R., Glässer, U., Ma, G.: Model Checking CoreASM Specifications. In: 14th International ASM Workshop (2007)
15. Gabrisch, W.: A Hoare-Style Verification Calculus for Control State ASMs. In: 5th Balkan Conference on Informatics, pp. 205–210 (2012)

16. Gargantini, A., Riccobene, E., Rinzivillo, S.: Using spin to generate testsfrom ASM specifications. In: Börger, E., Gargantini, A., Riccobene, E. (eds.) ASM 2003. LNCS, vol. 2589, pp. 263–277. Springer, Heidelberg (2003)
17. Gervasi, V.: An ASM model of concurrency in a web browser. In: Derrick, J., Fitzgerald, J., Gnesi, S., Khurshid, S., Leuschel, M., Reeves, S., Riccobene, E. (eds.) ABZ 2012. LNCS, vol. 7316, pp. 79–93. Springer, Heidelberg (2012)
18. Glausch, A., Reisig, W.: An ASM-characterization of a class of distributed algorithms. In: Abrial, J.-R., Glässer, U. (eds.) Rigorous Methods for Software Construction and Analysis. LNCS, vol. 5115, pp. 50–64. Springer, Heidelberg (2009)
19. Graf, S., Saidi, H.: Construction of abstract state graphs with PVS. In: 9th International Conference on Computer Aided Verification, pp. 72–83 (1997)
20. Gurevich, Y.: Sequential Abstract State Machines Capture Sequential Algorithms. ACM Transactions on Computational Logic **1**(1), 77–111 (2000)
21. Hopcroft, J.E., Ullman, J.D.: Introduction to Automata Theory, Languages, and Computation. Addison-Wesley (1979)
22. Klai, K., Desel, J.: Checking soundness of business processes compositionally using symbolic observation graphs. In: Giese, H., Rosu, G. (eds.) FORTE 2012 and FMOODS 2012. LNCS, vol. 7273, pp. 67–83. Springer, Heidelberg (2012)
23. Kindler, E.: Safety and Liveness Properties: A Survey. EATCS Bulletin **53**, 268–272 (1994)
24. Laplante, P.: Dictionary of Computer Science, Engineering and Technology. CRC Press (2000)
25. Luzzana, A., Rossetti, M., Righettini, P., Scandurra, P.: Modeling synchronization/communication patterns in vision-based robot control applications using ASMs. In: Derrick, J., Fitzgerald, J., Gnesi, S., Khurshid, S., Leuschel, M., Reeves, S., Riccobene, E. (eds.) ABZ 2012. LNCS, vol. 7316, pp. 331–335. Springer, Heidelberg (2012)
26. Perkins, C.E., Belding-Royer, E.M., Das, S.R.: Ad hoc On-Demand Distance Vector (AODV) Routing. RFC 3561 http://tools.ietf.org/html/rfc3561 (2003)
27. Reisig, W.: The Expressive Power of Abstract State Machines. Computing and Informatics **22**, 209–219 (2003)
28. Singhal, M.: Deadlock Detection in Distributed Systems. IEEE Computer **22**(11), 37–48 (1989)
29. Spielmann, M.: Automatic verification of abstract state machines. In: Halbwachs, N., Peled, D.A. (eds.) CAV 1999. LNCS, vol. 1633, pp. 431–442. Springer, Heidelberg (1999)

Implementation and First Evaluation of a Molecular Modeling Language

Alexander Andersson and John Krogstie[✉]

Norwegian University of Science and Technology – NTNU, Trondheim, Norway
alextommy89@gmail.com, krogstie@idi.ntnu.no

Abstract. Traditional conceptual modeling languages are oriented towards certain aspects of the problem domain e.g., data, processes, objects, actors, or goals. Although a distinct perspective can be beneficial when it is decided what perspective is most appropriate, there is no perspective that is best for all problem domains, and you usually have to integrate concepts from several perspectives to get a good understanding of the overall situation. In early analysis, it can be argued for a modeling languages that allow the modeler high freedom. An approach called GEMAL - Generic Enterprise Modeling and Analysis Language supporting these needs by building on a molecular modeling thinking, is described. An early evaluation of the language as implemented in the Troux Architect (METIS) tool is presented. It is illustrated how a modeling approach of this type can be an efficient way of capturing a large range of concepts by combining a small set of building blocks in a molecular manner, although more research is needed to investigate if this is a generally applicable approach or if it is only relevant to use for modeling experts for sense-making and early analysis.

1 Introduction

During problem analysis, conceptual models of the problem domain are developed, assessed and improved. Numerous modeling languages have been developed for this purpose. Each of them is oriented towards certain aspects of the phenomena in the real-world problem domain. Modeling languages can be divided into classes according to the core phenomena classes (concepts) that are focused on in the language. We have called this the *perspective* of the language. A classic distinction regarding modeling perspectives is between the structural, functional, and behavioral perspective [14]. Object-orientated analysis appeared as a particular way of combining the structural and behavioral perspective in the late eighties. Through other work, such as F3 [3], the NATURE project [7], and work within Enterprise Architecture [22], additional perspectives have been identified, include the goal, actor, communicational, and topological perspectives. Each of these perspectives has its pros and cons, and you can find critique of all in the literature. In [16] when presenting their facet-modeling approach, the authors argued for that strongly oriented models are not beneficial in early problem analysis, and that strong orientation may instead cause modeling-related problems during problem analysis because of

- limited freedom of choosing to represent different perspectives on the problem domain depending on the problems at hand;
- reduced possibility of co-representing multiple perspectives on the problem domain simultaneously;
- limited support for integrating these perspectives by representing relations between them;
- No way of extending the modeling language used to represent and visualize new perspectives which are discovered to be relevant first during problem analysis.

Although much modeling work is related to specific perspectives, earlier attempts for atomic and molecular languages exist, which will be presented briefly in section 2. In section 3, we will present GEMAL - Generic Enterprise Modeling and Analysis Language, a modeling language that is attempting to address the above challenges, and at the same time fulfilling later expressed requirements for language appropriateness [11, 13] balancing expressiveness and comprehensibility by supporting a molecular approach [9] to modeling. The following research questions are pursued

1. **RQ1:** Is it possible to implement a molecular modeling language such as GEMAL?
2. **RQ2:** Is GEMAL able to represent every relevant aspect of traditional modeling languages for analysis?
3. **RQ3:** Are all combinations of concepts possible in GEMAL meaningful?
4. **RQ4:** Is the comprehensibility appropriateness of a molecular modeling language like GEMAL better than that of traditional modeling languages with the same expressiveness?

The current implementation is described briefly, and a first evaluation of the approach is presented. The paper ends with an overview of plans for further work.

2 Related Work

Combined modeling languages that cover many perspectives in an integrated manner are often found to become overly complex. In [19] a detailed comparison using the complexity metrics devised by Rossi and Brinkkemper [18] reveals that for instance UML being a combination of several approaches consists of 3-19 times more object types, 2-27 more relationship types, and 2-9 times more property types than other object-oriented modeling approaches.

An approach to address such a problem is to go for a holistic or molecular modeling language. Kangassalo [9] distinguishes between holistic, molecular and atomic semantics of modeling languages. With atomic semantics, the meaning of a model element is completely specified by that element, while with holistic semantics the meaning depends on every other element in the model. Molecular semantics offer a compromise, where meaning of an element is given by some parts of the model. A language with atomic semantics requires a lot of specific constructs, while holism allows meaning to be constructed by combining a number of elements, like one combine letters into words, and words into a sentence. In its extreme this provides a large number of senseless statements (As also is the result of freely combining letters into words and words into sentences), thus we approach this using a molecular approach.

Most modeling languages have originally been developed according to a certain perspective. On the other hand already in [21] Sindre discussed pan-presumptionism, i.e. freeing yourself from conceptual presumptions as a guideline for modeling language design, work followed up in [16, 17] on real-world and facet-modeling. In facet-modeling different aspects can be co-represented whenever needed, and semantical relations between aspects are reflected in the created domain models. The modeler will have freedom from perspective, and can extend models with new kinds of aspects that are recognized as relevant to the problem at hand during analysis. Note that facet modeling is not built to prevent perspective and orientation in modeling. It is recognized that conceptual modeling should be based on a fundamental method or a set of guidelines, but these should be evolved through discoveries made in analysis instead of being implicit restrictions that come from the modeling language used. The presentation of this is not followed up with a visual modeling notation.

Ideas related to holistic and molecular modeling languages were proposed by Jørgensen [8]. Some of these aspects were included in the design of EEML [10] in particular to the process modeling domain there, and also in the work on active knowledge modeling (AKM) [12], in particular the underlying core EKA (Enterprise Knowledge Architecture) elements, originally developed by Jørgensen. The ideas behind GEMAL were originally discussed in connection to the POP* interchange format [23] and to the work on UEML [1], but neither these have a visual notation.

SeeMe [4] is a modeling notation specialized on modeling socio-technical systems that was originally developed in 1997 that has some similarities to our approach. The language is intended to be used in early stages of planning of socio-technical systems, that is, systems containing complex interdependencies between people and computers, and between technical components. It does not require the modelers to select a viewpoint, but allows the combination of roles, activities, and entities. Holism can be observed in SeeMe in the way concepts are built into complex composite structures bound together by relationships and events, exemplified in [6]. SeeMe has been applied to a number of process modeling case studies [5, 6], in which the researchers learned that incomplete models are simpler and easier to understand for the process participants. Also, SeeMe can be regarded as molecular because complex components can be decomposed freely into other perspectives. On the other hand, only role, activity and entity is supported, making it difficult to represent many other perspectives found useful in enterprise modeling such as goals and capabilities.

Over the last 20 years, the need to address modeling across perspectives have been addressed in a number of approaches such as UML and Archimate, but due to the amalgamation of a number of approaches, these kind of languages appear as very complex, with a large number of concepts with atomic semantics. In contrast, GEMAL have things with 7 perspectives, where any concept can be looked upon in the context of another concept, support the standard abstraction mechanisms found in conceptual modeling (generalization, classification, aggregation, and association) and combine this with 7 standard relationships in a fully orthogonal way, enabling a large number of concepts to be expressible by composing building blocks based on this limited number of core constructs.

3 Description of GEMAL

The language model (meta-model) in Fig 1 defines the components included in the current version of the GEMAL modeling language. We will go through the components briefly here;

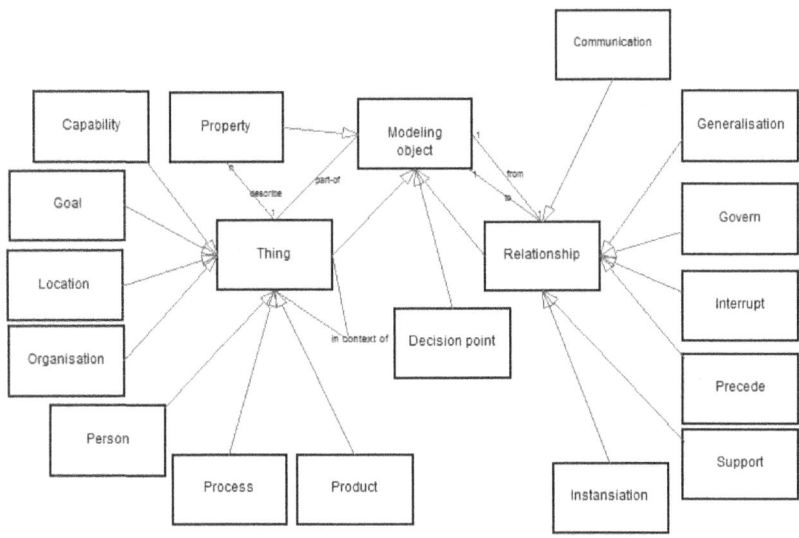

Fig. 1. GEMAL language model

Inspired by [24], Thing, Property, Decision Point and Relationship are on top level as *specializations of Modeling Object*. There are 7 types of things defined:

- **Goal:** A state that one wants to keep or achieve in some context. Generic goals apply for a type concept (Get paper accepted), and concrete goals apply to instances (Get this paper accepted for the EMMSAD 2015 conference).
- **Process:** A process is a transformation. Either it transforms things into other things, or changes their state: Both process types (write paper) and process instances (write the paper on GEMAL and send to EMMSAD 2015) exist
- **Product:** A thing produced through a process. Products also exist on type level (a scientific paper) and instance level (a concrete paper).
- **Capability:** A generic capability (everything needed to arrange a conference) is something useful for achieving a generic goal. A concrete capability (arranging the EMMSAD conference) makes it possible to change the state of the world in order to reach a concrete goal using a large set of means. Whereas a process more specifically take some input and transform it to some output, the capability makes it possible to address a less well-defined situation, using available people, processes, locations, products and systems to reach an emerging goal.
- **Person:** Type of person (person-role e.g. author) or concrete person (NN)

- **Organization:** A set of persons (more than one) where goal-oriented processes takes place. Types (organizational role e.g. program committee) and concrete organizations (EMMSAD 2015 Program Committee) can be modelled.
- **Location:** A type of location (place - e.g. a conference venue) or a concrete (physical) location (space - The room at the conference venue where EMMSAD 2015), where the state-changes actually take place.

Properties describe Things (a thing can have one to many properties). In addition things have the following predefined properties: ID, Name, Description, Start-time, End-time, Instantiation level (type or instance, showed visually on the concept), Modality (necessity, obligation, recommendation, permission, discouragement, prohibition, and contradiction), Current state (from a set of possible states of a thing), Vagueness: Indicate the state of the modelling process (inspired by SeeMe). This has the following possible values 'Must validate, Missing on purpose, Missing by lack of knowledge, Completed', Reference (to an external source).

A number of specific relationships are defined between modelling constructs

1. **Communicate** - A thing communicates with (informs) another thing.
2. **Govern** - A thing restricts and influences the state of another thing.
3. **Precede** - Used to connect a series of things that precede each other in time.
4. **Support** - A thing can support another thing in achieving a state change or goal. This can also be used to represent general dependencies
5. **Generalize** - Can be used to build generalization-hierarchies.
6. **Instantiate** - A thing can be an instance of another thing.
7. **Interrupt** - Can illustrate that a thing interrupts/works against another thing.

Any modeling object can be part of a thing, i.e. one can decompose a thing in anything (although the most usual use of this would be to decompose e.g. a task into subtasks. It can be discussed if one also would need a specific composition relationship if one what this to be expressed without using decomposition). One can also look upon any thing in context of another thing. When taking for instance a process-centric view, all other aspects are resources regarded to be relevant in the process. For a process-type exemplified with an instance of this type (e.g. arrange a conference), we have the following specific meanings.

- Process: A process needed to create and maintain the process:
  o Type: The process to set up the conference schedule.
  o Instance: The process for setting up the schedule for EMMSAD 2015
- Organization: As actor or resource in the process
  o Type: Program Committee (PC)
  o Instance: EMMSAD PC
- Person : An actor or role in the process
  o Type: Program Chair (i.e. a process role)
  o Instance: Program Chair of EMMSAD 2015
- Product: A resource (including data) used or produced in the process
  o Type: Proceedings
  o Instance: EMMSAD 2015 proceedings
- Goal: The goal of the process (what should be achieved)

- o Type: The conference should not lose money
- o Instance: CAiSE 2015 should have a small surplus
- Capability: The capabilities needed to perform the process
  - o Type: Skill, knowledge and competence needed to arrangement conferences
  - o Instance: Set up Easychair for reviewing EMMSAD papers
- Location: Space and place where the process is situated
  - o Type: Meeting venue, meeting rooms
  - o Instance: The room where EMMSAD 2015 is arranged

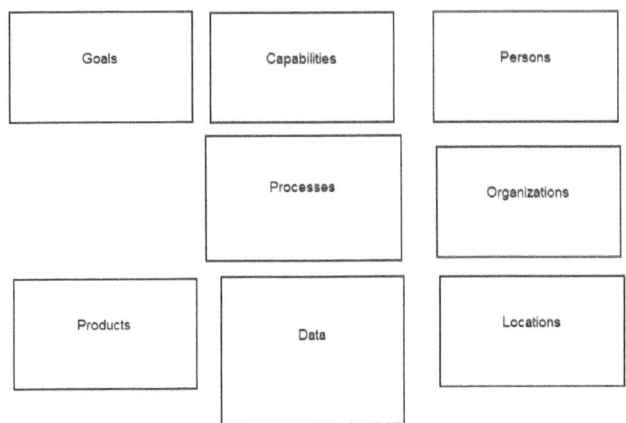

Fig. 2. Structure of a GEMAL model

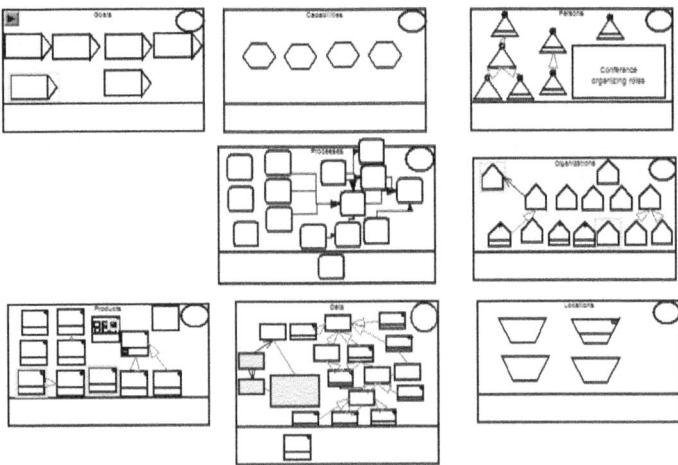

Fig. 3. GEMAL notation example

An example model from a larger case study (the conference arrangement case study described in [11]) is depicted in Fig. 2 and Fig. 3. Fig. 2 illustrates a possible overall structuring of the model, whereas Fig. 3 illustrates the notation of the main types of things. The concept with yellow border indicates that this is view objects (i.e. that one first has represented e.g. a goal as part of or in the context of another concept, and then also represents it as an individual concept.

4 Evaluation of Implementation

An early implementation of GEMAL is made available in the modelling tool METIS (now Troux Architect). To investigate the research questions in more detail, the following have been done

- Model cases originally depicted in other modelling languages for analysis such as DFD, ER, EEML, UML use case, class diagrams, and activity diagrams, BPMN descriptive modelling, i* (SD and SR), SeeMe and Archimate business modelling, to see if all important concepts in these languages can be represented.
- Perform a survey among non-experts of modelling investigating the semantic transparency of the chosen symbols for the main concepts.
- Do interviews with two modelling experts knowledgeable with different modelling languages for analysis and design.

Whereas details on the first area, relating to expressiveness, can be found in [2], the results from the two other areas are reported in more detail here. As part of the evaluation of GEMAL, an online survey was created. The survey was intended to provide some indication on the semantic transparency of the notation. The short survey was distributed to an arbitrary selection of people. To keep the survey's questions simple enough for untrained people to answer, it was focused on the connection between symbols and concepts, to find whether novice users can identify them correctly. 90.38% of the survey respondents (54 in total) had no training or previous experience with conceptual modeling or modeling languages used in enterprise modeling. The high percentage of novice users implies that the results are sound when analyzing them in order to show tendencies related to Semantic Transparency (ST) of symbols in GEMAL. Moody [13] claims that how novice users infer meaning from a symbol can help point out their level of ST.

4.1 Survey Results

The correct answer in each row is indicated by a grey background. The number of responses for each alternative is shown by the number in parentheses. To be able to point to a semantically immediate symbol in this survey, we could require that most respondents found the right name for it. This was not found for any symbols. 'Person' and 'Product' are the closest ones, with 44.23% correct answers. Semantically opaque concepts should be expected to display an arbitrary relationship between its symbol and name. We can infer that 'Person', 'Product', 'Goal', and 'Organization' can be classified somewhere between semantically immediate and semantically opaque symbols; the correct alternative got the most answers (or slightly lower than on other alternative in the case of 'Goal') for those symbols in the survey. 'Capability',

'Location', and 'Process' may be closer to a semantically perverse classification, since there was a noticeably higher number of respondents selecting a wrong alternative than there were respondents selecting the correct alternative. The survey indicates that 'Capability' is the concept closest to semantically perverse; as much as 50% of the respondents thought that the right symbol was the symbol used for 'Location'.

Table 1. Results for survey question 1: Each of the following 7 concepts is represented by one of the 7 symbols. Which symbol do you think represents each concept (one selection per concept)?

	Symbol A	Symbol B	Symbol C	Symbol D	Symbol E	Symbol F	Symbol G
Capability	9.61 % (5)	50.00 % (26)	9.61 % (5)	5.77 % (3)	0.00 % (0)	17.31 % (9)	7.69 % (4)
Goal	21.15 % (11)	13.46 % (7)	15.38 % (8)	5.77 % (3)	23.08 % (12)	15.38 % (8)	5.77 % (3)
Location	11.54 % (6)	15.38 % (8)	11.54 % (6)	23.08 % (12)	1.92 % (1)	9.61 % (5)	26.92 % (14)
Person	9.61 % (5)	1.92 % (1)	44.23 % (23)	1.92 % (1)	23.08 % (12)	9.61 % (5)	9.61 % (5)
Process	38.46 % (20)	9.61 % (5)	1.92 % (1)	1.92 % (1)	15.38 % (8)	25.00 % (13)	7.69 % (4)
Product	11.54 % (6)	7.69 % (4)	0.00 % (0)	44.23 % (23)	21.15 % (11)	5.77 % (3)	9.61 % (5)
Organisation	0.00 % (0)	0.00 % (0)	15.38 % (8)	19.23 % (10)	15.38 % (8)	21.15 % (11)	28.85 % (15)

4.2 Expert Evaluation

The expert evaluation was carried out to get indications on the quality of GEMAL as a modeling language from credible sources. The experts were provided with the following material:

- Models in the modeling languages DFD, ER, i*, UML, and Archimate.
- GEMAL models of the same situations, made with the Metis tool.
- A GEMAL model in Metis displaying some of the ways in which GEMAL can support conceptual and perceptual integration.
- A printout containing an overview of the GEMAL notation.

The material was sent via email to the experts a couple of days before the evaluation, along with a description of how the evaluation was intended to be carried out. They were also provided a link to an installation of the tool with support for modeling in GEMAL. The original models were brought to the evaluation as printouts, while the GEMAL models were presented on a laptop. The student took the initiative to present the models to the experts, but the experts were encouraged to choose models they wanted to have a closer look at, navigate themselves, or skip - depending on their interests and knowledge. Each evaluation session took about 50 minutes. The student and one expert took part in each session. Experts were encouraged to speak their mind about the models first, before the questions were answered. Questions were based on guidelines for modeling language design [11, 13].

We tried to find experts with a diverse set of specialization areas. Two experts have completed the evaluation; one focus on model-driven software development, while the other is more interested in topics like requirements engineering and information systems security.

4.2.1 Results
1. **About you**

   Do you have experience with DFD, ER, I*, UML, Archimate, other languages?

   Expert A: Fairly familiar with all the mentioned modeling languages.
   Expert B: Familiar with all the mentioned modeling languages, but has most experience with DFD, ER and i*. He knows some of the UML diagram types well, and has some experience with Archimate.

2. **Has GEMAL been successfully implemented?**

   With focus on the technical implementation of the language: What is done well? What could have been done better? Suggestions for improvements?

   Expert A: Larger fonts and more noticeable relationship lines could prove to be beneficial; it is hard to notice these things when the view is zoomed out a bit.
   Expert B: Hard to give a detailed answer here based on the evaluation walkthrough. GEMAL seems to respond fast and work well. Would have to try modeling with it himself to be able to provide suggestions for improvements.

3. **Is GEMAL able to represent every relevant aspect of traditional modeling cases?**

   3.1. *Based on the evaluation models, which notations were represented well in GEMAL? Which were not so well represented?*

   Expert A: In the UML Activity Diagram, he thinks it is a good solution to use a GEMAL 'Thing' in place of a UML'InitialNode', this makes it possible to add additional initial properties in GEMAL models. He does not think that GEMAL needs dedicated symbols to represent fork/join in UML (can use decision points). In the DFD model, he views the inline information flow of GEMAL as beneficial. However, it can be difficult to separate information flow from process flow (communicate vs. precede) in the diagram. For the Archimate model, he questioned what it means to have a process inside a product. He points out that the naming scheme of GEMAL should strive to be clear about if a word is in verb or noun form.
   Expert B: Based on the evaluation models, GEMAL seemed to cover them well enough. Sequence Diagrams may be problematic to show, as may for example special constructs in Activity Diagrams. The GEMAL equivalent to UML Use Case looked similar to a DFD diagram, since things could be nested. The GEMAL equivalent to the ER model looks very clean, but how can details for relationships be showed clearly (now represented similar to cardinality constraints in UML class diagrams)? i* has

two types of models ('Strategic Dependency Model' and 'Strategic Rationale Model'), while only one was showed in the evaluation case. Can both be supported well in concert?

3.2. Are there other traditional modelling cases that you can foresee will work well or not so well in GEMAL?

Expert A: When the notation has few concepts as in GEMAL, some things cannot be expressed as easily in the models. Models leave room for interpretation, and may be understood differently by different users. Better support for modeling deontic logic could be useful in some cases (now only as a property on things). He calls for a way to model the conditions of a modality property. Utilizing OCL (used together with UML) in GEMAL could help with this. For GEMAL relationship types, dotted lines could advantageously be given a meaning as seen in some notations; A hollow arrow-head could symbolize that information is sent, while a filled arrow-head means the opposite. The direction of a relation could be used to symbolize read/write operations
Expert B: Detailed events in BPMN may be tricky to represent. Also specialized concepts in for example BPMN and Archimate. To be complete, GEMAL would need a way to represent states in state diagrams, but this may not be required in a simple language. Also, could OCL be useful to model conditions in GEMAL?

3.3. Does GEMAL seem to adapt better to some modeling perspectives than others?

Expert A: [See some related answers under question 3.2.]
Expert B: Unsure about how well behavioral languages like Petri Nets and State-charts can be modeled, GEMAL may not have the concepts required for this. Communication languages like Speech Act Models can also be a problem, hard to say based on the evaluation cases. GEMAL seems to cover many modeling perspectives with few constructs, which is good. UML has many different symbols for the 'Actor' concept, while GEMAL has one. It may require some effort to put GEMAL to use for people that are accustomed to using existing modeling languages. Further case studies with modelers are needed to find out whether they would benefit from switching to GEMAL. Using GEMAL may be more tempting to new modelers than to experts.

3.4. Can you see advantages or disadvantages that come with the holistic/molecular approach of the GEMAL language?

Expert A: It provides the modeler freedom from perspectives, which can be an advantage in enterprise modeling. However, sometimes a more narrowly defined language can be easier to use, and this approach brings new user interface challenges.
Expert B: It makes it possible to model according to any perspective, which can be great, especially for high level enterprise models. May not be as useful for technical system models.

3.5. Does GEMAL fit large and complicated modeling tasks?

Expert A: [See the expert's answer to question 5.10.]
Expert B: For large tasks it is great to have the zoom and view/copy functionality that comes with Metis. It is also useful to have ways to search for components,
   Follow specific connections in a model, search by criteria (i.e. 'view all goals that

have not yet been decomposed'), and have tools to help get an overview of errors and incompleteness in a model, functionality also provided with Metis and GEMAL.

4. Are all notations when combining modeling concepts freely like in GEMAL meaningful?

4.1. **Expert A:** As mentioned, concepts in GEMAL leave room for interpretation. What the writer tries to convey may not be what every user receives. As an improvement that could help alleviate misunderstandings, a 'Method' concept could be added, that would be different from 'Process'.

4.2. **Expert B:** Meaningless constructs can be created, but it is not a big problem in a language like GEMAL. Modeling languages can have strict rules that may force modelers to follow a specific pattern when modeling (i.e. connection rules for an entity have to be fulfilled before more entities can be added), but in more loose languages, the modeler is given freedom to create the models how he wants. The expert thinks that this is right for GEMAL if the goal is to create a language where the modeler is free to follow any perspective and approach.

5. Is the comprehensibility appropriateness of a molecular modeling language like GEMAL better than what can be achieved by combining traditional modeling languages?

5.1. *Is the **number of concepts** reasonable?*

Expert A: Yes. If any more concepts are added, it can quickly become hard for users to keep aware of their separate meanings. However, a concept similar to BPMN 'event' could prove useful.

Expert B: The number of concepts in GEMAL seems reasonable. Perhaps one for 'state' could be added? The concepts needed in GEMAL would depend on what the intended use of the language is.

5.2. *Is it easy to understand the **meaning of the different concepts** in GEMAL? I.e. what a person represents when modeled as a sub-concept inside a process.*

Expert A: He thinks a more illustrative symbol for 'Product' could be found. 'Person' and 'Organization' could have been combined into 'Actor'.

Expert B: He would guess that a person inside an organization means that he works there. A person in a location means he's at the location or relates to the location. Location in a person could also mean that this is the current location of the person. He thinks that the meaning of sub-concepts is very open to interpretation. How to show if a person is the producer, recipient or has some other role related to a product? Sometimes this may cause people to interpret the diagram differently.

5.3. *(Perceptual Discriminability) - Is it easy enough to differentiate concepts graphically in GEMAL? The language use mostly shapes, dual coding (with shapes and text) and colors for this purpose.*

Expert A: He had no remarks for this question.

Expert B: Yes, the components are quite distinct without being too complex. One could experiment with more colors or icons in addition to shape, but this would add complexity to the notation.

5.4. (Semiotic Transparency) -Do you find the graphical representations of concept in GEMAL intuitive? I.e. the 'Person' symbol is meant to imitate a real person.

Expert A: Perhaps a 'Pin' symbol could replace the current 'Location' symbol. The pin would be familiar to people that use application like Google Maps.

Expert B: The 'Person' symbol is intuitive, the others less so. The 'Process' symbol is familiar to people that know DFD. The 'Decision Point' symbol is a common factor in many notations, but it is not very intuitive to understand. Icons could have been used in place of figures to increase transparency, but that would make the notation more complex.

5.5. (Semiotic Clarity) - Did you see examples of, or can you think of scenarios where, symbol overload (several symbols for the same concept) or symbol redundancy (the same symbol can refer to several concepts) can occur in GEMAL models?

Expert A: No examples.

Expert B: Saw no examples of this. However, the symbols for 'Product' and 'Thing' are maybe too similar.

5.6. (Graphical Economy/Visual Expressiveness) - The goal is to make GEMAL easy to learn, understand and use by utilizing few visual variables both at the concept level and in larger models. Shape, color, text, connectivity and texture are the variable categories used to differentiate concepts. Has GEMAL found a good balance between use of visual variables and the degree of expressiveness of its components?

Expert A: Hard to say, one could experiment with more visual variables, but in GEMAL's case, it seems like a good idea to keep a visually simple notation.

Expert B: Did not notice anything negative. Experiments with users would be needed to be able to say anything conclusive about this.

5.8. (Complexity Management) - GEMAL uses sub-concepts (thing-in-thing) to aid complexity management. A vagueness indicator (i.e. 'Missing by purpose') is meant to help show different levels of precision in a model. Things can also be annotated with a status (planned, waiting, ready...). Does GEMAL handle complexity in a good way?

Expert A: Few complaints about the functionality. Can the vagueness indicator be used to show shortcomings in the tool (Metis) as well as in the language (GEMAL)?

Expert B: This functionality seemed ok. Good to have the filter functionality, where one can choose to see only persons, only goals, or hide for example processes.

*5.9. Do you find GEMAL to be **flexible** in terms of what level of details and precision a model can have?*

Expert A: Yes, GEMAL seems to be more flexible than most current modeling languages. Textual differentiation is needed to model details, but this does not have to be an issue.

Expert B: Ability to model details seems good, but some concepts of languages with a large number of concepts like BPMN may be hard to model with good precision. Textual differentiation is an option here.

5.10. *(Cognitive Integration) - Do you think that GEMAL has good functionality for helping users understand how different models relate to each other (conceptual integration), and for navigating complex models (perceptual integration)?*

Expert A: It is good to for example be able to extract every 'Product' to a separate diagram. He would like if data generation based on a model was supported. It would be interesting to have the metamodeling functionality of Metis in a tool like CIRIUS for Eclipse; there the programmer would not be bound by settings in Metis, but be freer to customize how the tool should work. On a positive side, Metis brings a feeling of space when it zooms and animates between diagrams. It helps to show how the current view relates to the model as a whole. The functionality reminds him of the 'Prezi' presentation software. It would be nice to be able to 'tag' a thing, and then extract everything with a tag to a new view.

Expert B: I think that the navigation support from Metis is good, and it is useful to create different views with the same information. However, these views are not automatically created; the person creating the model has to build them as part of a model. Perhaps this could have been handled automatically using AI in other tools? It would be useful if the tool could suggest parts of the model that may be related.

6. Other

6.1. *Do you think creating two **view styles** that can be switched between to fit different tasks and users is a good idea? Would you add or remove anything from the basic and advanced view styles?*

Expert A: Different visualization for different users is a good idea. However, he would prefer to have the choices listed in a setting menu, where each user can choose what to display. Not a great solution to have to drag view styles into the model, but it works.

Expert B: Good idea. Different users may want different viewstyles, and they can be switched between for different tasks and purposes.

6.2. *Is it a good idea to show complexity, vagueness and modality visually as in GEMAL? Should other factors be displayed as well?*

Expert A: The functionality is good, but the values could perhaps have been visualized in a better way. Complexity could be indicated by folding a corner on a thing symbol, and vagueness could be shown by applying a dotted line with some texture on the border.

Expert B: This would depend on the use of a model. If risk and security is of concern, one may want to focus on those factors visually. In other situations one may want to see performance, time allocation and power consumption visually.

4.3 Summary and Analysis of Results

We will here summarize the feedback generated through the evaluations, and attempt to interpret what it can mean to future versions of the GEMAL language. Relative to research question # 1, the experts seem to agree that the implementation of GEMAL in Metis works in a satisfactory way. They had suggestions for improvements (including larger fonts, more noticeable lines, using updated software, which all can be addressed). For research question #2, experts provided quite rich feedback. They seemed to be generally pleased with the approach and expressiveness of GEMAL, but also noticed some of the same problems. There will be room for interpretation of GEMAL models, and advanced components of several mentioned notation may be challenging to express in GEMAL. They raised some concern towards complexity management, relationship scheme, and graphical economy. Expanding on the ideas shared for a revised relationship scheme seems like a good idea; not every visual variable in the current visualization of GEMAL relationships actually carry a meaning (i.e. dotted lines). Both expert mentioned more implementation oriented concepts such as OCL, detailed events and sequence diagrams, but one can question the usefulness of these kinds of mechanisms in a language primarily meant to be used for problem analysis. On the topic of research question #3, both experts agreed that there will be room for interpretation in GEMAL models, but they did not see this as a big issue for GEMAL as a molecular modeling language. Stricter rules may not be a good solution, because they would restrict the modeler's freedom of expression. Relative to research question #4, they seemed to agree that the number of concepts in GEMAL is reasonable; by adding more one would impair the graphical economy of the language, and thus make requirements for complexity management tougher to meet. Some changes to the current selection of concepts were suggested. It is suggested that symbols for concepts could be more intuitive, something also the survey pointed to. In particular the less established concepts such as location and capability might be visualized in a better way. On the other hand, with a limited number of shapes, it is likely that a modeler will be able to learn these quickly. When building composites of GEMAL concepts, users may interpret the meaning of the model differently depending on their experience level and the context of use. It seems one should experiment more with visual variables for the language, but we note that the experts were positive to the current selection of visual variables overall - even if there were some interesting suggestions for possible changes. GEMAL have a flexible notation, but it is dependent on textual differentiation. This can be attributed to the low number of concepts of the language. The experts were positive to the current functionality for complexity management, but possible additions were mentioned. On the topic of GEMAL view styles, the experts seemed positive to the idea. However, they outlined possible changes to visualization, and mentioned variables that are not currently shown visually in GEMAL that could be added, to make it easier to adapt GEMAL to additional mentioned modeling settings.

5   Conclusion and Future Work

The current tool-support for the approach has recently evolved, and we will experiment more on different cases, investigating the possible advantages and challenges with getting away from too early having to classify the concepts in the world, supporting the early analysis of a problem domain. In addition to the language itself it is evident that there is a need to provide more detailed style guidelines for how to best use the language (cf. work on the use of BPMN [15,20]). Further empirical evaluation of the language will consist in having a wider set of modelers use the language on practical modeling problems. A language like GEMAL is regarded to be most appropriate for early analysis work. How to transfer these models into more traditional models being more appropriate for supporting e.g. systems development is also to be investigated.

References

1. Anaya, V., Berio, G., Harzallah, M., Heymans, P., Matulevičius, R., Opdahl, A.L., Panetto, H., Verdecho, M.J.: The Unified Enterprise Modelling Language – Overview and Further Work. Computers in Industry **61**, 99–111 (2010)
2. Andersson, A.: The Molecular Modeling language GEMAL (2015)
3. Bubenko Jr., J.A., Rolland, C., Loucopoulos, P., De Antonellis, V.: Facilitating fuzzy to formal requirements modeling. In: Proceedings of ICRE 1994, pp. 154–157. IEEE Computer Society Press, Colorado Springs (1994)
4. Herrmann, T. et al.: SeeMe in a nutshell. The semi-structured socio-technical modeling method. Ruhr-Universität Bochum, Bochum (2006)
5. Herrmann, T., Hoffmann, M., Loser, K.-O., Moysich, K.: Semistructured models are surprisingly useful for user-centered design. In: Designing Cooperative Systems (Coop 2000), pp. 159–174 (2000)
6. Herrmann, T., Loser, K.-O.: Vagueness in models of socio-technical systems. Behaviour & Information Technology 18(5), 313–323 (1999)
7. Jarke, M.J.A., Bubenko Jr., C., Rolland, A.S., Vassiliou, Y.: Theories underlying requirements engineering: An overview of NATURE at genesis. In: Proceedings of RE 1993, pp. 19–31 (1993)
8. Jørgensen, H. D.: Interactive Process Models, PhD-thesis, NTNU, Trondheim, Norway (2004)
9. Kangassalo, H.: Are global understanding, communication, and information management in information systems possible? In: Chen, P.P., Akoka, J., Kangassalu, H., Thalheim, B. (eds.) Conceptual Modeling. LNCS, vol. 1565, pp. 105–122. Springer, Heidelberg (1999)
10. Krogstie, J.: Integrated Goal, Data and Process modeling: From TEMPORA to Model-Generated Work-Places. In: Johannesson, P., Söderström, E. (eds.) Information Systems Engineering From Data Analysis to Process Networks, pp. 43–65. IGI Publishing (2008)
11. Krogstie, J.: Model-based development and evolution of information systems: A Quality Approach. Springer (2012)
12. Lillehagen, F., Krogstie, J.: Active Knowledge Modeling of Enterprises. Springer (2008)
13. Moody, D.L.: The physics of notations: Toward a scientific basis for constructing visual notations in software engineering. IEEE TSE **35**(6), 765–779 (2009)

14. Olle, T.W., Hagelstein, J., MacDonald, I.G., Rolland, C., Sol, H.G., van Assche, F.J.M., Verrijn-Stuart, A.A.: Information Systems Methodologies. Addison-Wesley (1988)
15. OMG(2011) BPMN v2 Specification. Technical report, OMG (January). (http://www.omg.org/), http://www.omg.org/spec/BPMN/2.0/
16. Opdahl, A.L., Sindre, G.: Facet Modelling: An Approach to Flexible and Integrated Conceptual Modelling. Information Systems **22**(5), 291–323 (1997)
17. Opdahl, A.L., Sindre, G.A.: Taxonomy for Real-World Modelling Concepts. Information Systems **19**(3), 229–241 (1994)
18. Rossi, M., Brinkkemper, S.: Complexity Metrics for System Development Methods and Techniques. Information Systems **21**(2), 209–227 (1994)
19. Siau, K., Cao, Q.: Unified Modeling Language (UML) - A Complexity analysis. Journal of Database Management, January-March 2001
20. Silver, B.: BPMN method and style. Cody-Cassidy Press (2011)
21. Sindre, G.: HICONS: A general diagrammatical framework for Hierarchical modeling (1990)
22. Zachman, J.A.: A framework for information systems architecture. IBM Systems Journal **26**(3), 276–291 (1987)
23. Ziemann, J., Ohren, O., Jaekel, F.W., Kahl, T., Knothe, T.: Achieving enterprise model interoperability applying a common enterprise metamodel. In: INTEROP-ESA 2005, Bordeaux, France (2006)
24. Wand, Y., Weber, R.: On the ontological expressiveness of information systems analysis and design grammars. Journal of IS **3**(4), 217–237 (1993)

Requirements and Regulations

Analyzing Variability of Cloned Artifacts: Formal Framework and Its Application to Requirements

Iris Reinhartz-Berger, Anna Zamansky[(✉)], and Mark Kemelman

Department of Information Systems, University of Haifa, Haifa, Israel
{iris,annazam,mkemelman}@is.haifa.ac.il

Abstract. Software Product Line Engineering (SPLE) promotes systematic reuse through variability mechanisms, such as configuration, parameterization, and inheritance. In reality, however, such reuse is many times done ad-hoc, resulting in several clones of the same product artifact which need to be managed in all development stages. To address this need, we provide in this paper a formal framework to represent dimensions of variability, which can be applied for identifying and analyzing variability automatically. The framework is based on the assumption that software artifacts can be modeled as graphs, and variability can be analyzed through examining the properties of mappings between the elements of these graphs. We demonstrate the potential usefulness of our framework by applying it to identify and analyze variability of functional requirements written in a natural language.

Keywords: Software Product Line Engineering · Variability analysis · Variation points · Variability mechanisms · Systematic reuse · Requirements engineering

1 Introduction

The increase in the number and complexity of software products made reuse very essential in software development. Software Product Line Engineering (SPLE) further promotes systematic reuse, namely, reuse in prescribed ways, of software-related artifacts among different, yet similar, software products [8], [21]. The main way to achieve systematic reuse is through *variability mechanisms*, which are techniques that need to be applied for realizing differences among software products of the same line (family) [26]. To this end, *core assets*, i.e., reusable artifacts or resources that are used in the production of more than one product in a software product line (SPL) [8], are developed and variability mechanisms are applied on core assets in order to produce *product artifacts*.

Different variability mechanisms have been suggested over the years and corresponding catalogs have been published, e.g., [1], [13]. Examples of variability mechanisms are: (1) configuration, in which core asset elements are selected to be included in the product artifact; (2) parameterization, in which values are assigned to core asset parameters; and (3) inheritance, in which core asset elements are refined in the context of a particular product. Extension of structure or functionality over core assets in particular product artifacts may also be allowed. The variety of these variability mechanisms empowers creativity without compromising ability: each mechanism guides the reuse and supports verifiability, and in the same time the mechanisms are diverse and the product artifacts may be quite different from each other and from the core assets that generated them.

In reality, however, the reuse of artifacts may be done ad-hoc, without following a structured SPLE process. A recent survey in industry [4] reveals that in practice SPLE is commonly adopted using an extractive strategy in which existing product artifacts are re-engineered into a SPL. In these cases, the same product artifact may have several clones, which have been developed before the adoption of SPLE [24]. In order to manage these cloned artifacts, some studies suggested refactoring them into SPLs, e.g., [9], [18], [25], and [28]. Others suggest generating variability models that capture the differences among existing artifacts, e.g., [6], [20], and [27]. Overall, the studies concentrate on particular types of artifacts (e.g., code or requirements) and sometimes make assumptions on the cloned artifacts based on the project context or the application domain [24]. Even after adopting SPLE, the process of selecting variability mechanisms in order to reuse core assets in particular products is often done ad hoc, based on preferences and earlier experience of the developers [5], [14].

To address the need to systematically manage cloned artifacts in all development stages, we suggest here a formal two-dimensional framework for variability analysis. The first dimension – the *element* dimension – is based on a central way to model variability – *variation points* [4]. These are placeholders in the core asset that are associated with corresponding variants in the product artifact [21]. To realize variation points different variability mechanisms can be utilized, such as inheritance and parameterization. The second dimension takes a *product* view and refers to whether the elements of the core asset are optional or mandatory (conditions associated with the configuration mechanism) and whether extension in the product artifact is allowed. Our underlying assumption is that both core assets and product artifacts can be modeled as graphs, and variability can be analyzed through examining the properties of mappings between the elements of these graphs. Perceiving requirements as central drivers of many development stages and methods, we exemplify the ability of the formal framework to cluster similar requirements written in a natural language. The framework is further used to identify variable parts that can be later associated with variability mechanisms to promote systematic reuse of requirements. Although exemplified on requirements, the suggested framework has the potential to support other types of artifacts, such as design models and code.

The rest of this paper is structured as follows. Section 2 reviews related work, while Section 3 provides motivation through a running example. Section 4 elaborates on the framework and its theoretical foundations, and Section 5 exemplifies the application of the framework to support the reuse of textual requirements, as well as presents some preliminary results demonstrating the framework usefulness. Finally, Section 6 summarizes the work and presents future research directions.

2 Related Work

Managing cloned artifacts is dealt with in various fields. In the context of SPLE, some studies suggested refactoring cloned artifacts into SPLs (an extended list of such studies can be found at [24]). They mainly refer to code-related artifacts. Faust et al. [9], for example, propose a method for migrating multiple instances of code units of a "successful" single information system to a SPL. To this end, they introduce a grow-and-prune model which includes two phases: (1) grow, in which the code is copied and modified to implement additional similar functionality, and (2) prune, in which the different variants are merged to support easy percolation of changes. Mende et al. [18] further suggest a tool to support the maintenance of code developed following the grow-and-prune model. In order to identify similar functions that may be merged, token-based clone detection is used to detect pairs of functions sharing code. Then, textual similarity measures are utilized to lift sufficiently similar functions to the architectural level.

Ryssel et al. [25] introduce an automatic approach to recognize variants in a set of function-block-based models, namely, models that decompose the functionality of systems into components (function blocks). The variants are identified after mapping similar components, where similarity takes into consideration the components' names, their characteristics, and their relations to similar neighbor components. The approach further identifies the variation points and their dependencies within variants.

Yoshimura et al. [28] describe an approach to detect variability in a SPL from the change history of the software. The information on those changes is converted to vectors, and factor analysis – a multivariable analysis technique – is applied.

Another corpus of studies offers methods for generating variability models from existing artifacts. In the context of requirements artifacts (see [12] for an extended review of studies on this subject), Chen et al. [6] propose a semi-automatic approach for constructing feature diagrams, commonly used to model variability in SPLE [7]. For each requirements document, a relationship graph is constructed manually, and an application feature diagram is constructed applying a clustering algorithm. Those application feature diagrams are automatically merged into a domain feature diagram.

Niu and Easterbrook [20] introduce a semi-automatic method for constructing variability models. The method extracts functional requirements profiles (FRPs), represented as "verb – direct object" pairs, using expert knowledge and linguistic clues.

Weston et al. [27] introduce an automatic tool, named ArborCraft, for creating feature diagrams from textual requirements documents. The tool uses natural language processing techniques, as well as mining techniques, for finding potential variable elements within the documents.

Reinhartz-Berger et al. [23] suggest introducing ontological considerations when analyzing the variability between functional requirements. The method separately analyzes the differences between the initial states (pre-conditions), external events (triggers), and final states (post-conditions) of the behaviors specified in the functional requirements.

To summarize, existing studies that handle clones focus on specific types of artifacts (most notably code and requirements). Furthermore, they sometimes make assumptions on the cloned artifacts based on the project context or the application domain [24] and they do not directly relate the outcomes to existing variability mechanisms. In this work, we suggest a formal framework in which the relationships between core assets and product artifacts can be represented precisely. Due to its abstract nature, our framework is able to support different development artifacts, including both non-structured artifacts, such as requirements written in a natural language, and structured ones, such as models and code.

3 Running Example

We will use requirements and particularly functional requirements to exemplify the potential of the framework to analyze variability of cloned artifacts. Our example is taken from a domain of library management systems that support functionalities, such as borrowing, returning books, and searching. Table 1 presents four similar requirements of the searching functionality in libraries. These can be considered as clones of the same requirement, each being part of a different product in the SPL. To analyze the variability of these requirements, we have to identify first the common parts (non-bold text). Then, for the other (variable) parts, we have to identify whether they are product-specific parts (calling for core asset extension), optional parts (calling for configuration mechanisms), or variants (calling for applying other variability mechanisms, such as inheritance or parameterization). The first variable part in our example (borrower vs. librarian) should be handled as a variation point with two variants: borrower and librarian. In this case, borrowers and librarians can be considered specializations of a more abstract concept (e.g., users). This is also the case with "a list" and "details" which can be considered different types of information presentation. We further see that searching can be for a book "that has no available copies" (3, 4) or for a book "that has available copies" (1). In any case, this variable part is optional, as the second requirement does not refer to this possibility at all. This possibility is associated in all three cases (1, 3, 4) with presentation of book availability, which is another optional variable part. Finally, searching by author details is an extension, as it appears only in one requirement (4).

Table 1. Four similar requirements of the searching function in libraries

1	When **a borrower** searches for a book **that has available copies** by book name, the system retrieves **the details** of all books that have the requested name.
2	When **a librarian** searches for a book by its name, the system retrieves **a list** of all books with the requested name **and presents their availability**.
3	When **a borrower** searches for a book **that has no available copies** by book name, the system retrieves **a list** of all books with the requested name **and presents their availability**.
4	When **a librarian** searches for a book **that has no available copies** by book name **or author details**, the system retrieves **a list** of all books with the requested name **and presents their availability**.

To provide formal foundations for such variability analysis, we next suggest a framework based on mappings between graphs. Section 5 elaborates on how to use graphs and mappings to deal with textual requirements.

4 The Framework and Its Theoretical Foundations

4.1 Graph-Based Representation Languages

The processes of creating product artifacts from core assets are very diverse, may vary significantly from one domain to another and are language-dependent. In what follows, L is a representation language, on which we can base the definition of well-formed core assets and product artifacts. A prominent example of such a representation language is FD, the language of *feature diagrams*[1]. This language enables representing mandatory, optional, alternative, and OR-grouped features of a family of software products, in addition to dependencies in the form of logical constraints. Feature diagrams can represent the variability of different software development artifacts. The set of core assets represented by FD, CA(FD), is the set of all feature diagrams, while PA(FD) is the set of all feature diagrams instantiations, namely, the representations of software products configurations.

In this work we make further assumptions on the representation language and its associated well-formed core assets and product artifacts. First of all, observing that most of the representation languages used in SPLE can be transformed into some graph-based forms, we assume that L is a language consisting of labeled vertices V, potentially labelled arcs E and meta-data, usually expressed as annotations or constraints, on how the core assets should be reused. Thus our assumption is that the set of well-formed core assets of such language, CA(L), is a set of graphs with some meta-data attached to them, while the set of well-formed product artifacts, PA(L), are simply graphs.

Definition 1: We say that a representation language L is *graph-based* if:

1. Elements of CA(L) (i.e., core assets) have the form c = (G,MD), where G is a graph and MD are constraints represented by meta-data attached to vertices, arcs or the whole graph (G).
2. Elements of PA(L) (i.e., product artifacts) are graphs.

The language of feature diagrams, FD, can be considered a graph-based representation language. The elements of CA(FD) are graphs with metadata, such as "mandatory" and "optional", which are annotated on arcs using dedicated symbols, XOR and OR, which are attached to (parent) vertices, and "requires" and "excludes" constraints which can be either attached to vertices (in case of local constraints) or to the whole graph (in case of global constraints). Fig. 1(a) shows a concrete member c of CA(FD).

---

[1] Although there are many variants of the feature diagrams language, we will treat them as one language abstracting away their differences. A similar approach is taken in [11]. where the authors surveyed different feature diagrams notations and provided them with a generic formal syntax and semantics, called free feature diagrams (FFDs).

The elements of PA(FD) are graphs (without metadata) that represent specific configurations. On Fig. 1(b) we can see a graph p which belongs to the set PA(FD). p further satisfies all the constraints imposed by the metadata of c.

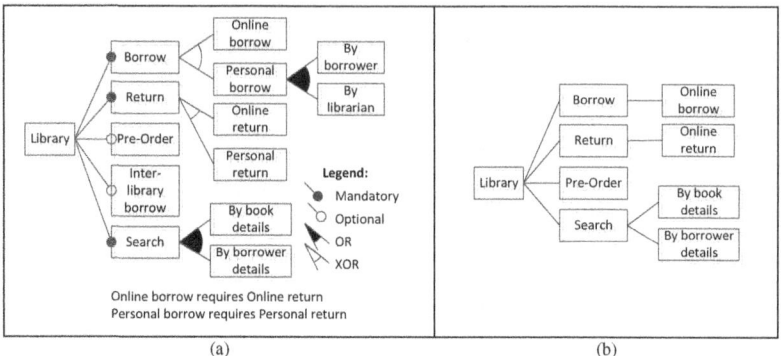

Fig. 1. A core asset (a) and a product artifact (b) specified as FDs

To capture the relation between members of CA(L) and PA(L) in a given graph-based representation language L, we associate L with a binary *reuse relation* which expresses for every $c \in CA(L)$ and $p \in PA(L)$ whether p can be created from c in L.

Definition 2: Let L be a graph-based representation language. The reuse relation of L is a binary relation, $\vdash^L$: CA(L) × PA(L), such that for each $c \in CA(L)$, $p \in PA(L)$, if $c \vdash^L p$, p satisfies the constraints imposed by the meta-data of c.

For instance, the reuse relation of the feature diagrams language FD is defined as follows: p can be created from c if p satisfies the mandatory, optional, OR-grouped, XOR-group, "requires", and "excludes" constraints of c. As noted, the diagrams in Fig. 1 satisfy $c \vdash^{FD} p$.

The reuse relation describes *what* product artifacts can be produced from a core asset, but not *how* the different product artifacts vary, i.e., which elements of the core asset are taken as they are, which are modified (creating different variants), and which are added in particular product artifacts. The notion of reuse relation thus needs to be refined to reflect these aspects. This is done next by introducing *similarity mappings*.

4.2 Similarity Mappings

Similarity mappings provide a concrete way of representing variability by mapping core asset elements to their product artifact counterparts, which are in some sense "similar" to them. To abstract away the details of this similarity, we assume the availability of a Boolean similarity metric, which determines whether a core asset element and a product artifact element are similar or not. Since similarity mappings do not use the core assets metadata, we refer in this section only to the graph part of core assets.

Definition 3: Let $G_1=(V_1,E_1)$ and $G_2=(V_2,E_2)$ be graphs respectively representing a core asset and a product artifact. Let S: $V_1 \cup E_1 \times V_2 \cup E_2 \rightarrow \{0,1\}$ be a Boolean similarity metric. The *similarity mapping* $M_S: V_1 \cup E_1 \rightarrow 2^{V_2} \cup 2^{E_2}$ is defined as follows:

1. For every $v \in V_1$ $M_S(v) = \{v' \in V_2 \mid S(v,v')=1\}$.
2. For every $e \in E_1$ $M_S(e) = \{e' \in V_2 \mid S(e,e')=1\}$.

In other words, the similarity mapping maps elements of the core asset to product artifact elements "similar" to them according to the metric S. Note that a core asset element may have no similar product artifact elements, or it may have more than one similar product artifact element, respectively corresponding to cases where the core asset element is not reused at all or is reused several times in the same product artifact.

For instance, one can define the following similarity metric from $G_1=(V_1,E_1)$ to $G_2=(V_2,E_2)$: for every $x_1 \in V_1 \cup E_1$ and $x_2 \in V_2 \cup E_2$, $S(x_1,x_2) = 1$ if $x_1=x_2$, and $S(x_1,x_2) = 0$ otherwise. We refer to this metric as drastic similarity (adapting the notion of drastic distance [2] to our context). We use the notation of $D(x_1,x_2)$ for drastic similarity. The drastic similarity is used for mapping the graph part of Fig. 1(a) to the product specified in Fig. 1(b). In particular:

M_D (Library) = {Library}, M_D (Borrow) = {Borrow},.., M_D (Inter library borrow)=∅
..., M_D ((Library,Borrow)) = {(Library,Borrow)}, ,...

The following definition introduces two useful notions with respect to similarity mappings: non-total and non-onto sets.

Definition 4: Let S be a similarity metric from $G_1=(V_1,E_1)$ to $G_2=(V_2,E_2)$ and M_S – the associated similarity mapping[2].

- The *non-total set of* M_s, NonTotal(M_s), is the set of all $x \in V_1 \cup E_1$ which has no similar counterparts in $V_2 \cup E_2$. Formally expressed:
  NonTotal(M_s) = {$x \in V_1 \cup E_1 \mid M_S(x)=\emptyset$}.
- The *non-onto set of* M_s, NonOnto(M_s), is the set of all $y \in V_2 \cup E_2$ which has no similar counterparts in $V_1 \cup E_1$. Formally expressed:
  NonOnto(M_s) = {$y \in V_2 \cup E_2, \mid \neg \exists x \in V_1 \cup E_1$ such that $y \in M_S(x)$}.

The similarity mapping defined above between Figure 1(a) and Figure 1(b), which was based on the drastic similarity metric D, satisfies {Inter library borrow, Personal borrow, Personal return} $\subseteq$ NonTotal(M_D) and NonOnto(M_D) = ∅. We would like to relate such a mapping to notions associated with the configuration mechanism.

4.3 Relating Properties of Similarity Mappings to Dimensions of Variability

Consider again a core asset, c, and a set of product artifacts, $p_1,...,p_n$, represented in some graph-based representation language, L, such that $c \vdash^L p_i$. We look at similarity mappings $M_{S1},..., M_{Sn}$ associating c with $p_1,...,p_n$, respectively, in two dimensions: product and element.

1. *Product dimension:* Comparing c against $p_1,...,p_n$, we focus on the questions *which* elements of a core asset are reused, and *which* elements of a product artifact are obtained by reuse. We distinguish between the following cases (see Table 2):

---

[2] The terminology of non-total and non-onto sets is due to their direct relations to properties of total and onto functions: M_s is total if NonTotal(M_s)=∅, and M_S is onto if NonOnto(M_s)=∅.

a. *Mandatory* – a core asset element x, which is reused in *all* product artifacts. This holds whenever for every i: $x \notin NonTotal(M_{Si})$ and $x \notin NonOnto(M_{Si})$.
   b. *Optional* – a core asset element which is not mandatory (i.e., reused only in some of the product artifacts). This holds when for some i: $y \in NonTotal(M_{Si})$.
   c. *Extension* – a product artifact element y which is a *product-specific addition*, i.e., is not derived (reused) from the core asset. This holds whenever $z \in NonOnto(M_{Si})$ for some i.
2. *Element dimension*: comparing a core asset element against a single counterpart in the product artifact, we can distinguish between the following cases of reuse:
   a. *Common* – the product artifact element is identical to the core asset element.
   b. *Variant* – the product artifact element is a variant of the core asset element.

   This dimension is characterized by the properties of the chosen similarity metric. Here we only distinguish between two types of relationships between a core asset element and a product artifact element: identical or similar (see Table 3). If none of them holds (i.e., the elements are different), we assume that no reuse is feasible and the product artifact element is product-specific (see the extension option in the product dimension).

The framework presented above provides a concrete way of representing the association between a core asset and its product artifacts while abstracting away the details of the representation language. This allows us to handle both structured and unstructured representation languages in a uniform way.

Table 2. The product dimension

#	NonOnto	NonTotal	Visualization	Variability
1	$\forall i:$ $x \notin NonOnto(M_{Si})$	$\forall i:$ $x \notin NonTotal(M_{Si})$	CA (x) → PAi (x)	Element x is **mandatory**
2	---	$\exists i:$ $y \in NonTotal(M_{Si})$	CA (y) PAi	Element y is **optional**
3	$\exists i:$ $z \in NonOnto(M_{Si})$	---	CA PAi (z)	Element z is an **extension**

Table 3. The element dimension

#	Similarity	Visualization	Definition	Variability
1	Identical	CA (x) → PA (x)	Element x is used as is.	Common
2	Similar	CA (x) → PA (x')	Element x is modified, creating variants in the product artifacts.	Variant

5 Application of the Framework to Textual Requirements

To demonstrate the applicability of the suggested framework, we utilized it to textual functional requirements, which are common and available development artifacts. Each requirement is composed of one or more sentences, and expresses a different behavior. For simplicity, we currently assume that the different requirements in the same product are independent: a behavior does not influence the execution or the outcome of another behavior. One challenge in applying our framework for this type of artifacts is that graph-based representations are not immediately available, and should be obtained using natural language processing (NLP) techniques.

In terms of our framework, the nodes we refer to are textual phrases of the functional requirements, each describing an atomic action. We name such phrases *behavioral vectors*[3]. The arcs relate the different phrases belonging to the same requirement and their order in the requirement. This way we are able to analyze variability within requirements and not only across requirements, as is the case with existing studies partially reviewed in Section 2. Furthermore, this fine-grained variability analysis sets the ground for automatic generation of "core requirements", namely requirements at the SPL level. The problem we try to tackle is: given a set of requirements, what are the most similar requirements and what are their variable parts in terms of optionality, extension, and variants? Note that in this case, we do not have the core asset explicitly, but different product artifacts from which we eventually would like to reconstruct similarity mappings to support generation of core requirements for systematic reuse. We assume that the given requirements were cloned from the same source, which may be part of the requirements or not. The aim is to analyze the variability of the requirements in order to improve their future management.

To this end, we propose a three stage process, depicted in Fig. 2. In this process, similar requirements are first clustered. Then, the potential variable parts of the requirements in the same cluster are identified, using the common parts as anchors. Finally, the variable parts are categorized, using syntactic techniques, to identify the appropriate types of variability. The following three sub-sections elaborate these stages, while sub-section 5.4 presents some preliminary results demonstrating the potential usefulness of the suggested framework to variability analysis of requirements.

5.1 Clustering Similar Requirements

In the first stage, the different requirements are clustered using a hierarchical agglomerative clustering algorithm [15]. Different semantic similarity measures [19] can be used for this purpose, including corpus-based approaches that are based on information derived from large corpora, e.g., Latent Semantic Analysis (LSA) [16], and knowledge-based measures that use information drawn from semantic networks, such as the method suggested by Mihalcea, Corley, and Strapparava (MCS) in [19]. We further use a similarity measurement, named SOVA, which introduces ontological considerations to overcome some shortcomings of pure semantic measurements and is tailored to analyze variability of functional requirements [23].

---

[3] Section 5.2 elaborates how such behavioral vectors are extracted from text.

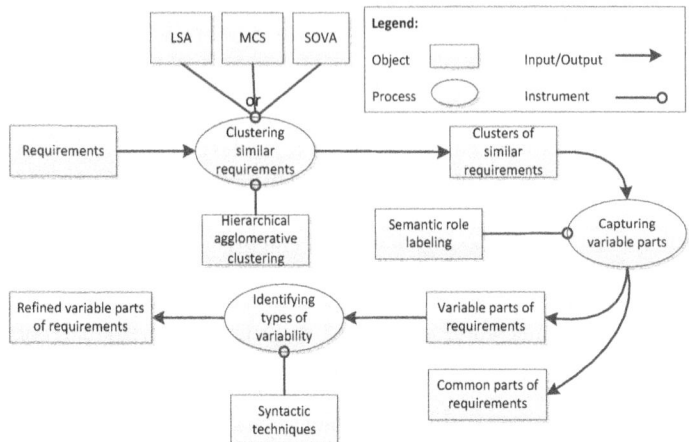

Fig. 2. Applying the suggested framework for identifying and analyzing variability in textual requirements

The emerging clusters may be associated with the elements of the core asset: each cluster represents a core requirement which may appear in different variants in a specific product. The requirements presented in Table 1 can be considered as an example of requirements that were clustered together using MCS or SOVA measures.

5.2 Capturing Variable Parts

In order to capture the variable parts of the requirements in the same cluster, we use NLP techniques. Particularly, we use semantic role labeling (SRL) [10][4] which labels constituents of a phrase with their semantic roles in the phrase. Currently, we refer to the following four semantic roles which are of special importance to functional requirements: (1) Agent – Who performs? (2) Action – What is performed? (3) Object – On what object is it performed? and (4) Instrument – How is it performed? A (functional) requirement is composed of one or more behavioral vectors, where each behavioral vector is composed of the aforementioned roles. Some of the roles may be missing in certain behavioral vectors (e.g., when the requirement is phrased in passive).

Returning to our example, Fig. 3 presents the first two requirements in Table 1 as graphs. Each requirement appears in a separate product and, hence, is considered a separate product artifact (surrounded by an ellipse). Each requirement is further composed of behavioral vectors (depicted in circles). The behavioral vectors are linked according to their appearance order in the sentence or to the occurrence of their associated events in reality (following semantic temporal ordering algorithms, e.g., [17]). Each behavioral vector is further divided into roles (the instrument role is missing in all the demonstrated behavioral vectors). The behavioral vectors are then compared based on their roles. In our example, the first behavioral vector of the first requirement (R1) is similar to the first behavioral vector of the second requirement (R2), and the second behavioral vector of R1 is similar to the second behavioral vector of R2. The third behavioral vector of R2 has no similar counterpart in R1.

---

[4] The online demo is available at http://barbar.cs.lth.se:8081/.

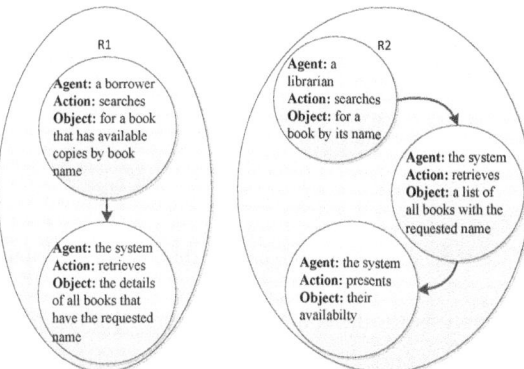

Fig. 3. Graph-based representation of the first two requirements in Table 1

Similar behavioral vectors may include roles whose values are identical, e.g., the actions in the first pair of similar behavioral vectors and the agents and the actions in the second pair of similar behavioral vectors. These common parts are used as anchors for analyzing the variable parts – roles which are similar (but not identical) or different. The identical roles are classified as "common" in the element dimension, while the other roles (the variable parts) are further analyzed in the next stage.

5.3 Identifying Types of Variability

The variable parts of the requirements are now analyzed using syntactic techniques, including tokenization, Part Of Speech (POS)-tagger, lemmatizer, morphological tagger, and dependency parser[5]. This kind of parsing enables examination of variability within a single role and separation between identity and similarity in the element dimension. Consequentially, we can identify variants and associate them to variation points. Examining the two requirements from Fig. 3, Fig. 4 shows the syntactic parsing of these requirements. Based on these parsing outcomes (for the four requirements), Table 4 presents the results of the variability analysis along the lines of the formal framework[6]. The first, third, and sixth rows include non-identical terms and hence are classified as variant in the Element dimension. The terms in the third, eighth, and ninth rows appear in most products (three out of the four used in our example) and thus are classified as optional in the Product dimension. The term in the fifth row, on the other hand, appears only in one product and can be considered as a (particular) extension.

---

[5] These techniques are used within the semantic role labeler at http://barbar.cs.lth.se:8081/.

[6] Note that we currently use in the comparison only key POS, such as nouns, verbs, adjectives, and adverbs. Communicational POS, such as determiners (DT, WDT) and prepositions and conjunctions (IN), are ignored. We further use the dependency parser outcome to extract terms (rather than individual words) that can be compared.

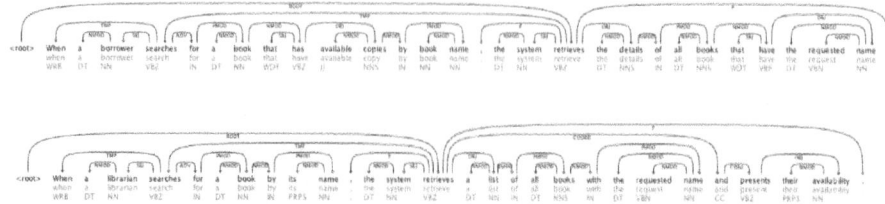

Fig. 4. Syntactic parsing of R1 (top) and R2 (bottom) presented in Table 1

The result of our analysis for the four given requirements is the following core requirement:

When VP1 searches for a book VP2 by book name VP3, the system retrieves VP4 of all books with/that have the requested name VP5.
VP1 (mandatory) – a borrower, a librarian
VP2 (optional) – that has available copies, that has no available copies
VP3 (extension) – or author details
VP4 (mandatory) – a list, the details
VP5 (optional) – and presents their availability

Table 4. Analyzing the variability of four library requirements according to the product and element dimensions

R1	R2	R3	R4	Element	Product
borrower	librarian	borrower	librarian	Variant	Mandatory
book	book	book	book	Common	Mandatory
available copies		no available copies	no available copies	Variant	Optional
book name	its book[7] name	book name	book name	Common	Mandatory
			author details	Common	Extension
details	list	list	list	Variant	Mandatory
books	books	books	books	Common	Mandatory
requested name	requested name	requested name	requested name	Common	Mandatory
	presents	presents	presents	Common	Optional
	availability	availability	availability	Common	Optional

5.4 Preliminary Results

In order to demonstrate the potential usefulness of the suggested framework and its application to functional requirements, we implemented a tool that follows the process presented in Fig. 2. We further compared the tool's outcome to the variability

---

[7] A coreference algorithm, such as that in [22], is used to replace pronouns with their anaphors (i.e., the nouns to which they refer). Here, "its" was replaced with "book".

analysis of a human expert with 25 years of experience in requirements engineering and software development. To this end, we used a questionnaire with 10 requirements of typical functionalities in library management systems[8]. For each requirement, four alternatives were presented. We asked the expert to identify the variability between each one of the four alternatives and the original requirement in terms of changes needed to adapt the alternatives to the requirement. We further asked him to categorize the variability into three categories: adding functionality, removing functionality, and changing functionality (e.g., changing roles or conditions). In terms of our framework, the alternatives can be considered as potential core requirements and the requirements – product artifacts. Optionality, extension, and variants correspond to removing, adding, and changing functionality, respectively[9].

We observed that the tool found all variable parts identified by the expert and categorized them in a similar way. In addition, the tool found further variable parts which were not identified by the expert. While some of the parts originated from syntactic differences and need to be further processed and eliminated by our tool, others were meaningful additions to the expert's variability analysis. This was extremely noticeable when the differences between the original requirement (the product requirement) and the alternative (the core requirement) were substantial, e.g., when the alternative both added functionality and removed another functionality. In those cases, the expert perceived only the most noticeable change – addition of functionality (extension), neglecting the other types of variability and especially change of functionality (variants).

This preliminary evaluation establishes the ability of our framework to capture the main variability types, demonstrating its usefulness for analyzing variability. Based on this, we plan to refine the framework to represent further dimensions of variability.

6 Summary and Future Work

In this paper we have presented a formal framework to represent dimensions of variability. This framework, which has two dimensions – product and element, can be used as a basis for managing cloned artifacts in all development stages and automatically identifying and analyzing variability. We demonstrated the usefulness of our framework by applying it to functional requirements written in natural language. We further demonstrated how the framework can support fine-grained variability analysis and generation of core assets.

As for future research directions, we plan to associate dimensions represented in our framework with concrete variability mechanisms. This way we hope to support the selection of mechanisms to realize variability, thus promoting a systematic reuse of core assets. Another issue deserving further exploration in this context is the relation of the dimensions of our framework to the model projection principle and the generic reference model adaptation mechanisms, presented in [3]. We also plan to extend the

---

[8] The questionnaire can be accessed at http://mis.hevra.haifa.ac.il/~iris/research/OA/QuestionnaireEng.pdf

[9] We deliberately did not use the framework terminology in order to avoid bias.

application of the framework to different development stages, such as architecture design and code implementation, as well as to define an abstract language to specify core assets based on the dimensions of the suggested framework. Finally, we plan to improve and extend the evaluation of the framework usefulness for different software development activities, applying both quantitative and qualitative approaches. In the context of natural language requirements, we plan to extend the support to other roles relevant to functional requirements, such as when an action is performed (temporal modifiers) and in what conditions it is performed (adverbial modifiers).

References

1. Anastasopoules, M., Gacek, C.: Implementing product line variabilities. In: Proceedings of the 2001 Symposium on Software Reusability: Putting Software Reuse in Context, SSR 2001, pp. 109–117 (2001)
2. Arieli, O., Denecker, M., Bruynooghe, M.: Distance semantics for database repair. Annals of Mathematics and Artificial Intelligence 50(3–4), 389–415 (2007)
3. Becker, J., Delfmann, P., Knackstedt, R.: Adaptive Reference modeling: integrating configurative and generic adaptation techniques for information models. In: Becker, J., Delfmann, P. (eds.) Reference Modeling – Efficient Information Systems Design Through Reuse of Information Models. Physica-Verlag HD, pp. 27–58 (2006)
4. Berger, T., Rublack, R., Nair, D., Atlee, J.M., Becker, M., Czarnecki, K., Wasowski, A.: A survey of variability modeling in industrial practice. In: Proceedings of the 7th International Workshop on Variability Modelling of Software-Intensive Systems, VaMoS 2013, pp. 7:1–7:8 (2013)
5. Dannenberg, R.B., Verslype, K., Greefhorst, D., Kuusela, J., Obbink, H., Pohl, K.: Variability issues in software product lines. In: van der Linden, F.J. (ed.) PFE 2002. LNCS, vol. 2290, pp. 13–21. Springer, Heidelberg (2002)
6. Chen, K., Zhang, W., Zhao, H., Mei, H.: An approach to constructing feature models based on requirements clustering. In: Proceedings of the 13th IEEE International Conference on Requirements Engineering, RE 2005, pp. 31–40 (2005)
7. Chen, L., Ali Babar, M., Ali, N.: Variability management in software product lines: a systematic review. In: Proceedings of the 13th International Software Product Line Conference, SPLC 2009, pp. 81–90 (2009)
8. Clements, P., Northrop, L.: Software Product Lines: Practices and Patterns. Addison-Wesley (2001)
9. Faust, D., Verhoef, C.: Software Product Line Migration and Deployment. Journal of Software Practice and Experiences 30(10), 933–955 (2003)
10. Gildea, D., Jurafsky, D.: Automatic Labeling of Semantic Roles. Computational Linguistics 28(3), 245–288 (2002)
11. Heymans, P., Schobbens, P.Y., Trigaux, J.C., Bontemps, Y., Matulevicius, R., Classen, A.: Evaluating formal properties of feature diagram languages. IET Software 2(3), 281–302 (2008)
12. Uslar, M., Specht, M., Rohjans, S., Trefke, J., Gonzalez, J.M.V.: Perspective. In: Uslar, M., Specht, M., Rohjans, S., Trefke, J., Vasquez Gonzalez, J.M. (eds.) The Common Information Model CIM. POWSYS, vol. 2, pp. 189–196. Springer, Heidelberg (2012)
13. Jacobson, I., Griss, M., Jonsson, P.: Software Reuse: Architecture, Process, and Organization for Business Success. Addison-Wesley Longman, Reading, MA (1997)

14. Jaring, M., Dannenberg, R.B.: Representing variability in software product lines: a case study. In: Chastek, G.J. (ed.) SPLC 2002. LNCS, vol. 2379, pp. 15–36. Springer, Heidelberg (2002)
15. Kurita, T.: An efficient agglomerative clustering algorithm using a heap. Pattern Recognition **24**(3), 205–209 (1991)
16. Landauer, T.K., Foltz, P.W., Laham, D.: Introduction to Latent Semantic Analysis. Discourse Processes **25**, 259–284 (1998)
17. Mani, I., Verhagen, M., Wellner, B., Lee, C.M., Pustejovsky, J.: Machine learning of temporal relations. In: The 21st International Conference on Computational Linguistics and the 44th Annual Meeting of the Association for Computational Linguistics, pp. 753–760 (2006)
18. Mende, T., Koschke, R., Beckwermert, F.: An Evaluation of Code Similarity Identification for the Grow-and-Prune Model. Journal of Software Maintenance and Evolution **21**(2), 143–169 (2009)
19. Mihalcea, R., Corley, C., Strapparava, C.: Corpus-based and knowledge-based measures of text semantic similarity. In: The 21st National Conference on Artificial intelligence, AAAI 2006, vol. 1, pp. 775–780 (2006)
20. Niu, N., Easterbrook, S.: Extracting and modeling product line functional requirements. In: The 16th IEEE International Requirements Engineering Conference, RE 2008, pp. 155–164 (2008)
21. Pohl, K., Böckle, G., van der Linden, F.: Software Product-line Engineering: Foundations, Principles, and Techniques. Springer, Heidelberg (2005)
22. Raghunathan, K., Lee, H., Rangarajan, S., Chambers, N., Surdeanu, M., Jurafsky, D., Manning, C.: A multi-pass sieve for coreference resolution. In: The Conference on Empirical Methods in Natural Language Processing, EMNLP 2010, pp. 492–501 (2010)
23. Reinhartz-Berger, I., Itzik, N., Wand, Y.: Analyzing variability of software product lines using semantic and ontological considerations. In: Jarke, M., Mylopoulos, J., Quix, C., Rolland, C., Manolopoulos, Y., Mouratidis, H., Horkoff, J. (eds.) CAiSE 2014. LNCS, vol. 8484, pp. 150–164. Springer, Heidelberg (2014)
24. Rubin, J., Czarnecki, K., Chechik, M.: Managing cloned variants: a framework and experience. In: Proceedings of the 17th International Software Product Line Conference, SPLC 2013, pp. 101–110 (2013)
25. Ryssel, U., Ploennigs, J., Kabitzsch, K.: Automatic variation-point identification in function-block-based models. In: Proceedings of GPCE 2010, pp. 23–32 (2010)
26. Schnieders, A., Puhlmann, F.: Variability mechanisms in e-business process families. In: International Conference on Business Information Systems, BIS 2006 (2006)
27. Weston, N., Chitchyan, R., Rashid, A.: A framework for constructing semantically composable feature models from natural language requirements. In: Proceedings of the 13th International Software Product Line Conference, SPLC 2009, pp. 211–220 (2009)
28. Yoshimura, K., Narisawa, F., Hashimoto, K., Kikuno, T.: FAVE: factor analysis based approach for detecting product line variability from change history. In: Proceedings of MSR 2008, pp. 11–18 (2008)

Solving Semantic Disparity and Explanation Problems in Regulatory Compliance- A Research-In-Progress Report with Design Science Research Perspective

Sagar Sunkle[✉], Deepali Kholkar, and Vinay Kulkarni

Tata Consultancy Services, 54B, Hadapsar Industrial Estate, Pune 411013, India
{sagar.sunkle,deepali.kholkar,vinay.vkulkarni}@tcs.com

Abstract. Modern enterprises increasingly face the challenge of keeping pace with regulatory compliances. Semantic disparity between regulation texts, their interpretations, and operational specifics of enterprise often leads enterprises to situations where it becomes difficult for them to establish what compliance means, how they are supposed to affect it in the operational practices, and how to prove that they comply when asked for explanations of (non-)compliance. We take a step toward reducing the semantic disparity by using semantic vocabularies to map regulations with available operational details of enterprise and utilize them in enacting compliance. We also propose to provide explanations of proofs of (non-)compliance. We report our ongoing work in this regard using the design science research (DSR) paradigm. Initial iterations of design cycle from DSR have been useful to us in identifying and matching stakeholder-specific goals in solving these problems.

Keywords: Regulatory compliance · SBVR · Business process · Enterprise data · GRC · Design science research · Design cycle

1 Introduction

Modern enterprises need to face many change drivers effectively and efficiently in order to stay competitive. *Regulatory compliance* is a unique change driver that enterprises find more difficult to deal with in a cost effective manner than other change drivers because a) it impacts operational practices of enterprise substantially and b) non-compliance is penalized severely in most countries and across various domains of operations [1,2]. In most enterprises, regulatory compliance is still largely a document-oriented function [3] supported by industry governance, risk management, and compliance (GRC) solutions. GRC solutions make use of in-/semi-formal techniques and are in high demand not only in US but also in Canada, Japan, India, Australia, South Africa, and members of EU [1]. With spend on compliance slated to rise to many billions of dollars in next 10 years, the need for improving the state of practice in regulatory compliance is all the more evident.

In contrast to industry GRC solutions, state of the art in regulatory compliance research mostly uses formal models of both regulations and operational details of enterprise to check and remediate non-compliance [4–10]. Based on underlying formal logics, it may also be possible to obtain proofs of (non-) compliance which are critical in the real world compliance determination [11]. Compliance function of enterprises can thus benefit from industry GRC solutions improved with ideas from research in a number of ways.

In this paper, we present a research-in-progress report on our attempt to bring the best of both worlds of industrial GRC solutions and compliance research approaches in industry setting. We are using design science research (DSR) paradigm for researching, implementing, and reporting our approach. Based on our observations of problematic phenomena in our interactions with clients, we propose to a) reduce semantic disparity between regulations and operational details of enterprise and b) improve explanations of proof of (non-) compliance. We utilize the design/regulative cycle [12,13] in DSR with particular focus on artifacts [14] to explicate the practical problems in a) and b). Our specific contributions in this paper are twofold- 1) we present initial solutions to problems of semantic disparity and stakeholder-specific proof explanations in regulatory compliance where meanings of concepts expressed using vocabularies form the core and 2) we report the implementation of DSR design cycle for both the problems.

The paper is arranged as follows. In Section 2, we briefly discuss DSR and steps of design cycle. We have early results for which we report a) an iteration of all the steps of design cycle in Section 3, b) the first two steps in Section 4. Section 5 concludes the paper.

2 Note on DSR

Design Science Research (DSR) is a problem-solving paradigm that aims to resolve problems using innovative artifacts against criteria of utility within an operating context such as an enterprise, a domain, a social setting, etc. [12]. It is well suited to solve practical problems which are identified as *"intended differences between the way stakeholders experience the world and the way they would like to experience it"* [15]. The development and evaluation of specific problem solutions can be viewed as having a special emphasis on sufficiently new or decisively better *artifacts* [14].

A widely accepted classification of types of artifacts is: (1) constructs, (2) models, (3) methods, and (4) instantiations [12]. *Constructs* are abstract conceptualizations of problem and solution spaces with varying levels of formalism. Natural language documents such as regulation texts are informal constructs. *Models* are a representation of reality for varying purposes such as communication and analysis. Formal models of regulations and business process models are examples of models. *Methods* are a structured, goal-oriented, repeatable set of activities where other types of artifacts may be generated and utilized. Steps in using Semantics of Business Vocabularies and Rules (SBVR) as a semantic bridge between regulations and operational details explained in Section 3 constitute a method. The *instantiation* artifact is the operationalization of prior construct, model, and method artifacts as a proof-of-concept to demonstrate the feasibility and effectiveness of the proposed solution.

The DSR process itself is divided into various steps such that based on problem analysis, specific artifacts are constructed, evaluated, and interpreted. Artifacts are improved with insights from evaluation in iteration till the problem-specific goal is met. From the next section onward, we report our research using five distinct steps for each area of improvement referred to in Section 1: practical problem investigation, artifact design, artifact validation, artifact implementation, and implementation evaluation [15]. *Investigation of a practical problem* examines the involved stakeholders and their goals, causes of problematic phenomenon that stakeholders want to resolve, and criteria applicable to the solution. *Artifact design* studies available solution artifacts and proposes design for new set of artifacts for solving the problem under consideration. *Artifact validation* tests properties of suggested artifacts against criteria laid out earlier and explores in which ways conclusions drawn can be applied to real solution behavior as expected by stakeholders. *Artifact implementation* attempts to implement the artifacts in their respective tooling environment as applicable in an integrated manner. Finally, *Implementation evaluation* is carried out using methods such as assisted interviews, questionnaires, and case studies among others [12–14] and may lead to insights which can be incorporated into artifacts in the next iteration of the design cycle.

DSR process/design cycle steps may be conducted in any order depending on situations with several nested iterations [13]. Next section elaborates how we attempt to solve the practical problem of semantic disparity in regulatory compliance using DSR.

3 Reducing Semantic Disparity

3.1 Problem Investigation

In attempting to solve a practical problem of enabling formal approaches in an industry offering of a compliance solution, we found that the need for reconciling the distinct conceptual realms of regulations and enterprises' operations is often sidestepped in formal compliance checking solutions and emulated via taxonomies in industrial GRC solutions. We show this in a high level depiction of formal and industrial GRC approaches to compliance checking in Figure 1.

Academic compliance checking approaches tend to utilize business process (BP) models as explicit representation of operational details of enterprise and check compliance of regulations against these [4–10]. While formal models of regulations and business processes may differ in these approaches, each approach ideally requires to *relate* regulations to business process entities. A terminological mapping would essentially tell *where* in the business process a rule from the regulation becomes applicable. Surprisingly, business process compliance checking approaches implicitly assume such mapping to exist without describing how to arrive at it.

Solving Semantic Disparity and Explanation Problems   329

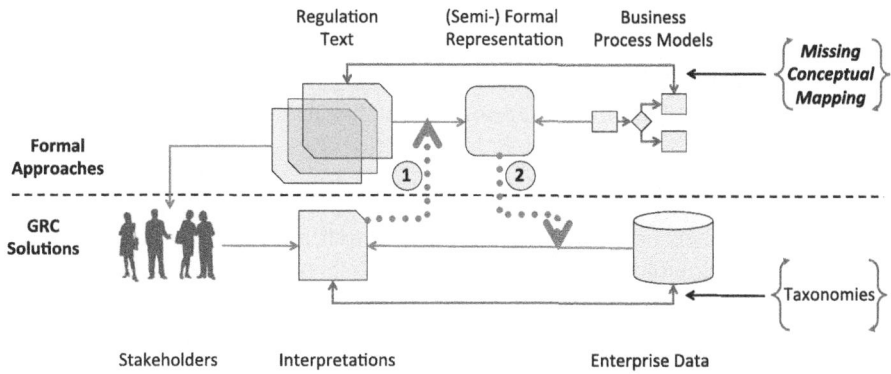

Fig. 1. State of the Art and Practice in Compliance Checking

In contrast, industrial governance, risk, and compliance (GRC) solutions tend to include taxonomy tagging mechanisms in their content management tools. Experts trained on GRC solution platforms tag regulations and use these to search and validate documents. Compared to academic compliance approaches, GRC solutions are document/artefact-oriented and rely on enterprise data to prove compliance to regulations [3].

As shown above the horizontal dashed line in Figure 1, formal approaches represent both rules derived from the text of regulations and facts derived from business processes in a given formal language for compliance checking. Steps followed in industrial GRC solutions are shown below the horizontal dashed line in Figure 1. Various stakeholders interpret regulations in the current context of enterprise and tag these interpretation with enterprise taxonomies. The tags indicate what enterprise data should be pulled and checked against interpreted regulations.

With regards semantic disparity or differences in the meanings of concepts used in regulation texts and operational details, two pointers are relevant. ① in Figure 1 indicates that interpretations of regulation text by such stakeholders as enterprise legal advisors, compliance experts, CxO level business stakeholders, and operational managers, which are prevalent and even necessary in industry are ignored in formal approaches to a large extent [16]. Formal approaches often show direct translation of regulations to rules in the formal language used by the approach. A terminological mapping forms the first step of supporting interpretations of regulations when using formal approaches.

② in Figure 1 indicates that industrial GRC solutions hardly leverage formalisms available in research. Complexity of legal text of regulations and frequent amendments by regulatory bodies make it a demanding task to check and re-validate compliance. Formal methods to compliance checking can be very useful in industry solutions for this very reason.

The problematic phenomenon in the observation depicted in Figure 1 is that of semantic disparity between regulations and operational details with stakeholder-specific interpretations adding to that disparity. The criteria for a solution to this problem is that it should be possible to formally relate regulations to operational details including business process models and enterprise data in general and also support interpretations in further iterations of this solution.

3.2 Artifact Design

As described earlier, academic compliance research approaches often implicitly assume that a terminological mapping already exists between regulations and the BP models. The locations where regulations become applicable are then found by constructing an execution trace as in [4], finding paths in process structure tree as in [7], or placed manually on a business property specification language diagram as in [5]. Labels from business process in such traces, paths, or other representations are often presumed to map to labels of formal models of regulations. Ontologies are suggested in [6] to tackle the semantic disparity- *"Semantic consistency of the data and of the definition of regulations must be achieved. We see semantics in the form of ontologies as the solution to this challenge."*, but the actual mapping is not explained nor clarified with an example. Use of Semantics of Business Vocabularies and Rules (SBVR) or vocabularies is suggested in [17–19], but these approaches use SBVR to describe process rules, transform SBVR process vocabulary to a formal compliance language, and represent legal rules using SBVR respectively without using SBVR/vocabulary for *both* legal rules and processes/enterprise data.

Based on the existing work in process compliance checking and the way industry GRC solutions *relate* regulations to enterprise data, it is clear that a vocabulary/ontology based approach is required that makes the terminological mapping intuitive. Compared to domain-specific ontologies, SBVR provides a semantic model for formal terminology which is the SBVR vocabulary, as a cohesive set of interconnected concepts, with behavioral guidance in terms of policies and rules to govern the actions of subject of the formal terminology [20]. Also SBVR supports *disambiguation* of concepts via semantic communities that enable cross-discipline / cross-subject field capability with support for adopting concepts. Concepts like speech community and subject field provide the necessary context to *disambiguate* multiple uses of the same signifier to designate concepts. Based also on the criteria in the problem investigation, we need vocabularies for both regulations and business process/enterprise data. Once vocabularies are created based on SBVR, we need to elaborate how concepts can be mapped from these vocabularies. Furthermore, since our approach needs to work for both business process models (when they are available) and enterprise data, we also need to arrive at a data model of some sort from the mapping which the enterprise data can map to.

Some iterations of this line of thought over the design/regulative cycle led us to various steps shown in Figure 2 that make use of the aforementioned artifacts. In the following, we describe what these artifacts are and how they are validated by implementing them in the context of a case study of Reserve Bank of India's (RBI) Know Your Customer (KYC) regulations for a typical Indian bank BankA.

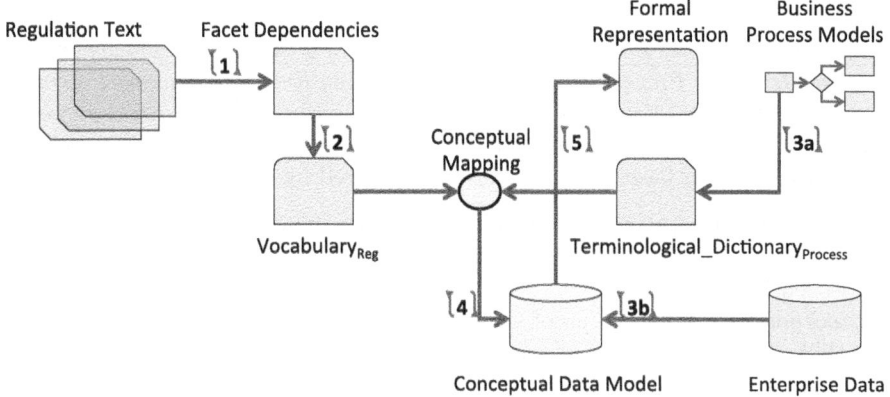

Fig. 2. Steps in using SBVR as a Semantic Bridge Between Regulations and Operational Details

3.3 Artifact Validation

Step [1] in Figure 2 shows that we obtain what we refer to as *facet depedencies* from the regulation text. Facets are key concerns of a regulation in that often (pre-)conditions on some facets obligate/permit (post-)conditions on some other facets. Facet dependencies also help to scope the target of the vocabularies, since regulation text may often have number of interrelated concepts which need not be considered for compliance. The facets that we showcase here are from RBI's KYC regulations. KYC is a process by which banks obtain information about the identity and address of the customers[1]. KYC processes are aimed at identifying *different types of customers*, accepting them as customers of given bank when they fulfill certain identity and address documentation criteria laid out in various regulations and annexes in the most recent RBI KYC master circular[2], and categorising them into various risk profiles for periodic updating.

The following shows how facet dependencies are manually extracted and represented as if-then-[else] statements.

KYC Regulation for Salaried Employees [RBI KYC Customer Identification 2014 §2.5 (vii)]

... for opening bank accounts of salaried employees some banks rely on a certificate / letter issued by the employer as the only KYC document ..., banks need to rely on such certification only from corporates and other entities of repute and should be aware of the competent authority designated by the concerned

---

[1] See RBI KYC Guidelines for Customers http://www.rbi.org.in/commonman/english/scripts/FAQs.aspx?Id=840.

[2] See RBI KYC 2014 Master Circular http://www.rbi.org.in/scripts/BS_ViewMasCirculardetails.aspx?id=9074#23.

employer to issue such certificate/letter. Further, *in addition to the certificate from employer, banks should insist on at least one of the officially valid documents as provided in the Prevention of Money Laundering Rules (viz. passport, driving license, PAN Card, Voters Identity card etc.) or utility bills for KYC purposes for opening bank account of salaried employees of corporates and other entities.* The corresponding facet dependencies are captured like so-

> **If** customer_type=individual and customer_subType=private_salaried_employee and employer_approved_corporate = true and accept_certificate_issued_by
> _corporate = true and no_of_additional_documents_for_verification ¿= 1
> **then** open_account_for_customer = true

Facets here are *customer type, customer sub-type, whether employer of the salaried employee is an approved corporate, whether to accept certificates issued by this employer and additional verification documents*. We split customers types into main and sub-types because KYC regulations distinguish between single (individual) customer and group of customers and further kinds of individual and group customers.

The vocabulary for regulations (shown as Vocabulary$_{Reg}$) in step [**2**] in Figure 2 is defined by focusing on these key terms and various facets from the dependencies pertaining to given rules. First, *business vocabulary* is created to capture semantic communities, in our case the banking industry, with sub-communities as RBI and BankA. Each semantic community is unified by a shared understanding of an area, i.e., body of shared meanings such as RBI_Regulations for RBI as illustrated in Figure 3 by Vocabulary$_{Reg}$. This in turn can comprise smaller bodies of meanings, e.g., RBI_KYCRegulation, each containing a body of shared concepts that captures concepts and their relations, and a body of shared guidance containing business rules.

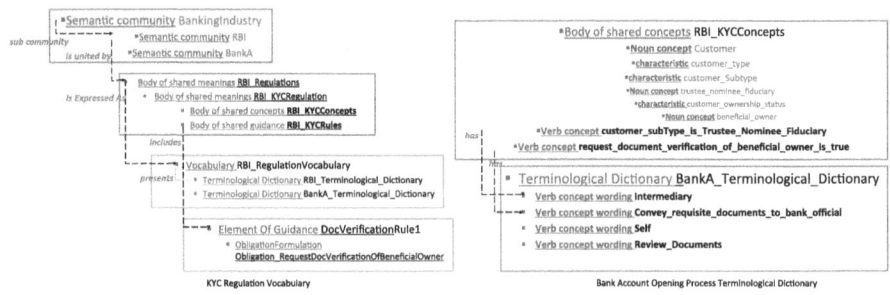

Fig. 3. Vocabulary$_{Reg}$ and Terminological_Dictionary$_{Process}$

Second, *meaning and representation vocabulary* concepts are captured as noun and verb concepts. Noun concepts, e.g., Customer, can have characteristics such as customerType. Each condition in the facet dependencies shown earlier becomes a verb concept in the regulation vocabulary of the form *characteristic = value*. These verb concepts such as *customer_type is individual* form the basis for atomic logical formulations.

Third, we build *business rule vocabulary* that represents rules in the KYC regulations using *logical formulation of semantics vocabulary* which consists of atomic and compound formulations such as conjunctions, implications, obligation and permission formulations etc. Each rule is defined as an element of guidance which embeds a logical formulation. For instance, DocVerificationRule1 is defined as an obligationFormulation for the facet dependency shown earlier.

In order to map an enterprises business process to regulation concepts as shown by step [3] in Figure 2, we use the notion of a terminological dictionary to map terms used in the enterprise business process as representations of concepts in the regulations body of concepts. Each community can define its own terminological dictionary containing expressions to represent common concepts from the shared body of concepts, in its own vocabulary. E.g., BankA_Terminological Dictionary is created to contain business terminology used by BankA as illustrated in Figure 3 by Terminological_Dictionary$_{Process}$.

BankA's process model is shown in Figure 4. A general bank official interacts with a client while KYC documents are managed by content management official. The compliance official is in charge of compliance function. This BP model is traversed to generate BankA Terminological Dictionary which is generated in

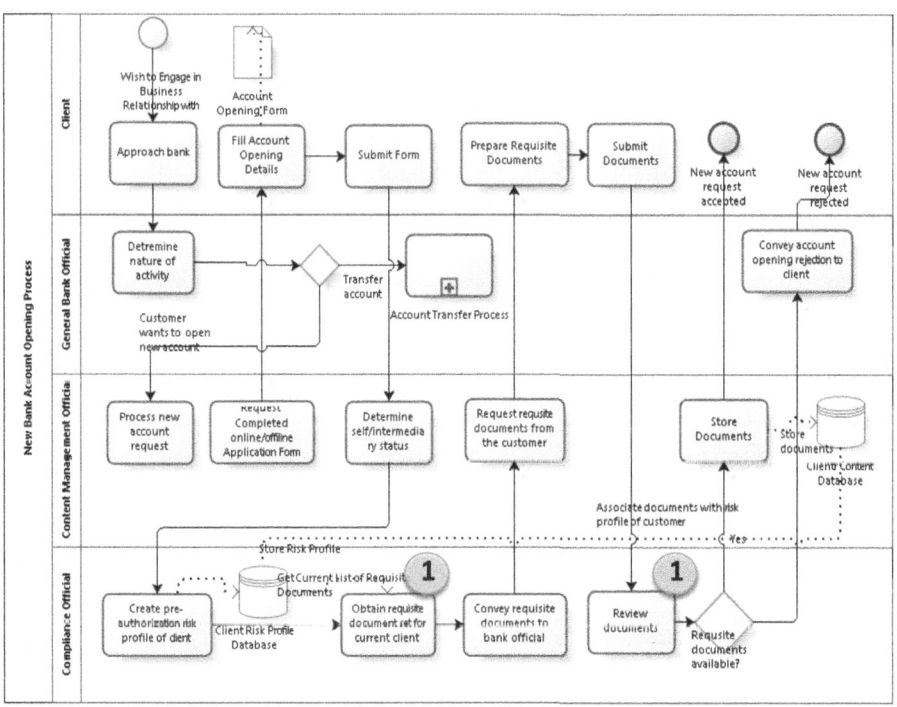

Fig. 4. Account Opening Business Process

the form of a list of Verb concept wordings corresponding to a) each Task/SubProcess from the process, e.g., Approach Bank, Process New account Request and b) each object and condition label/phrase in the process, e.g., Client Risk Profile Database, Self, Intermediary.

Mapping Regulations to Business Process Models The noun concept Customer from Vocabulary$_{Reg}$ maps to the designation Client in the Terminological_Dictionary$_{Process}$. SBVR defines verb concept wordings as representations of verb concepts in their most general form. Every verb concept in the Vocabulary$_{Reg}$ is mapped to corresponding verb concept wording from the Terminological_Dictionary$_{Process}$. Note that *this activity is performed by domain experts*, here legal advisors and operational managers in charge of the business process. If this mapping is stored in a table, it can be used as lookup table for consequent terms in the eventual rules formed in step [5]. In this step, the process entity that corresponds to given term according to the mapping is treated as a placeholder.

Step [4] in Figure 2 creates a conceptual data model using Vocabulary$_{Reg}$ and Terminological_Dictionary$_{Process}$. For each designation requisite schema is created from characteristics. From the schema, **client_account_data, client_data, pse_data,** and **pse_KYC_document_data** are tables of the conceptual data model which can also be treated as relations in logic programming. This helps us in both steps [3b] and [5] in Figure 2. Regarding step [3b], ideally a mapping similar to shown earlier in step [3a] would exist between regulation vocabulary and enterprise data dictionary. A compliance official would pull requisite data from enterprise data-stores as per conceptual data model to check compliance.

Step [5] in Figure 2 can be implemented by using any given formalism to represent the rules and facts for compliance checking. Here, we showcase rules and facts for private salaried employee using DR-Prolog [21] syntax as shown in Listing 21.1. The schema relations are highlighted in Listing 21.1. The regulation is complied with for individual 17 whereas conditions for individuals 18 and 19 result in non-compliance.

Listing 21.1. Case Theory in DR-Prolog

```
1  defeasible(r3,obligation,client_account_data(Client_ID,open_account),
2  [client_data(Client_ID,ind,pse),pse_data(Client_ID,approvedCorporate),
3  pse_KYC_document_data(Client_ID,acceptApprovedCorpCertificate,
       pse_kyc_document_set)]).
4  /* Everything is OK, so account can be opened.*/
5  fact(client_data(17,ind,pse)).
6  fact(pse_data(17,approvedCorporate)).
7  fact(pse_KYC_document_data(17,acceptApprovedCorpCertificate,
       pse_kyc_document_set)).
8  /* Corporate is not approved, account cannot be opened*/
9  fact(client_data(18,ind,pse)).
10 fact(pse_data(18,not(approvedCorporate))).
11 fact(pse_KYC_document_data(18,acceptApprovedCorpCertificate,
       pse_kyc_document_set)).
12 /* Requisite documents not submitted, account cannot be opened*/
13 fact(client_data(19,ind,pse)).
14 fact(pse_data(19,approvedCorporate)).
15 fact(pse_KYC_document_data(19,acceptApprovedCorpCertificate,not(
       pse_kyc_document_set))).
```

Other compliance checking approaches such as [4] can similarly adopt this schema by incorporating it into their rules and facts and check for compliance.

This rule is then annotated to the BP model at the placeholders found earlier in Figure 4. It is thus attached to various process entities as shown in Figure 4 by number 1 in coloured circle ① to various tasks of the compliance official. Note that annotation ① ends up on two tasks; in the first task where compliance official obtains requisite documentation criteria for private salaried employee, she needs to check whether the corporate employer is approved by the bank or not. Depending on this, customer is asked to submit one or more documents as required in RBI KYC guideline for private salaried employee.

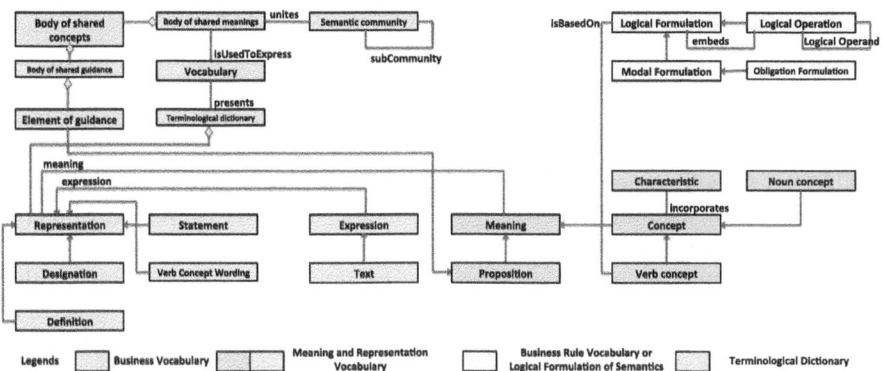

Fig. 5. SBVR Meta-model used for Modelling Vocabularies

3.4 Artifact Implementation

To implement vocabulary artifacts, we imported elements shown in Figure 5 from the consumable XMI of SBVR meta-model available at OMG site[3] into Ecore model. The BP model is created and traversed using an in-house tool described in [22]. Both facet dependencies and data model are textual artifacts. For DR-Prolog compliance checking, we used TuProlog[4]

3.5 Implementation Evaluation

We have already demonstrated artifacts created for solving the problem of semantic disparity in regulatory compliance using *case study* method for evaluation where RBI's KYC regulations are related to BankA's business process model. Apart from *private salaried employee* type of customer, we have modeled vocabularies for regulations about *trustee/nominee/fiduciary, politically exposed person,*

---
[3] See under *Normative Machine Consumable Files* at http://www.omg.org/spec/SBVR/20130601/SBVR-XMI-Metamodel.xml
[4] See http://apice.unibo.it/xwiki/bin/view/Tuprolog/

and *intermediary* types among others and mapped them to BankA's business process model. These are all *individual customer* types. RBI's KYC also refers to *groups of individuals* such as *accounts of companies and firms* which we are yet to model. An internal evaluation of *artifact implementation* by partner stakeholders and also our own led to following remarks:

1. The proposed approach is better than taxonomies used currently as it has the provision for describing semantics or meaning of concepts for both regulations and enterprise data.
2. The approach still maps regulations directly from legal text which is never the case as far as banking domain is considered. Question was raised as to whether the approach can accommodate interpretations by stakeholders. Suggestion was made to represent legal and business context interpretations via vocabularies. Since our problem investigation had already revealed that regulation texts are interpreted, we responded by planning to include interpretation in the next iteration of the design cycle.
3. The mapping still requires domain experts on legal and operations sides. Suggestion was made to use *semantic similarity* measurement techniques [23,24] to help experts with meaning and disambiguation of concepts. We responded by planning to investigate semantic similarity techniques and creating requisite artifacts according to design cycle in the next iteration.
4. A suggestion was made to recheck the formulation of rules as some rules were annotated on number of process entities (as in case of rule described in this paper). If the regulation denoted that certain information was not already available at a given process entity then suggestion was made to split the rule accordingly so that requisite information was available and a rule in its entirety would map to only a single process entity as is the practice. We realized that there could be two reasons why a rule may be annotated to more than one process entity- a) either the rule is in fact trying to capture more than one set of dependencies or b) the mapping needs to be improved by further disambiguation so that it gets annotated to a single correct process entity. We also attributed the problem to the level of granularity of business process as well, since BP models in practice were often at a high level of granularity compared to BP models showcased in research.
5. Facet dependencies and DR-Prolog rules and facts are currently textual artifacts created manually. A suggestion was made to represent these two artifacts in such a way that consistency of specification of rules and facts could be assured. Further discussion within internal reviewers resulted in suggestion to use some sort of simple model transformation where such checks could be easily implemented.

The above evaluation led to refining the design of the artifacts such as interpretation texts and also to the inclusion of other lines of investigation such as similarity measurement techniques for mapping and consistent transformation of rules and facts using model-driven techniques. Currently, we are iterating over the design cycle for the inclusion of these. Overall, the solution was deemed as matching the desired criteria of being able to formally relate regulations to operational details including business process models and enterprise data in general.

We now report the progress in research on the second practical problem of stakeholder-specific proof explanations, for which we found that vocabulary artifacts designed and implemented for the first problem could be reused.

4 Stakeholder-Specific (Non-)Compliance Proof Explanation

4.1 Problem Investigation

Study of trends in industry compliance reporting reveals that auditors increasingly expect consistent evidence of compliance whereas enterprise management expects an accurate and succinct assessment of risks associated with compliance [2]. Furthermore, explanation of proofs of (non-)compliance are increasingly expected to include which regulations a given operational practice of enterprise is subject to and what parts of a regulation does the practice depart from and why [25]. The last part is especially relevant for shareholders since it forces an enterprise to give business reasons for non-compliance.

The criteria for solution to this problem is that it should be possible to generate basic explanations that can be enhanced with stakeholder-specific perspectives and in the long run include ability to explicate enterprises' business reasons behind (non-)compliance in specific operational practices.

4.2 Artifact Design

In compliance research, in spite of formal techniques in use for compliance checking, proof explanation has received less attention [11,26] in general. The proof explanations are constructed by obtaining inference trace and processing it to indicate which rules matched and which failed based on which facts. This certainly provides the starting point in providing explanations of (non-)compliance. But this basic functionality needs to be extended with artifacts that can at least help compliance team to create stakeholder-specific explanations.

Industry GRC solutions often market reporting abilities for risk adjusted decision making where control objectives are described based on views of enterprise data and risk heat maps [1,2]. These functionalities are widely used in reporting, but using formal techniques, they can be enhanced because formal techniques could help in automated generation of at least basic proof explanation.

The design cycle for the problem of semantic disparity already provides some useful artifacts for stakeholder-specific proof explanations of (non-)compliance. Figure 6 shows how artifact design for the problem of semantic disparity can be leveraged for enhanced proof explanations.

Basic proof explanation generation can be implemented easily on top of a given formalism especially if there are ways of constructing inference proofs using those formalisms already in place. We demonstrated DR-Prolog rules and facts for a RBI KYC regulation in Section 3.3. Proof generation in DR-Prolog

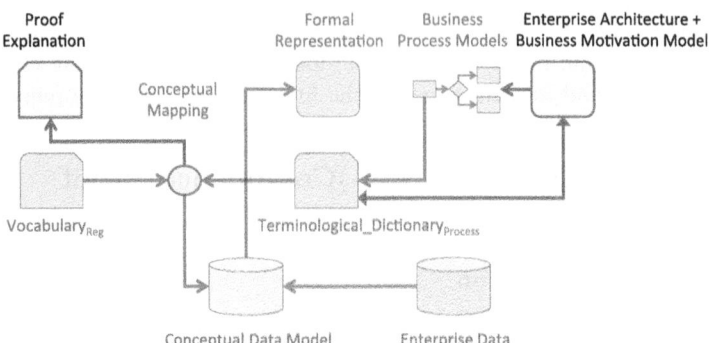

Fig. 6. Artifact Design for Stakeholder-specific Proof Explanations

as in [11] can be leveraged to obtain basic proof of (non-)compliance. Existing vocabulary artifacts Vocabulary$_{Reg}$ and Terminological_Dictionary$_{Process}$ can be used to generate a more detailed proof from the basic proof by inserting relevant meanings from the vocabularies for current set of terms in the proof.

Also, business objectives which the existing operational practices serve can be modeled with enterprise architecture and business motivation models. In [27], we showed how to incorporate directives such as internal policies and external regulations into enterprise to-be architecture. In a nutshell, our approach in [27] enabled querying directives given operational details such as business process specifics and also querying business process specifics given directives. If the high level motivations of enterprise were expressed in the Terminological_Dictionary$_{Process}$, it might be possible to relate these to terms in the proof explanation. This would help in bringing business reasons for (non-)compliance of specific regulations. It would also help in generating business-stakeholder-specific explanations.

Finally, as in the next iteration of problem of semantic disparity, if vocabulary artifacts pertaining to interpretations were included, it might be possible to enhance basic proof explanation with these artifacts for the concerned stakeholders.

Other Steps in Design Cycle We have not yet conducted artifact validation, artifact implementation, and implementation evaluation for the problem of stakeholder-specific (non-)compliance proof explanations and it constitutes our ongoing work along with the next iteration of design cycle for the problem of semantic disparity.

5 Conclusion

We investigated the problems of semantic disparity and stakeholder-specific proof explanations in regulatory compliance in Section 3 and 4 respectively. First iteration of design cycle for the problem of semantic disparity led us to using

SBVR to express meaning of regulations and operational details. The conceptual mapping between the vocabularies of regulation and operational specific and data model enables using BP models when they are available and enterprise data alike. Various suggestions from the internal evaluation indicated that further investigations are needed in terms of incorporating vocabularies for interpretations of regulations and using semantic similarity and model-driven techniques to enhance proposed conceptual mapping. We also found that the artifacts created for the problem of semantic disparity are useful in solving the problem of stakeholder-specific proof explanations. We believe that further iterations of steps of design cycle for both the problems will yield solutions that take us closer to defined criteria for each.

The DSR and the design cycle has been useful in our work to clarify what goals are important to which stakeholders and which artifacts can help in reaching those goals. DSR also helped us in positioning our solution approaches with existing state of the art and practice. While the actual implementation of various steps of design cycle is messier than the linear manner in which we report them here, focus on specific activities at each step helped us in keeping the holistic view of the problem and lines of attack.

References

1. French Caldwell, J.A.W.: Magic Quadrant for Enterprise Governance, Risk and Compliance Platforms (Gartner) (2013)
2. English, S., Hammond, S.: Cost of Compliance 2014 (Thomson Reuters Accelus) (2014)
3. Racz, N., Weippl, E., Seufert, A.: Governance, risk & compliance (GRC) software - an exploratory study of software vendor and market research perspectives. In: Proceedings of the 2011 44th Hawaii International Conference on System Sciences. HICSS 2011, Washington, DC, USA, pp. 1–10. IEEE Computer Society (2011)
4. Sadiq, W., Governatori, G., Namiri, K.: Modeling control objectives for business process compliance. In: Alonso, G., Dadam, P., Rosemann, M. (eds.) BPM 2007. LNCS, vol. 4714, pp. 149–164. Springer, Heidelberg (2007)
5. Liu, Y., Müller, S., Xu, K.: A Static Compliance-checking Framework For Business Process Models. IBM Systems Journal **46**(2), 335–362 (2007)
6. El Kharbili, M., Stein, S., Markovic, I., Pulvermüller, E.: Towards a framework for semantic business process compliance management. In: The Impact of Governance, Risk, and Compliance on Information Systems (GRCIS). CEUR Workshop Proceedings, vol. 339, pp. 1–15. Montpellier, France, June 17, 2008
7. Awad, A., Smirnov, S., Weske, M.: Resolution of compliance violation in business process models: a planning-based approach. In: Meersman, R., Dillon, T., Herrero, P. (eds.) OTM 2009, Part I. LNCS, vol. 5870, pp. 6–23. Springer, Heidelberg (2009)
8. Ly, L.T., Rinderle-Ma, S., Knuplesch, D., Dadam, P.: Monitoring business process compliance using compliance rule graphs. In: Meersman, R., Dillon, T., Herrero, P., Kumar, A., Reichert, M., Qing, L., Ooi, B.-C., Damiani, E., Schmidt, D.C., White, J., Hauswirth, M., Hitzler, P., Mohania, M. (eds.) OTM 2011, Part I. LNCS, vol. 7044, pp. 82–99. Springer, Heidelberg (2011)

9. Hashmi, M., Governatori, G.: A methodological evaluation of business process compliance management frameworks. In: Song, M., Wynn, M.T., Liu, J. (eds.) AP-BPM 2013. LNBIP, vol. 159, pp. 106–115. Springer, Heidelberg (2013)
10. Fellmann, M., Zasada, A.: State-of-the-art of business process compliance approaches. In: Avital, M., Leimeister, J.M., Schultze, U. (eds.) 22st European Conference on Information Systems, ECIS 2014. Tel Aviv, Israel, June 9–11, 2014
11. Bikakis, A., Papatheodorou, C., Antoniou, G.: The DR-prolog tool suite for defeasible reasoning and proof explanation in the semantic web. In: Darzentas, J., Vouros, G.A., Vosinakis, S., Arnellos, A. (eds.) SETN 2008. LNCS (LNAI), vol. 5138, pp. 345–351. Springer, Heidelberg (2008)
12. Hevner, A.R.: Design science research. In: Topi, H., Tucker, A. (eds.) Computing Handbook, Third Edition: Information Systems and Information Technology, vol. 22, pp. 1–23. CRC Press (2014)
13. Wieringa, R.: Design science methodology: principles and practice. In: Kramer, J., Bishop, J., Devanbu, P.T., Uchitel, S. (eds.) Proceedings of the 32nd ACM/IEEE International Conference on Software Engineering, ICSE 2010, vol. 2, pp. 493–494. Cape Town, South Africa, May 1–8, 2010. ACM (2010)
14. Mettler, T., Eurich, M., Winter, R.: On the Use of Experiments in Design Science Research: A Proposition of an Evaluation Framework. Communications of the Association for Information Systems **34**(10) (April 2014)
15. Wieringa, R.: Design science as nested problem solving. In: Vaishnavi, V.K., Purao, S. (eds.) Proceedings of the 4th International Conference on Design Science Research in Information Systems and Technology, DESRIST 2009. Philadelphia, Pennsylvania, USA, May 7–8, 2009. ACM (2009)
16. Boella, G., Janssen, M., Hulstijn, J., Humphreys, L., van der Torre, L.: Managing legal interpretation in regulatory compliance. In: Francesconi, E., Verheij, B. (eds.) International Conference on Artificial Intelligence and Law, ICAIL 2013, pp. 23–32. Rome, Italy, June 10–14, 2013. ACM (2013)
17. Goedertier, S., Mues, C., Vanthienen, J.: Specifying process-aware access control rules in SBVR. In: Paschke, A., Biletskiy, Y. (eds.) RuleML 2007. LNCS, vol. 4824, pp. 39–52. Springer, Heidelberg (2007)
18. Kamada, A., Governatori, G., Sadiq, S.: Transformation of SBVR compliant business rules to executable FCL rules. In: Dean, M., Hall, J., Rotolo, A., Tabet, S. (eds.) RuleML 2010. LNCS, vol. 6403, pp. 153–161. Springer, Heidelberg (2010)
19. Abi-Lahoud, E., Butler, T., Chapin, D., Hall, J.: Interpreting regulations with SBVR. In: Fodor, P., Roman, D., Anicic, D., Wyner, A., Palmirani, M., Sottara, D., Lévy, F. (eds.) Joint Proceedings of the 7th International Rule Challenge, the Special Track on Human Language Technology and the 3rd RuleML Doctoral Consortium, Seattle, USA, July 11–13, 2013. CEUR Workshop Proceedings, vol. 1004. CEUR-WS.org (2013)
20. OMG: Semantics of Business Vocabulary and Business Rules (SBVR), v1.0 (November 2013)
21. Antoniou, G., Dimaresis, N., Governatori, G.: A Modal and Deontic Defeasible Reasoning System For Modelling Policies and Multi-agent Systems. Expert Syst. Appl. **36**(2), 4125–4134 (2009)
22. Kholkar, D., Yelure, P., Tiwari, H., Deshpande, A., Shetye, A.: Experience with industrial adoption of business process models for user acceptance testing. In: Van Gorp, P., Ritter, T., Rose, L.M. (eds.) ECMFA 2013. LNCS, vol. 7949, pp. 192–206. Springer, Heidelberg (2013)

23. Resnik, P.: Semantic Similarity in a Taxonomy: An Information-Based Measure and its Application to Problems of Ambiguity in Natural Language. CoRR abs/1105.5444 (2011)
24. Batet, M., Harispe, S., Ranwez, S., Sánchez, D., Ranwez, V.: An Information Theoretic Approach To Improve Semantic Similarity Assessments Across Multiple Ontologies. Inf. Sci. **283**, 197–210 (2014)
25. FRC: What Constitutes an Explanation Under 'Comply or Explain'? Report of Discussions between Companies and Investors (February 2012)
26. Kravari, K., Papatheodorou, C., Antoniou, G., Bassiliades, N.: Reasoning and proofing services for semantic web agents. In: Walsh, T. (ed.) IJCAI 2011, Proceedings of the 22nd International Joint Conference on Artificial Intelligence, Barcelona, Catalonia, Spain, July 16–22, 2011, pp. 2662–2667. IJCAI/AAAI (2011)
27. Sunkle, S., Kholkar, D., Rathod, H., Kulkarni, V.: Incorporating directives into enterprise TO-BE architecture. In: Grossmann, G., Hallé, S., Karastoyanova, D., Reichert, M., Rinderle-Ma, S. (eds.) 18th IEEE International Enterprise Distributed Object Computing Conference Workshops and Demonstrations, EDOC Workshops 2014, Ulm, Germany, September 1–2, 2014, pp. 57–66. IEEE (2014)

Enterprise and Software Ecosystem Modelling

On the Support of Automated Analysis Chains on Enterprise Models

Andrés Ramos, Juan Pablo Sáenz[✉], Mario Sánchez, and Jorge Villalobos

Systems and Computing Engineering Department,
Universidad de los Andes, Bogotá, Colombia
{am.ramos260,jp.saenz79,mar-san1,jvillalo}@uniandes.edu.co

Abstract. Enterprise Architecture (EA) is being widely used across medium and large companies to document, analyze, plan, and manage the state of business and IT in order to successfully achieve organizational goals described in the enterprise strategy. During EA management projects, enterprise-level models and artifacts such as catalogs, matrices and diagrams are produced and used by architects to analyze a variety of enterprise situations. These analysis processes are fairly complex but not in all cases they rely entirely on the architect's expertise. In fact, it is possible to automate many analysis techniques, partially or in full. This reduces the necessary manual work and allows experts to concentrate their efforts in the tasks where their expertise is more valuable. This paper presents an approach to describe, validate, and execute analysis chains composed by both automated and non-automated analysis methods. This work is based on our previous characterization of automated analysis methods.

Keywords: Enterprise architecture · Enterprise modeling · Automated analysis · Analysis tools · Analysis chains

1 Introduction

Many approaches of Enterprise Architecture (EA) have been proposed, each one of them having their own definitions and concerns. For instance, [1] introduces two approaches within the EA context: EA-Frameworks and Enterprise Modeling. An EA-Framework is the set of structures that can be used to develop a broad range of different architectures [2]. Enterprise Models refers to the use of a modeling language to coherently specify and describe components of an organization along with their relationships [3]. Both approaches lead to the fact that models are essential into the EA development process as they largely satisfy visualization and communication needs in order to enable analysis capabilities.

In this context, analysis is *"the process of transforming mere facts into reasoned facts in order to provide the information needed to resolve a problem using solid arguments"* [4]. Therefore, architects performing Enterprise Model Analysis transform mere facts (contained in the model) into reasoned facts that support their observations and conclusions. These are typically presented using artifacts such as catalogs, matrices and diagrams [5][6].

However, Enterprise Model Analysis is a complex task because of the inherent characteristics of enterprise models. These models are typically large, have a high complexity, and are built using a number of disconnected modeling languages. Given this situation, we started the construction of a specialized tool for Enterprise Architecture Analysis (ArchiAnalysis) [7] which is capable of running a variety of automated analysis methods. This tool also offers a framework for the development of further analysis methods.

Based on initial experiments, we identified several characteristics of the original approach which justified the introduction of a new concept: that of Analysis Chain. First of all, we saw that some sets or analysis methods were frequently used together, and they had to be adjusted to make them compatible. These sequences of analysis methods were also repeated several times, using slightly different parameters. Furthermore, we identified a need to allow the participation of experts in particular points during the execution of a sequence of analysis methods. And finally, we discovered that it was possible to reduce the effort required to build new analysis methods by building on top of existing ones. The solution to these issues is Analysis Chains: configurable sequences of analysis methods with a common analysis goal, assembled and configured to guarantee consistency in their inputs and outputs.

As an example, consider an architect that wants to identify bottlenecks in processes due to relying on a single employee. In the proposal presented in this paper, the corresponding analysis chain connects four automatable steps: 1) filtering the organization business model by actors, processes/tasks and relations between the actors and these tasks. 2) creating a matrix showing the relations between actors and tasks. 3) counting the tasks assigned to each actor. 4) sorting the catalog of actors based on the number of assigned tasks and performing a filter to select those with a value that represents overload in the assignation (e.g., 25% of the total number of tasks). Hereafter we will refer to this example as the process Bottleneck Analysis Chain.

The rest of the paper is structured as follows. Section 2 briefly describes what Enterprise Model Analysis is, as well as the main characteristics of automated analysis methods. Section 3 is the core of this paper: it describes Analysis Chains including a full detail of the Bottleneck Analysis Chain, a formalization strategy, and the presentation of ArchiAnalysis with Analysis Chains. Section 4 illustrates Analysis Chains by applying them on the TOGAF ADM. Finally, Section 6 concludes the paper.

2 Manual and Automated Enterprise Model Analysis

In the domain of Enterprise Model Analysis, an analysis method is a repeatable sequence of steps performed to reason over a model and obtain information to support, among others, diagnosis, optimization, and decision-making processes across diverse areas of the organization [8–10]. Analysis methods are typically applied by architects, which explore the models to produce recommendations and assessments, by combining a predefined methodology (the aforementioned steps) with their own accumulated expertise. The outcome of analysis methods can assume different representations: it can be an artifact (a Catalog, a Matrix, or a Diagram) or it can be a set of additional facts integrated into the original enterprise model.

For example, consider Gap Analysis. This is an analysis method that an architect can perform in order to plan a roadmap and requires him to go over the baseline and the target architecture models looking for differences. The outcome of this method can be twofold: on the one hand, the architect builds a Matrix showing which elements to keep, which to remove, and which to add; on the other hand, elements can also be enriched with this information in the architecture models.

Automated Analysis Methods are the means for extracting, manipulating and reasoning about data contained in a model, using automated mechanisms [11]. By implementing automated analysis methods, some of the responsibilities that architects have regarding analysis are revoked, especially those responsibilities concerning manual tasks that add very little value to the overall analysis processes. By implementing this, architects earn a lot of extra time that they can devote to perform more advanced and valuable analyses.

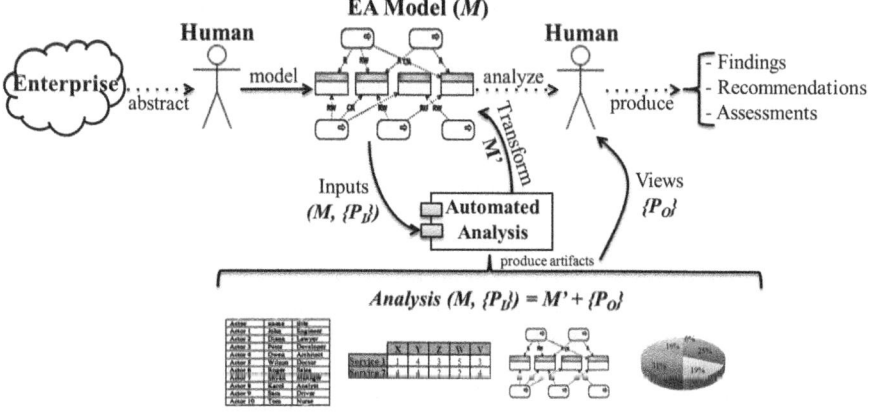

Fig. 1. Human-based analysis supported with an automated analysis tool

Figure 1 illustrates how automated analysis methods and expert-base analysis are not exclusive, but are rather complementary. From left to right, the figure shows that 1) humans are responsible for observing and abstracting the reality (the enterprise); 2) humans materialize their observations in a model; 3) automated analysis methods are capable of extracting information from the model and producing artifacts; 4) humans can also participate in the analysis process by studying the model to produce the final deliveries (findings, recommendations and assessments).

A problem that appears with automated analysis methods is that of metamodel conformance: an automated analysis method is created to process models that conform with a given model; however, if the output of the analysis method is not an artifact but an updated model, it is not necessarily true that it is going to conform to the same metamodel. For example, an analysis method may operate over an ArchiMate model, but the output may be an enriched model where ArchiMate concepts are accompanied by risk concepts which are not part of the original metamodel. This situation creates a number of conceptual and technical problems, especially for Analysis Chains. These problems are addressed later on this paper.

In a previous work [7] we identified the need to characterize automated analysis methods in order to be able to implement them in concrete tools and be able to use them effectively. The characterization included the following required attributes: identifier, name, description, analysis dimension, analysis type, layer, entities and relations, structural attributes and algorithm. Moreover, to deal with the problem described in the previous paragraph it is also necessary to know the input and the output metamodel of the method.

In [7] we also presented a catalog of analysis functions or methods which summarizes and organizes the methods currently described in the literature. Table 1 presents one method from the catalog: Data/Information Architecture Gap Analysis - FGP002. This method is based in the Gap analysis proposed by TOGAF.

3 Analysis Chains

3.1 Definitions and Requirements

By observing the usage of single analysis methods and functions, we have been able to see that there is a lot of potential beyond their independent usage. In the first place, we saw that analysis methods are typically used in groups, either sequentially or in parallel. Furthermore, they are not applied just once and it is not precisely known a priori which methods are going to be used: in the face of uncertainty, architects *explore* the available functions and then apply them tentatively, extracting bits of new information, until they are satisfied with the results. We also noticed that the simpler the functions, the more likely they are to be used in conjunction with other ones. This also means that relatively advanced functions do not have to be implemented from scratch, but instead they can be formed by composing basic functions.

Given this situation, we decided to study the problem of analysis function composition in order to offer to architects better tools to support the processes that they were naturally doing. This entailed, among other things, developing the mechanisms to describe these chains of functions, and also to manage the possible inconsistencies between the primitive functions.

Table 1. Analysis Function FGP002: Data/Information Architecture Gap Analysis

ID: FGP002	Dimension: Functional	Type: Gap
Name: Data/Information Architecture GAP Analysis		**Layer:** Application
Description Gap Analysis detects and documents changes from an As-Is to a To-Be view, where elements could have been modified, deleted or added to the architecture. Gap Analysis requires identifying information from the elements to recognize the modifications, as well as the removal or addition of elements. For specifically doing Data/Information Architecture Gap Analysis, the analysis requires an As-Is and a To-Be (baseline and target) version of Data Objects, as well as information about access permissions (Read/Write) from the Application Components in the architecture. The outcome of this analysis is a Catalog specifying which data elements are new, which were removed, and what changes occurred to data permissions level.		
Entities and relations Application components/functions/interactions access data objects for write or read operations.		

Structural Attributes
 	
- Data object
  - Name
- Application Component
  - Name
- Access Relationship
  - Access level: (Read, Write)		

Input metamodel ArchiMate.		
Algorithm There are three different kinds of gaps: New, Modify and Deleted. Both Models will be matched to identify each gap and a gap element will be associated to each element that meets any of the following rules: - Deleted: The data object exists in the As-Is but doesn't exist in the To-Be. - New: The data object exists in the To-Be but not in the As-Is. - Modify: data object exist in the As-Is and To-Be but the access level from the application components is different.		
Output metamodel Catalog. In particular for this analysis function it is a catalog or inventory of gaps. Each gap consists in the following attributes: - GAP-ID: A unique identifier for each gap. - Dimension: As this catalog can be reused for different gap analysis, it has a domain for identify the architectural domain (Business, Application and Technological). - Description: Gap general description. For example, CRM no longer writes Customer information, only reading access is allowed.		

An Analysis Chain is thus the composition of several Analysis Functions. An Analysis Chain can be sequential (the output of one function is the input to the next) or they can also have segments that happen in parallel. Furthermore, an analysis chain may not be completely automated and they may require the manual intervention in certain specific points. An analysis chain is characterized by the following attributes: *identifier*, *name*, *description*, *parameters*, *analysis functions involved* and the *input* and *output metamodels* for the entire chain. Table 2 presents the characterization of the Process Bottleneck Analysis Chain.

Supporting Analysis Chains creates a number of requirements for the underlying platform. On the one hand, it is necessary to have the mechanisms to describe the actual structure of the chain and the way in which inputs and outputs of the functions are connected. This requirement is not really complicated and thus the rest of the section will not concentrate on it. The complicated requirements come from the need to support inputs and outputs in different formats (models, catalogs, matrices, diagrams). On top of that, models may conform to different metamodels, making the composition of functions even more complex both from the conceptual and the technological point of view. The following section describes how these issues were tackled in our approach.

3.2 A Strategy to Support Analysis Chains

An Analysis Chain can be seen as an organized sequence of transformation functions which operate over a model and produce updated models or other kinds of artifacts. The problem for supporting this are the compatibility issues that may exist between the analysis functions selected for a given chain. In order to address this, we adopted a strategy based on three main ideas. The first idea is that, on top of the analysis functions, an Analysis Chain can also integrate transformation functions with the aim of eliminating the compatibility problems. The second idea is that a basic metamodel can be used as a canonical language along the entire analysis chain, leaving to the transformation functions the responsibility of transforming between this metamodel and the ones expected by the concrete analysis functions. The third idea is about managing artifacts as models by means of transformations, and only manipulate them as artifacts (diagrams, catalogs, etc.) when the user requires it.

Figure 2 shows how this strategy is materialized in an Analysis Chain. On the top part it can be seen that users interact with two languages or formats: firstly, the input format which is used to create the initial models; secondly, the output format (or formats) which is used to deliver information to the user. In the middle, a canonical metamodel called GIMM (Generic Intermediate Metamodel) [12] is used to eliminate the compatibility problems between functions. The second layer of the figure presents a sequence of transformation tasks which provide the structure of the chain by implementing transformation functions. The *Receive* tasks receive a model belonging to any modeling language which may be the initial model, or may be the results produced by an analysis function. *Map* tasks are responsible for mapping the received models into GIMM models. *Invoke* tasks consume Analysis Function using as input a GIMM model. Finally, the *Reply* tasks are responsible for returning the output to the Analysis Chain's invoker.

Figure 3 and Figure 4 illustrates how an ArchiMate model is transformed into a GIMM instance. For reasons of space we cannot explain the process fully in here, but complete information can be found in [12].

Table 2. Process bottleneck Analysis Chain

ID: ACH001	Name: Process bottleneck Analysis Chain
Description	This Analysis Chain is intended to identify business process bottlenecks due to the high number of tasks executed by a single employee
Parameters	**Overload limit percentage**: represents a percentage used to calculate the maximum number of tasks in which an employee can be assigned. **Overload limit value**: if the above percentage is not provided, this parameter can be used to represent the maximum number of tasks in which an employee can be assigned.
Analysis functions	**VWP001 Create Viewpoint**: This function performs a filter over the elements of the model and produces a viewpoint composed just by some elements and relationships specified by the invoker. *For example, filter all Business Objects, Data Objects and Realization relationships.* The input and output metamodel is ArchiMate. **CAT001 Create Catalog**: This function creates a catalog of elements based on the input parameter, which defines which element is going to be inventoried. The input metamodel is expressed in ArchiMate and the output metamodel is expressed as a Catalog. *For example, the Catalog of Business Actors and Catalog of Business Functions* **STR002 Create Matrix**: Taking two different catalogs as input (*Catalog of Business Actors and Catalog of Business Functions*), the function creates a matrix in which the rows corresponds to the elements of the first catalog, the column to the elements of the second catalog, and the intersection between them represents the corresponding value. The input metamodel is a Catalog and the output metamodel is a Matrix. *For example, Actors vs. Business Functions matrix* **STR006 Counting Matrix elements**: This function is responsible for counting the elements of a matrix, and groups them according to some specific value. The input metamodel is a Matrix and the output metamodel is a Catalog. *For example, Actors vs. Business Functions counting Matrix* **CAT002 Sort and Filter a catalog by**: This function is intended to sort and filter a Catalog based on a single attribute and a specific value. For instance, execute a filter over all the actors who have more than 5 tasks assigned. The input and output metamodels is a Catalog. For example, *Actors and Business Functions catalog sorted by number of tasks and constrained by an Overload limit percentage of 25%*
Input metamodel	ArchiMate
Output metamodel	Catalog

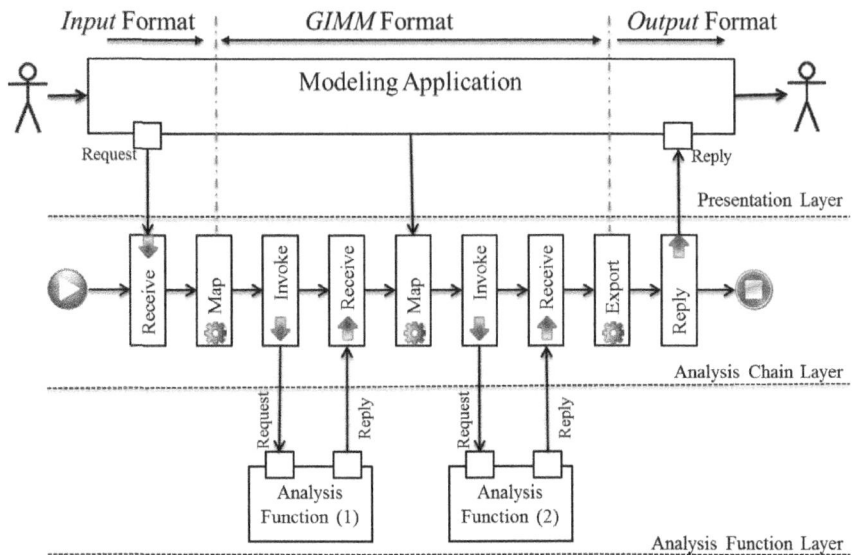

Fig. 2. The structure of an Analysis Chain

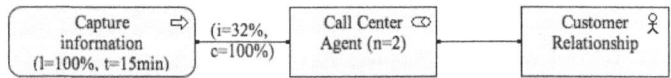

Fig. 3. ArchiMate model

As anticipated, the final point in the strategy is handling all artifacts as models conforming to GIMM. The goal behind this was the same as the one behind the usage of GIMM: to reduce the compatibility problems between analysis functions by offering mechanisms to maintain homogeneity in formats. To support this, transformations between each kind of artifact and GIMM were developed. The following are just the basic transformations that were defined and implemented. More powerful ones are required as artifacts become more complex.

Transformation of Catalogs Figure 5 illustrates the transformation of a Catalog into a GIMM instance.

Transformation of Diagrams and Charts Diagrams or charts such as bar charts, line charts, pie charts, radar charts among others represent tabular numeric data. In this case rows represent elements just like the previous catalog, and the columns represent dimensions that are mapped in GIMM as attributes.

Transformation of Matrices For this kind of artifacts, columns and rows are mapped into Elements and the crossing cells into relations between them.

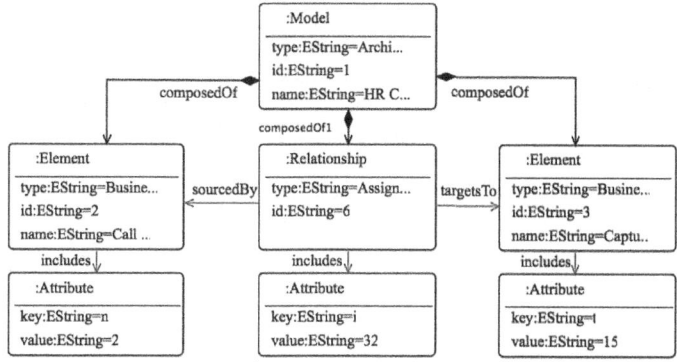

Fig. 4. Input Parameter: ArchiMate Model into GIMM

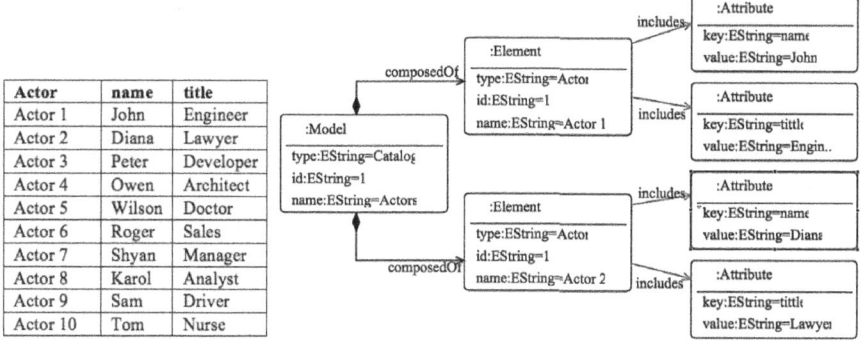

Fig. 5. A catalog and its reification as a model

Accordingly, the information inside the cells is represented as attributes of those relationships.

3.3 ArchiAnalysis

ArchiAnalysis is a framework for supporting analysis functions automation and execution. It was developed as an Archi plug-in. Archi is a cross-platform tool and editor to create ArchiMate models, based on the Eclipse Rich Client Platform, and open sourced under an MIT-type license. The main elements of ArchiAnalysis are shown in figure 6.

The *Analysis Chains Engine* component is responsible for supporting the Analysis Chain creation and execution scripts. It is uses the *ArchiAnalysis Core* component to access to the analysis functions catalog, and uses *analysis interfaces* that expose services for discovering and executing functions. Additionally, the Analysis Chains Engine also implements the features required to map models into GIMM and vice versa. Figure 7 presents a screenshot of ArchiAnalysis after executing the Gap Analysis Function. It can be noticed that the results are returned as a catalog displayed in the lower part of the screen.

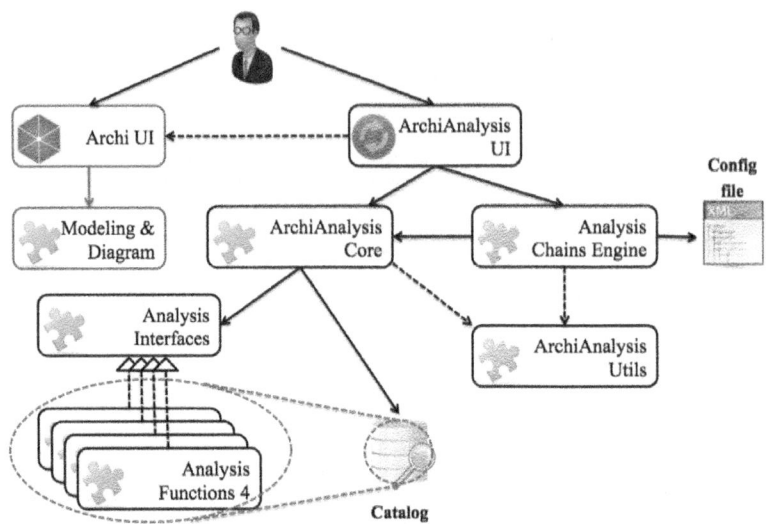

Fig. 6. ArchiAnalysis framework conceptual model

4 Analysis Chains Applied On TOGAF ADM Process

The TOGAF Architecture Development Method (ADM) involves several repeatable and sequential processes focused on the development of Enterprise Architecture Models. For this reason, it represents a suitable scenario to exemplify the composition of an Analysis Chain. The ADM is composed by ten phases during which the architects develop an Enterprise Architecture following an iterative and cyclic process. Each phase is divided into steps. For example, the phase titled *Information System Architectures* is divided into the following nine steps: (1) Select reference models, viewpoints, and tools, (2) Develop Baseline Architecture Description, (3) Develop Target Architecture Description, (4) Perform gap analysis, (5) Define candidate roadmap components, (6) Resolve impacts across the Architecture Landscape, (7) Conduct formal stakeholder review, (8) Finalize the Architecture, and (9) Create Architecture Definition Document. Based on this observation, and taking into account that the steps are sequential and repeatable, an Analysis Chain can be introduced into this scenario to perform specific analyses and support the architecture process.

Below, an example of the ADM Information Systems Architecture phase, applied over the ArchiSurance case study [13], will be presented. For this example it will be assumed that ArchiSurance is facing a Master Data Management Project involving two core systems. The first of them is the *Policy Management System*. It is responsible for managing The Policy Issuing Process, in which beneficiaries of a policy are registered, considering that if they do not exist already,

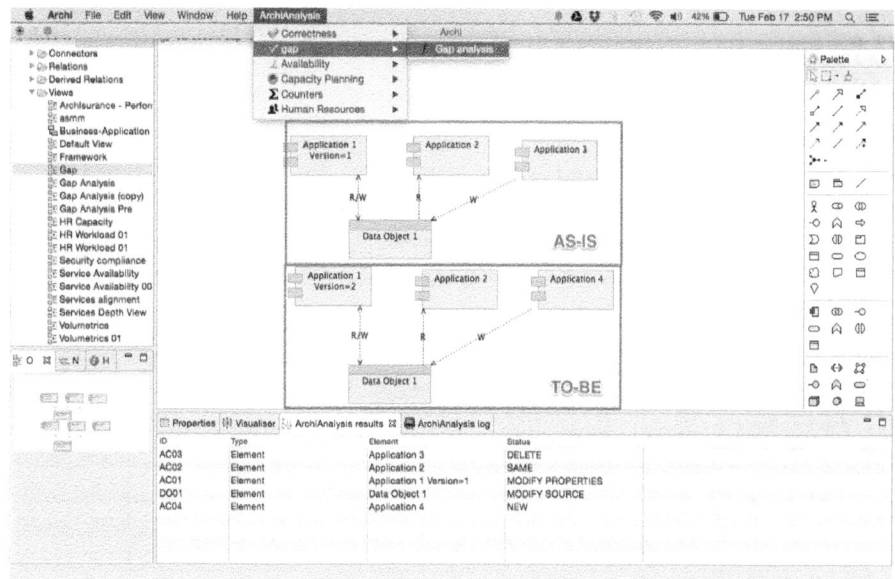

Fig. 7. ArchiAnalysis screenshot

they have to be created before. The second system is the *Claims Management System*. This component is used during a Claim registration process, where the information of the policy beneficiaries is validated and updated if necessary.

In the current architecture, the beneficiary information resides in both systems and it is kept consistent by a program that makes copies during the night of all the new beneficiaries from the Policy Management System into the Claims Management System. Nevertheless, this system does not synchronize the data from the beneficiaries that are already created. For that reason, Architecture team is carrying out a data governance initiative and they are visualizing a target architecture which integrates a Master Data Management System.

What an Architect would typically do is to determine that the application and information viewpoint will be taken as the initial inputs. Then, he designs an application and information viewpoint to represent the current and the target architecture as shown in Figure 8.a and Figure 8.b, respectively. Once the views are constructed, the architect creates a catalog of current and target application components and data objects. Based on the catalog, and following the TOGAF suggestions, he creates a cross matrix with the aim of identifying the gaps between both architectures as illustrated in Figure 8.c. Finally, the architect identifies and documents a set of solutions, as if they were building blocks, which are required to close all the gaps. At this point he proposes a roadmap where he includes all the identified initiatives as presented in Figure 8.d.

The purpose of this section is to expose that some of the steps explained above can be automatized through the use of ArchiAnalysis [7]. Consequently,

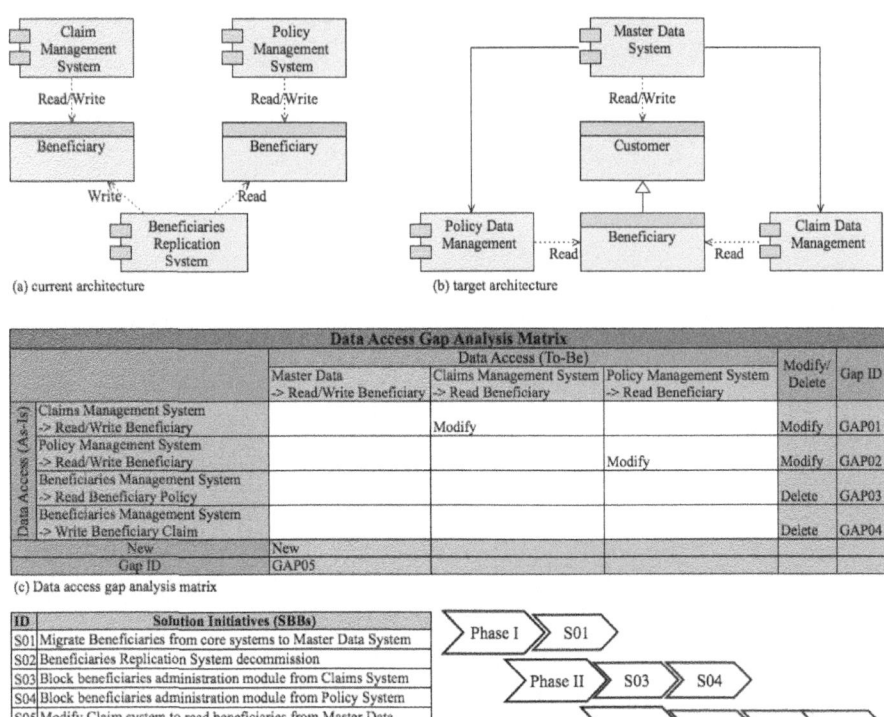

Fig. 8. TOGAF ADM Process example

an Analysis Chain may be built in order to enable the architect to repeat the same process as many times as he needs, even with different Data Objects and Applications. Then, based on the fact that an Analysis Chain can be composed considering each architect experience, Figure 9 exemplifies a possible Analysis Chain built over the ADM Information System Architecture phase.

In this process the Architect takes entities and interfaces models from the current and target architecture and uses them as inputs for a Gap Analysis function. Later, he generates a gap catalog that will be mapped against the Architecture Solutions Building Blocks (SBBs) catalog in order to identify whether all gaps are covered by at least one SBB. Finally, the SBB catalog will be used to prioritize the initiatives, creating a candidate roadmap. Figure 9 shows how entire process is expressed as an Analysis Chain by composing different Analysis Functions. It also reveals the way in which human intervention is required during the whole Analysis Chain process.

- **FGP002 – Data/Information Architecture GAP Analysis**: Analysis performed over the Application entities taking into account their access level (Read/Write). The input metamodel is expressed in ArchiMate and the

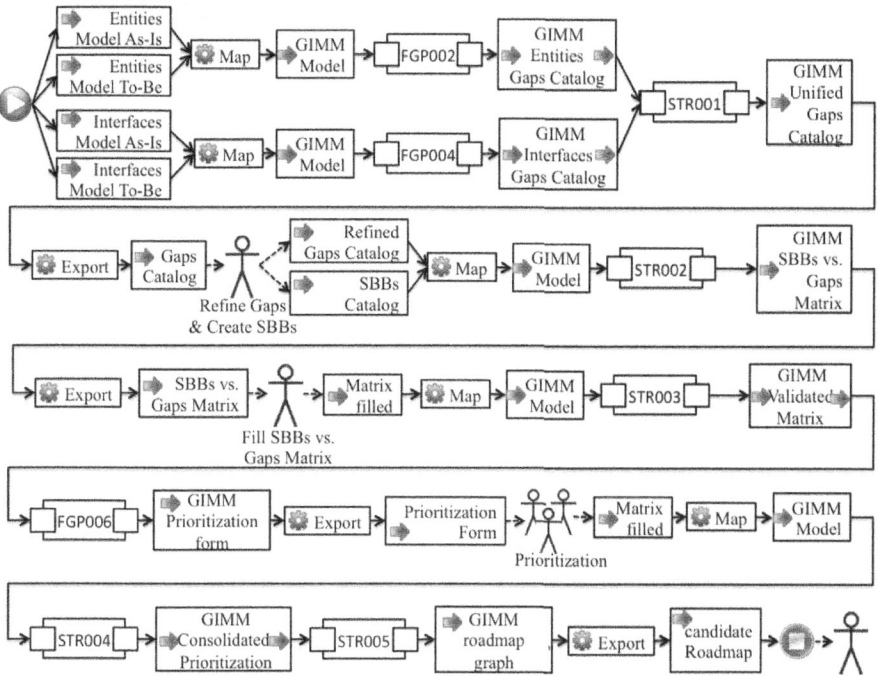

Fig. 9. ADM Information System Architecture Analysis Chain example

output metamodel is a Catalog containing a list that indicate which data objects are new, which data objects have changed in their data access level, and which data objects were deleted in the Target (TO-BE) Architecture.

- **FGP004 – Application-Integration Architecture GAP Analysis**: The analysis is performed based on the integration attributes. It uses the Application Interface View to place all the integration attributes and compares them. The input metamodel is ArchiMate. The output metamodel is a Catalog. The catalog will have a list of gaps such as: new, dropped and modified interfaces.
- **STR001 – Catalogs Merge**: This function able to merge/unify two different catalogs into a single inventory. It performs structural validation in order to determine whether the incoming catalogs are valid or not for merging. The input and output metamodels are both Catalogs.
- **STR002 – Create Matrix**: Consistent with the definition given in Table 2.
- **STR003 – Validate Matrix**: This function is responsible for validating whether a matrix is completely filled. For instance, all the columns must have a relation with at least one row. The input and output metamodel of this analysis function is a Matrix.
- **FGP006 – Generate Prioritization Form**: This function is able to create a priorization form with certain default scoring criteria. It has several

inputs including a catalog with initiatives to be prioritized and the criteria catalog. If the second one is not defined, the function will use the default scoring criteria. Input metamodel is a Catalog and the output metamodel is a Matrix.
- **STR004 – Consolidate Matrix**: This function receives a set of matrices with the same structure and executes a join between the crossed information. For this analysis function it is required to specify the kind of join that will be performed. Functions such as average, count, max, min and sum are used to consolidate the values. The input and output metamodel for this function are Matrices.
- **STR005 – Create Dependencies Graph**: This function is used to create dependency graphs. It requires as input a catalog with a specific column indicating the order and dependency between other elements. For this analysis the input metamodel is a Catalog and the output metamodel is a Diagram/Graph.

5 Conclusions

In this paper a solution to create Analysis Chains was proposed based on the automated analysis functions detailed in [7]. Analysis Chains are used for accompanying enterprise architecture processes, helping architects to deal with the complexity of performing strategic analysis over large enterprise models. Besides, it enables architects to create a base of knowledge with several analysis chains and consequently, to reuse them for facing new scenarios. Moreover, the paper presents an implementation strategy used to overcome the inconsistencies between the outputs and inputs of analysis functions which may prevent the creation of chains. This strategy uses a Generic Intermediate Metamodel (GIMM) which enables transformations from any metamodel to GIMM and vice versa. In addition the paper presents a series of considerations within the ArchiAnalysis characterization to fit into the Analysis Chain specification. Finally it is important to remark that analysis chains are instruments to facilitate analysis work by means of automated processes, but it must be driven by architects or analyzers capable of rationalizing the analysis outputs.

References

1. Schöenherr, M.: Towards a common terminology in the discipline of enterprise architecture. In: Feuerlicht, G., Lamersdorf, W. (eds.) ICSOC 2008. LNCS, vol. 5472, pp. 400–413. Springer, Heidelberg (2009)
2. Haren, V.: TOGAF Version 9.1, 10th edn. Van Haren Publishing (2011)
3. Jonkers, H., Lankhorst, M.M., van Buuren, R., Hoppenbrouwers, S., Bonsangue, M.M., van der Torre, L.: Concepts for modeling enterprise architectures. International Journal of Cooperative Information Systems **13**, 257–287 (2004)
4. Byrne, P.H.: Analysis and Science in Aristotle: New Essays on Auto/Biography. SUNY series in Ancient Greek Philosophy. State University of New York Press (1997)

5. van Ramshorst, E.: Application portfolio management from an enterprise architecture perspective (2013)
6. Speckert, T., Rychkova, I., Zdravkovic, J., Nurcan, S.: On the changing role of enterprise architecture in decentralized environments: state of the art. In: 2013 17th IEEE International Enterprise Distributed Object Computing Conference Workshops (EDOCW), pp. 310–318, September 2013
7. Ramos, A., Gomez, P., Sánchez, M., Villalobos, J.: Automated enterprise-level analysis of archimate models. In: Bider, I., Gaaloul, K., Krogstie, J., Nurcan, S., Proper, H.A., Schmidt, R., Soffer, P. (eds.) BPMDS 2014 and EMMSAD 2014. LNBIP, vol. 175, pp. 439–453. Springer, Heidelberg (2014)
8. Kohlhammer, J., May, T., Hoffmann, M.: Visual analytics for the strategic decision making process. In: Amicis, R., Stojanovic, R., Conti, G. (eds.) GeoSpatial Visual Analytics. NATO Science for Peace and Security Series C: Environmental Security, pp. 299–310. Springer, Netherlands (2009)
9. Johnson, P., Johansson, E., Sommestad, T., Ullberg, J.: A tool for enterprise architecture analysis. In: 11th IEEE International Enterprise Distributed Object Computing Conference, EDOC 2007, pp. 142–156, October 2007
10. Lankhorst, M.: Enterprise architecture at work: modelling, communication and analysis. The enterprise engineering series. Springer, Heidelberg (2013)
11. Benavides, D., Segura, S., Ruiz-Cortés, A.: Automated analysis of feature models 20 years later: A literature review. Inf. Syst. **35**, 615–636 (2010)
12. Gómez, P., Sánchez, M., Villalobos, J.: GraCoT, a tool for co-creation of models and metamodels in specific domains. In: Proceedings of the Workshop on ACademics Tooling with Eclipse, ACME 2013, New York, NY, USA, pp. 5:1–5:10. ACM (2013)
13. Jonkers, H., Band, I., Quartel, D.: The ArchiSurance case study. The Open Group (2014)

Designing Software Ecosystems:
How Can Modeling Techniques Help?

Mahsa H. Sadi[1(✉)] and Eric Yu[2,1]

[1] Department of Computer Science, University of Toronto, Toronto, ON, Canada
{mhsadi,eric}@cs.toronto.edu
[2] Faculty of Information, University of Toronto, Toronto, ON, Canada

Abstract. It has become an increasingly common practice for software companies to collaborate with external developers in order to develop software platforms for a shared market, constituting software ecosystems. Creating and sustaining a software ecosystem is a challenging problem that involves numerous technical, organizational, and business concerns. To support the systematic design of software ecosystems, modeling is a crucial tool. In this paper, we (a) identify a set of descriptive and analytical requirements raised in the design of software ecosystems; (b) review several modeling techniques used for describing and examining software ecosystems; and (c) assess the support of the reviewed techniques towards addressing the identified requirements. The results provide insight into the gaps between the issues raised in the design of software ecosystems, and the coverage of the studied techniques, suggesting an agenda for future research.

Keywords: Software ecosystems · Design · Modeling · Analysis · Review

1 Introduction

Collaboration has become an increasingly critical factor to the success of software companies [1-2]. There are various forces driving software companies to collaborate [3-4]: the shift towards the development of software platforms; the urge to share the costs of production; and the need to satisfy the varying demands of market, which usually fall outside the domain expertise of one software company. These forces have led to a recent software development practice, referred to as software ecosystem in which a keystone software company collaborates with other companies and developers to develop and extend a software platform for a shared market [5-6].

Two well-known examples of software ecosystems are the Google Android and the Apple iOS ecosystems. The main goal of both ecosystems is to provide a software platform, and complementary software applications and services for the market. Hence, both Google and Apple companies have established a network of collaborators (or partners) consisting of application developers, software companies, and content providers. While Google and Apple develop the key software platform (i.e. the mobile operating system), other application developers and software development companies provide complementary software applications and services for the related operating systems. Moreover, content providers supply content such as data, music and game for the applications. Google and Apple make the applications and content developed by external parties visible to the market via the online stores of App Store and Google Play [7-8].

Designing and organizing a sustainable collaboration in software ecosystems is a challenging problem for a keystone software company. First, it involves various boundary decisions, made at the legal and economical borders of the keystone software company and its environment [4, 9]. To open up the software platform and development activities to external companies and developers, the scope of the decisions transcends the organizational boundaries of the company and raise serious concerns and risks about control, ownership, intellectual property, security, privacy, trust, and quality. [9-13]. Second, collaboration in software ecosystems is multi-faceted, spanning various technical, business, and organizational concerns that must be addressed simultaneously [10, 14-15]. A successful software ecosystem needs to have a viable business model, a well-organized inter-organizational interaction model [7, 10], a well-designed collaborative software development process [13], and a software platform that enables the collaboration [16].

For instance, Google and Apple each pursues different approaches to organize their collaboration with external developers and content providers [17]. Each strategy has its own advantages and disadvantages. Google licenses Android for free. This makes the Android platform openly accessible to external developers in order to extend it with complementary applications and services. In contrast, the Apple iOS is proprietary and is accessible to a limited community of software developers directly controlled or owned by Apple. One advantage of Google's strategy in organizing its software ecosystem is that it attracts more software developers and companies to adopt Google Android as a platform. Its disadvantage is that the open strategy results in higher uncertainty in the quality of the final set of software products and makes Android platform and its complementary services and application loosely integrated. On the other hand, Apple's strategy, while limiting the adoption of the platform to a smaller number of developers, leads to a tight integration between platform and its complementary services and applications, and thus a software platform of higher quality [17].

The above example illustrates the complexity of the issues raised in organizing collaboration in a software ecosystem. The pivotal role of a well-organized collaboration among partners in the success software ecosystems [1-3] demands concentrated effort on developing systematic methods and techniques to support the design of software ecosystems. To address this need, a small but growing strand of recent research efforts have specifically focused on providing model-based approaches to describe and analyze software ecosystems. To this objective, two main strategies are pursued: (a) developing new modeling techniques [18], (b) using or adapting the available modeling techniques to describe and analyze software ecosystems [7, 10, 19-21]. However, these modeling approaches vary widely in the terminology that they use, and the analytical capabilities they provide. Moreover, due to the short time since the widespread adoption of the practice of software ecosystem by the software community, there is as yet no rigorous study on analyzing the needs raised in the design of software ecosystems.

In a preliminary attempt to address the above issues, in this paper, we identify what descriptive and analytical requirements are raised in the design of software ecosystems, and to what extent they are currently supported by a set of modeling techniques used to describe and analyze software ecosystems. The results identify the gaps between the descriptive and analytical needs raised in the design of software ecosystems and the current coverage of model-based approaches, suggesting an agenda for future research.

The rest of this paper is organized as follows: Section 2 identifies a set of descriptive and analytical requirements in designing software ecosystems. Section 3 reviews and summarizes a set of modeling techniques used to examine software ecosystems. Section 4 evaluates the support of the reviewed techniques for the specified requirements. Section 5 concludes the paper and discusses how to improve the support of modeling techniques for designing software ecosystems.

2 Designing Software Ecosystems: A Set of Descriptive and Analytical Requirements

In this section, we identify what issues are raised in the design of software ecosystems and what descriptive and analytical capabilities are required to address these issues. The requirements are developed based on the analysis and synthesis of the available literature on software ecosystems from a design-oriented perspective.

Table 1. A set of requirements to describe software ecosystems

P-1	**Collaborator**: Identifying the members of a software ecoytem and their roles – *Example:* Specifying the keystone software company; content providers; software developers; software companies. RR*: [10, 18, 22, 23]
P-2	**Interaction**: Identifying the relationships among members – *Example:* Specifying the business or technical relationships among the members of a software ecosystem. RR*:[10, 18, 22, 23]
P-3	**Activity (or Responsbility)**: Specifying the resources, activities and commitment of the members – *Example:*The specific business or software development activity performed by a member. RR*:[10, 22]
A-1	**Type**:Specifying different types and categories of collaborators, interactions, and responsibilities– *Example:* Identifying a financial or knowledge exchange relationship RR*: [18, 22]
A-2	**Constraint**: Specifying constraints and rules on collaborators, interactions, and activitiess – *Example:*Describing the conditions and rules on an interaction between a keystone software company and external software developer; or Describing the level of access of one software developer to the platform. RR*:[18]
A-3	**Attribute:** Specifying attributes and characteristics of collaborators, interactions, and their activities – *Example*: Identifying an important or a reliable collaborator or a critical interaction. RR*: [18, 24]
A-4	**Characteristic of Collaboration:** Specifying the characteristics and attributes of a collaboration (i.e. the configuration of collaborators, their activities and their interactions) – *Example:* Identifying a healthy, productive or secure collaboration between two or more collaborators. RR*: [1, 25]

*RR: Related Resources

2.1 A Set of Requirements for Describing Software Ecosystems

The first step in designing software ecosystems is to describe them. The description should provide a clear view of the structure of collaboration or partnership in software ecosystems; i.e., members and the interactions among them. For this purpose, a modeling technique needs to at least be able to describe and represent the following primary elements (either textually or graphically): (a) *Collaborator*; (b) *Interaction*; (c) *Activity (or Responsibility)*. To further delineate the structure of a collaboration, each of the above concepts can be augmented with the following ancillary information: (i) Type; (ii) Constraint; (iii) Attribute; and (iv) Characteristic of Collaboration. The ancillary information enables more elaborate description of a software ecosystem. The description for each of these features is provided in Table 1.

2.2 A Set of Analytical Requirements in Software Ecosystems

To identify what analysis issues are raised in the design of software ecosystems, we adopt a top-down domain analysis approach: we first explain general steps in the development of a software ecosystem; we then identify a set of the analysis concerns raised in each of these steps. A modeling technique used to represent a software ecosystem needs to support answering these analysis concerns.

Generally, from the perspective of a keystone software company who is in charge of the software platform, three main phases can be considered in the development of a software ecosystem [3, 6]:

1. *Setting up the software ecosystem* (**S**): The main activities in this step include: (a) to identify the objectives of developing the software ecosystem, and (b) to motivate external stakeholders (including software developers and software companies) to collaborate and contribute to the software platform.
2. *Organizing collaboration and opening up the software development processes and the software platform to collaborators* (**O**): In this stage, the keystone software company needs (a) to organize and configure the collaboration in the software ecosystem by specifying the collaborators, their roles and activities, and configuring the interactions among them; (b) to decide about how to distribute, decentralize and share access, information, activities, resources, products, responsibilities, and control among the collaborators.
3. *Monitoring and governing the software ecosystem* (**M**): The main activities in this stage include: (a) monitoring the health and sustainability of the collaboration, (b) orchestrating collaborations among the members, and (c) maintaining and evolving the collaboration and the platform.

In the above phases, specifically phase 1 and 2, several concerns are raised which require elaborate analysis. In Table 2, we identify a set of these concerns by providing example questions that can be raised for a keystone software company. It should be mentioned that the analysis concerns are generic. Therefore, these concerns can be raised for software business managers in the business and organizational context, or for the software project managers and software developers in the software development context.

Table 2. A set of analytical requirements in designing software ecosystems

S-1	**Analyzing incentives and motivations of collaborators:** – *Example questions:* **Q1.** How to foster collaboration and how to motivate external developers and companies to participate and contribute to the platform? / **Q2.** What are the intrinsic and extrinsic motivations of software developers and software companies for joining the software ecosystem? RR*: [17]
S-2	**Analyzing for trust and reliability:** – *Example questions:* **Q1.** How to create and ensure trust between the collaborators? / **Q2.** How reliable is the collaboration? / **Q3.** How reliable are the collaborators? RR*:[6]
O-1	**Analyzing for risk, vulnerability, tolerance, costs and benefits:** – *Example questions:* **Q1.** What risks are involved in the collaboration? / **Q2.** What are the costs and benefits of opening up software platform towards external stakeholders? / **Q3.** What dependencies are created between the collaborators and how critical are these interactions and dependencies? / **Q4.** What if the collaborators do not fulfill their commitments? / **Q5.** How tolerant is the keystone company and other collaborators against potential failures in collaboration? RR*:[12]
O-2	**Analyzing for distributing and decentralizing responsibilities and resources:** – *Example questions:* **Q1.** How to distribute the activities, responsibilities, and resources of software development and service provision among collaborators? RR*: [2, 3, 9, 11]
O-3	**Analyzing for distributing control, authority, decision making, and access:** – *Example questions:* **Q1.** How to distribute the control and authority of decision making over software development among collaborators? / How much control and access should be given to each collaborator over software platform? RR*: [9, 11, 13]
O-4	**Analyzing for distributing ownership and power:** – *Example questions:* **Q1.** How to distribute the ownership of software products and services among collaborators? RR*: [9]
O-5	**Analyzing for openness and sharing in collaboration:** – *Example questions:* **Q1.** What is the acceptable level of openness of the keystone software company in collaboration? / **Q2.** What information, products, and resources need to be shared between the collaborators? / **Q3.** How to open up software development processes and platforms to external collaborators? RR*: [2, 3, 7, 9, 13,15]
O-6	**Analyzing for security and privacy:** – *Example questions:* **Q1.** How to preserve the security and privacy of the platform and processes of the keystone software development organization in collaboration? / **Q2.** Is the collaboration secure? / **Q3.** Is the privacy of collaborators preserved? RR*: [6]
O-7	**Analyzing for health, productivity, robustness, performance:** – *Example questions:* **Q1.** Is the configuration of collaboration productive and robust? / **Q2.** Will the relationships among members lead to a productive collaboration? RR*: [1,2, 9, 25]
O-8	**Analyzing for alignment and conflict resolution:** – *Example questions:* **Q1.** How to resolve conflicts between the collaborators and their contributions to the platform? / **Q2.** How to align the objectives of the collaborators with the keystone software company? RR*: [5, 10]

*RR: Related Resources

3   Several Techniques Used for Modeling Software Ecosystems

In this section, we review and summarize a set of modeling techniques that have been used to examine software ecosystems. To include the modeling techniques in the review, two steps have been performed:

Collecting the Modeling Techniques. To gather the modeling techniques, two steps were taken: (a) An extensive search was conducted to collect the available literature on software ecosystems. To perform the search, two recent systemic reviews (published in 2013) [6, 26] were used as an initial catalog for collecting the resources. Then, the collection was updated and extended with more recent literature. The study [6] contains a categorized list of the resources that propose procedures or techniques, qualitative or descriptive models, tools or notations, analytical models, and empirical models for software ecosystems. The study [26] identifies a set of resources on software ecosystem modeling. (b) From the collected set, those research efforts are selected that use a model-based approach and a modeling technique to describe or analyze a software ecosystem.

Selecting the Modeling Techniques. The criteria for the inclusion of the modeling techniques in this review are as follows: (i) The collected resource must use a modeling technique to represent the structure of software ecosystems. (ii) The members of a software ecosystem and the relationships among them must be explicitly modeled. (iii) The modeling technique must have a well-defined and well-documented syntax, semantics, and notation. The notation can be either graphical or textual.

In the collected literature, a few modeling techniques have been used to describe software ecosystems including Product Deployment Context (PDC) Diagram, a component of Software Ecosystem Meta-model (SEM) [18], Technical Ecosystem Modeling Notation (TECMO) meta-model [27], and UML Deployment Diagram [10], but were omitted according to the second criterion. These techniques focus on modeling the software platform, but do not deal with the involved actors and the relationships among them. Another group of work offers various meta-models such as Associate Models [22], the SPO software ecosystem meta-model [28], and SPEM meta-model [29]. This group is excluded according to the third criterion. The focus of these efforts is on the meta-model level and not on the technique. There were also a few models, such as Graph Representations [30], and Food-web models [8] which lack a well-defined semantics and syntax, and are also omitted from this study.

Ultimately, five modeling techniques were selected that met the above criteria, namely, SSN [18], i* [15, 21], BMC [10, 20], VN [19], and e^3Value [7]. From among the selected techniques, only SSN is specifically proposed for modeling software ecosystems. The other techniques are generic and widely used in various domains. In the following, we briefly review how each technique can help describe and examine software ecosystems.

3.1 An Overview of the Selected Techniques

Software Supply Network Diagram (SSN). SSN is one component of the Software Ecosystem Meta-model (SEM), the formally proposed meta-model for describing and analyzing software ecosystems [18]. SSN is used to represent the structures of

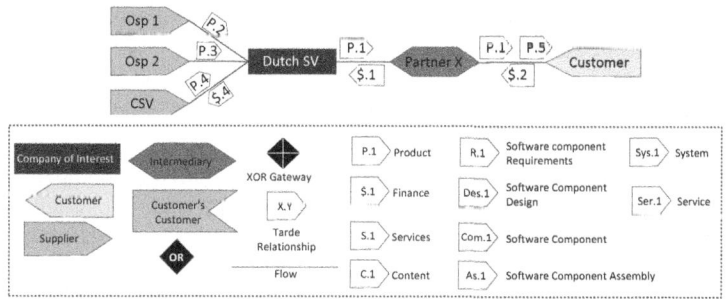

Fig. 1. An example of a SSN Diagram (originally developed in [18])

software supply chains in software ecosystems [24]. SSN explicates the business relationships among the members of a software ecosystem in terms of input and output flows between actors. One specific characteristic of SSN is that its terminology is developed based on the terms used in software development activities. Therefore, it is understandable by software developers. The main elements of SSN modeling language are "Actors", "Trade Relationships", "Flows", and "Gateways". An "Actor" is an organization or company that participates in a software ecosystem and can be a "Company of Interest", "Supplier", "Customer", "Intermediatory" or "Customers' customer". A "Trade Relationship" connects two actors, and is comprised of one or more flows. A "Flow" represents an artifact or service from one actor to another and is of different types of: "Products", "Services", "Finance", and "Content". A "Gateway" represents a logical relationship between flows and can be "OR" or "XOR". A SSN diagram is comprised of nodes and edges (see Fig. 1 as an example). Nodes represent the members and their roles in a software ecosystem. Edges represent input / output flows between the members. In SSN, participant actors (organizations) are represented by their names. Trade relationships are depicted in the format of X.Y. X represents the types of flow and Y represents the ID of flow.

The i* Modeling Technique. i* is a generic social modeling technique describing intentional relationships among actors from different business, technical and organizational perspectives [31]. i* explicates the relationships among the members of a software ecosystem in terms of strategic dependencies among strategic actors. The model can be used to explicate the objectives and reasoning of the members for developing or joining a software ecosystem [15, 21]. In i*, "actors" are intentional, set "Goals" or "Soft Goals", and can come up with different alternatives ("Task", "Resource", "Strategic Dependency") to achieve their goals. Strategic dependencies indicate that one actor relies on another actor to have a goal achieved, a task performed, or a resource be furnished. Different types of open, committed and critical strategic dependencies explicate different degrees of control and vulnerability of the depender in the relationship between two actors. As Fig. 2 illustrates, the motivations of developers for adopting a software platform are captured in terms of soft goals such as "Easy to use", and "Profitability from platform". Moreover, as shown by a sample evaluation of the objectives, although the platform is not easy to use, it adequately satisfies the developers to join the software ecosystem because it has a big market.

Business Model Canvas (BMC). BMC is a structured textual technique, developed based on Business Model Ontology (BMO) – a generic ontology to represent the business model of an organization [32]. This technique can be used to illustrate a high-level business view of a software company and its collaborators, and to describe how a member creates value in a software ecosystem [10, 20] (see Fig. 3 for an example). BMC describes which products and services a software company provides and lists who are its collaborators (partners) in the software ecosystem. For this purpose, the following building blocks are used: The products and services a company

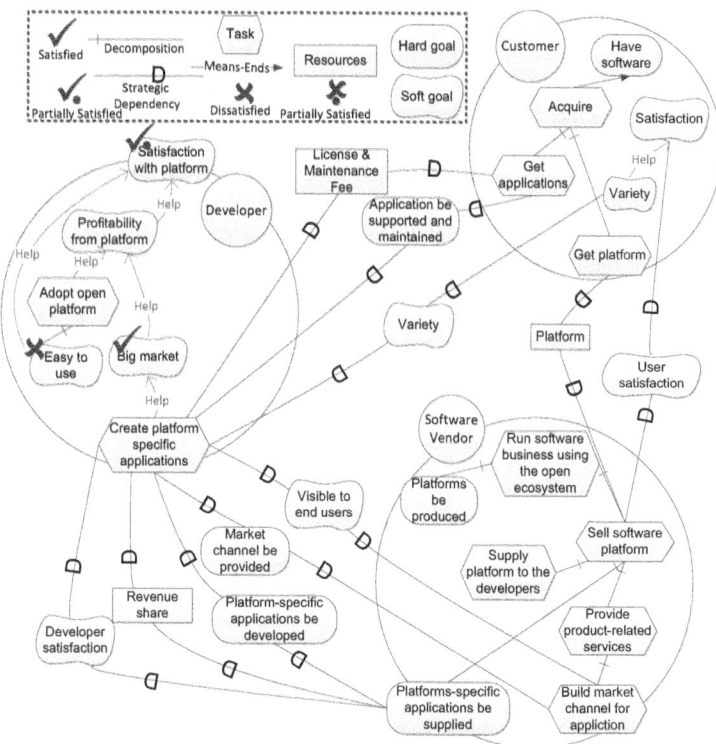

Fig. 2. An example of software ecosystem modeling using i* (excerpted from [21])

provides are listed in the "Value Proposition". The activities and the resources that are necessary to provide the services and products are listed in the "Key activities" and "Key resources" section. Customers to whom the organization offers services and products are listed in the "Customer Segments" section. Collaborators and partners of the software company are listed in the "Key Partners" section. "Channels" identifies the means by which the company gets in touch with its customers, and "Relationship" describes the type of link a company establishes between itself and its customer.

Key Partners	Key Activities	Value Proposition	Customer Relationships	Customer Segments
• Manufacturer	• Production	Service types:	• Product life cycle	• Broader customer reach
• Supplier	• Product line man-	• Product	• Liability	• Improved customer feedback
• Content Provider	agement	• Process	• Ownership of information and privacy	
• Communication Provider	**Key Resources**	• Life Cycle	**Channels**	• Customer differentiation
• Service Operator	• Development	• Extended	• Sales	
• Add-on developer	Environment	• Open interfaces	• Distribution	
• Owner	• Human capital	• Quality attributes	• Configuration	
• End user	• Branding		• Information	
• Regulatory Agency	• Standards		• Channel ownership	
• Information Broker				
Cost Structure			**Revenue streams**	
• Development cost			• Volume increases	
• Product cost			• Recurring sales	
• Operating cost			• Direct sales of software	
• Information cost			• Subscription fees	
			• Revenue sharing	

Fig. 3. An example of a BMC developed for a software ecosystem (excerpted from [20])

Value Network Diagram (VN). VN is a generic technique to describe the value exchange relationships between a set of human actors. The language of VN is comprised of "Actor", "Transaction", "Value exchange" and "deliverable" concepts, and is mainly used to explicate the business and inter-organizational relationships between a set of organizations [33]. VN can be used to describe and analyze how members create value in a software ecosystem [19] (see Fig. 4 as an example). In VN, the members of a software ecosystem are represented by ovals and the relationships among the members are represented by uni-directional or bi-directional arrows. Arrows represent the exchange of tangible and intangible deliverables (such as goods, services, knowledge, and revenue, or benefit) between the members.

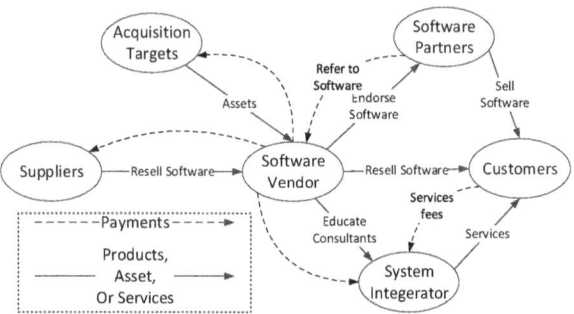

Fig. 4. An example of a VN developed for a software ecosystem (excerpted from [19])

e³Value Modeling. e³Value modeling technique explicates how economic value is created and exchanged within a network of actors [34]. This technique is used to illustrate the economically valuable activities of the members of a software ecosystem and the inter-organizational relationships among them [7] (see Fig. 5 as an example). In e³value modeling, the relationships among the actors are captured in terms of activity flows and input/output flows. The main modeling elements are "Actor", "Market Segment", "Value Activity", "Value exchange", and "Value Object". An Actor" is an economically independent (and often a legal) entity and represents a company or a customer. A group of actors (companies or customers) that share common properties are identified as a "Market Segment". For example, in Fig. 5, "Testing and verification party" is identified as a market segment representing a group of companies which collaborate with the operating system manufacturer to test the operating system.

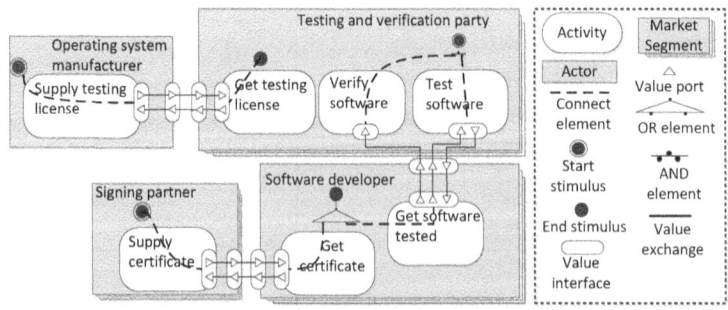

Fig. 5. An example of an e³value developed for a software ecosystem (excerpted from [7])

Activities in e^3value modeling are "value activities" meaning that they are economically profitable for the actors. Interactions among actor are captured in terms of "Value exchange" and "Value object". "Value exchange" represents trade relationships between actors. "Value object" represents the exchanged object and can be of the types "Services", "Product", "Money", or "Experiences".

3.2 Summary of the Reviewed Modeling Techniques

Table 3 provides an overall summary of the reviewed techniques for modeling software ecosystems based on several general criteria. These criteria investigate three main characteristics of the modeling techniques, namely (a) the objective and the intended users, (b) general usability features, and (c) the specific support and maturity of the techniques for modeling software ecosystems. These characteristics are labeled with A, B and C in Table 3.

Table 3. Summary of the reviewed techniques for modeling software ecosystems

		SSN	i*	VN	BMC	e^3V
A-1 Focal viewpoint	Business	+	+	+	+	+
	Inter-organizational	+	+	+	−/−[*]	+
	Technical (Software Development)	*[1]/+	*/+	−/−	−/−	−/−
A-2 Intended users	Software Business Manager	+	+	+	+	+
	Software Project Manager	−/−	−/+	−/−	−/−	−/*
	Software Developer	−/−	−/+	−/−	−/−	−/−
B-1 Support for analysis	Qualitative	+	+	+	+	+
	Quantitative	−/*	−/×	−/−	−/−	*[2]
B-2 Representation mode	Textual	−	−	−	+	−
	Visual	+	+	+	−	+
B-3 Refinement and traceability		−	+	−	−	−
B-4 Multiple views		*[3]	−	−	−	−
B-5 Formal syntax and semantics		+	+	−[4]	+	+
B-6 Tool support		−	+	−	+	+
C-1 Experimentation maturity for software ecosystems[5]		+	×	×	*	×
C-2 Methodology support for software ecosystems		×	−	−	−	−
C-3 Documentation support for software ecosystems[6]		*	*	×	×	×

Legend: (−): Not supported.   (×): Poorly supported.   (*): Partially supported.   (+): Supported.
[*] In pair evaluations, the first symbol shows current support of the modeling language and the second symbol shows the potential of the language for supporting the criteria

Notes:
[1] The main focus of SSN is on modeling the business relationships among the members of a software ecosystem; however, the terminology is based on software development activities and can support the technical relationships to some extent.
[2] e^3V supports quantitative analysis on the financial aspect of interactions among actors.
[3] Although SSN does not support multiple views of a software ecosystem, it is accompanied by another component in the SEM meta-model, named Product Deployment Context (PDC). PDC provides the architectural viewpoint of a software ecosystem.
[4] VN lacks a formal definition of syntax. It is not clearly defined what information should be represented in the models and how the information should be represented.
[5] This criterion is assessed as follows: (×) Poorly supported: the model is experimented in examples.   (*) Partially supported: The model is experimented in real-world case studies by researchers.   (+) Supported: the model is experimented by practitioners and intended users in real settings.
[6] This criterion is assessed as follows: (×) Poorly supported: the number of available documentation; i.e. publications, technical papers and websites is less than 2.   (*) Partially supported: the number of available documentation is more than 2 and less than 5.   (+) Supported: the number of available documentation is more than 5.

As the criteria of group A demonstrate, the focus of the modeling techniques is mainly on addressing the business and inter-organizational aspects of collaboration in software ecosystems, reflecting the viewpoint of software business managers. Group B criteria show that all of the techniques mainly support qualitative analysis, and the majority of the techniques do not support or poorly support two features of "refinement and traceability", and "multiple views". Refinement criterion evaluates the support of the modeling technique for developing a hierarchy of models with different levels of details and different levels of information in them and the ability to trace between the models. The multiple views criterion evaluates the ability of the technique to represent different views from a software ecosystem for different stakeholders. Finally, the criteria of group C reveal that currently, the majority of modeling techniques have not received adequate experimentation in modeling software ecosystems and there is not yet enough documentation, methodology, or guidelines available for using these techniques in the practice of software ecosystems.

4 Applying the Requirements to Assess the Reviewed Techniques

In this section, we analyze and assess the nature and extent of support offered by the reviewed modeling techniques towards the design of software ecosystems based on the descriptive and analytical requirements identified in Section 2.

4.1 Support for Describing Software Ecosystems

Evaluation Procedure. To evaluate the support of each modeling technique, it has been checked whether the technique supports representing the primary and ancillary elements, (P-1 to P-3) and (A-1 to A-4), introduced in Table 1.

Table 4. Assessment of the modeling techniques for describing software ecosystems

	P-1: Collaborator			P-2: Activity (or Responsibility)			P-3: Interaction			A-4
	A-1	A-2	A-3	A-1	A-2	A-3	A-1	A-2	A-3	
SSN	+/(Actor)*			— [1]			+/(I-O Flow)			—
	+**	—	+	—	—	—	+	—	—	
i*	+/(Actor/ Role/Agent)			+/(Tasks/Resources/Goals/ Softgoals)			+ /(Strategic Dependency)			—
	—	—	—	—	—	—	—	—	+	
VN	+/(Actor)			—			+/(Activity Flow)			—
	—	—	—	—	—	—	+	—	—	
BMC	+/(Actor)			*[2] /(Activities / Resources)			—			—
	—	—	—	—	—	—	—	—	—	
e^3V	+ / (Market Segment /Actor)			+ (Activities)			+/(I-O Flow/Activity Flow)			—
	—	—	—	—	—	—	+	—	—	

Legend: (—): Not supported.   (*): Partially supported.   (+): Supported.

* The first row in front of each modeling technique shows the support for describing the primary elements. The related element of the modeling technique that support the represnetation of the primary element is also identified.

** The second row in front of each modeling technique shows the support for describing the anciliary concepts.

Notes:

[1] SSN does not support representing the repsonbilities of collaborators. However, the other component of SEM, Product Deployment Context (PDC), identifies the architectural components of a software ecosystem and SEM enables linking the actors in SSN to the relevant components in PDC. [18]

[2] BMC only identifies the key activities and resources of one colabrator (software company).

Evaluation Results. The assessment of the reviewed modeling techniques based on the descriptive requirements is presented in Table 4. As Table 4 demonstrates, except for BMC, all the modeling techniques support representing collaborators and their interactions in software ecosystems. In all the studied techniques, collaborators are represented in terms of actors (but in each technique, actor has a different meaning.), and the interactions among collaborators are mostly represented in terms of input-output flows. However, the notion of activity is not supported by most of the modeling techniques. Moreover, the majority of techniques do not support describing ancillary information about the collaborators, the activity and the interactions among them.

4.2 Support for Analyzing Software Ecosystems

Evaluation Procedure. To evaluate the support of the modeling techniques for each type of analysis identified in Table 2, the following four criteria are used:

(1) Information Support (IS). This criterion evaluates whether the model expresses the required information to draw conclusion about an analysis concern.
(2) Analysis Representation (AR). This criterion identifies whether the information related to the analysis is captured inside the model or outside the model.
(3) Alternative Analysis and Comparison (AAC). This criterion identifies whether the modeling technique enables representing and comparing the consequences of two or more alternatives to address one analysis concern.
(4) Type of Analysis (TA). This criterion identifies whether the modeling technique enables descriptive analysis or predictive analysis.

Table 5. Assessment of the modeling techniques for analyzing software ecosystems

	S-1	S-2	O-1	O-2	O-3	O-4	O-5	O-6	O-7	O-8
SSN	–/O *	–/O	+/O	–/O	–/O	–/O	+/O	–/O	–/O	–/O
	O/D **									
i*	+/I	+/I	+/I	+/I	+/I	–/I	+/I	+/I	–/I	+/I
	I/D									
VN	–/O	–/O	+/O	–/O	–/O	–/O	+/O	–/O	–/O	–/O
	O/D									
BMC	–/O	–/O	–/O	–/O	–/O	–/O	–/O	–/O	–/O	–/O
	O/D									
e^3V	–/O	–/O	+/O	+/O	+/O	+/O	+/O	–/O	–/O	–/O
	O/D									

Legend: (–): Unable to support. (+): Able to support. I: Information captured inside the model. O: Information captured outside the model. D: Descriptive analysis. P: Predictive or Prescriptive analysis.

* The pair evaluations in the first row in front of each modeling technique identify the following information: The first symbol evaluates the *information support*. The second symbol evaluates *analysis representation capability*.

** The pair evaluations in the second row in front of each modeling technique identify the following information: The first symbol evaluates *alternative analysis and comparison* capability. The second symbol identifies the *type of analysis* supported.

Evaluation Results. Table 5 evaluates the capabilities of the reviewed modeling techniques to support the analytical requirements of software ecosystems (S-1 to O-8 in Table 2). As the results demonstrate, the majority of the reviewed techniques do not provide enough information (IS) to address the analysis questions raised in the design of a software ecosystems. From among those techniques that provide adequate information support for one type analysis, the majority do not capture adequate information related to performing the analysis (AR). Representing and comparing alternatives to address the analysis concerns (AAC) is covered by only one technique (i*). Finally, all of the models merely enable descriptive analysis and do not support prediction and prescription capabilities for designing a software ecosystem.

5 Conclusion

Software ecosystem is a recent software development practice in which various software companies, application developers, and content providers collaborate to develop software platforms, and complementary software applications and services for a shared market. Herein, we identified a set of descriptive and analytical requirements raised in the design of software ecosystems, and investigated to what extent these requirements are addressed by a set of techniques used to model software ecosystems. In the following, we identify the gaps and suggest how to enhance the modeling support for designing software ecosystems:

- *Lack of support for representing the technical aspect of collaboration*: Designing and organizing a sustainable collaboration in software ecosystems involves technical concerns as well as business concerns. However, the focus of the studied modeling techniques is mainly on describing and analyzing the business aspect of collaboration. They mainly reflect the viewpoint of software business managers. Modeling techniques that reflect the viewpoint of software project managers and software developers, and the technical relationships among the members can complement the reviewed techniques.
- *Lack of alignment between the business and organizational viewpoints and the technical viewpoints*: Technical aspect of collaborations in a software ecosystem should follow the rules and restrictions in the business and organizational aspects [6]. Models are the main tools for aligning and tracing between these dimensions. Specifically, two features of "refinement and traceability", and "multiple views" in modeling techniques enable aligning between different viewpoints. Enriching these features in the studied modeling techniques alleviates the issue of alignment in designing software ecosystems.
- *Weak representation support*: SSN is the formally proposed technique for modeling the relationships among the members of a software ecosystem. However, SSN does not support describing the activities of the collaborators in software ecosystems. Moreover, the majority of the studied techniques provide very little support for describing constraints and the attributes of collaborators, their activities, and the interactions among them. Enriching the syntax and semantics of the studied techniques to support these features, or using the techniques that already support these features enhance the representation of software ecosystems.

- *Weak methodological support*: There is not enough methodology and documentation support on how to model and analyze software ecosystems using the reviewed techniques. Moreover, most of the studied modeling techniques have not received enough experimentation and evaluation in real case studies and by practitioners. Addressing these weaknesses leads to further clarification of the descriptive and analytical needs in software ecosystems as well as how to improve the effectiveness of each technique.
- *Weak analysis support*: The majority of the studied techniques do not include enough information to support the analysis concerns raised in the design of software ecosystems. Alternative analysis and comparison is not supported by most of the modeling techniques. No support is provided for the predictive and prescriptive analysis on software ecosystems. Enriching the support of each individual technique for analysis, or providing guidelines to use a group of the modeling techniques in combination facilitates the design and development of software ecosystems.

Limitations of the Study. In this study, an initial set of modeling requirements for designing software ecosystems is identified through the analysis of the published literature. Confirming these requirements with practitioners and/or other active research groups in software ecosystems strengthens the results of this study. Moreover, the assessment of the modeling techniques against the identified analytical requirements needs to be further complemented by elaborating the evaluation criteria and conducting empirical studies on the modeling techniques.

References

1. Iansiti, M., Levien, R.: Strategy as ecology. Harvard business review **82**(3), 68–81 (2004)
2. Jansen, S., Finkelstein, A., Brinkkemper, S.: A sense of community: a research agenda for software ecosystems. In: 31st International Conference on Software Engineering-Companion, vol 2009, pp. 187–190 (2009)
3. Bosch, J.: From software product lines to software ecosystems. In: Proceedings of the 13th International Software Product Line Conference, pp. 111–119. Carnegie Mellon University (2009)
4. Bosch, J.: Software ecosystems: Taking software development beyond the boundaries of the organization. Journal of Systems and Software **85**(7), 1453–1454 (2012)
5. Jansen, S., Cusumano, M.: Defining software ecosystems: a survey of software platforms and business network governance. In: Proceedings of Fourth International Workshop on Software Ecosystems, pp. 41–58 (2012)
6. Manikas, K., Hansen, K.M.: Software ecosystems–a systematic literature review. Journal of Systems and Software **86**(5), 1294–1306 (2013)
7. Müller, R.M., Kijl, B., Martens, J.K.: A comparison of inter-organizational business models of mobile app stores: there is more than open vs. closed. Journal of theoretical and applied electronic commerce research **6**(2), 63–76 (2011)
8. Lin, F., Ye, W.: Operating system battle in the ecosystem of smartphone industry. In: International Symposium on Information Engineering and Electronic Commerce. IEEC 2009, pp. 617–621 (2009)

9. Jansen, S., Brinkkemper, S., Souer, J., Luinenburg, L.: Shades of gray: Opening up a software producing organization with the open software enterprise model. Journal of Systems and Software **85**(7), 1495–1510 (2012)
10. Christensen, H.B., Hansen, K.M., Kyng, M., Manikas, K.: Analysis and design of software ecosystem architectures–Towards the 4S telemedicine ecosystem. Information and Software Technology **56**(11), 1476–1492 (2014)
11. Boudreau, K.: Open platform strategies and innovation: Granting access vs. devolving control. Management Science **56**(10), 1849–1872 (2010)
12. Franch, X., Susi, A., Annosi, M.C., Ayala, C.P., Glott, R., Gross, D., Siena, A.: Managing risk in open source software adoption. In: ICSOFT, pp. 258–264 (2013)
13. Hanssen, G.K.: A longitudinal case study of an emerging software ecosystem: Implications for practice and theory. Journal of Systems and Software **85**(7), 1455–1466 (2012)
14. Jarke, M., Loucopoulos, P., Lyytinen, K., Mylopoulos, J., Robinson, W.: The brave new world of design requirements. Information Systems **36**(7), 992–1008 (2011)
15. Sadi, M.H., Yu, E.: Analyzing the evolution of software development: from creative chaos to software ecosystems. In: IEEE Eighth International Conference on Research Challenges in Information Science (RCIS), pp. 1–11. IEEE (2014)
16. Cataldo, M., Herbsleb, J.D.: Architecting in software ecosystems: interface translucence as an enabler for scalable collaboration. In: Proceedings of the Fourth European Conference on Software Architecture: Companion Volume, pp. 65–72. ACM (2010)
17. Koch, S., Kerschbaum, M.: Joining a smartphone ecosystem: Application developers' motivations and decision criteria. Information and Software Technology **56**(11), 1423–1435 (2014)
18. Boucharas, V., Jansen, S., Brinkkemper, S.: Formalizing software ecosystem modeling. In: Proceedings of the 1st International Workshop on Open Component Ecosystems (2009)
19. Popp, K., Meyer, R.: Profit from Software Ecosystems: Business Models, Ecosystems and Partnerships in the Software Industry. Books on Demand (2010)
20. Axelsson, J., Papatheocharous, E., Andersson, J.: Characteristics of software ecosystems for federated embedded systems: A case study. Information and Software Technology **56**(11), 1457–1475 (2014)
21. Yu, E., Deng, S.: Understanding software ecosystems: a strategic modeling approach. In: Proc of 3rd IWSECO, pp. 65–76 (2011)
22. van Angeren, J., Kabbedijk, J., Jansen, S., Popp, K.M.: A survey of associate models used within large software ecosystems. In: IWSECO@ ICSOB, pp. 27–39 (2011)
23. Jansen, S., Brinkkemper, S., Finkelstein, A.: Business network management as a survival strategy: a tale of two software ecosystems. In: IWSECO@ ICSR (2009)
24. Handoyo, E., Jansen, S., Brinkkemper, S.: Software ecosystem modeling: the value chains. In: Proceedings of the Fifth International Conference on Management of Emergent Digital Ecosystems, pp. 17–24 (2013)
25. van den Berk, I., Jansen, S., Luinenburg, L.: Software ecosystems: a software ecosystem strategy assessment model. In: Proceedings of the Fourth European Conference on Software Architecture: Companion Volume, pp. 127–134 (2010)
26. Werner, C., Jansen, S.: A systematic mapping study on software ecosystems from a three-dimensional perspective, pp. 59–81. Analyzing and Managing Business Networks in the Software Industry, Software Ecosystems (2013)
27. Seidl, C., Aßmann, U.: Towards modeling and analyzing variability in evolving software ecosystems. In: Proceedings of the Seventh International Workshop on Variability Modelling of Software-Intensive Systems, p. 3 (2013)

28. Lungu, M., Lanza, M., Gîrba, T., Robbes, R.: The small project observatory: Visualizing software ecosystems. Science of Computer Programming **75**(4), 264–275 (2010)
29. Pettersson, O., Svensson, M., Gil, D., Andersson, J., Milrad, M.: On the role of software process modeling in software ecosystem design. In: Proceedings of the Fourth European Conference on Software Architecture: Companion Volume, pp. 103–110 (2010)
30. Manikas, K., Hansen, K.M.: Characterizing the Danish telemedicine ecosystem: making sense of actor relationships. In: Proceedings of the Fifth International Conference on Management of Emergent Digital Ecosystems, pp. 211–218 (2013)
31. Yu, E., Giorgini, P., Maiden, N., Mylopolous, J. (eds.). Social modeling for requirements engineering. MIT Press (2011)
32. Osterwalder, A.: The business model ontology: A proposition in a design science approach. Ph.D. Dissertation. Institut d'Informatique et Organisation. Lausanne, Switzerland, University of Lausanne, Ecole des Hautes Etudes Commerciales HEC, p. 173 (2004)
33. Allee, V.: Value network analysis and value conversion of tangible and intangible assets. Journal of Intellectual Capital **9**(1), 5–24 (2008)
34. Gordijn, J., Akkermans, H., van Vliet, H.: Business modelling is not process modelling. In: Mayr, H.C., Liddle, S.W., Thalheim, B. (eds.) ER Workshops 2000. LNCS, vol. 1921, pp. 40–51. Springer, Heidelberg (2000)

Information and Process Model Quality

Dealing with Information Quality Requirements

Mohamad Gharib[(✉)] and Paolo Giorgini

University of Trento - DISI, 38123 Povo, Trento, Italy
{gharib,paolo.giorgini}@disi.unitn.it

Abstract. Information Quality (IQ) has been a growing concern for most organizations, since they depend on information for managing their daily tasks, make important decisions, etc., and relying on low quality information may negatively influence their overall performance. Despite this, most of the Requirements Engineering approaches either ignore or loosely define such requirements, i.e., they deal with them as generic non-functional requirements. In this paper, we propose a goal-oriented framework that is based on an extended version of secure Tropos methodology for modeling and reasoning about IQ requirements since the early phases of the system development, and refine these requirements until reaching their operational specifications. Moreover, the framework offers a methodological process along with several reasoning techniques to help designers during the different phases of the system design. We illustrate our framework with an example concerning a stock market crash.

Keywords: Information quality · Requirements engineering · Modeling · Reasoning

1 Introduction

The importance of Information Quality (IQ) for organizations is out of discussion, since they depend on information for managing their daily tasks, make important decisions, etc., and relying on low-quality information may result in undesirable outcomes [1], or even disasters in the case of critical systems (e.g., Air Traffic Management). Despite this, most existing Requirements Engineering (RE) approaches either loosely define IQ requirements, or simply ignore them (e.g., UMLsec [2], i^* [3], etc.). In particular, like all non-functional (quality) requirements, IQ requirements use to be represented as generic qualitative properties of the system, without specific methods for their analysis [4].

Although there exists many technical solutions for dealing with IQ related concerns in storage, network and database systems related literature (e.g., integrity constraints), such solutions are not able to satisfy the needs of current complex systems, such as socio-technical systems [5]. Existing solutions are able to solve IQ related issues at the technical level, but seem to be limited in solving IQ issues that may rise at social or organizational levels [6]. The Flash Crash (a main U.S market crash) is an example about the limitation of such solutions for addressing IQ needs for socio-technical systems, where the crash was not caused by a mere technical failure, but it was due to undetected vulnerabilities in socio-technical interactions [7].

In particular, the literature offers several RE approaches that are able to capture the social and organizational aspects of the system-to-be (e.g., secure Tropos [8], etc.), but their main focus is on the functionality of the system, and usually they ignore IQ needs. An integrated analysis of both functional and non-functional requirements is essential, since as highlighted in [4], some functional requirements might not be useful without their necessarily related non-functional requirements. Consider for example, an actor who wants to send an order to a stock market (functional requirements), and it requires its order to be send in an already defined period of time (non-functional requirements). If the system fails to satisfy time related aspects, requirements concerning the send of the order might not be achieved.

In [9], we have proposed a RE framework for modeling IQ requirements in terms of four IQ dimensions: accuracy, completeness, timeliness, and consistency. However, the framework does not provide a systematic process to guide the analysis of the four dimensions, and does not cover possible relations among them based on the actual information use. In this paper, we propose an extension to our previous framework [9] extending and refining its modeling concepts, and providing mechanisms for capturing IQ requirements, and then gradually refining them in terms of their different IQ dimensions until reaching their operational specifications. Moreover, the framework offers a methodological process along with several reasoning techniques to help designers during the different phases of the system design.

The paper is organized as follows; in Section (§2) we describe our motivating example concerning the Flash Crash scenario that is used to illustrate our framework. While in Section (§3), we propose our multi-dimensional model for analyzing IQ. Section (§4) introduce our proposed extensions for modeling IQ requirements, and in Section (§5), we briefly discuss the reasoning support that our framework offers. We implement and evaluate the proposed framework in Section (§6). Finally, we discuss the related work in Section (§7), and we conclude the paper and discuss future work in Section (§8).

2 Motivating Example

Our motivating example describes the May 6, 2010 Flash Crash, in which the Dow Jones Industrial Average (DJIA) dropped about 1000 points (9% of its value). Based on [10,11], several stakeholders of system can be identified, including: *stock investors* are individuals or companies, who have a main goal of "making profit from trading securities". While *stock traders* are persons or companies involved in trading securities in *stock markets* either for their own sake or on behalf of their *investors* with a main goal of "making profit by trading securities". *Traders* can be classified under several main categories, including: *Market Makers*: facilitate trading on a particular security in the market, and they have the capability to trade very large number of securities; *High-Frequency Traders*

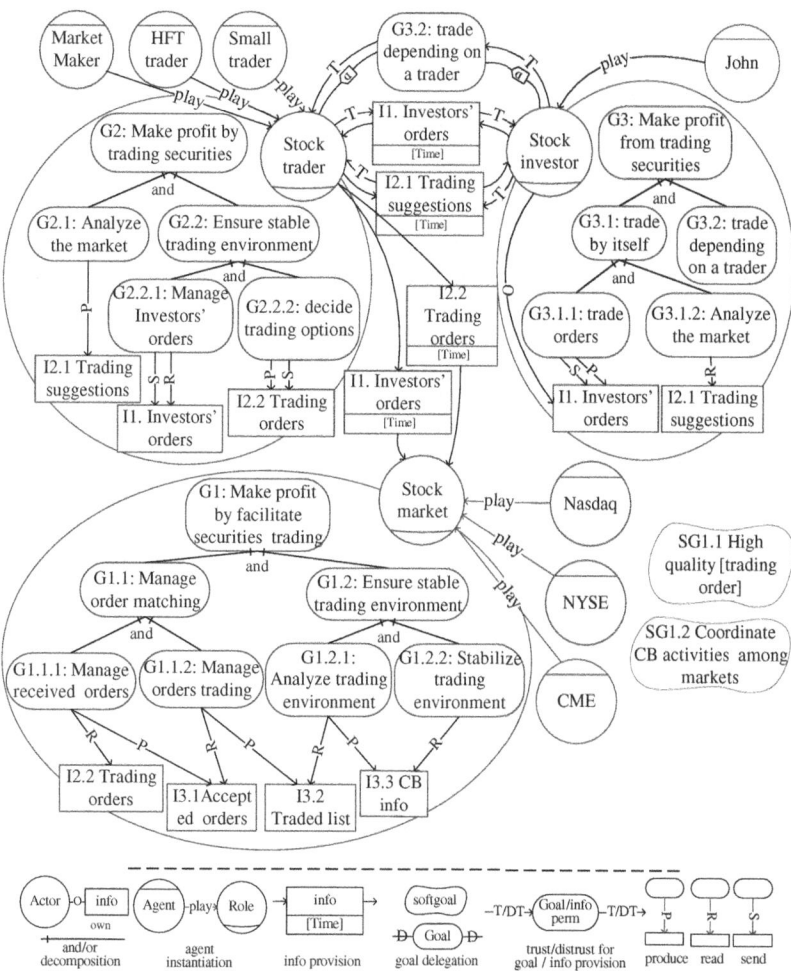

Fig. 1. A partial goal model concerning the Flash Crash scenario

(*HFTs*): are able to trade with very high trading frequency; and *small traders*: trade small amount of securities with very low trading frequency.

Stock markets are places where *traders* gather and trade securities, and they can be a physical trading place (e.g., New York Stock Exchange (NYSE), Chicago Mercantile Exchange (CME)), or electronic systems (e.g., NASDAQ). *Markets* have a main goal of "making profit by facilitating security trading". Usually, they "manage traders' order matching", and they should "ensure stable trading environment", which can be done by depending on their Circuit Breakers (CBs), where a CB is a technique that is used to slow down or halt trading activities to prevent a potential market crash.

Figure 1 shows a partial goal model of the Flash Crash scenario represented in the extended secure Tropos modeling language [9]. An actor covers two concepts a role and an agent, where the first is an abstract characterization of the behavior of an actor (e.g., Stock Market), and the last is an actor within a concrete manifestations (e.g., CME). Moreover, an agent can *play* a role or more (e.g., CME *play* a Stock Market). Actors can have a set of goals, they intend to achieve (e.g., G1: Make profit by facilitate security trading). When a goal is too coarse to be achieved, it can be refined through AND/ OR-decompositions of a root goal into finer sub-goals (e.g., G1: is AND-decomposed into G1.1: and G1.2:), where an AND-decomposition means that in order to achieve the root goal, all of its sub-goals must be achieved, while for an OR, achieving only one of its sub-goals is enough.

Moreover, softgoals (e.g., SG1.1) are used to represent non-functional requirements, and they do not have clear-cut criteria for achievement [4]. An actor may own information, which gives it a full control concerning the usage of information it owns (e.g., stock investor (O)wn I1). A goal may (P)roduces, (R)eads, and (S)ends information (e.g., G3.1.1: (P)roduces and (S)ends I1). Actors may depend on one another for information to be provided, where information provision has a time attributes to describe the provision time, and they may trust/ distrusts one another for the provided information (e.g., investor depend on trader for I2.1, and trusts it for its provision). Finally, actors may delegate goals to one another, and trust/ distrusts one another for the achievement of the delegated goals.

3 Multi-dimensional Model for Analyzing IQ

There is a general consensus that IQ is a hierarchical multi-dimensional concept that can be characterized by different dimensions/ sub-dimensions (e.g., accuracy, completeness, consistency, etc. [12–14]). That is why deciding whether information is high or low quality is not an easy task, and it became harder for socio-technical systems, since intentional, social and organizational aspects might underlie some of these dimensions. Although there exist many models for analyzing IQ (e.g., [13,15,16]), yet most of them can be criticized by their ambiguity [14], inconsistency among the dimensions they consider (e.g., completeness is a sub-dimension of believability in [15], while it is a sub-dimension of integrity in [13]), and most of them were not designed to capture the needs of socio-technical systems, i.e., they do not consider intentional, social and organizational aspects of IQ.

Our multi-dimensional model (Figure 2) for analyzing IQ is based on 7 IQ dimensions: accessibility, accuracy, believability, trustworthiness, completeness, timeliness and consistency, and it consider the intentional, social and organizational aspects that might underlies these dimensions. We define and discuss each of these dimensions along with their interrelations as follows:

Accessibility: the extent to which information is available, or easily and quickly retrieved [12]. In this paper, accessibility is defined as having the required permission over information to perform a task at hand.

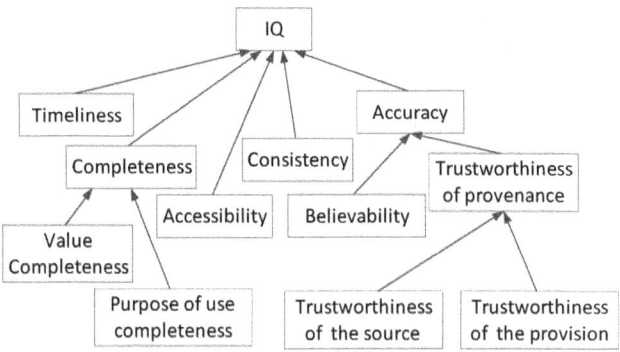

Fig. 2. IQ Model: the hierarchy of IQ dimensions

Accuracy: means that information should be true or error free with respect to some known, designated or measured value [13]. Accuracy is the most important and studied dimension, yet without clear standards, estimating accuracy is not an easy task. However, Dai et al. [17] stated that information accuracy is highly influenced by the *trustworthiness of its provenance*. While Wang and Strong [18] argued that accuracy can be analyzed based on several dimensions including *believability*. Thus, we analyze accuracy based on these two sub dimensions.

Believability: can be defined as the extent to which information is accepted or regarded as true and real [12,13]. Concerning our motivating example, in order to fulfill their obligation (facilitate trading), Market Makers provide what is called "stub quotes", which are orders with prices far away from the current market prices, i.e., such orders can be considered as fraud (inaccurate, falsified) orders. During the Flash Crash, over 98% of the trades were executed at prices within 10% of their values because of such orders [11]. However, such failure could be avoided, if *markets* apply a mechanism to verify the *believability* of the trading orders.

Trustworthiness: can be defined as the extent to which information is credible [16]. We rely on the *trustworthiness of the provenance* to analyze information trustworthiness, i.e., trustworthiness of information is analyzed depending on the (*trustworthiness of its source*), and the (*trustworthiness of its provision*) [17]. Concerning our example, some HFTs provide fraud/ falsified orders to affect the prices of some securities before starting their real trades (e.g., flickering quotes that are orders last very short time, which make them unavailable for most traders). If markets analyze the *trustworthiness of the provenance* of the trading orders, they will be able to detect such orders and apply the required mechanisms to mitigate their harmful effect.

Completeness: means that all parts of information should be available, and information should be complete for performing a task at hand [13]. Thus, completeness can be analyzed depending on two sub dimensions: *Value Completeness*: information is preserved against corruption or lost that might endanger its integrity (e.g., during its storage/ transfer); and *Purpose of use completeness*: information is complete for performing a task at hand, i.e., all the required information for performing a specific task should be available. Usually, the *purpose of use completeness* is harder to be analyzed and it requires a domain knowledge. For example, a main reason of the Flash Crash was the lack of coordination among the CBs of the trading markets. In particular, markets depend only on their own CBs information to stabilize their trading environment. However, such information is enough for each market alone, but when it comes to coordinate the CBs activities among all the markets, such information can be considered as incomplete. For instance, during the Flash Crash CME employs its CB, but NYSE did not [19], since each of them depends only on its CB information.

Timeliness: means to which extent information is valid in term of time (e.g., sufficiently up-to-date) [12]. According to [20], information timeliness can be analyzed depending on information *currency (age)* that is the time interval between its creation (or update) to its usage time [12,18]), and information *volatility* that is the change frequency of information value [18], i.e., information is not valid, if its currency is bigger than its volatility interval, otherwise it is valid. Note that timeliness can be subject to the needs of their stakeholders, i.e., stakeholders might define the timeliness of their own information. For example, stock investor can define the timeliness (validity time) of its sell/buy orders.

Consistency: means all multiple records of the same information should be the same across time and space [13]. In this paper, *consistency* is a time related aspect, i.e., the value of information among its different users might became inconsistent due to time related aspects (e.g., currency). For example, the lack of coordination among CB activities of the trading markets will not be resolved unless markets depend on consistent information for their CB activities.

4 Extended Concepts for Capturing IQ Requirements

Our previous framework [9] proposes concepts for modeling and analyzing 4 IQ dimensions, namely: accuracy, competence, timeliness, and consistency. However, it does not provide a systematic process that justifies why a certain IQ dimension should be considered or not for analyzing IQ, or how the considered IQ dimensions may contribute to one another. Thus, we extend our previous framework with mechanisms for capturing IQ requirements based on the actual purpose of use, and then gradually refining them in terms of 7 different IQ dimensions until reaching their operational specifications. In particular, we introduce two sets of modeling extensions: *(I) Basic IQ concepts*: that adopts and refine the concepts proposed in [9] for modeling IQ dimensions; and *(II) Top-level IQ concepts*: that are used to capture the IQ requirements of the stakeholders based on the actual information usage, and gradually refining them until reaching their operational specifications. More specifically, this set is used to identify how top-level IQ requirements can be captured and refined in terms of their different IQ dimensions, which can be modeled by the basic IQ concepts.

(I) Basic IQ concepts: first we extend and refine IQ modeling concepts proposed in [9] to accommodate the new IQ dimensions we consider along with their interrelations. In particular, our previous framework introduces concepts for capturing IQ requirements in terms of their different dimensions such as accuracy, timeliness, consistency, etc. For instance, it introduce *trusted provision* concept that enables for capturing information accuracy, and it provides *information volatility*, *read timeliness* and *send timeliness* concepts for capturing information timeliness. Moreover, it proposes *interdependent readers* and *read-time* for capturing information consistency [9,21]. However, some of the proposed concepts are at high abstraction level, and need to be refined to a level that enables for identifying detailed IQ specifications. For instance, it proposes *trusted provision* that helps in analyzing the accuracy of transferred information, yet we cannot rely on such concept to derive detailed IQ specifications. Moreover, we need to propose new constructs to accommodate the new IQ dimensions we consider (e.g., believability and accessibility). In what follows, we propose concepts to address these limitations.

Accessibility: can be influenced by the permissions that an actor has over information, which might enable or prevent it from using information as intended. However, our previous framework does not support the notion of permissions. Thus, we refine the modeling language by proposing 4 types of permissions concerning the 4 types of information usage that our framework supports (e.g., (P)roduces, (R)eads, (M)odifies and (S)ends). Moreover, we extend the language to model permission delegation among actors, and to model trust/ distrusts concerning the delegated permissions.

Completeness: completeness can be subject to (1) *value completeness* for which we rely on *Integrity Preserving provision* (IP-provision) that preserves the integrity of the provided information against corruption or lose [22], i.e., the *value completeness* of information is guaranteed during its transfer; and (2) *purpose of use completeness* we rely on the "Part of" concept to model the relation between an information item and its sub-items.

Trustworthiness: is subject to (1) *trustworthiness of the source* that can be captured by *trust/distrust produce* relations between information consumer and its producer concerning the produced information; and the (2) *trustworthiness of the provision* that can analyzed based on the way information arrives to its destination (e.g., P/IP provision), and the operations (e.g., modify) that have been applied to it taking into consideration if such operations were authorized or not (e.g., permissions and trust).

Believability is considered in both *read* and *produce*, since only these two relations can be influenced by information believability. Thus, we extend these two

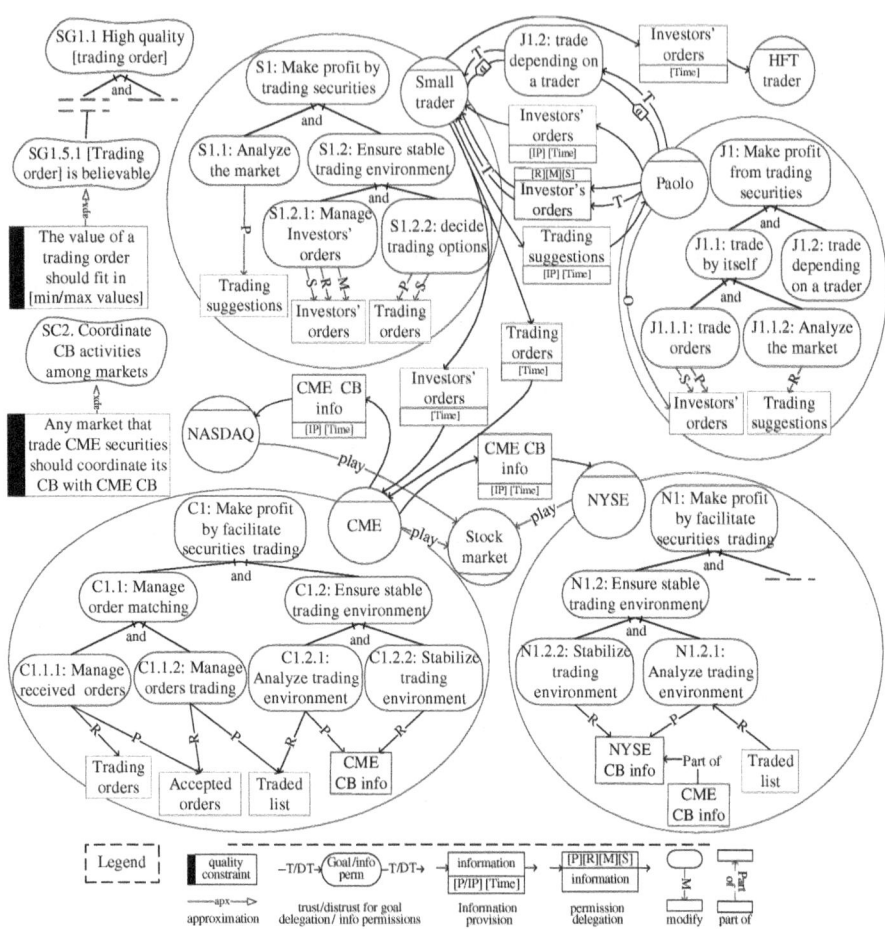

Fig. 3. A partial goal model of the Flash Crash extended with new IQ constructs

concepts to accommodate a *believability* check for read/produced information respectively, i.e., produced/read information is believable from the perspective of its producer/ reader, if the produce/ read operation apply a believability check.

Accuracy: can be analyzed based on : (1) *Accuracy of produced information* that can be analyzed based on its *believability*, which enables to avoid producing unintended information (e.g., fat finger mistakes), and its *trustworthiness of the production process*, if the producing goal has been delegated; (2) *Accuracy of provided information* can be analyzed based on the *trustworthiness of the provision*; and (3) *Accuracy of read information*: can be analyzed based on its *believability* and its *trustworthiness of the provenance*. Figure 3 shows an example of the graphical representation of the modeling concepts introduced above. A second set of modeling extensions are presented below to capture the rational of IQ requirements.

(II) High-level IQ concepts: enable for capturing the stakeholders' high-level IQ requirements, and then gradually refining them until reaching their operational specifications.

Top-level IQ softgoals: are softgoal concerning IQ requirements, and they are used as a starting point for identifying the stakeholders' needs concerning information they use/own. For example, in Figure 1 we have two softgoals of the stock market SG1.1: high quality [Trading order] that is defined based on information usage, and we have SG1.2: Coordinate CB activities among markets that is defined based on information owner needs. In particular, in stock market domain the same security can be traded in different markets, but it will always have only one primary listing market that is the main market for trading the security, such market has full authority over the way its listed security can be traded in other markets. In our motivating example, CME is the primary listing market and it requires that all CB activities related to its securities to be coordinated with its own activities, i.e., if CME turn to slow trading mode or stop trading, all markets trade the same security should do the same. However, at this point IQ softgoals are likely to be informal and imprecise, but they became more precise during the latter refinement activities.

And-decomposition for IQ softgoals refinement: a softgoal can be refined into more specific sub softgoals, if the joint satisfaction of these softgoals is considered equivalent to the satisfaction of the refined softgoal [23]. Usually, softgoals refinement into more specific sub softgoals can be done based on some related taxonomy. For example, Mylopoulos et al. [24] propose taxonomy for refining accuracy and performance softgoals. While Antón et al. [25] introduce taxonomy for refining privacy softgoals. The same can be applied to IQ softgoals, i.e., they can be refined based on different IQ dimensions (taxonomy proposed in Section §3). In particular, we introduce *and-decomposition* relation between an IQ softgoal and its sub IQ softgoals instead of *contribution links* proposed in [24], since such requirements will reach a point that enables for clearly deciding whether they can be achieved or not (operational specifications). For example, the IQ softgoal of stock market (SG1.1: high quality [Trading order]) can be *and-decomposed* into softgoals concerning accessibility, accuracy, completeness, timeliness, and consistency etc. based on the actual needs of the stock market. Note that IQ softgoals cannot be refined more that the leaf IQ softgoals, which are used to represent leaf IQ dimensions (shown in Figure 2), where a leaf IQ dimension is an IQ dimension that does not has sub dimensions (e.g., leaf IQ believability softgal is used to represent believability (leaf IQ dimension)).

Approximating leaf IQ Softgoals: as previously mentioned, softgoals are difficult to be expressed in a measurable way. Yet Jureta et al. [23] introduce the *approximation* relation through which a softgoal can be satisfied by a Quality

Table 1. IQ softgoal classification & approximation into IQC

Leaf IQ softgoals	Kind	Satisfaction	Representation	Approximated into IQC
Believability	Functional	Hard	Operational	Operational IQC
Trustworthiness	Constraint	Hard	Declarative	Declarative IQC
Completeness	Constraint	Hard	Declarative	Declarative IQC
Timeliness	Performance	Hard	Quantitative	Quantitative IQC
Consistency	Performance	Hard	Quantitative	Quantitative IQC

Constraint (QC), i.e., a QC can provide clear-cut criteria for the satisfaction of a softgoal. However, leaf IQ softgoals are used to capture different IQ dimensions (e.g., accuracy, completeness, etc.), i.e., each of them is used to describe different aspects of IQ. Thus, leaf IQ softgoals might not have the same nature/type, and in turn, they may need to be approximated in different ways. To tackle this problem, we rely on Glinz [26] work[1], to get better understanding of the nature/type of the leaf IQ softgoals, and to define the appropriate Information Quality Constraints (IQC)[2] for their approximation. Moreover, for the approximation to be consistent with the different types of leaf IQ softgoals, we define 3 different types of IQCs:

1. *Operational IQC:* are constraints that define the required actions to be performed in already determined situations. For example, IQ softgoal concerning information *believability* can be approximated into operational IQC that define a mine and max values to determine the believability of produced/read information.
2. *Declarative IQC:* are constraints used to define properties of the system that should hold. For example, IQ softgoal concerning the *trustworthiness of provision* can be approximated into declarative IQC stated that information should be transferred only through IP provision.
3. *Quantitative IQC:* are constraints used to specify properties of the system that should hold, and can be measured on an ordinal scale. For example, IQ softgoal concerning *consistency* can be approximated into quantitative IQC stated that *interdependent readers* should rely on information that has the same currency (age).

Table 1 shows how leaf IQ softgoals can be classified, and how they can be approximated into the appropriate IQCs. Finally, in order for the approximation relation between IQ softgoal and its related IQC to hold, a well-defined quality space should exist [23], where a quality space can be defined as a certain conceptual space that can be used to describe the quality value [27]. The main purpose of the quality space is providing a general consensus among the stakeholders of the system on how quality aspects (e.g., IQ dimensions) can be measured, which removes any ambiguity related to the verification of IQCs, i.e., determining whether a certain IQC is satisfied or not. For instance, both information

---
[1] Glinz classify requirements based on their kind, satisfaction and representation
[2] We use IQC to refer to QC, since no other type of constraints is used in this paper

timeliness and consistency are time related aspects. Thus, how time can be represented and measured should be clear to all stakeholders of the system, i.e., the allowed number of digits along with the value they represent (e.g., seconds, milliseconds, etc.).

Our framework is equipped with an engineering methodological process that is based on the extended secure Tropos methodology [9], and extends it with the required activities to accommodate the new extensions. The process provides the required guidance to designers during the different phases of the system design, and it consists of 6 main steps: (1) *Actors modeling*: in which the stakeholders of the system are identified and modeled along with their top level goals, and then these goals might be refined through And/ Or-decompositions if required. Finally, based on the actors' capabilities, they may delegate goals to one another; (2) *Information modeling*: the different relations among goals and information are modeled, and then information provisions / permissions delegation among actors are modeled as well; (3) *Identifying top IQ softgoals*: IQ softgoals are defined by the stakeholders based on their needs concerning information they use/own, and then they are refined through AND-decomposition until reaching their leaf IQ softgoals; (4) *Leaf IQ softgoals approximation*: leaf IQ softgoals are approximated into their corresponding IQC; (5) *Trust modeling*: trust among actors concerning goals/ permissions delegation and information producing are modeled; (6) *Analyzing and refining the model*: at this step the model is analyzed to verify whether all the stakeholders' requirements are achieved or not, if some requirements were not achieved, the analyst tries to find appropriate solutions.

5 Reasoning About Information Quality Requirements

We use Datalog [28] to formalize all the concepts along with the related axioms (reasoning rules) that have been introduced in this paper. Further, we define a set of properties of the design that can be used to verify the correctness and consistency of the IQ requirements model, i.e., such properties define constraints that designers should consider during the system design[3]. In particular, our framework offers several reasoning techniques that enable for detecting different design vulnerabilities (e.g., IQ related issues such as accuracy, competence, timeliness and consistency, etc.), and how such vulnerabilities may influence the achievement of the actors' requirements. Further, it enables to detect if social dependencies among actors (e.g., information provision) holds, and whether they satisfy the actors' needs or not. More specifically, the reasoning enables for checking whether stakeholders' IQ requirements are achieved or not, and identify the reason(s) preventing their achievement (if any). Further, it is able to identifies which functional requirements is influenced if certain IQ needs were not achieved.

---

[3] The formalization of the concepts, axioms and properties of the design are omitted due to space limitation

6 Implementation and Evaluation

Evaluation is an important aspect of any research proposal; it aims to demonstrate the utility, quality, and efficacy of a design artifact. Hevner et al. [29] classify evaluation methods in design science under five categories: observational, analytical, experimental, testing, and descriptive. Since our framework belongs to the design science area, we evaluated its applicability and effectiveness depending on simulation method (experimental), i.e., execute artifact with artificial data. To this end, we developed a prototype implementation of our framework[4] to test the applicability and effectiveness of our framework for modeling and reasoning about IQ requirements. In what follows, we briefly describe the prototype, discuss its applicability and effectiveness over the Flash Crash scenario, and then test the scalability of its reasoning support.

Implementation: our prototype consist of 3 main parts: (1) a graphical user interface (GUI) developed using Sirius[5], which enable designers for drawing the requirements model; (2) model-to-text transformation that supports translating of the graphical models into Datalog formal specifications depending on Acceleo[6]; and (3) automated reasoning support (DLV system[7]) takes the Datalog specification that result from translating the graphical model into Datalog along with the reasoning axioms, and helps in verifying the correctness and completeness of the stakeholders' IQ requirements.

Applicability and effectiveness: we evaluated our framework by showing its applicability in capturing the IQ requirements along with its effectiveness to detect inconsistencies/ conflicts of the stakeholders' requirements by applying it to the Flash Crash motivating example. We used our extended modeling language to model the Flash Crash motivating example, and then we translate the requirements diagram into Datalog formal language. Finally, we depend on the reasoning support technique that our framework provides to check whether the requirements model is correct and consistent. The analysis captured several inconsistencies in the design, including:

Incomplete information: the CB information of both *NYSE* and *NASDAQ* were identified as incomplete information, since they miss *CME* CB information that is considered as *part of* them (sub item), such information were not provided to them.

Inconsistent information: the incompleteness problem can be solved by providing "CME CB information" to both of *NYSE* and *NASDAQ*. However, "CME CB information" is provided to them with two different provision times. According to [30], provision time from *CME* to *NASDAQ* was 13 (ms), while provision time from *CME* to *NYSE* was 14.65 (ms). Thus, we

---

[4] https://github.com/disi-unitn-RE-IQ/RE-IQ
[5] https://projects.eclipse.org/projects/modeling.sirius
[6] https://projects.eclipse.org/projects/modeling.m2t.acceleo
[7] http://www.dlvsystem.com/dlv/

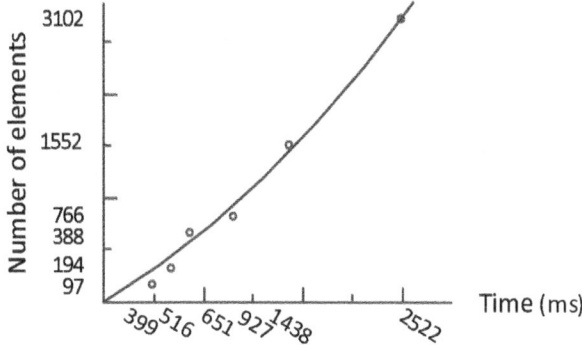

Fig. 4. Scalability results with increasing the number of modeling elements

face another problem that is inconsistency among *NYSE* and *NASDAQ* concerning "CME CB information", since they are *interdependent readers*.

Inaccurate information: *CME* market considers information received from *Market Marker* inaccurate information, since no believability check was applied to orders received from it. At the other hand, *CME* considers information received from *HFT trader* as inaccurate information, since no trust in information production holds between *CME* and *HFT trader*.

Experiments on scalability: to test the scalability of the reasoning technique, we expanded the model shown partially in Figure 3 by gradually increasing the number of its modeling elements from 97 to 3104 through 6 steps, and investigate the reasoning execution time at each step. The result is shown in Figure 4, and it is easy to note that the relation between the size of the model and execution time is not exponential. We have performed the experiment on laptop computer, Intel(R) core(TM) i3- 3227U CPU@ 190 GHz, 4GB RAM, OS Window 8.1, 64-bit.

7 Related Work

Requirements engineering community did not appropriately support capturing IQ requirements. For example, UMLsec [2] propose concepts for modeling information integrity (IQ related aspect) as a constraint, which can restrict unwanted modifications of information, yet IQ still can be compromised in several other ways. Abuse frame [31] addresses integrity related issues (modification) by preventing unauthorized actors from modifying information, or prevent authorized actors from doing unauthorized modifications. Finally, secure Tropos [8] seems to be sufficient to capture the functional, security and trust requirements of system, yet it provides no primitives for capturing IQ needs.

At the other hand, we find several approaches for improving IQby design, but they were not designed to capture neither the organizational nor the social aspects of the system-to-be, which are very important aspects in current complex systems. For instance, Ballou et al. [20] presented an information manufacturing system that can be used to determine the data quality in terms of timeliness, quality, cost, and value of information products. While Shankaranarayanan et al. [32] extend Ballou 's work to develop a formal modeling method for creating an IP-MAP. While Scannapieco et al. [33] propose IP-UML approach that relies on the IP-MAP framework, which is a software engineering approach developed to improve information quality in a single organization.

8 Conclusion and Future Work

In this paper, we discussed the importance of capturing IQ requirements of the system starting from the early design phase. Moreover, we introduced a novel RE framework that proposes an extended language for modeling and reasoning about IQ requirements, and offers a systematic process for refining these requirements until reaching their operational specifications. We illustrated our framework by example concerning a U.S stock market crash, and showed how many of the reasons that led to the crash could be avoided if the IQ requirements of the system were addressed properly during the system design. For the future work, we aim providing more expressive analysis for IQ related aspects rather than the binary one, which use only two values to evaluate IQ related concepts (e.g., accurate or inaccurate, believable or not believable, etc.). Another track under investigation involves extending the relations between stakeholders and information they use, and enriching the proposed IQ model by considering other dimensions along with refining the already considered ones.

Acknowledgments. This research was partially supported by the ERC advanced grant 267856, "Lucretius: Foundations for Software Evolution", http://www.lucretius.eu/.

References

1. Redman, T.: Improve data quality for competitive advantage. Sloan Management Review **36**, 99–99 (1995)
2. Jürjens, J.: Secure systems development with UML. Springer-Verlag New York Incorporated (2005)
3. Yu, E.S.K.: Modelling strategic relationships for process reengineering. PhD thesis, University of Toronto (1995)
4. Chung, L., do Prado Leite, J.: On non-functional requirements in software engineering. Conceptual Modeling: Foundations and Applications, 363–379 (2009)
5. Emery, F., Trist, E.: Socio-technical systems. In: Churchman, C.W., et al. (eds.) Management Sciences, Models and Techniques (1960)
6. Fisher, C., Kingma, B.: Criticality of data quality as exemplified in two disasters. Information & Management **39**(2), 109–116 (2001)
7. Sommerville, I., Cliff, D., Calinescu, R., Keen, J., Kelly, T., Kwiatkowska, M., Mcdermid, J., Paige, R.: Large-scale complex it systems. Communications of the ACM **55**(7), 71–77 (2012)

8. Mouratidis, H., Giorgini, P.: Secure tropos: A security-oriented extension of the tropos methodology. International Journal of Software Engineering and Knowledge Engineering 17(2), 285–309 (2007)
9. Gharib, M., Giorgini, P.: Modeling and reasoning about information quality requirements. In: Fricker, S.A., Schneider, K. (eds.) REFSQ 2015. LNCS, vol. 9013, pp. 49–64. Springer, Heidelberg (2015)
10. Securities and Exchange Commission: Findings regarding the market events of may 6, 2010. Report of the Staffs of the CFTC and SEC to the Joint Advisory Committee on Emerging Regulatory Issues (2010)
11. Kirilenko, A., Kyle, A.S., Samadi, M., Tuzun, T.: The flash crash: The impact of high frequency trading on an electronic market. Manuscript, U of Maryland (2011)
12. Pipino, L.L., Lee, Y.W., Wang, R.Y.: Data quality assessment. Communications of the ACM 45(4), 211–218 (2002)
13. Bovee, M., Srivastava, R.P., Mak, B.: A conceptual framework and belief-function approach to assessing overall information quality. International Journal of Intelligent Systems 18(1), 51–74 (2003)
14. Jiang, L.: Data quality by design: agoal-oriented approach. PhD thesis, University of Toronto (2010)
15. Wang, R.Y., Reddy, M.P., Kon, H.B.: Toward quality data: An attribute-based approach. Decision Support Systems 13(3), 349–372 (1995)
16. Liu, L., Chi, L.: Evolutional data quality: A theory-specific view. In: IQ, pp. 292–304 (2002)
17. Dai, C., Lin, D., Bertino, E., Kantarcioglu, M.: An approach to evaluate data trustworthiness based on data provenance. In: Jonker, W., Petković, M. (eds.) SDM 2008. LNCS, vol. 5159, pp. 82–98. Springer, Heidelberg (2008)
18. Wang, R., Strong, D.: Beyond accuracy: What data quality means to data consumers. Journal of Management Information Systems, 5–33 (1996)
19. Subrahmanyam, A.: Algorithmic trading, the flash crash, and coordinated circuit breakers. Borsa Istanbul Review 13(3), 4–9 (2013)
20. Ballou, D., Wang, R., Pazer, H., Tayi, G.K.: Modeling information manufacturing systems to determine information product quality. Management Science 44(4), 462–484 (1998)
21. Gharib, M., Giorgini, P.: Detecting conflicts in information quality requirements: the may 6, 2010 flash crash. Technical report, Università degli studi di Trento (2014)
22. Gharib, M., Giorgini, P.: Analysing information integrity requirements in safety critical systems. In: The 3rd International Workshop on Information Systems Security Engineering WISSE 2013 (2013)
23. Jureta, I.J., Mylopoulos, J., Faulkner, S.: Revisiting the core ontology and problem in requirements engineering. In: 16th IEEE International Requirements Engineering, RE 2008, pp. 71–80. IEEE (2008)
24. Mylopoulos, J., Chung, L., Nixon, B.: Representing and using nonfunctional requirements: A process-oriented approach. IEEE Transactions on Software Engineering, 483–497 (1992)
25. Antón, A.I., Earp, J.B., Reese, A.: Analyzing website privacy requirements using a privacy goal taxonomy. In: Proceedings of the IEEE Joint International Conference on Requirements Engineering, pp. 23–31 IEEE (2002)
26. Glinz, M.: Rethinking the notion of non-functional requirements. In: Proc. Third World Congress for Software Quality, vol. 2, pp. 55–64 (2005)

27. Masolo, C., Borgo, S., Gangemi, A., Guarino, N., Oltramari, A., Schneider, L.: Dolce: a descriptive ontology for linguistic and cognitive engineering. WonderWeb Project, Deliverable D17 v2 **1** (2003)
28. Abiteboul, S., Hull, R., Vianu, V.: Foundations of databases. Citeseer (1995)
29. Hevner, A.R., March, S.T., Park, J., Ram, S.: Design science in information systems research. MIS Quarterly **28**(1), 75–105 (2004)
30. Lewis, M.: Flash boys: a Wall Street revolt. WW Norton & Company (2014)
31. Lin, L., Nuseibeh, B., Ince, D., Jackson, M., Moffett, J.: Introducing abuse frames for analysing security requirements. In: Proceedings of the 11th IEEE International Requirements Engineering Conference, pp. 371–372. IEEE (2003)
32. Shankaranarayanan, G., Wang, R., Ziad, M.: Ip-map: representing the manufacture of an information product. In: Proceedings of the 2000 Conference on Information Quality, pp. 1–16 (2000)
33. Scannapieco, M., Pernici, B., Pierce, E.: Ip-uml: towards a methodology for quality improvement based on the ip-map framework. In: 7th Intl. Conf. on Information Quality (ICIQ 2002), pp. 8–10 (2002)

Understanding Model Quality Concerns When Using Process Models in an Industrial Company

Merethe Heggset[1], John Krogstie[1(✉)], and Harald Wesenberg[2]

[1] Norwegian University of Science and Technology - NTNU, Trondheim, Norway
merethhe@gmail.com, krogstie@idi.ntnu.no
[2] Statoil ASA, Stavanger, Norway
hwes@statoil.com

Abstract. Modelling has been used as a general technique in many companies for the last decades. Some already started using modelling in the eighties, trying out the first industrial CASE-tools. Their usage of modelling techniques has evolved over the years, finding new uses, and thus using the modelling techniques for supporting new goals. In our case company semi-formal modelling techniques have been taken into use on a large scale as a backbone for the company' quality system. In this paper we report on the use of process modelling in particular on the aspects found necessary to emphasise to achieve the right quality of the models in this organisation, and how the understanding of needed quality has evolved as the usage of modelling has evolved. A recent evaluation of the use of models in the company is reported, using the SEQUAL framework as an analytical lens for understanding and assessing the quality of models. Whereas earlier a focus has been on objective quality characteristics, with detailed guidelines for empirical and syntactic quality of models, the later investigations have identified the importance of also supporting the *process* to achieve and keep higher level quality on the semantic, pragmatic and social level.

Keywords: Enterprise process modelling · Case study · Model quality

1 Introduction

Process modeling has been performed relative to IT development and organizational development at least since the 70ties. The interest has been going through phases with the introduction of different approaches, including Structured Analysis in the 70ties [6], Business Process Reengineering in the late eighties/early nineties [7], and Workflow Management in the 90ties [41]. Lately, with the proliferation of BPM (Business process management) [8], interest in and use of process modeling has increased even further.

Models of work processes have long been utilized to learn about, guide and support practice also in other areas. In software process improvement [4], enterprise modeling [5] and quality management, process models describe methods and standard working procedures. Simulation and quantitative analyses are also performed to improve efficiency [22]. In process centric software engineering environments [2] and workflow systems [40] model execution is automated.

Statoil is a company which has used process modelling and other type of modelling for many years. It is a Norwegian oil company with more than 23 000 employees and around the same number of external contractors. Statoil operates in 36 different countries all over the world and has in particular in the last decade been using process modelling in order to structure their vast amount of organizational knowledge.

A lot of research has been done in the field of enterprise process modelling, as well as on the subject of how to judge the appropriateness of models. Much work is done regarding the use and creation of models on a theoretical level, but in order to better understand the mechanisms at work in the application of enterprise process models, real-life cases can provide interesting insights. How enterprise process models are actually used within an organisation will vary from case to case, so collecting as much information as possible about this from several sources seems appropriate and useful.

This paper presents some of the results from an ongoing case study on the use of enterprise process models in Statoil, in particular analysing how we can understand quality of models in an industrial setting. The main research question we have investigated in connection to this paper is;"How can we use existing frameworks of quality of models for structuring our understanding of the important aspects of model quality as it appears and evolves in practice"

Background on the evolution of the use of modelling in Statoil is found in section 2. An overview over the main analytical lens, the SEQUAL framework, is provided in section 3. In section 4 we discuss the current means for developing high quality models, and the challenges identified in this regard using the categories of SEQUAL.

Discussion of results, concluding remarks and ideas on further work on understanding the trade-off on different quality aspects are found in section 5.

2 Modelling in Statoil Through the Last Decades

As an advanced technology company, Statoil has a long tradition of taking new approaches into use for IT and organizational development. Back in the eighties, they experimented with the use of process and data modelling in connection to the use of CASE tools [32]. In the nineties, modelling where used for a broader set of tasks. As summarized in [3] the usage of process and enterprise models in Statoil was divided into three categories, based on their purpose, in what they called "The PAKT taxonomy":

1. Construction of reality: Modelling as a technique for creating a common understanding among people whose cognitive models do not necessarily coincide
2. Analysis and simulation: Making changes to simulated enterprise models and monitoring the consequences, in order to decide if a change should be put into action.
3. Model deployment and activation: An enterprise model being used for controlling and performing work. The operation of the enterprise is being done through and in the enterprise model.

Detailed case-studies particular on the first usage area were done and reported in [38], analyzing four case-studies in detail

1. VPT – Creation of value across organizational and disciplinary borders. Enterprise modeling of the Statoil value chain related to the Norwegian continental shelf, arguing for new and improved ways-of-working across existing boundaries.
2. PA30 – Process Plant 30+: Enterprise modeling as an activity in a large restructuring project at a Statoil-operated gas processing plant, conducted in order to have an overall view of business processes before changing them.
3. Gazz – Gas logistics – Development of an enterprise modeling software tool to be used for holistic and strategic thinking concerning Statoil's gas business.
4. TEK-s Technology strategy: Enterprise modeling as an aid in both development and dissemination of a corporate technology strategy.

Although the notations used in the different cases differed and partly covered larger part of the enterprise than the business processes, a standardized process modeling notation appeared. This was evaluated and compared with other notations in 2001 [20] using the current version of SEQUAL, but was at that stage kept with some changes. Some years later when BPMN arose as a standard this was also chosen and adapted by Statoil. Statoil decided to use enterprise process models as part of their corporate management system in 2004. The introduction of models has had a positive effect on operations. The models contribute to reducing risk, from an operational, environmental and safety perspective [39]. To illustrate, the number of serious incidents per million work hours have been reduced from 6 to around 0.8 since the introduction of enterprise models. Statoil employees and sub-contractors perform around 2 million work hours per week in total, so the reduction is important also in absolute terms. While other aspects certainly have contributed to this reduction, enterprise modelling has played a large role in changing the way of working in Statoil during the last decade. The current enterprise model is realized through the Statoil management system. The Statoil Book [37], which is the foundation the management system is built upon, describes the management system as "the set of principles, policies, processes and requirements which support our organisation in fulfilling the tasks required to achieve our goals". It defines how work is done within the company, and all employees are required to act according to relevant governing documentation.

The Management System consists of three main parts:

- Process models using a restricted subset of BPMN represented in ARIS, the modelling solution from which all governing documentation (GD) is accessed by the end users.
- Docmap, used for handling and publishing textual governing documentation.
- Disp, a tool which supports the process of handling applications for deviation permits in cases where compliance with a requirement is difficult or impossible to achieve.

The three main objectives of the Statoil Management System are:

1. Contributing to safe, reliable and efficient operations and enabling compliance with external and internal requirements.
2. Helping the company incorporating their values, people and leadership principles into everything they do.
3. Supporting business performance through high-quality decision-making, fast and precise execution and continuous learning.

Governing documentation (GD) describes what is to be achieved, how to execute tasks, and ensures standardisation. Each process area has governing documentation in the form of documents and/or process models, accessible from the ARIS start page.

The management system function is responsible for creating and improving the management system based on business needs and ensuring that the governing documentation is understood and used, as well as monitoring compliance with work requirements. The work of the function follows a five-step cycle; Assess and plan, design, implement, use, and monitor and control. This is done in close collaboration with line management and owners of the governing documentation.

The enterprise process model is created according to a set of rules for structuring and use of notation, and can be used for a variety of purposes, such as compliance management, competence management, portfolio management, decision making and performance analysis. There are three levels of abstraction in the enterprise model: The contextual level, the conceptual level and the logical level, including the following interrelated diagrams as illustrated in Fig. 1:

- The top-level diagram is a mandatory navigational diagram visualizing core value chain processes, management processes, and support processes, capturing what they in Statoil term the contextual level. This is similar to what is known as a process map [15], depicting the core, support, and management processes at the highest level.
- The navigation diagram(s) are optional diagrams to support more tailored access to the processes than the top-level diagram.
- Model diagram: Is a mandatory diagram that visualizes the model of one process area in the organization.
- Process navigation diagram is an optional model for navigational support on the conceptual level.
- Workflow diagram - Contains BPMN models [1, 31] on the descriptive level. The quality system contains around 2000 BPMN models at this level, qualifying the case to be an example of BPMN in the large [12]. An example of a simple workflow diagram is found in Fig. 2 (labels in Norwegian). Note that this example follows the specific version of BPMN made by Statoil, which differs a bit from the official BPMN-definition, e.g. including special semantics to the grouping mechanism.

Understanding Model Quality Concerns when Using Process Models 399

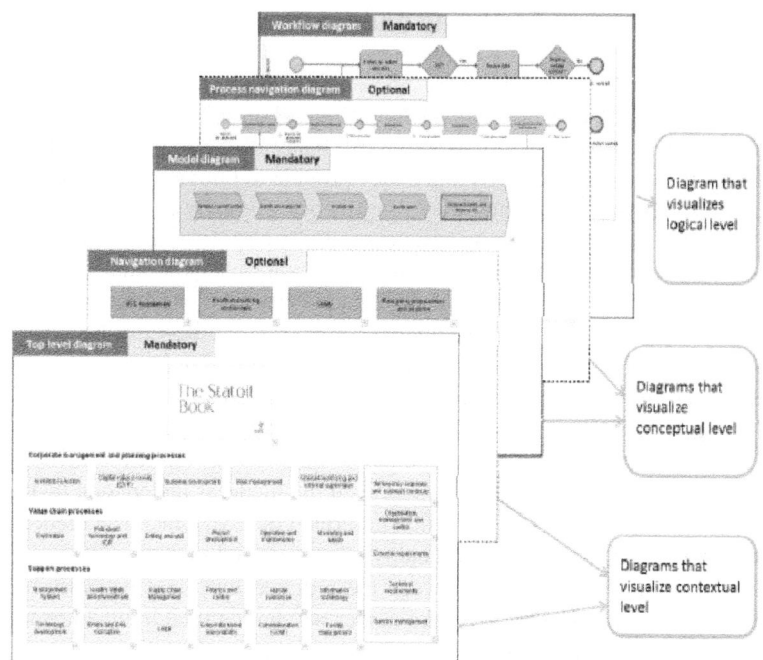

Fig. 1. Structure of models in STATOIL Management System

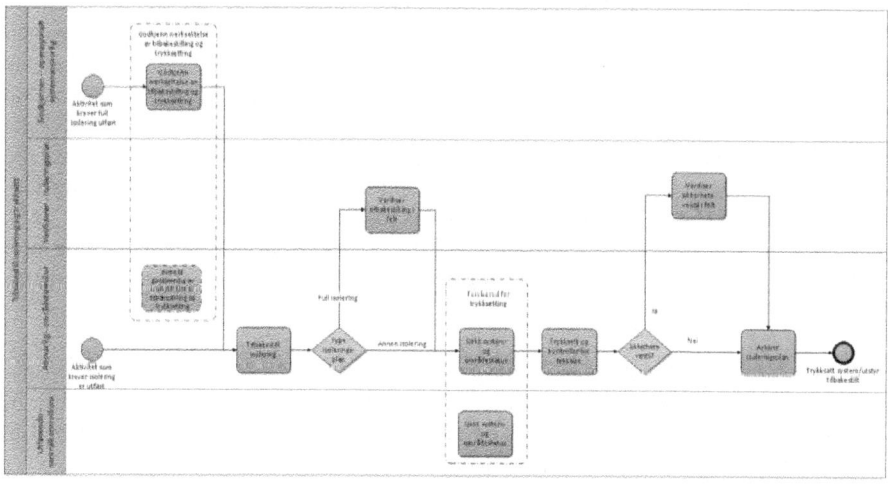

Fig. 2. Example of BPMN model from Statoil

3 Background on Modeling and Quality of Models

Model quality has been discussed by many researchers over the years, and many frameworks and methods have been developed based on scientific theories from various fields. However, as stated by Moody [26], many of these methods suffer from a lack of adoption in practice, partly since they provide too limited guidance. While the main goal of applying such frameworks in practice normally is providing a detailed evaluation of model quality in a specific case, it can also give indications of the usefulness of the framework and, based on the results improve the framework, and possibly enforce its position in the field which again may lead to a wider adoption.

From the start of the modelling initiative supporting the new management system in 2004, Statoil has been aware of the need to balance different levels of quality of the models. According to [34, 39] Statoil have found that it is useful to differentiate between at least three dimensions of model quality: Syntactic quality (how well the model uses the modelling language), semantic quality (how well the model reflects the real world) and pragmatic quality (how well the model is understood by the target audience), building upon distinctions first described in [23], which is a predecessor to the current SEQUAL framework on quality of models and modelling languages [15]. In enterprise models the balance between these dimensions becomes very important based on the goal of modelling; else the model will not be used by its intended target audience in the right way. In our analysis, we have thus applied the current version of SEQUAL.

SEQUAL is a quality framework used for assessing the quality of models and modeling languages. The choice of using SEQUAL as an analytical lens for studying the Statoil enterprise model is mainly based on that the company has applied the three core quality levels of SEQUAL (syntactic, semantic and pragmatic) as also reported in earlier work [39]. Krogstie and Arnesen [20] used parts of SEQUAL to evaluate various enterprise modelling languages for use in Statoil. SEQUAL builds on early work on quality of models, but has been extended based on theoretical results [26, 27, 30] and practical experiences [15, 21, 25] with various versions of the framework. It has earlier been used for evaluation of modelling and modelling languages of a large number of perspectives, including data [17, 18], ontologies [11], process [16, 19], enterprise [20], topological [28] and goal-oriented modelling [13, 14]. Quality has been defined referring to the correspondence between statements belonging to the following sets:

- G, the set of goals of the modelling task. The goals of modelling can be many, and may vary greatly. Nysetvold and Krogstie outlines five main usage areas of enterprise models [29] (partly inspired by the PAKT taxonomy [3] described in section 2 and general model theory [33]):

  1. Human sense-making and communication: Actors can use the enterprise model to make sense of various aspects of the enterprise, and best practices and requirements can be communicated throughout the organisation to create a common understanding.

2. Computer-assisted analysis: Models can be used e.g. for simulation of process changes.
3. Business process management and quality assurance: Models can be used for quality assurance of work processes (e.g. ensuring compliance to regulations).
4. Model deployment and activation: The model can be deployed directly to be used for controlling, supporting and performing work. The activation can either be manual, automatic or interactive.
5. To give context for other tasks such as supporting system development.

- **D**, the domain, i.e., the set of all statements that can be stated about the situation. The goal of modelling restricts the domain to only those things relevant to achieve this/these goal(s).
- **L**, the language extension, i.e. what can be expressed by the modelling language chosen.
- **M**, the externalized model itself.
- **K**, the explicit knowledge that the audience (both modelers and model interpreters) have of the domain.
- **I**, the social actor (human) interpretation of the model
- **T**, the technical actor (tool) interpretation of the model

The main quality types relates these sets and are depicted in Fig. 3:

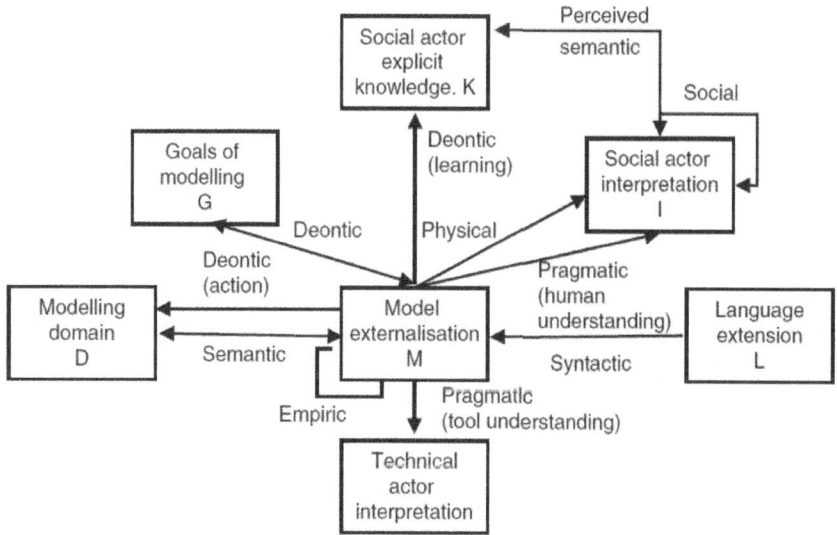

Fig. 3. SEQUAL Framework for quality of models

- Physical quality: The basic quality goal is that the relevant parts of the externalized model **M** is available to the relevant actors (and not others) for interpretation (**I** and **T**).
- Empirical quality deals with the comprehensibility of the model **M**.

- Syntactic quality is the correspondence between the model M and the language extension L. Is the language used correctly in the model?
- Semantic quality is the correspondence between the model M and the domain D.
- Perceived semantic quality is the similar correspondence between the social actor interpretation I of a model M and his or hers current knowledge K of domain D.
- Pragmatic quality is the correspondence between the model M and the actor interpretation (I and T) of it. Thus whereas empirical quality focus on if the model is understandable according to some objective measure that has been discovered empirically in e.g., cognitive science, we at this level investigate to what extend the model has actually been understood.
- The goal defined for social quality is agreement among social actor's interpretations of the models.
- The deontic quality of the model relates to that all statements in the model M contribute to fulfilling the goals of modelling G, and that all the goals of modelling G are addressed through the model M.

When we structure different quality aspects according to these levels, one will find that there might be conflicts between the levels (e.g. what is good for semantic quality such as completeness might be bad for pragmatic quality), thus it is important to make a trade-off between achieving the different quality levels for achieving the most important goals of modelling.

4 Evaluation of Quality of Models in Statoil

As reported in [9], the detailed guidelines for how modelling should be done to support high-quality models in Statoil, the TR0002, has earlier been analysed. As described in section 2, modelling has been used for a number of years in Statoil. The requirements for modelling to achieve a balance of syntactic, semantic and pragmatic quality has through this period evolved based on concrete needs identified. Thus, although we in [9] in particular analysed the current requirements (Version 3, valid from Dec. 5 2013) [35] we also investigated the development, in particular relative to version 1 of the requirements, that was made available Feb. 12 2009 [34]. Although the levels of syntactic, semantic, and pragmatic quality are emphasized, the existing requirements are not structured according to these levels. Also other levels of SEQUAL are relevant, partly since the original SEQUAL-categories have been divided in sub-areas in the later versions of the framework (e.g. splitting pragmatic quality into empirical quality (for aspects that at least in theory can be evaluated objectively by a tool) and pragmatic quality for aspects of understanding that have to take the human interpreter into account). Looking first at the sets of SEQUAL in the light of the case of the Statoil management system, we have the following:

- M: The models we analyse here are in particular the workflow-models (bottom level) of the overall model-framework. Relative to the description of purpose of modelling in Section 3 the models are meant to be as-is models, to support communication on the current process, manual activation (i.e. supporting human

action in the organization according to the models), and checking of compliance (area 3: quality assurance).
- *G*: Whereas the general requirements for the quality system were described in Section 3, five main more concrete usage areas have been defined:
  1. **Compliance management:** To monitor and control that the way of working is compliant with the standards set for the way to work. This enables producing predictable output from work.
  2. **Competence management:** Document the competency profiles needed to perform tasks, compare required competency profiles with competence represented in the organization, and therefore manage the competency gap.
  3. **Portfolio management:** Gain an overview of the current portfolio of e.g. processes, information systems, and technologies. This gives opportunities for analysing whether the existing portfolio/installed base will meet future needs, and to plan the roadmap to get from the current to the future portfolio.
  4. **Analysis and decision making:** The model enables analysis of the relationships between different objects in the models and how changes to one object (e.g. a process) will impact other objects (e.g. the information systems used by that process or relations between different work processes)
  5. **Performance analysis:** Monitoring of these results to get experience and data on the quality. This information can be used to analyse if the way of working produce the best possible result.

Even if several possible purposes are listed, one model always has one primary purpose, with potentially a (set of) secondary purpose (s). The current primary purpose of the enterprise model is compliance management therefore the priority is given on achieving the right quality of governing documentation models with corresponding governing elements, roles and responsibilities. We notice that two of these goals were not in version 1 of the requirements (competency management and performance analysis). Rather than being an example of 'goal creep' (that models used for one purpose over time is used also for other things not originally envisioned [21]) it is because the models have to be current as-is models (due to focus on compliance). First recently the underlying infrastructure to support competency management and performance analysis has been put in production.

- *D*: Domain: The work processes in Statoil.
- *L*: The language for workflow modelling is a subset of BPMN 2.0. In the original version of the requirements [34] it was a similar sub-set of BPMN 1.
- *A*: The target audience comes from the whole company. It is therefore necessary to do a stakeholder analyses to ensure that models have the right abstraction level, complexity, with a terminology suitable for the target audience.
- *K*: The relevant explicit knowledge of the actors (A).
- *T*: The tool currently used is ARIS. We note that two other tools (APOS and QLM/BPM) were used at the time version 1 of the requirements was made, ARIS being introduced at a later stage.
- *I*: Relates to how easy it is for the different actors to interpret the data as it can be presented in ARIS.

4.1 Classification of Existing Model Requirements

In the analysis of the current version of the guidelines in [9], we found requirements relative to the following levels

Physical quality

- The models should be available to the right people in a physical form (through the ARIS-tool) when needed for interpretation. Only the relevant parts should be available.
- Both the current and previous versions of the model should be available.
- It should be possible to store relevant meta-data e.g., on purpose and validity (a Statoil-term for what part of the organization the model applies for).

Empirical quality

- Focus on naming conventions for labels – a large number of detailed rules has been defined for this.
- Detailed rules for readability of textual parts
- Rules for colour usage in models

Syntactic quality

- The language used should be a well-defined subset of BPMN
- Some specific additions to BPMN defined in the language
- Many strict usage rules and guidelines for the use of the selected elements

Semantic, Pragmatic, Social, Deontic quality

- Fewer concrete guidelines on these levels, much supported in the modelling process adhering to regulations (cf. compliance goal)
- Pragmatic quality regarded as most important (including using the model to act correctly), but this level is mostly supported through rules on the empirical and syntactic level
- Mechanisms for providing input on models by users if errors are found
- Different parts of the model relevant for different parts of the organization – limiting the visibility is limiting the problem of social quality.
- More emphasis on safety and compliance than business performance. The relation between the different goals of modelling, different quality aspects and the goals of the quality systems are not explicit. Neither are the cost/benefit trade-offs between effort used and sufficient quality achieved.

4.2 Evaluation of the Management System

During the end of 2013 and the beginning of 2014, a large-scale user survey [36] was conducted in Statoil in order to better understand users' experiences and opinions related to the management system and governing documentation. A similar survey was also conducted in 2012. 4828 employees took part in the survey, which was about half of those invited. Many challenges were identified from the survey, related to the management system itself, learning processes and work practice, all of which contribute in some way to the management system goals of safety, reliability and efficiency (relative to objective 1 described in Section 2). The survey is seen as very useful, due to the large amount of quantitative data as well as the amount of detailed feedback given by the participants. Statoil is using the survey results as a basis when planning and implementing changes to the management system, and will use a similar survey in 2015 to hopefully be able to see a measurable improvement. Many of the issues discovered can be connected to model quality, and below the most important findings are summarised using the quality-levels of SEQUAL.

4.2.1 Physical Quality Issues
The survey showed that many of the employees have trouble finding what they need when they look for governing documentation. Moreover, when they do find the relevant documentation, more than half of the respondents are unsure that they have found all relevant documentation. Some describe ARIS as a "maze", in which it is hard to keep track of where the displayed page is situated in the hierarchy. According to the respondents, the search function often does not produce the desired result. Each user has a personal space called "MyPage", accessible from the menu at the top left of each page. From a workflow model page, users can click the "Subscribe" tab, and confirm that they want to subscribe to this particular model. The familiarity with this functionality is unfortunately low among many respondents.

Many are not satisfied with the way changes to GD affecting their work are communicated, which makes it difficult to know if the information they possess is current. Employees are not aware of the possibility for staying updated on changes, and when they do, they experience that the reasoning behind the changes are not clearly communicated. 14% of the respondents report using paper copies to access GD, so unless employees are clearly notified of changes they might keep using old versions.

4.2.2 Empirical Quality Issues
Users feel that governing documentation suffers from lack of clarity, and 42% of the survey respondents often do not understand abbreviations used in text and models. Note that the guidelines for modeling discourage explicitly the use of abbreviation, thus here it seems that it is not necessarily the guidelines that are the problem, but the lacking adherence to the guidelines.

4.2.3 Syntactic Quality Issues
Although there are many guidelines on this level, there is many examples that these are only partly adhered to, as also will be reported in [10]. Still this was not explicitly mentioned as an issue.

4.2.4 Semantic Quality Issues
The possibility for users with hands-on experience with the process to add improvement suggestions could improve the semantic quality of workflow models, as it could impose an improved correspondence between model and domain. However, the process of handling improvement proposals appear to be too slow and inconsistent, as many users experience waiting a long time to get feedback on their suggestions, and often the reasoning behind the outcome is not clear. Almost half of the respondents have experienced not receiving any feedback at all on their proposals. This could lead to lack of motivation for posting suggestions in the first place, even though they might

be useful. In addition, even though 68% feel that the governing documentation has the right amount of detail, it is also often seen as too rigid and general to account for local needs and variations, which leads to a lot of requests for deviations as the models are not experienced to fit the domain properly. 17% of survey respondents report often seeing gaps between what is described in the GD and what is being done in practice.

4.2.5 Pragmatic and Social Quality Issues

The survey uncovered challenges regarding understanding and processing. About half of the respondents feel that governing documentation is easy to understand. By others, governing documentation is perceived as vague and ambiguous, especially when it comes to authorities and responsibilities. This ambiguity often causes interpretations by different users to differ from each other. One in five of the respondents often or always experience this within their department or unit. A good support system for learning could improve users' understanding of the models and the system in general, but only 44% report being satisfied with the support they are given. About half of the respondents have participated in organised training related to the use of GD. These have a higher score for confidence in, use of and compliance with GD than the ones who have not participated in a training program. The survey showed that good leadership support has a strong positive effect on use, but in general, leaders do not sufficiently encourage better use of the governing documentation, and are often not able to answer questions related to the management system that they receive from their employees.

4.2.6 Deontic Quality Issues

Considering how governing documentation contribute to the goals of the management system, the results from the survey indicate that it contributes a lot to high safety (as confirmed by 75% of the respondents) and moderately to high reliability, but not to high efficiency (37%). One in five of the respondents feel that safety and efficiency is not properly balanced. Reasons for this imbalance are given as:

- The GD is too focused on safety, and this slows down execution of tasks
- Requirements are too rigid and complying with them is time-consuming
- Low user-friendliness. The relevant GD can be hard to find
- Differing interpretations lead to time-consuming discussions
- Local best practice is not always reflected in the GD
- Lack of cost awareness
- Competitiveness is not addressed, the emphasis is put on meeting formal requirements to assure compliance

5 Discussion, Conclusion and Further Work

The quality system of Statoil is developed supporting in particular compliance to requirements to reduce risk, an area where large improvements have been observed over the last decade. Still one find challenges with among other things the comprehension of some of the models as described above. Whereas the requirements given in TR0002 are quite detailed and structured providing guidelines on most quality levels of SEQUAL, they focus mostly on empirical and syntactic quality. Although quality on these levels is also important for pragmatic quality, the guidelines are not always followed [10].

Through the user survey [36], interviews and conversations have provided valuable insights into how users experience the management system. The use of SEQUAL to structure this discussion points to issues also on higher levels (semantic, pragmatic, social, and deontic) where the following of concrete objective guidelines for the quality of models is not sufficient. Some measures can be taken to achieve higher quality. Some users feel that governing documentation is hard to understand. Increased understanding is a necessity if 100% compliance is the goal. Measures that can contribute to this include applying the language guidelines and naming conventions more strictly and tailoring the complexity of models according to the needs of its target audience. Processes for taking the knowledge of the employees more directly into the loop, and for clearer model governance is also pointed to as important. Interestingly it can also be noted that also different emphasis in the organization (more towards efficiency and not only safety and compliance) seems to influence the perception of quality. As we saw in section 2, the use of modelling within the company has evolved over the years, and models and modelling practices that were regarded as good on an earlier stage might no longer be looked upon as being sufficient.

We pointed earlier to that theoretical frameworks such as SEQUAL is little used in practice. We hope this and other case studies of its use can make it easier to apply the framework for quality assessment tasks also in other situations and organizations. That all aspects pointed to in the investigation can be categorized according to one of the quality levels also support the completeness of the framework.

There are several possibilities for further work related to the Statoil enterprise process model. Based on the internal evaluation, a new modelling standard and improved tool support is being developed and implemented. When the new functionality developed has been implemented in full-scale, the actual effect of these changes on model quality and the effect of the models in practice can be analysed. A new user survey, similar to the one carried out in 2013/2014 will be distributed by Statoil when these changes have been put into effect. Studying the results based on the new standards and tools and comparing them to the old may give important insight into the real value of such changes. In particular, following the implementation of the new TR0002 document in practice, and how it impacts model quality and use is an interesting possibility for future research.

References

1. Aagesen, G., Krogstie, J.: Analysis and design of business processes using BPMN. In: vom Brocke, J., Rosemann, M. (eds.) Handbook on Business Process Management. Springer (2010)
2. Ambriola, V., Conradi, R., Fuggetta, A.: Assessing Process-Centered Software Engineering Environments. ACM Transactions on Software Engineering and Methodology 6(3) (1997)
3. Christensen, L.C., Johansen, B.W., Midjo, N., Onarheim, J., Syvertsen, T., Totland, T.: Enterprise modeling-practices and perspectives. Computer in Engineering, 1071–1084 (1995)

4. Derniame, J.C. (ed.): Software Process. LNCS, vol. 1500. Springer, Heidelberg (1998)
5. Fox, M.S., Gruninger, M.: Enterprise modeling. AI Magazine (2000)
6. Gane, C., Sarson, T.: Structured Systems Analysis: Tools and Techniques. Prentice Hall (1979)
7. Hammer, M., Champy, J.: Reengineering the Corporation: A Manifesto for Business Revolution. Harper Business (1993)
8. Havey, M.: Essential Business Process Modelling. O'Reilly (2005)
9. Heggset, M., Krogstie, J., Wesenberg, H.: Ensuring quality of large scale industrial process collections: experiences from a case study. In: Frank, U., Loucopoulos, P., Pastor, Ó., Petrounias, I. (eds.) PoEM 2014. LNBIP, vol. 197, pp. 11–25. Springer, Heidelberg (2014)
10. Heggset, M., Krogstie, J., Wesenberg, H.: The Influence of Syntactic Quality of Enterprise Process Models on Model Comprehension. Accepted for publication at CAiSE forum 2015 (2015)
11. Hella, L., Krogstie, J.: A structured evaluation to assess the reusability of models of user profiles. In: Bider, I., Halpin, T., Krogstie, J., Nurcan, S., Proper, E., Schmidt, R., Ukor, R. (eds.) BPMDS 2010 and EMMSAD 2010. LNBIP, vol. 50, pp. 220–233. Springer, Heidelberg (2010)
12. Houy, C., Fettke, P., Loos, P., van der Aalst, W.M.P., Krogstie, J.: Business Process Management in the Large. Business & Information Systems Engineering **6** (2011a)
13. Krogstie, J.: Using quality function deployment in software requirements specification. Paper Presented at the Fifth International Workshop on Requirements Engineering: Foundations for Software Quality (REFSQ 1999), Heidelberg, Germany, June 14-15 (1999)
14. Krogstie, J.: Integrated goal, data and process modeling: from TEMPORA to model-generated work-places. In: Johannesson, P., Søderstrøm, E. (eds.) Information Systems Engineering From Data Analysis to Process Networks, pp. 43–65. IGI (2008)
15. Krogstie, J.: Model-based development and evolution of information systems: A quality approach. Springer, London (2012)
16. Krogstie, J.: Quality of business process models. In: Sandkuhl, K., Seigerroth, U., Stirna, J. (eds.) PoEM 2012. LNBIP, vol. 134, pp. 76–90. Springer, Heidelberg (2012)
17. Krogstie, J.: Quality of conceptual data models. In: Proceedings 14th ICISO, Stockholm, Sweden (2013)
18. Krogstie, J.: Capturing Enterprise Data-integration Challenges using a Semiotic Data Quality Framework. Business & Information Systems Engineering **57**(1), 27–36 (2015)
19. Krogstie, J., Jørgensen, H.D.: Quality of interactive models. In: Olivé, À., Yoshikawa, M., Yu, E.S. (eds.) ER 2003 Ws. LNCS, vol. 2784, pp. 351–363. Springer, Heidelberg (2003)
20. Krogstie, J., de Flon Arnesen, S.: Assessing enterprise modeling languages using a generic quality framework. Information Modeling Methods and Methodologies, (1537–9299), 63–79 (2005)
21. Krogstie, J., Dalberg, V., Jensen, S.M.: Process modeling value framework. In: Manolopoulos, Y., Filipe, J., Constantopoulos, P., Cordeiro, J. (eds.) ICEIS 2006. LNBIP, vol. 3, pp. 309–321. Springer, Heidelberg (2008)
22. Kuntz, J.C., Christiansen, T.R., Cohen, G.P., Jin, Y., Levitt, R.E.: The virtual design team: A computational simulation model of project organizations. Communications of the ACM **41**(11) (1998)
23. Lindland, O.I., Sindre, G., Sølvberg, A.: Understanding Quality in Conceptual Modelling. IEEE Software **11**(2), 42–49 (1994)
24. Malinova, M., Leopold, H., Mendling, J.: A meta-model for process map design. In: CAISE Forum 2014, June 16-20, Thessaloniki, Greece (2014)

25. Moody, D.L., Sindre, G., Brasethvik, T., Sølvberg, A.: Evaluating the quality of process models: empirical testing of a quality framework. In: Spaccapietra, S., March, S.T., Kambayashi, Y. (eds.) ER 2002. LNCS, vol. 2503, pp. 380–396. Springer, Heidelberg (2002)
26. Moody, D.L.: Theorethical and practical issues in evaluating the quality of conceptual models: Current state and future directions. Data and Knowledge Engineering **55**, 243–276 (2005)
27. Nelson, H.J., Poels, G., Genero, M., Piattini, M.: A conceptual modeling quality framework. Software Quality Journal **20**, 201–228 (2012)
28. Nossum, A., Krogstie, J.: Integrated quality of models and quality of maps. In: Halpin, T., Krogstie, J., Nurcan, S., Proper, E., Schmidt, R., Soffer, P., Ukor, R. (eds.) BPMDS 2009 and EMMSAD 2009. LNBIP, vol. 29, pp. 264–276. Springer, Heidelberg (2009)
29. Nysetvold, A.G., Krogstie, J.: Assessing business process modeling languages using a generic quality framework. In: Siau, K. (ed.) Advanced Topics in Database Research, vol. 5, pp. 79–93. Idea Group, Hershey (2006)
30. Price, R., Shanks, G.: A semiotic information quality framework: Development and comparative analysis. Journal of Information Technology **20**(2), 88–102 (2005)
31. Silver, B.: BPMN Method and Style. Cody-Cassidy Press (2012)
32. Solum, P.E., Østerud, M.: Integreret CASE-verktøy. Kartlegging av teknologien og problemer i forhold til tradisjonell systemutvikling. Master Thesis NTNU, Trondheim Norway (1989)
33. Stachowiak, H.: Allgemeine Modelltheorie. Springer, Wien (1973)
34. Statoil: TR0002 Enterprise Structure and Standard Notation. version 1 (2009)
35. Statoil: TR0002 Enterprise Structure and Standard Notation. version 3 (2013)
36. Statoil. Management system user survey u13 (2014)
37. Statoil: Statoilboken (2014). http://www.statoil.com/no/About/TheStatoilBook/Downloads/Statoil-Boken.pdf
38. Totland, T.: Enterprise Modeling as a Means to Support Human Sense-making and Communication in Organizations, PhD Thesis, NTNU Trondheim, Norway (1997)
39. Wesenberg, H.: Enterprise modeling in an agile world. In: Johannesson, P., Krogstie, J., Opdahl, A.L. (eds.) PoEM 2011. LNBIP, vol. 92, pp. 126–130. Springer, Heidelberg (2011)
40. Weske, M.: Business Process Management: Concepts, Languages, Architectures. Springer-Verlag New York Inc. (2007)
41. WfMC Workflow Handbook 2001. Workflow Management Coalition. Future Strategies Inc., Lighthouse Point (2000)

Meta-Modeling and Domain Specific Modeling and Model Composition

Towards Metamodelling-In-The-Large: Interface-Based Composition for Modular Metamodel Development

Srđan Živković[✉] and Dimitris Karagiannis

Faculty of Computer Science, University of Vienna, Vienna, Austria
{srdjan.zivkovic,dimitris.karagiannis}@univie.ac.at

Abstract. Modelling language engineering approaches based on metamodelling provide powerful concepts to define metamodels, pivotal constructs for language definition. With increasing popularity of domain-specific, hybrid and evolving modelling languages, the necessity for efficient and flexible metamodelling becomes apparent. Modularisation and composition techniques can reduce effort and improve efficiency and flexibility in metamodel development. Existing metamodelling languages and standards provide means for metamodel modularisation and composition, however based on pure white-box packages and inheritance-like composition operators. In this paper, we propose an approach based on interfaces and interface-based composition operators. Inspired by component-oriented concepts of programming languages, we introduce black-box metamodel fragments with explicit provided and required interfaces that can be combined to systematically and flexibly build new metamodels. We discuss the realisation of our approach as an extension to existing metamodelling language concepts and demonstrate its applicability by modularising the metamodel of the BPMS method, a hybrid BPMN-centred language for enterprise modelling.

Keywords: Metamodelling · Metamodel composition · Metamodel modularisation · Metamodelling tools

1 Introduction

Until recently, modelling language engineering has been reserved for a handful of language experts, since the number of (general-purpose) modelling languages was rather low and the complexity of creating new languages high. This situation changed dramatically with the rise of domain-specific modelling languages (DSML) and appropriate tools for language development. Instead of sticking to a handful of one-size-fits-all languages, a *plethora of DSMLs* are being designed to facilitate model-based analysis, design, and development of systems and software. DSMLs are usually restricted to narrow domains. Hence, DSMLs are combined into s.c. *hybrid languages*, to address the complexity of the system from different perspectives holistically. Furthermore, languages undergo changes during their life cycle, i.e. they evolve. Such *evolving languages* may be adapted, customised or extended for problem and project-specific needs. Increasing number of DSMLs, their hybrid and evolving nature are three phenomena that imply systematic, flexible and efficient approaches for language definition.

Metamodelling has been recognised as a practical yet rigorous formalism for language definition. A metamodel is used to define the abstract syntax of a language. As a pivotal element in language definition, metamodel defines language concepts for which precise semantics and one or more concrete syntaxes may be defined [1]. Nowadays, a multitude of *metamodelling languages* for language definition exist such as MOF [2], MetaEdit+'s GOPPRR [3], ADOxx's Meta^2-Model [4,5], GME's MetaGME [6], Eclipse EMF Ecore [7] or GrUML [8]. Referring to the issues of programming-in-the-small vs. programming-in-the-large [9], existing metamodelling languages focus largely on concepts for "metamodelling-in-the-small", i.e. they provide constructs for developing metamodels as monolithic design artefacts, but offer less support for modularity common to modern component-oriented programming languages. Even though the standard metamodelling language MOF offers comprehensive modularisation and composition concepts such as *Package*, *packageImport*, *elementImport* and *packageMerge*, concepts for specifying component-like metamodel modules that support information hiding as well as flexible composition operators are missing. It has been recognised that modular, component-oriented, compositional development of software systems [10,11] addressed the issues of programming-in-the-small and pushed the software industry into a new era of software factories [12] and component-based software assembly. Hence, to contribute to the idea of metamodelling-in-the-large, we propose a modular approach to metamodelling based on the idea of metamodel fragment interfaces and corresponding interface-based composition operators. While existing modularisation concepts allow for creating white-box fragments that can be combined based on fragile class inheritance and intrusive merge operators, explicit interfaces and interface-based composition operators contribute to systematic and flexible composition of reusable, self-contained black-box metamodel components.

The remainder of the paper is organised as follows. We introduce the basic notions of interface-based modularisation and composition in Section 2. In Section 3 we discuss how such concepts can be realised as a part of a metamodelling language. Section 4 illustrates the applicability of the approach in the case of modular BPMS. Finally, in Section 5 we discuss related work on metamodel modularisation and composition. Section 6 concludes the paper.

2 Interface-Based Modular Metamodel Development

This section introduces the idea of interface-based modular metamodel development (I-MMD). After we have motivated the need for better support towards metamodelling-in-the-large, we introduce the concepts for interface-based modularisation and composition.

Table 1. Comparison of metamodelling language capabilities

Capability	ADOxx Meta²-Model	EMF Ecore	GME MetaGME	GrUML	MetaEdit+ GOPPRR	MOF 2.0
Core capabilities for Metamodeling-in-the-small						
Class	Class	EClass	Atom	Vertex Class	Object Type	Class
Attribute	Attribute	EProperty	Attribute	Attribute	Property	Property
Relation	Relation Class	EReference	Connection	Edge Class	Relationship	Association
Relation End	Endpoint	-	Connection Role	Incidence Class	Role, Port	Property
Model Type	Model Type, Mode	EPackage	Model, Aspect, Role	Package	Graph Type	Package
Supporting capabilities for Metamodeling-in-the-large: Modularisation						
Encapsulation	Library	Package	Project	Package	Graph Type	Package
Info. hiding	-	-	-	-	-	-
Supporting capabilities for Metamodeling-in-the-large: Composition						
White-box	Inheritance, Aggregation	Inheritance	Proxy Reference, Impl. inheritance, Interface inheritance	Inheritance	Inheritance, Inclusion	Inheritance, Redefinition, packageImport, packageMerge, elementImport
Black-box	-	-	-	-	-	-

2.1 Metamodelling-In-the-Small vs. Metamodelling-In-the-Large

To paraphrase Szyperski [10], there is nothing that can be done with components, that cannot be done without them. Clearly, while a metamodel may be developed from scratch as a monolithic design artefact, the same result may be achieved in a more productive and flexible way through the composition of prefabricated metamodel fragments. Analysing the capabilities of metamodelling languages we can distinguish between, what we call, core and supporting metamodelling capabilities. *Core capabilities* provide constructs to define fundamental elements of a metamodel, i.e. to enable metamodelling-in-the-small. Constructs such as class, attribute, relation, relation end and model type belong to core capabilities as they contribute to the core expressive power of a metamodelling language. On the other side, *supporting capabilities* enrich the metamodelling language with constructs for efficient metamodelling. Supporting capabilities contribute to an increased reuse of core metamodel artefacts, via *modularisation* and *composition*. A typical example of a supporting composition construct is inheritance, which enables reuse of attributes along the class hierarchy.

The analysis of the state-of-the-art metamodelling languages shows that both core capabilities as well as supporting capabilities for white-box modularisation and composition are well supported (Table 1). However, existing encapsulation concepts such as *Package* which are used for the compartmentalisation of elements into reusable modules are pure white-box components. It can be observed that none of the existing modularisation concepts support the principle of *information hiding*, which is crucial to separate the component internal implementation from its external interfaces. Furthermore, existing composition operators such as inheritance and packageMerge are designed to operate purely on white-box components. They require full white-box visibility for property inheritance and property merge. It is known that white-box composition leads to highly coupled modules with strong dependency. To further support metamodelling-in-the-large and to overcome deficiencies of white-box composition, we introduce the interface concept, already established in GPLs, for metamodel definition. Interfaces contribute to information hiding and definition of black-box components and allow for flexible composition based on black-box composition operators.

2.2 Metamodel Modularisation Based on Black-Box Metamodel Fragments

Referring to the definition of a software component [10], we define *a metamodel fragment as a unit of composition with contractually specified provided interfaces and required interfaces only. A metamodel fragment can be deployed independently and is subject to composition by third parties*. This definition introduces several key properties of metamodel fragments such as the existence of provided interfaces and required interfaces as a form of explicit dependencies to other fragments. The independent deployment refers to the fact that a fragment must be self-contained, and must not have any other dependencies than those explicitly declared.

Encapsulation. As a unit of composition and reuse, a *metamodel fragment* encapsulates metamodel elements that contribute to either fragment implementation or fragment interface definition. *Fragment implementation* defines the actual metamodel structure and consists of one or more *core metamodel elements* such as classes, relations, etc. A fragment may internally nest other owned fragments. Elements of a nested fragment are indirectly and recursively owned by a nesting fragment. If a fragment consists only of direct elements it is called *atomic*. If it includes other fragments it is called *composite*.

Information hiding. To support information hiding, a fragment defines a set of interfaces, which hide internal implementation of a fragment. The concept of an interface is the cornerstone for the flexible black-box metamodel composition. Metamodel fragment with explicit interfaces is treated as a black-box and is therefore subject to composition by other fragments in various contexts. The coupling of fragments based on interfaces is loose such that fragments may easily be replaced or modified as long as the interfaces remain intact. A *provided interface* of a fragment exposes a subset of the internal metamodel implementation to other fragments. A *required interface* of a fragment specifies explicit context dependencies to other components. A required interface is always implemented by another fragment. Both provided and required interfaces may either be *owned* by a fragment or *imported* from another fragment. Ability to have a required interface that is owned is crucial in order to realise *extension points* of a fragment, where other fragments may hook into. An owned required interface independently specifies a contract, which has to be fulfilled by other fragments that may realise that interface. A fragment that consists of at least one owned required interface is considered abstract. Fig. 1 illustrates a glass-box view of a reusable black-box fragment of the BPMN metamodel based on a simplified combined notation of UML class and composite structure diagrams. We explain this sample fragment in detail in Section 4.

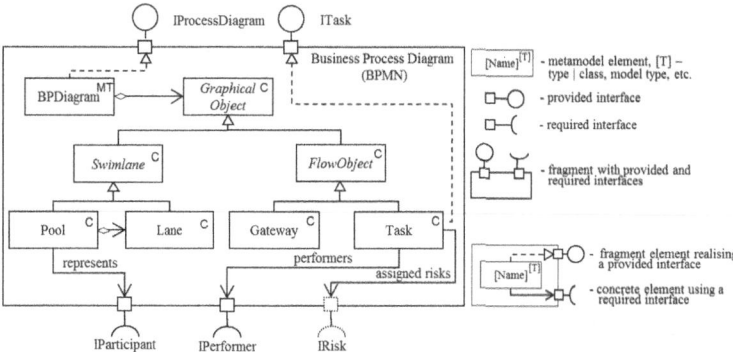

Fig. 1. Example of a black-box metamodel fragment of the BPMN metamodel

2.3 Metamodel Composition Based on Black-Box Operators

Black-box metamodel composition combines black-box fragments based on explicitly defined interfaces and appropriate flexible composition operators. For this purpose, two fundamental interface-based composition operators are introduced: *interface realisation, interface subtyping*.

Interface realisation. Interface realisation connects a concrete core metamodel element with its interface to denote that the realising metamodel element conforms to the interface by contract. A concrete metamodel element may realise many interfaces. In turn, an interface may be realised by many concrete elements. Interface realisation is the key composition operator for black-box modular metamodel development. On the one side, it allows for the *provisioning of provided interfaces* of a fragment, by stating that a concrete element realises an interface of a fragment that is exposed to the outside. On the other side, it allows for the *realisation of required interfaces* (explicit extension points) of a fragment by an arbitrary number of other fragments.

Interface subtyping. Interface subtyping connects a base interface and a derived sub-interface to denote the substitutability of the base interface by its subtype. A subtype interface may extend the base interface with additional members. An interface may extend and be extended by many other interfaces. Unlike class inheritance, interface subtyping inherits only member definitions instead of their implementations. Finally, the ability to extend existing interfaces and to declare the compatibility of a subtype with its supertype, interface subtyping allows for highly flexible metamodel composition scenarios on the level of black-box fragments.

3 Implementation of I-MMD

This section elaborates on the implementation of the I-MMD proposed in Section 2. We extend the existing concepts of metamodelling languages with constructs for interface-based metamodel modularisation and composition. We call such extended meta-language the Modular Metamodel Development Language (MMDL). MMDL consists of a core module, a modularisation module, and a composition module. We follow this structure to introduce the abstract syntax and basic semantics of MMDL.

3.1 The Core Metamodelling Language (CML)

The core metamodelling language (CML) provides concepts for defining the core metamodel structure, i.e. the fragment implementation. As we discussed in Section 2, existing metamodelling languages provide extensive support for core metamodelling, such that any of the mentioned meta-languages may be used as a basis for the implementation of the I-MMD. However, to abstract from any specific implementation, we introduce a generic core meta-metamodel that we then extend with modularisation and composition capabilities.

In CML, a *metamodel* consists of metamodel elements, which may either be attributable elements or attributes. An *attributable element* contains attributes and supports *inheritance* for the purpose of specialisation and attribute reuse along the same type element hierarchies. An *attribute* represents a property of a metamodel element and is of some attribute type. For the sake of simplicity, we do not consider attribute types any further. A *class* is an attributable element and the central meta-construct used to specify entities of a modelling language. A *model type* is another attributable element and a meta-construct used to typify models (diagram types). Model type may contain classes and relations. A *relation* is an attributable construct that connects classes and/or model types. A relation connects to other elements indirectly using the concept of a relation end. The number of relation ends defines the arity of the relation. A relation must have at least one relation end of type From and one of type To in order to be directed. A *relation end* specifies how a target of a relation participates in a relation (e.g. multiplicity, direction type). Usually, a relation end allows for only one explicit possible target type. However, by virtue of inheritance all subtypes may become valid targets. The same may be achieved by explicitly allowing that more than one target type is possible. This way, by the virtue of aggregation, one can define allowed targets that are otherwise not possible via single inheritance. Finally, a *connectable element* is a helper construct to generalise classes and model types as types that may be targets of relation ends. Note that for each of the core concepts we introduce an *abstract supertype*. This is needed as, we will see, a core concept may be either an implementation element of a fragment or an interface element. Fig. 2 illustrates the CML meta-metamodel. Note that containment relationships between elements are weak denoting the fact that intra-level reuse between elements is supported. For example, the relationship *classes* between the class and the model type allows for reusing classes in multiple model types. The same is true for relations such as *attributes*, *relations*, *relationEnds* and *targets*. This kind of weak containment that enables inter-level reuse is a powerful concept found in metamodelling languages such as GOPPRR and ADOxx. For the comparison and mapping of the introduced constructs to existing metamodelling languages refer to Table 1.

3.2 The Interface-Based Metamodel Modularisation Language (I-MML)

I-MML extends CML with modularisation constructs for encapsulation and information hiding.

Encapsulation. The central encapsulation construct for decomposing metamodels into reusable, modular units is *fragment*. A fragment is a container of metamodel elements. A contained element may either be *owned* or *imported*. Owned elements are existential members of that fragment and may be core elements or interfaces. Imported elements are those referenced from other fragments. For black-box modules, only interfaces are importable elements. Imported elements contribute to inter-fragment reuse and provide a basis for arbitrary composition operators. As a prerequisite to import elements from other fragments, an explicit *dependency* to that fragment must be established. Fragment dependency is acyclic, transitive relation. Further, fragments may participate in nested structures, i.e. a fragment may contain other fragments, as its internal packages. Fragment dependency is acyclic, transitive and reflexive relation. Fig. 3 illustrates the metamodel of the I-MML.

Interfacing. An *interface* as we now it from GPLs, is a class-based interface that specifies properties and method signatures, an implementing class should support. In a modular metamodelling approach, it is sensible to support interfaces for all first-order reusable elements of a metamodelling language. Hence, not only a class may be exposed via interfaces, but also attributes, model types, relations and relation ends. For example, a fragment may implement a model type and expose via interface only a subset of classes of that model type. To allow for such interfacing, each abstract metamodel element type is specialized by an adequate

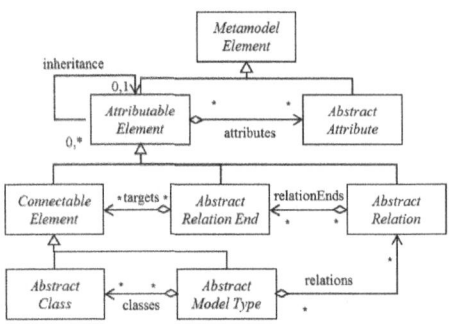
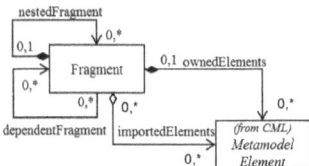

Fig. 2. Metamodel of the abstract core metamodelling language (CML)

Fig. 3. Metamodel of the modularisation language (I-MML)

interface type counterpart. Furthermore, by subtyping the interface from the core abstract elements, we basically mimic the abstract syntax structure of the CML to allow for interfacing of elements.

Fig. 4 illustrates the interfacing part of the I-MML. For the sake of brevity we introduce only the class interface and the attribute interface. A *class interface* specifies exposable elements of a class. Class interface specifies which attribute interfaces a concrete class has to support. An *attribute interface* represents an interface of the core attribute construct. Attribute interface as a contract specifies the data type that implementation attribute must support. Finally, for all interfaces hold that they cannot aggregate implementation constructs as member elements. For example, an attributable interface such as class interface can only aggregate attribute interfaces as members but no concrete, implementation attributes. Also, inheritance is not applicable for interfaces.

3.3 The Interface-Based Metamodel Composition Language (I-MCL)

I-MCL extends both CML and I-MML with constructs for interface-based metamodel composition. We introduce two composition operators that allow for flexible, black-box composition of metamodel fragments.

An *interface realisation* ensures the contractual binding of an implementing core element to an interface. In other words, it allows a single fragment to implement provided interfaces and to fill the explicit fragment holes defined by required interfaces with implementations from other fragments. A core element is said to realise an interface only if it realises all of its member interfaces. A concrete element may realise many different interfaces, which allows for *interface segregation*, an important aspect in modular development. In contrast, an interface may have different implementations. Multiple interface implementation is desirable to enable flexible compositions. An *interface subtyping* allows for declaring the compatibility between interfaces based on Liskov substitution semantics. It also has extensional semantics in that the subtype interface may extend the supertype by defining additional member elements. Interface subtyping is acyclic, transitive and reflexive relation. Finally, subtyping contributes to an additional flexibility of interface realisation. An element that realises an interface, also realises all of the interface supertypes. Fig. 4 shows the metamodel of the I-MCL reduced to class and attribute interfaces.

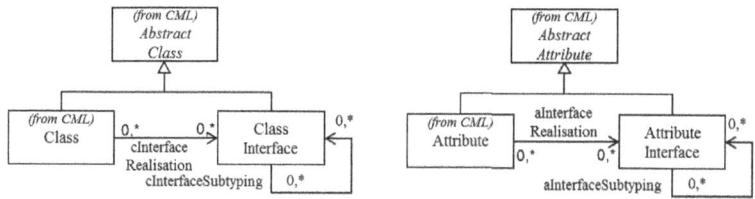

Fig. 4. Metamodel of the composition language (I-MCL)

Note that the semantics of interface realisation and interface subtyping may vary based on the core element type (class, attribute, etc.) it is applied on. Hence, MCL introduces specific realisation and subtyping operators for each core type from CML. In the following, we do this for classes and attributes.

An *attribute interface realisation* denotes the implementation of an attribute interface by its concrete counterpart. *Given the attribute A_1, its data type T_1, the attribute interface AI_1 and its data type T_2, A_1 realises AI_1 only if T_1 is a compatible type to T_2.* As for *attribute interface subtyping*, an attribute interface AI_2 is a valid subtype of AI_1 only if the compatibility of data types between AI_1 and AI_2 is given.

A *class interface realisation* denotes the implementation of a class interface by a concrete class. In order to fulfill the interface contract, the concrete class must realise all attribute interfaces of that class interface. *Given the class C_1, with a set of attributes Sa_1 and a class interface CI_1 with a set of attribute interfaces Sai_1, C_1 realises CI_1 only if for each of the attribute interfaces in Sai_1 a matching realisation attribute in Sa_1 exists.* In the case of *class subtyping, given the class interfaces CI_1 and CI_2 with corresponding sets of attribute interfaces Sai_1, Sai_2, CI_1 is a subtype of CI_2 only if for each of the attribute interfaces in Sai_2 the same or valid subtype attribute interface in Sa_1 exists.*

4 Case Study: Modular BPMS

To demonstrate the applicability of the introduced approach for I-MMD, we use MMDL to modularise the BPMS method. The BPMS method is based on the Business Process Management System (BPMS) framework [13], and is a modelling method for the systematic modelling of core enterprise areas such as business processes, products, organisations and IT. BPMS method is the central modelling formalism of the business process management tool ADONIS and has been successfully used in research and industry [4]. As a hybrid DSML, BPMS features the process modelling standard BPMN 2.0 [14] and extends it with sub-DSMLs for product, organisation and IT modelling. The necessity of extending the BPMN with more business-oriented concepts has been extensively discussed in [15]. Analysing the hybrid structure of the BPMS, a practical way to modularise the metamodel is to follow the decomposition based on enterprise areas and existing model types. For example, having organisation modelling DSML as a reusable module allows us to recombine it with other modules for process modelling or IT modelling. Furthermore, BPMS as a modular system can easily be extended by new modules. Likewise, new versions of fragments such as BPMN may be flexibly replaced as long as interfaces remain compatible. The outcome of the modularisation is illustrated in Fig. 5. Here we adopt UML package diagram notation to visualise metamodel fragments and their structure.

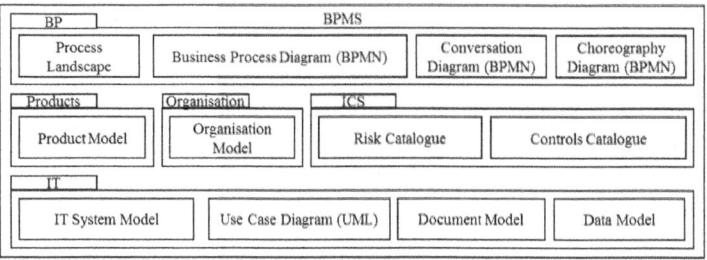

Fig. 5. Reusable metamodel fragments of modular BPMS

Application of I-MML. For the sake of brevity, we explicate the metamodel modularisation based on the pivotal fragment, the Business Process Diagram fragment (BPMN fragment). In Fig. 1, we showed the glass-box view of BPMN fragment. The core of the BPMN fragment consists of constructs for process flow modelling. Hence, two central elements may be provided for the integration with other languages such as the model type *BPDiagram* and the class *Task*. Two interfaces make sense, model type interface *IProcessDiagram*, and class interface *ITask*. BPMN fragment also defines explicit required interfaces to other fragments. Since the *Pool* element is a graphical representation of the participants (business entities, roles, etc.), we may define a required class interface *IParticipant*, which other metamodel fragments may implement. Another appropriate required interface seems to be the actual performer of the tasks. Instead of hard-coding it to a concrete class, we simply define the class interface *IPerformer* and let other fragments provide an appropriate implementation. We can continue to foresee integration points for other aspects in the same fashion. For example, while not part of the BPMN standard, to allow for the integration of risk management within process management, we may define a required interface for risk modelling *IRisk*, that is used by the class *Task* to denote assigned risks.

Application of I-MCL. Having defined explicit black-box metamodel fragments, we can combine them using the composition operators of I-MCL.

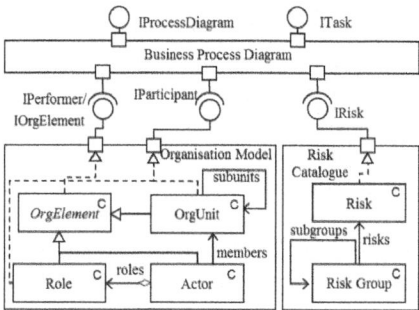

Fig. 6. Applying the interface-based metamodel composition for modular BPMS

We explicate the composition by extending the BPMN fragment with the *Organisation Model (OM)* fragment and the *Risk Catalogue (RC)* fragment, based on the integration points defined in BPMS. The OM fragment contributes the constructs for the modelling of organisational units, actors and roles. Classes *OrganisationUnit* and *Role* are appropriate candidates to implement the *IParticipant* interface required by the BPMN fragment. In order to perform this composition, we use the *cInterfaceRealisation* composition operator. Further, the role of the task performer may be fulfilled by organisational units (*OrgUnit*), roles (*Role*) and actors (*Actor*). The OM fragment already provides the interface *IOrgElement* which is realised by the common class *OrgElement*. Hence, we use the class interface subtyping to declare the compatibility of the interface *IOrgElement* with the required interface *IPerformer* from the BPMN fragment. On the other side, the RC fragment encapsulates the *Risk* class that is used to model company risks. Accordingly, we combine the RC fragment with the BPMN fragment by declaring the realisation of the interface *IRisk* by the class *Risk*. Fig. 6 shows the aforementioned fragment composition based on I-MCL.

5 Related Work

Modularisation and composition are active research topics in the context of metamodel-based modelling language definition. In more broader area of method engineering, the modularisation principle has long been the cornerstone of situational method engineering (SME) [16]. In SME, methods are created on-the-fly based on the assembly of existing method fragments [17], method chunks [18], method components [19], method services [20]. For composition, assembly-based association techniques are applied when method fragments are complementary, and integration techniques when method fragments are overlapping [21]. In [22], a pattern-based approach for integrating process metamodels is proposed. Our idea of metamodel modularisation is in-line with that of general method modularisation, with a difference that we call for explicit definition of required and provided interfaces already during fragment design. Furthermore, composition techniques require white-box view of fragments and combine elements intrusively, by creating new combined fragments with modified element structures.

On the language definition level, the idea of language interfaces as an ability of languages to inter-work has been discussed in [23] and its relevance underlined in [1]. In the area of grammar-based textual language definition, an interface-based approach for compositional domain-specific language (DSL) development has been proposed in [24]. While there are other works that deal with compositional specification of textual DSLs, they focus on the concrete syntax as a primary language artefact. In the metamodel-based modelling language definition, the focus of our work, the idea of interfaces hasn't been sufficiently embraced. In the following, we discuss existing metamodel-based approaches in detail.

To support modularisation, MOF and MOF-like meta-languages such as Ecore or GrUML feature the notion of a package. However, elements in packages may only be defined as white-boxes. For inter-package reuse MOF defines operators such as packageImport and elementImport. Imported elements may be combined using white-box operators such as inheritance or redefinition. In addition, only MOF introduces the packageMerge operator. Package merge is similar to the semantics of inheritance in the sense that the merging element receives the properties of the merged element. The restriction is that merge works based on name matching and it is applied on the package level. PackageMerge is a handsome operator to support efficiency of building internals of metamodel fragments defined in our work. Finally, MOF doesn't mention the notion of interfaces.

To support modularisation, GME's meta-language MetaGME [25–27] introduces aspects on the intra-metamodel level and projects on the inter-metamodel level. Elements are reused between metamodels by importing them as proxies. There are no concepts for explicit information hiding, hence metamodels are white-boxes. For composition, MetaGME reuses white-box constructs based on class inheritance such as implementation inheritance and class equivalence. To allow for non-intrusive inheritance, MetaGME introduces interface inheritance that operates between classes with subtyping semantics. This concept is similar to our interface subtyping, with a difference that it works on classes instead of dedicated interface concepts. Template instantiation is another white-box composition operator supported. Instead of importing an existing metamodel part, a metamodel is instantiated based on a template.

MetaEdit+' GOPPRR [28,29] supports white-box modularisation in the sense of graph types. Since all atomic elements such as object types and relationship types are available globally for reuse, graph types group metamodel elements to form a metamodel. For inter-metamodel reuse the notion of element inclusion is used. Hence, metamodel elements may be reused between multiple graphs from the global pool of meta-elements. Once the elements are available for reuse, they are combined using inheritance, aggregation, or association. GOPPRR supports two special associative composition operators, to allow for inter-model references. As in other approaches, the supporting interface concept for richer encapsulation and interface-based composition is not supported.

The metamodelling language of ADOxx [5], Meta^2-Model encapsulates metamodels within a library. Metamodel elements may be packaged into model types to depict single aspects of an integrated modelling language. Similar to inclusion in GOPPRR, ADOxx supports global intra-metamodel reuse of classes, attributes, relations and endpoints by aggregation. However, the concept of a reusable black-box module for inteface-based inter-metamodel reuse is not provided. Metamodels are therefore white-boxes. As for the composition, elements may be composed using inheritance, aggregation and association/inter-reference.

In addition to standard and tool-specific metamodelling languages, there are few other approaches towards interface-based metamodel development.

In [30], an extension for MOF 2.0 is proposed, suggesting concepts for the specification of metamodel components having import and export interfaces. Import interfaces are bound to export interfaces using interface binding, resulting in a new composite component based on graph morphisms. This approach is similar to ours in that it supports the definition of black-box metamodel components. However, instead of interface realisation and subtyping it uses interface binding to map interfaces. Furthermore, it differs substantially in the way that the composition requires derivative techniques to produce a composite fragment. In our approach, composition logic is part of the fragment design.

In [31] the idea of having place-holders in metamodels is proposed based on the Ecore. A metamodel is called parametric if it has "holes" (required interfaces). In contrast, a metamodel is called effective if it "fills the holes" of the parametric metamodel. Parameterisation operator is based on the substitution operation which replaces an existing element in the metamodel by a compatible one. The modularisation concept is a white-box Ecore package. As in the previous approach, metamodel modules are kind of templates out of which prototype metamodels are instantiated.

In [32] a role-based approach for modular language engineering is proposed. In addition to classes, the concept of role is used as a required interface for other classes that may play a role. The roleBinding operator declares a subtype relationship between a class and a role. Role playing is comparable to our interface realisation, whereas for subtyping a separate operator is used. The target composite module is generated out of the external composition specificaiton using generative derivation by translating the composition operations to constructs known to Ecore. Eventually, the subtyping operator is realised using the class inheritance in Ecore. In our approach, we incorporate composition constructs natively in the meta-language, to avoid transformative derivation step which may lead to conflicts that authors mention.

6 Conclusion

In this paper, we introduced an interface-based approach to modular metamodel development. The I-MMD approach advises extensions to existing metamodelling concepts towards black-box modularisation and composition based on interfaces and corresponding composition operators. We elaborated on these concepts deliberately based on an abstract metamodelling language such that ideas may be applied for various meta-languages. Furthermore, we demonstrated the usefulness of the approach by modularising the metamodel of the BPMS modelling method. We showed how a large metamodel can be decomposed into modular, reusable modules that can be flexibly combined to build hybrid and evolving metamodels. Unlike approaches that are either white-box-based or suggest interface-based composition based on the external composition and the usage of derivative techniques, we suggest compositionality to be native, but optional part of a metamodelling language, such that modular development becomes more forethought than afterthought in metamodelling. This requires that instead of building external transformation engines, the operational semantics of modularisation concepts must be an integral part of metamodelling tools. We experimented with such an approach within the ADOxx[1] metamodelling platform.

One may argue that I-MMD adds to the complexity as one needs to cope with concepts that do not contribute to the actual expressiveness of the modelling language. While this argument is valid for metamodelling-in-the-small, the benefit of systematic modularisation is achieved through economies of scale, by reusing and combining pre-fabricated modules to build new metamodels in the sense of metamodelling-in-the-large. Finally, we believe that a metamodel with clean interfaces, as a pivotal element, may contribute to the greater flexibility in overall modelling method definition. Multiple notations, mechanisms, algorithms and method procedures that are built around a metamodel may refer to metamodel constructs via interfaces only, in the way that greatly increases the substitutability and reuse of both metamodels and corresponding modelling method parts. A part of our future work at OMiLab[2] will go also in this direction.

References

1. Selic, B.: The theory and practice of modeling language design for model-based software engineering—a personal perspective. In: Fernandes, J.M., Lämmel, R., Visser, J., Saraiva, J. (eds.) Generative and Transformational Techniques in Software Engineering III. LNCS, vol. 6491, pp. 290–321. Springer, Heidelberg (2011)
2. OMG: Meta Object Facility (MOF) Version 2.4.2 (2014). http://www.omg.org/spec/MOF/2.4.2/
3. Kelly, S., Lyytinen, K., Rossi, M.: Metaedit+ a fully configurable multi-user and multi-tool CASE and CAME environment. In: Advanced Information Systems Engineering. pp. 1–21. Springer (1996)
4. Junginger, S., Kühn, H., Strobl, R., Karagiannis, D.: Ein Geschäftsprozessmanagement-Werkzeug der nächsten Generation - ADONIS: Konzeption und Anwendungen. Wirtschaftsinformatik **42**(5), 392–401 (2000)
5. Kühn, H.: The ADOxx metamodelling platform. In: Workshop on Methods as Plug-Ins for Meta-Modelling, Klagenfurt, Austria (2010)
6. Ledeczi, A., Maroti, M., Bakay, A., Karsai, G., Garrett, J., Thomason, C., Nordstrom, G., Sprinkle, J., Volgyesi, P.: The Generic Modeling Environment. In: Workshop on Intelligent Signal Processing, Budapest, Hungary, vol. 17 (2001)

---

[1] http://www.adoxx.org
[2] http://www.omilab.org

7. Steinberg, D., Budinsky, F., Merks, E., Paternostro, M.: EMF: Eclipse Modeling Framework. Pearson Education (2008)
8. Ebert, J., Winter, A., Dahm, P., Franzke, A., Süttenbach, R.: Graph-based modeling and implementation with EER/GRAL. In: Thalheim, B. (ed.) Conceptual Modeling ER '96. LNCS, vol. 1157, pp. 163–178. Springer, Heidelberg (1996)
9. DeRemer, F.L., Kron, H.H.: Programming-in-the-Large versus Programming-in-the-Small. In: Schneider, H.-J., Nagl, M. (eds.) Programmiersprachen. IF, vol. 1, 80th edn, p. 89. Springer, Heidelberg (1976)
10. Szyperski, C.: Component Software: Beyond Object-Oriented Programming, 2nd edn. Addison-Wesley Longman Publishing Co. Inc, Boston, MA (2002)
11. Aßmann, U.: Invasive Software Composition. Springer (2003)
12. Greenfield, J., Short, K., Cook, S., Kent, S., Crupi, J.: Software Factories: Assembling Applications with Patterns, Frameworks, Models and Tools. Wiley, (2004)
13. Karagiannis, D.: BPMS: Business Process Management Systems. ACM SIGOIS Bulletin **16**(1), 10–13 (1995)
14. OMG: Business Process Model and Notation (BPMN) Version 2.0.2 (2013). http://www.omg.org/spec/BPMN/2.0.2/
15. Rausch, T., Kuehn, H., Murzek, M., Brennan, T.: Making BPMN 2.0 Fit for Full Business Use. BPMN 2.0 Handbook Second Edition p. 189 (2011)
16. Harmsen, A.F.: Situational Method Engineering. Moret Ernst & Young Management Consultants (1997)
17. Brinkkemper, S., Saeki, M., Harmsen, F.: Meta-modelling Based Assembly Techniques for Situational Method Engineering. Inf. Systems **24**(3), 209–228 (1999)
18. Ralyté, J., Rolland, C.: An Approach for Method Reengineering. Springer (2001)
19. Wistrand, K., Karlsson, F.: Method components – rationale revealed. In: Persson, A., Stirna, J. (eds.) CAiSE 2004. LNCS, vol. 3084, pp. 189–201. Springer, Heidelberg (2004)
20. Deneckère, R., Iacovelli, A., Kornyshova, E., Souveyet, C.: From Method Fragments to Method Services. In: Proceedings of EMMSAD. p. 81 (2008)
21. Ralyté, J., Rolland, C., Deneckère, R.: Towards a meta-tool for change-centric method engineering: a typology of generic operators. In: Persson, A., Stirna, J. (eds.) CAiSE 2004. LNCS, vol. 3084, pp. 202–218. Springer, Heidelberg (2004)
22. Hug, C., Front, A., Rieu, D., Henderson-Sellers, B.: A Method to Build Information Systems Engineering Process Metamodels. Journal of Systems and Software **82**(10), 1730–1742 (2009)
23. Kleppe, A.: Software Language Engineering: Creating Domain-Specific Languages Using Metamodels. Addison-Wesley Professional (2009)
24. Krahn, H., Rumpe, B., Völkel, S.: MontiCore: A Framework for Compositional Development of Domain Specific Languages. International Journal on Software Tools for Technology Transfer **12**(5), 353–372 (2010)
25. Ledeczi, A., Nordstrom, G., Karsai, G., Volgyesi, P., Maroti, M.: On Metamodel Composition. In: Proc. of the 2001 IEEE International Conference on Control Applications, pp. 756–760 (2001)
26. Karsai, G., Maroti, M., Lédeczi, Á., Gray, J., Sztipanovits, J.: Composition and Cloning in Modeling and Meta-Modeling. IEEE Transactions on Control Systems Technology **12**(2), 263–278 (2004)
27. Emerson, M., Sztipanovits, J.: Techniques for Metamodel Composition. In: OOPSLA 6th Workshop on Domain Specific Modeling, pp. 123–139 (2006)
28. Tolvanen, J.P.: Incremental Method Engineering with Modeling Tools: Theoretical Principles and Empirical Evidence. University of Jyväskylä (1998)

29. Kelly, S., Tolvanen, J.P.: Domain-specific Modeling: Enabling Full Code Generation. Wiley (2008)
30. Weisemöller, I., Schürr, A.: Formal definition of MOF 2.0 metamodel components and composition. In: Czarnecki, K., Ober, I., Bruel, J.-M., Uhl, A., Völter, M. (eds.) MODELS 2008. LNCS, vol. 5301, pp. 386–400. Springer, Heidelberg (2008)
31. Pedro, L., Amaral, V., Buchs, D.: Foundations for a Domain Specific Modeling Language Prototyping Environment: A Compositional Approach. In: Proc. 8th OOPSLA Workshop on Domain-Specific Modeling. University of Jyväskylän (2008)
32. Wende, C., Thieme, N., Zschaler, S.: A role-based approach towards modular language engineering. In: van den Brand, M., Gašević, D., Gray, J. (eds.) SLE 2009. LNCS, vol. 5969, pp. 254–273. Springer, Heidelberg (2010)

Towards Static Analysis of Executable DSMLs Using Model Typing

Reza Gorgan Mohammadi[✉] and Ahmad Abdollahzadeh Barforoush

Department of Computer Engineering and Information Technology,
Amirkabir University of Technology, Tehran, Iran
{reza_gorgan,ahmad}@aut.ac.ir

Abstract. Executable domain-specific modeling languages (xDSMLs) are enriched with sufficient details which allows their direct execution. An important challenge hampering the use of executable models is the potentiality of run-time misbehavior. In this paper we present a static analysis technique for xDSMLs based on the notion of model typing. The abstract syntax of the modeling language is represented using a directed type graph. The operational semantics of the language is defined using graph transformation rules with priorities. The model typing is provided with a type graph enriched by a collection of graph constraints. We discuss the soundness of the analysis technique with respect to the operational semantics. The proposed approach allows the enforcement of the desired properties in xDSMLs and will eliminate run-time misbehavior by preventing ill-typed models.

Keywords: Executable model · Static analysis · Type graph · Model typing

1 Introduction

Domain specific modeling languages (DSMLs) raise the level of abstraction by specifying the solution directly using the concepts in the problem domain. This approach can improve agility and productivity of software development in the corresponding domain. DSMLs are mainly concerned with a certain domain with a set of concepts which allows a greater expressive power on their focused domain.

Model executability is now an emerging trend in model driven engineering [1]. An executable model is comprehensible by humans and at the same time machine executable and eliminates the intermediate code generation steps. Executable DSMLs (xDSMLs) enjoy the intuitiveness of modeling elements and simultaneously are equipped with behavioral specifications allowing direct execution of models.

The general case concerned in this paper is illustrated in Fig. 1. The language designers (a group of domain and technical experts) define an xDSML for a certain domain in a collaborative manner [2]. The end users employ the graphical editor for drawing their intended models based on the visual elements.

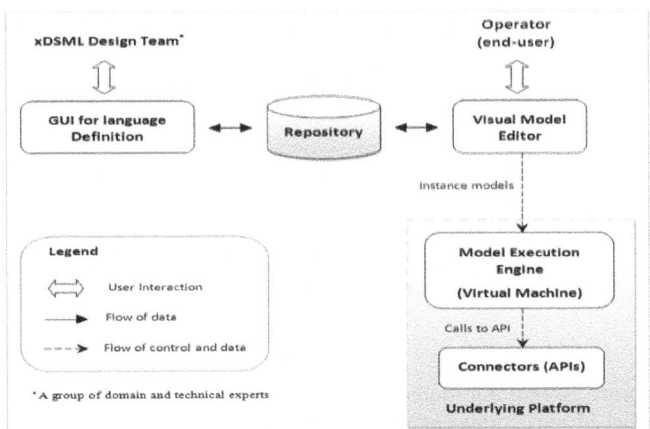

Fig. 1. The high-level design in the general case

The model is then executed on a virtual machine (model execution engine) which can have direct effects on the corresponding real-world system. There are connectors (APIs) which allow reifying the engine's inputs/outputs via the underlying host platform.

The potentiality of run-time misbehaviors is a major challenge hindering the application of executable models. In this paper we provide a static analysis technique which investigates an input model in order to prevent some undesirable run-time behaviors. We employ graph theory as the formalism for definition of the language and extend the notion of model typing which was first introduced by Steel et al. [3]. The metamodel is represented using a directed type graph which defines the abstract syntax of the language. The operational semantics is defined using graph transformation rules with priorities. Model typing is provided by enriching the type graph with a set of graph constraints. An instance model is called well-typed if it conforms to its metamodel and satisfies the graph constraints.

As programming languages enjoy the power of type systems, there is a rising interest to employ typing in modeling languages. The use of types for improving reusability in model transformation has motivated several studies in recent years [3,4]. The notion put emphasis on the reusability and subtyping in model manipulation operators. A model type is considered as a substructure of the metamodel class diagram. The use of OCL constraints for safe substitutability is reported in [5]. Contrary to all the previous works, in this paper we introduce a graph-based approach to model typing for modeling languages taking into account the desired properties of the domain. The use of directed graphs for definition of the xDSML provides a uniform formalism which allows static analysis of the instance models. The soundness of the static analysis technique is also investigated with respect to the operational semantics of the modeling language. The use of types allows to restrict the set of possible instance models.

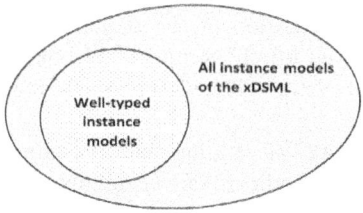

Fig. 2. Well-typed instance models are a subset of all possible models

The inner set in Fig. 2 is confined to instance models of the xDSML which are well-typed and therefore well-behaved at run-time.

In order to illustrate the important aspects of the proposed approach we follow a running example inspired by a railway control system throughout the paper. The example involves a visual editor for a special industrial rail control system (Fig. 3). The end user can define the desired movement of wagons between stations through the visual notations provided in the xDSML. The models are directly executed upon the underlying platform which result in the physical movement of wagons towards the specified targets. We extend a type system for the modeling language in the example in order to prevent the run-time unsafe conditions which are the collision of wagons. The running example has been implemented using AGG [6] which provides a development environment and execution engine for attributed graph transformation systems.

The structure of this paper is as follows. Section 2 introduces the running example which comprises an xDSML for a special rail control system. The state of the art are elaborated in Section 3. The static analysis technique based on model typing is introduced in Section 4. The soundness property of the static analysis technique is investigated in Section 5. Section 6 gives discussions and future works. Finally, Section 7 concludes the paper.

2 Motivation and Running Example

The running example concerns an xDSML developed for a special industrial rail control system. We assume that there are several stations in the manufacture line and each wagon carries a special material from a certain station to a target station. The movement is limited from left to right and there may be several wagons running simultaneously. Any collision of the wagons can bring about a catastrophe. Also when a wagon leaves its current station it should eventually reach the destination point otherwise it may cause unsafe conditions. The control system puts the wagon on the initial station and moves it towards the target station.

The end user is provided with a visual editor and can draw models conforming to the definition of the xDSML. The user can determine which wagon should move to which station using the visual notation then he/she can press the "Run" button to commence the execution of the model. The execution is performed via a virtual machine hosted upon the underlying platform which connects the machine to the real-world system. The virtual machine performs based on model interpretation.

For the sake of simplicity, we assume that the number of stations are given and their icons are fixed in the editor. The visual notations constitute a "Wagon" icon which indicates a wagon and two relation icons "On" and "Goto". The former indicates the current position of the wagon and the latter defines the target station which the wagon should be relocated (Fig. 3).

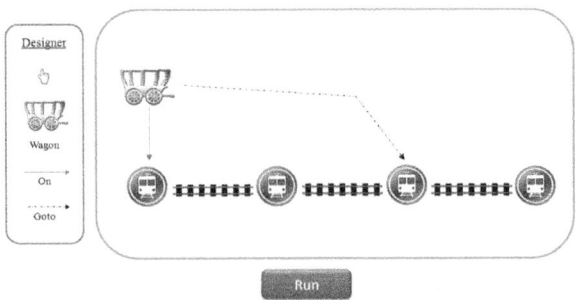

Fig. 3. The visual editor for the special rail-control system

Informally, The semantics of the language of the running example can be explained as follow: the execution engine moves the wagon from the current position one step to the successive station and continues until it reaches the station determined by "Goto" relation. An unsafe situation may arise when the end-user draws a model like the one depicted in Fig. 4 which bring about the collision of two wagons.

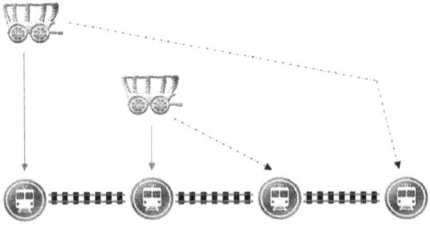

Fig. 4. A sample ill-typed model

The formal definition of the xDSML for the running example has been elaborated in [7] which is based on graph theory [8]. Using a graph formalism, a metamodel can be recognized as a type graph. A model is a directed graph G typed over TG_M by a graph morphism $type : G \rightarrow TG_M$. Fig. 5-a illustrates the metamodel of the language and part b of the figure depicts the graph-based representation of the metamodel. Note that the AGG tool allows the definition of type graph with multiplicity constraints. A "next" relation is defined between two consecutive stations and a "successor" relation indicates that a station is transitively reachable from a station next to the current one. The graph representation of the instance model in Fig. 4 is given in Fig. 5-c. (the stations and their relations are auto-generated by the visual editor).

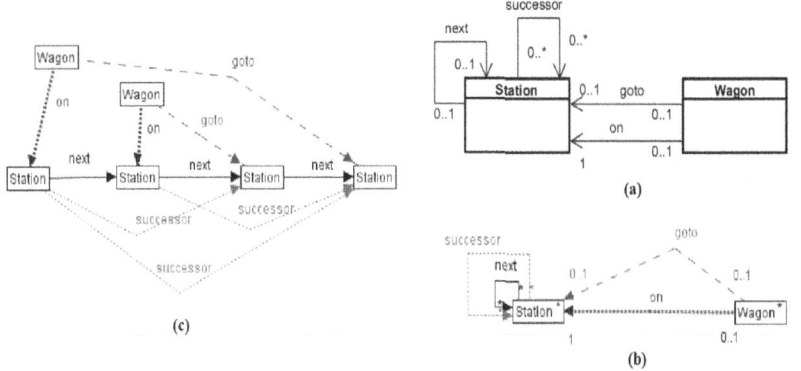

Fig. 5. (a) Metamodel of the xDSML (b) The graph representation of the metamodel (c) graph representation of instance model in Fig. 4

The behavior of the executable model can be specified by the operational semantics assigned to the modeling language. In this paper we employ graph transformations for the definition of the operational semantics of the language. A graph transformation $G \stackrel{p}{\Rightarrow} H$ converts a given graph G to a graph H by applying a transformation rule $p : (L \stackrel{l}{\leftarrow} K \stackrel{r}{\rightarrow} R)$. Graphs L, K, and R are typed over a type graph TG. L and R are called left-hand side (LHS) and right-hand side (RHS), respectively. The rules can be assigned priorities which defines a partial order on rules. A rule is applied to a host graph if there is no rule with higher priority applicable to the graph.

The two graph transformation rules illustrated in Fig. 6 are employed to specify the operational semantics of the example language. The first rule (Rule 1) has the highest priority and is applied when the target station is not the consecutive station. The second rule (Rule 2) is applied when there is a *goto* relation between two consecutive stations. A graph transformation engine like AGG can be employed as the model execution engine which performs based on the definition of graph transformation rules.

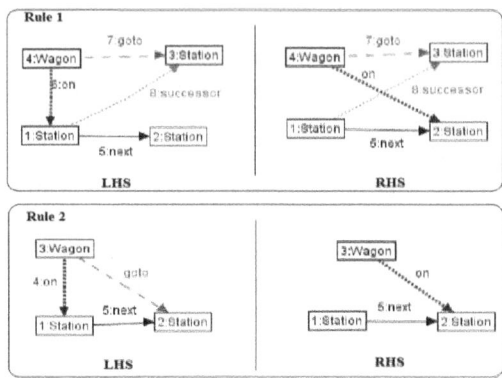

Fig. 6. The operational semantics of the xDSML using graph transformation rules

3 State of the Art

The correctness of software models in the context of model-based engineering (MBE) is a broad research area. The analytical methods applicable to software verification can be categorized to static methods and dynamic methods. The former is performed at compile (design) time which investigates a program/model without executing it. The dynamic methods are performed by executing the model/program and observing its execution [9]. The major formal static verification techniques are Data-Flow Analysis [10], Constraints-Based Analysis [11], and Abstract Interpretation [12]. Model checking [13] is an automatic technique for dynamic verification. The major drawback of model checking methods is state-explosion problem.

An executable model is composed of a structural model and a behavioral model. In order to specify executable models, OMG has provided a new standard called fundational UML (fUML) [14]. The fine-grained behavior can be specified using fUML which is an action-based behavior specification. Alf [15] is a concrete and text-based syntax for concise and readable specification of UML executable models. The verification of UML behavioral models has been previously addressed in [16,17]. In [18] a light-weight verification framework is provided for UML based executable models which employs fUML and Alf. The analysis mainly concerns strong executablility of the operations in the sense that by applying the operations the initial state of the system evolves to a state satisfying all the integrity constraints of the structural model.

In this paper, we propose a static analysis technique for executable DSMLs. The modeling language is formally defined using graph theory which allows the application of the existing concepts and theories to the models. A major contribution of our work in comparison with the existing works especially [18] is that it provides a sound static analysis technique for executable models which can prevent unsafe run-time behaviors. The soundness of the analysis technique is formally proved based on preservation and progress properties. This technique extends the notion of model typing to determine valid instance models. The soundness guarantees that all the instance models of the xDSML are well-behaved at run-time. Using graph theory provides a uniform formalism which considers the integration of graph transformation rules (operational semantics) and metamodel constraints.

There are many resources concerning analysis of graph transformation rules which can be classified into two main categories. The first category employs SPO and DPO theory. The second one translates the graph transformation rules to other domains (mainly OCL) [19]. The integration of graph transformation with meta-modeling based on type graph has been previously addressed in [20,21]. However, they do not provide analysis capabilities. The work in [19] is based on transformation of graph rules to OCL which allows analysis based on facilities available for OCL. Approximated unfolding [22] is an analysis technique for graph transformation systems which can be employed for concurrent systems.

Model types were introduced by Steel et al. [3] as an extension of object typing for enabling the reuse of model manipulation operators. They defined a model type as a substructure of the metamodel class diagram. The model typing definition and implementation in their work is limited to MOF-based metamodels. Based on the notion of model type, the work in [4] defined four subtyping relations between model types. The subtyping relation in [4] has been extended in [5] which considers OCL contracts for a safe substitutability of models conforming to a metamodel. The OCL invariants are added to MOF classes to specify additional structural properties. Contracts are the invariants and are expressed using the first order logic using OCL.

Contrary to the previous works concerning the model typing, this paper presents a graph-based approach to model typing which is extended by enriching the type graph with a set of graph constraints. The model typing allows static analysis of the instance models of the xDSMLs. Due to the use of graph formalism, the static analysis is proved to be sound which means that execution of models preserve constraints and eventually reach a final state (progress property).

4 The Static Analysis Technique

The use of type in programming languages allows static semantics to be applied to a program and provides a lightweight formalism which enables the enforcement of the desired properties in the programs. The input programs are restricted using types in order to prevent run-time errors and illegal behaviors. It may be conservative but it guarantees that well-typed programs cannot go wrong at run-time.

In this paper we extend the notion of model typing for an xDSML which allows static verification of the correctness properties of the model. Since a model can be represented as a directed graph, it seems rational to define the type of a model as a type graph enriched with a set of constraints. In this paper we enrich the type graph in order to encompass the typing constraints.

Graph g corresponding to an instance model is well-typed if it conforms to the type graph TG and satisfies the constraint C denoted by g▷[TG,C]. (g▷[TG,C] ⇔ g : TG and g ⊨ C). The diagram illustrated in Fig. 2 can be redrawn as depicted in Fig. 7. The outer ellipse encompass all the graphs which can be mapped to the type graph using a graph morphism. The inner circle includes all the graphs conforming to the type graph and satisfying the constraints.

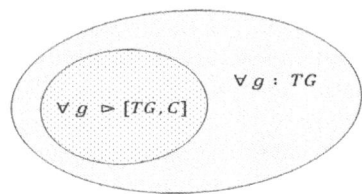

Fig. 7. Graph-based well-typed instance models

If graph g is well-typed then we can conclude that predicate X(g) is true meaning that property X holds for the graph (g▷[TG,C] ⇒ X(g)). Therefore, the desired invariant properties of the domain can be encoded in the typing constraints which are enforced through the typing mechanism. The correctness property of the running example can be defined as follows:

- *(Correctness property): Two wagons should never crash and should eventually reach their desired destinations.*

In this paper we specify model typing by means of a type graph enriched by a set of graph constraints. A graph constraint formulates the condition that a graph G must (or must not) contain a certain subgraph G' [8].

Definition 1 (Atomic Graph Constraint). *A graph constraint is a typed graph morphism $c : P \to C$. A graph G satisfies a graph constraint c, written $G \models c$, if for every injective graph morphism $p : P \to G$, there exists an injective graph morphism $q : C \to G$ such that $q \circ c = p$.*

It is also possible to define a Boolean formula over graph constraints which makes a composite constraint.

Definition 2 (Composite Graph Constraint). *A composite graph constraint can be defined as follows:*

- *$G \models \sim c$ iff G does not satisfy c;*
- *$G \models C$, $(C = \wedge_{i \in I} c_i)$ iff G satisfies all c_i with $i \in I$;*
- *$G \models C$, $(C = \vee_{i \in I} c_i)$ iff G satisfies some c_i with $i \in I$;*

The major atomic graph constraints for the running example are given in Fig. 8. The multiplicity constraints can also be specified using graph constraints (The complete set of constraints are accessible in [7]). The constraint "GotoOn" prevents a situation where a wagon should be relocated to another station having a wagon without "goto" relation. The constraint "CrGotoOn" obviates a situation where the movement of a wagon is encompassed by the movement of another one. A composite constraint has been defined based on the atomic constraints which is illustrated in Fig. 9.

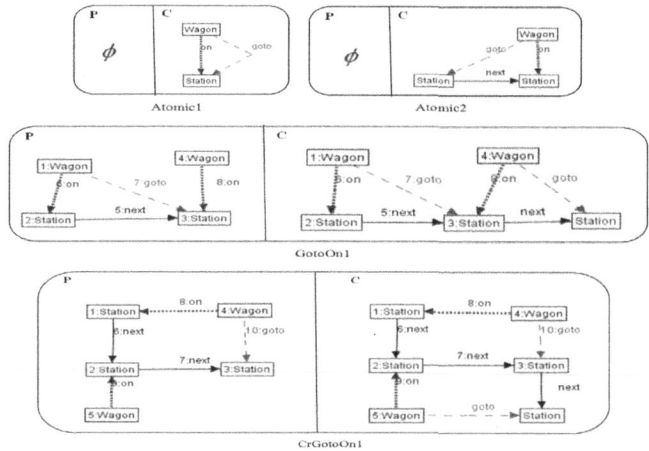

Fig. 8. Atomic constraints as typing rules for the xDSML

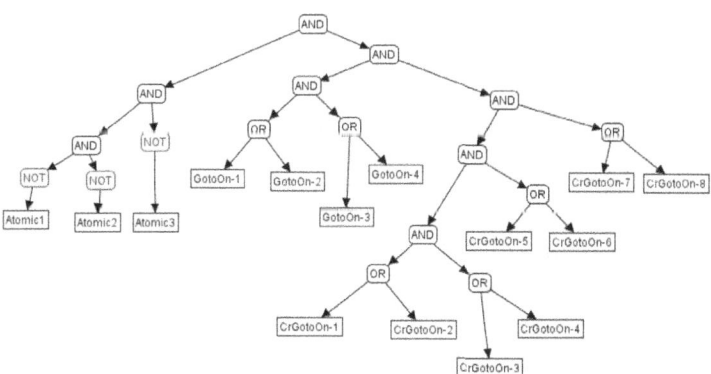

Fig. 9. Composite constraint defined based on atomic constraints

5 Soundness of the Static Analysis

The static analysis technique is performed without executing the models. Therefore, the soundness of the analysis should be investigated to ensure that the analysis gives correct approximation of the behavior. The soundness property guarantee that no run-time error occurs when executing a well-typed model. The soundness property can be proved formally with regard to the definition of the operational semantics of the modeling language.

In programming languages the soundness property of a type system can be verified using a standard approach which investigates preservation and progress. The first property also known as subject reduction indicates that types are invariant under reduction. In other words, types describe the invariant properties of the system under consideration. The second property implies that well-typed programs/models cannot get stuck in an ill-defined state and always reach a final state (value).

5.1 Subject Reduction

We define the subject reduction property for the proposed analysis technique as given in Definition 3.

Definition 3 (Subject Reduction). *If $g \triangleright [TG, C]$ and $g \Rightarrow h$ then $h \triangleright [TG, C]$.*

In order to support subject reduction property the transformation rules should be constraint preserving. The preservation can be provided by defining application conditions for the transformation rules. An application condition allows to restrict the application of graph transformation rules to a host graph.

Definition 4 (Negative Application Condition). *For a graph transformation rule $p: (L \xleftarrow{l} K \xrightarrow{r} R)$, a negative application condition is of form NAC(x) where $x: L \to N$ is an arbitrary graph morphism. A morphism $g: L \to G$ satisfies NAC(x), written $g \models NAC(x)$ if $\nexists q: N \to G$ such that $q \circ x = g$.*

We employ the previous work by Ehrig et al [23] which provides a mechanism to integrate constraints into the production rules in order to ensure constraint preservation. The mechanism is advocated by theorem 1. Using the mechanism and the accompanying theorem we can find the left application condition for the two rules based on the definition of graph constraints. The negative application conditions which preserve the defined constraints are depicted in Fig. 10. They prevent the application of Rule 1 and Rule 2 while there is a wagon on a consecutive station. The subject reduction property follows directly from the definition of graph transformation and theorem 1.

Theorem 1 (Constraints Preservation). *Given a graph transformation system (TG,R) and a set of constraints C, for every graph transformation rule p and graph constraint c, there is a transformation Pre such that Pre(c) is the left application condition for p and (p, Pre(c)) is c − preserving i.e. $G \models c$ and $G \Rightarrow H$ then $H \models c$ [23].*

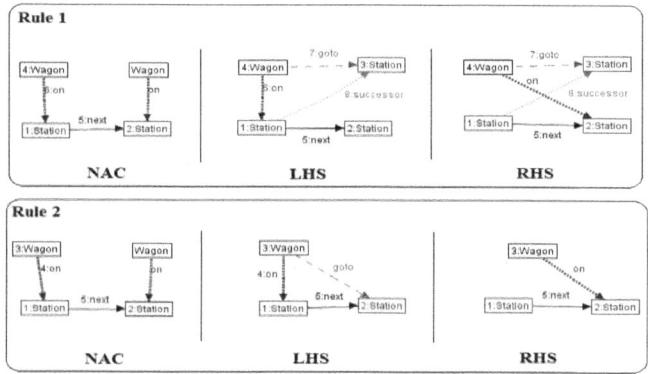

Fig. 10. Negative application conditions for graph transformation rules

5.2 Progress

The progress property indicates that a well-typed model which is not final can be executed and eventually reach a final state. Due to the use of graph representation, a final state in our approach is a graph in the corresponding graph transformation system called a final graph. The definition of a final graph is domain dependent. A graph $g : TG$ is a final graph if it satisfies the final graph constraints $g \models C_f$.

In the running example, a final state is where there is no wagon in the model or there is no wagon with "goto" relations meaning that all wagons are on their corresponding destinations. The atomic final constraints are shown in Fig. 11-a and the corresponding composite constraint is given in part b of the figure. The formal definition of progress property for the proposed type system is given in Definition 5.

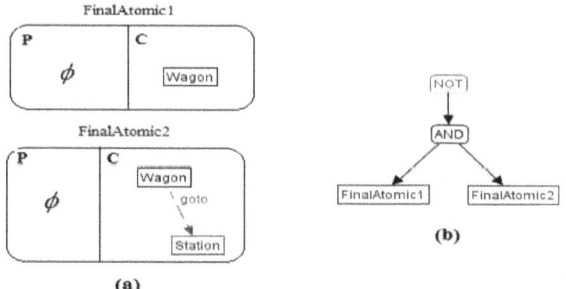

Fig. 11. Final constraint which specifies the final state (graph)

Definition 5 (Progress). *If $g \triangleright [G, C]$ then either g is a final graph or there exists a graph h such that $g \Rightarrow h$.*

The progress property indicates that if a well-typed graph g is not final there is always a terminating transformation sequence $G_0 \stackrel{*}{\Rightarrow} G_n$ such that $G_n \models C_f$. The progress property can be investigated from different aspects.

Applicability. The applicability property indicates that there is exactly one rule (due to the rule priority) applicable to a non-final well-typed graph. In other words, if $g \triangleright [G, C]$ then there exists exactly one rule $p : (L \stackrel{l}{\leftarrow} K \stackrel{r}{\to} R)$ such that $\exists m : L \to g$ and $m \models AC$ where AC is the application condition of the rule p.

In the language developed in the running example the applicability property holds. A well-typed graph g satisfies the composite constraint (Fig. 9). Therefore, if graph g is not final then there is at least one wagon and one "goto" relation. Hence there is always a match $m : L \to g$ for the two graph transformation rules (Rule 1 and Rule 2). Furthermore, graph g satisfies constraints "GotoOn" and "CrGotoOn" which prevent crash situations. Therefore, $m \models AC_1$ and $m \models AC_2$ where AC_1 and AC_2 are application conditions of Rule 1 and Rule 2, respectively. Hence, any well-typed model in the running example is always executable.

Termination. The termination property ensures that the transformation sequence is always terminating. A graph transformation sequence $G \stackrel{*}{\Rightarrow} H$ is terminating if no graph production in the graph transformation system (GTS) [8] is applicable to H anymore. A GTS is terminating if all transformation sequences are terminating for any start graph G. In general, it is undecidable whether a graph transformation system is terminating [24]. However, in certain cases it is possible to guarantee termination [25].

The aim is to guarantee the termination of the graph transformation after a finite number of steps for every input valid finite graph. A general idea for proving termination is to consider a termination criterion (e.g. a measure function) which can end the transformation sequence. In our case, it is possible to consider the number of remaining stations to the target station as the metric. By applying Rule 1 a wagon is relocated one station towards the target station. We define the termination metric as the number of remaining stations which is evidently decreasing. Application of Rule 2 removes the "goto" relation, hence preventing more application of the rule.

Reaching a final state. At last, we need to show that all the well-typed graphs representing models of the sample language eventually reach a final state. Up to now, we have shown that in the running example if graph $g_0 \triangleright [G, C]$ and g_0 is not a final graph there is a terminating graph transformation sequence $g_0 \stackrel{*}{\Rightarrow} g_n$ based on rule sequence $p_0, \ldots, p_n$ ($n \neq 0$). Now we should investigate that graph g_n is always a final graph i.e. $g_n \models C_f$.

In order to show that the property holds for all graphs of type $g \triangleright [G, C]$ we employ an inductive approach. Assume that set S contains all the well-typed graphs. We can partition the set based on the number of times the graph transformation rules can be applied (based on the termination property the number is finite). For example S_1 is the set of all well-typed graphs such that only one rule can be applied.

Assume that P(n) indicates that all the well-typed graphs in S_n eventually reach a final graph. In the running example, If P(1) and P(k) holds we should investigate if the property holds for P_{k+1}. Any $g \in S_{k+1}$ can be transformed to another graph h (the applicability property). Since the transformation rules are constraint preserving we have that $h \triangleright [G, C]$. The termination metric is decreasing and hence resulting graph h always is a member of S_k which is assumed to eventually reach a final state.

6 Discussions and Future Works

The proposed approach in this paper extend model typing for xDSMLs which detects ill-typed instance models, thus preventing unsafe run-time situations. In this section we discuss the major issues and future works in three categories:

- **Formal proof of the soundness property**: the formal proof of the soundness property of the proposed technique is required for its general applicability. The work by Ehrig [23] provides a mechanism which ensures the preservation property. However, the case is more complicated for the progress property. The sufficient criteria that guarantees the termination of the graph transformation system should be determined. Moreover, the Church-Rosser theorem [8] is required to be investigated in the general case. The approach selected for the specification of an xDSML conducts the formal verification of the soundness property. It is possible to define the abstract syntax of an xDSMLs using graph grammar. Furthermore, the Plotkin's structural operational semantics [26] which is extensively employed in programming languages can be mapped to graph transformation rules [27].
- **Definition of the xDSML**: the syntax and semantics of an xDSML are the basic ingredients of its definition. In this paper, we have exclusively employed graph formalism for specification of the xDSML. Another alternative is to abide by the OMG's specifications. Metamodels can be defined using the commonly accepted and well-established language of OMG standard MOF [28]. The OMG's specification for executable UML (fUML) [14] and the corresponding textual notation Alf [15] allows to specify the behavioral semantics of the system. Although several works have addressed the static verification of properties in models based on OMG's specifications [29], the proposed approach in this paper provides an intuitive and uniform graph-based formalism to verify safety and liveness properties in xDSMLs.
- **Constraint specification**: the typing rules in the proposed approach have been specified using graph constraints. Another alternative for constraint specification is OCL which is employed as the de-facto option with MOF and fUML. Although it is possible to transform a subset of OCL constraint into graph constraints, the comparison of the two options needs more investigation.

– **Industrial application**: although the proposed analysis technique has been explained using a toy example, we believe that it is applicable in industrial contexts and real-world models comprising hundreds or thousands of elements. The technique is scalable since the analysis is performed by detection of some graph patterns in the graph representation of the model.

7 Conclusion

In this paper we have presented a graph-based static analysis technique for executable domain specific modeling languages. The approach has been exemplified using a running example concerning a visual editor for a special rail control system. We have extended the notion of model typing by means of type graph and graph constraints which restrict the possible instance models of the xDSMLs.

The proposed technique allows the enforcement of the correctness properties in a domain and prevents unpleasant run-time behaviors during the model execution. The soundness of the static analysis has been investigated with respect to the definition of the operational semantics of the modeling language. Therefore, if there are no type-level violations in the model it guarantees the invariant properties of the domain.

References

1. Combemale, B., Crgut, X., Pantel, M.: A Design pattern to build executable DSMLs and associated V and V tools. In: 19th Asia-Pacific Software Engineering Conference (APSEC), pp. 282–287 (2012)
2. Cánovas Izquierdo, J.L., Cabot, J.: Enabling the collaborative definition of DSMLs. In: Salinesi, C., Norrie, M.C., Pastor, Ó. (eds.) CAiSE 2013. LNCS, vol. 7908, pp. 272–287. Springer, Heidelberg (2013)
3. Steel, J., Jézéquel, J.M.: On Model Typing. Software and Systems Modeling **6**(4), 401–413 (2007)
4. Guy, C., Combemale, B., Derrien, S., Steel, J.R.H., Jézéquel, J.-M.: On model subtyping. In: Vallecillo, A., Tolvanen, J.-P., Kindler, E., Störrle, H., Kolovos, D. (eds.) ECMFA 2012. LNCS, vol. 7349, pp. 400–415. Springer, Heidelberg (2012)
5. Sun, W., Combemale, B., Derrien, S., France, R.B.: Using model types to support contract-aware model substitutability. In: Van Gorp, P., Ritter, T., Rose, L.M. (eds.) ECMFA 2013. LNCS, vol. 7949, pp. 118–133. Springer, Heidelberg (2013)
6. AGG. http://user.cs.tu-berlin.de/gragra/agg/ (last visited 1 December 2014)
7. Mohammadi, R.G., Barforoush, A.A: Formal Definition of Executable DSMLs: A Graph-based Approach. Technical Report, Computer Engineering and Information Technology Faculty, Amirkabir University of Technology (2015). http://ceit.aut.ac.ir/islab/researches/modeling/TR.pdf
8. Ehrig, H., Ehrig, K., Prange, U., Taentzer, G.: Fundamental Theory for Typed Attributed Graphs and Graph Transformation based on Adhesive HLR Categories. Fundamenta Informaticae **74**(1), 31–61 (2006)
9. Ernst, M.D: Static and dynamic analysis: synergy and duality. In: Proceedings of the ACM-SIGPLAN-SIGSOFT Workshop on Program Analysis for Software Tools and Engineering, pp. 35–35 (2004)

10. Nielson, F., Nielson, H.R., Hankin, C.: Principles of Program Analysis. Springer-Verlag New York Inc, Secaucus (1999)
11. Agesen, O.: Constraint-based type inference and parametric polymorphism. In: LeCharlier, B. (ed.) SAS 1994. LNCS, vol. 864, pp. 78–100. Springer, Heidelberg (1994)
12. Cousot, P., Cousot, R.: Abstract interpretation: a unified lattice model for static analysis of programs by construction or aproximation of fixpoints. In: Proceedings of the 4th ACM SIGACT-SIGPLAN Symposium on Principles of Programming Languages, pp. 238–252 (1977)
13. Clarke Jr., E.M., Grumberg, O., Peled, D.A.: Model Checking. MITPress, Cambridge (1999)
14. Object Management Group. Semantics of a Foundational Subset for Executable UML Models (fUML), Version 1.1, Auegst 2013. http://www.omg.org/spec/FUML/1.1/
15. Object Management Group. Action Language for Foundational UML (Alf), Version 1.0.1, September 2013. http://www.omg.org/spec/ALF/1.0.1/
16. Abdelhalim, I., Sharp, J., Schneider, S., Treharne, H.: Formal verification of tokeneer behaviours modelled in fUML using CSP. In: Dong, J.S., Zhu, H. (eds.) ICFEM 2010. LNCS, vol. 6447, pp. 371–387. Springer, Heidelberg (2010)
17. Graw, G., Herrmann, P.: Transformation and Verification of Executable UML Models. Electronic Notes in Theoretical Computer Science **101**, 3–24 (2004)
18. Planas Hortal, E.: Lightweight and static verification of UML executable models. Universitat Politcnica de Catalunya, PhD Thesis (2013). http://www.tdx.cat/handle/10803/116449
19. Cabot, J., Claris, R., Guerra, E., De Lara, J.: A UML/OCL Framework for the Analysis of Graph Transformation Rules. Software and Systems Modeling **9**(3), 335–357 (2010)
20. De Lara, J., Bardohl, R., Ehrig, H., Ehrig, K., Prange, U., Taentzer, G.: Attributed Graph Transformation with Node Type Inheritance. Theoretical Computer Science **376**(3), 139–163 (2007)
21. Taentzer, G., Rensink, A.: Ensuring structural constraints in graph-based models with type inheritance. In: Cerioli, M. (ed.) FASE 2005. LNCS, vol. 3442, pp. 64–79. Springer, Heidelberg (2005)
22. Baldan, P., Corradini, A., König, B.: Unfolding graph transformation systems: theory and applications to verification. In: Degano, P., De Nicola, R., Meseguer, J. (eds.) Concurrency, Graphs and Models. LNCS, vol. 5065, pp. 16–36. Springer, Heidelberg (2008)
23. Ehrig, H., Ehrig, K., Habel, A., Pennemann, K.H.: Theory of Constraints and Application Conditions: From Graphs to High-level Structures. Fundamenta Informaticae **74**(1), 135–166 (2006)
24. Plump, D.: Termination of Graph Rewriting is Undecideable. Fundamental Informatica **33**(2), 201–209 (1998)
25. Varró, D., Varró-Gyapay, S., Ehrig, H., Prange, U., Taentzer, G.: Termination analysis of model transformations by Petri Nets. In: Corradini, A., Ehrig, H., Montanari, U., Ribeiro, L., Rozenberg, G. (eds.) ICGT 2006. LNCS, vol. 4178, pp. 260–274. Springer, Heidelberg (2006)
26. Plotkin, G.D.: A Structural Approach to Operational Semantics. J. Log. Algebr. Program. **60**(61), 17–139 (2004)

27. Dorman, A., Heindel, T.: Structured operational semantics for graph rewriting. In: ICE (2011), pp. 37–51 (2011)
28. Object Management Group. OMG Meta Object Facility (MOF) Core Specification, Version 2.4.1, August 2011. http://www.omg.org/spec/MOF/2.4.1
29. Hvid Hansen, H., Ketema, J., Luttik, B., Mousavi, M.R., van de Pol, J., dos Santos, O.M.: Automated verification of executable UML models. In: Aichernig, B.K., de Boer, F.S., Bonsangue, M.M. (eds.) Formal Methods for Components and Objects. LNCS, vol. 6957, pp. 225–250. Springer, Heidelberg (2011)

Modeling of Architecture and Design

Real-Time Design Patterns: Architectural Designs for Automatic Semi-Partitioned and Global Scheduling

Amina Magdich[1(✉)], Yessine Hadj Kacem[2], Adel Mahfoudhi[3], Mickaël Kerboeuf[4], and Mohamed Abid[1]

[1] CES Laboratory, University of Sfax, ENIS, Sfax, Tunisia
{amina.magdich,mohamed.abid}@ceslab.org
[2] College of Computer Science, King Khalid University, Abha, Saudi Arabia
yaalhaj@kku.edu.sa
[3] College of Computers and Information Technology, Taif University, Taif, Saudi Arabia
adel.mahfoudhi@yahoo.fr
[4] Lab-STICC, MOCS Team, University of Brest, Brest, France
kerboeuf@univ-brest.fr

Abstract. The scheduling problem is becoming an important topic for different fields especially for Real-Time applications. Considering the complexity of Real-Time Embedded Systems (RTES) coupled with the variety of scheduling approaches and algorithms, the designer task is becoming increasingly hard. Few approaches have investigated design patterns to perform an automatic scheduling at a high-level of abstraction. However, only the partitioned scheduling that prevents task migrations has been taken into account. In this context, this paper proposes two design patterns maintaining an automatic choice of semi-partitioned and global scheduling algorithms. The Unified Modeling Language (UML) profile for the Modeling and Analysis of Real-Time Embedded systems (MARTE) is used to annotate the proposed design patterns with functional and non-functional properties.

Keywords: Semi-partitioned scheduling · Global scheduling · Automatic scheduling · MDE · Design patterns · MARTE

1 Introduction

RTES complexity is increasingly growing due to the variety of required constraints. We believe that this complexity requires high-level development methodologies that facilitate the development steps and ameliorate the software quality. In this context, the use of models reuse is emerging as a major trend in high-level systems development. This concept allows minimizing development cost/ maintenance and responding to «time to market» constraint. In fact, models reuse may be supported through various techniques such as the Model Driven Engineering (MDE) [18].

While most of the researchers' interest is focused on the use of MDE, other researchers have proposed to use design patterns [6].

It is obvious that both MDE and design patterns support the models reuse concept, but each concept has its own goals. Indeed, MDE is used mainly to build frameworks and automate the development flows. The corresponding metamodels and models are conceived through a specific language (Ecore). For design patterns, they are used to model solutions for given problems, as they may be conceived using different languages. Using both MDE and design patterns in the same development approach is beneficial. Actually, MDE is based mainly on the use of generic artifacts that support the models reuse concept. The fact of using design patterns as generic artifacts in an MDE-based approach guarantees the obtention of relevant results.

Owing to the impact of design patterns on systems development, they have been investigated in various application domains to overcome electronic systems complexity such as specifying RTES functionalities [12], domain space exploration [4], scheduling analysis [5] [7], etc.

While developing RTES, the temporal determinism remains a goal to achieve and satisfy. Indeed, the allocation of tasks on the available computing resources while satisfying deadlines represents an important issue. This step is typically performed while using the scheduling theory [8].

The scheduling of tasks on processors depends not only on the application and the architecture criteria, but also on the scheduling approach and algorithm choice. In fact, three multiprocessor scheduling approaches are available in the literature [2]; namely the partitioned, the semi-partitioned and the global approach.

The partitioned approach allocates each task to be executed on a particular computing resource and prevents task migrations. Consequently, some execution resources may be free for a while and delays may be caused. The semi-partitioned and the global approaches aim to reach CPU (Central Processing Unit) optimality by allowing task migrations. Although most algorithms of the global approach achieve CPU optimality by enabling full migrations, a context switch may be costly. The semi-partitioned approach enables restricted migrations to reach optimality while performing a trade between the context switching and the CPU occupations.

Along with the complexity of RTES, the variety of scheduling approaches and algorithms makes the scheduling step harder especially for new designers who are required to choose the appropriate ones. To address this issue, researchers in [7] have proposed a design pattern-based approach for automatic scheduling. Yet, they have supported only partitioned scheduling that prevents task migrations. It is in this context that we propose in this paper two design patterns intended to support the automatic scheduling that enables task migrations. To model and annotate these design patterns, the UML/MARTE profile [9] is adopted.

The remainder of this paper begins with related works in section 2. Section 3 highlights the impact of the design patterns use on the scheduling analysis.

In sections 4 and 5, the proposed design patterns are presented. In section 6, a case study is performed and finally, a summary is given in section 7.

2 Motivation and Related Works

Nowadays, the scheduling of electronic systems remains a challenge especially for the new designers since it requires a huge knowledge and expertise about the scheduling theory. As an attempt to facilitate the scheduling step, some approaches have proposed an automatic scheduling. In [17], an automatic scheduling has been suggested to deal with distributed systems. In this context, the MONARC tool has been adopted for systems modeling and simulation. In [20], a hybrid classifier has been proposed to maintain an automatic selection of scheduling algorithms.

Despite the fact that an automatic scheduling has been performed in [17] and [20], the authors have dealt only with partitioned scheduling. To deal with such a problem, the authors of [3] have proposed automatic adaptive scheduling supporting task migrations.

The cited research works have dealt with automatic scheduling to facilitate the scheduling step, but no attention has been given to the use of high-level techniques.

In fact, complex systems development at a high-level abstraction is one of the most promising software engineering approaches. Such methods facilitate the development stages and overcome the RTES growing complexity challenge. In the same vein, some research works proposing a high-level scheduling have been documented in the literature such as [11], [7], [10].

In [11] and [10], a model-based approach for scheduling analysis has been proposed. It has taken advantage of the use of MDE and UML/MARTE to overcome RTES complexity and perform a high-level scheduling analysis. However, only the partitioned scheduling has been supported, and no attention has been given to the automatic choice of scheduling approach or algorithm. In this context, a design pattern-based approach for automatic scheduling has been proposed in [7]. To perform this approach, the authors have proposed five design patterns modeled through Architecture Analysis Design Language (AADL). Although the introduced methodology supports the automatic multiprocessor scheduling, it addresses only the partitioned approach that prohibits task migrations.

In this paper, two UML/MARTE design patterns for an automatic semi-partitioned and global scheduling are proposed. These design patterns are used in the context of a model-based approach. This work represents an extension of the proposal highlighted in [14].

3 Design Patterns for Automatic Scheduling

Nowadays, the impact of design patterns on systems development has been the interest of many researchers since they allow them to overcome systems complexity, promote the models reuse and improve the software quality. That is why

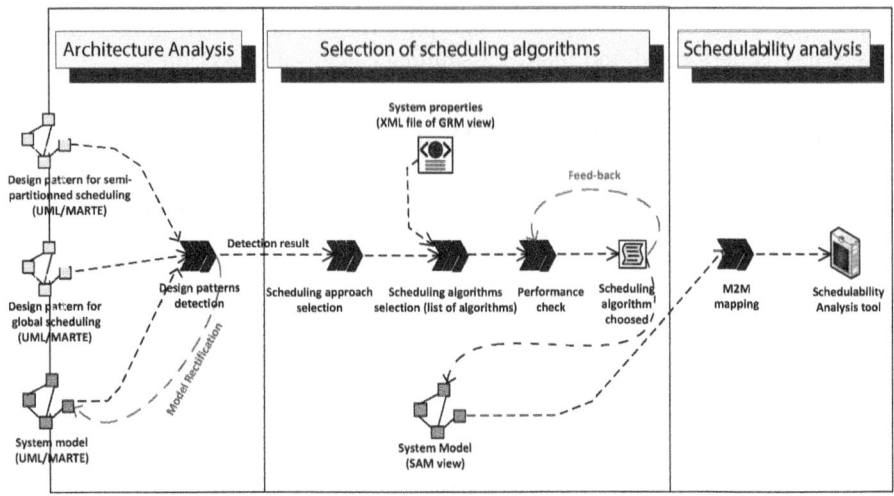

Fig. 1. A design patterns-based approach for automatic scheduling

we propose two design patterns intended to support semi-partitioned and global scheduling to be used to fulfill our MDE-based approach, which is presented in Figure 1.

The proposed design patterns are modeled through the UML/MARTE profile that aims to support a high-level modeling and produce a unified model rich in annotations. During a semi-partitioned and global scheduling, tasks are allowed to migrate between executions hosts. Then, various attributes (deadline, period, scheduler, processor, execution time, etc) may change due to the context switching. Consequently, to model properties for semi-partitioned and global scheduling, the extensions of MARTE [16][15] have to be used.

The proposed approach begins by an architecture analysis step, which relies on a detection step to detect design patterns. It uses three models: a UML/MARTE view that represents the studied system modeling and two design patterns for semi-partitioned and global scheduling. The detection step is based on a similarity scoring algorithm [19] that allows detecting design patterns and checking whether the system model is correct and complete. In case of problems, the designer will be notified to correct or complete the system model.

Concerning the second step, it uses the detection result for an automatic choice of the scheduling approach. Based on this result coupled with the studied system properties, a list of applicable scheduling algorithms will be generated. Relying on a performance check, only one scheduling algorithm will be selected from this list.

The schedulability analysis, which allows the establishment of an early schedulability analysis, is the last step of our approach. It is based on a model to model mapping to translate the dynamic view of the detected design pattern to the model of the schedulability analysis tool.

4 Design Pattern for Automatic Semi-Partitioned Scheduling: DP-ASPS

Each design pattern is used in a specific context to present a solution for a given issue. DP-ASPS aims at supporting the choice of semi-partitioned scheduling approach and algorithms.

4.1 Specification

- Context:
  The main intention of this design pattern is to help designers specify the criteria used for semi-partitioned scheduling, which allows restricted task migrations.
- Problem:
  The design patterns documented in the literature support only the partitioned scheduling approach. Moreover, no attention has been given to make only one design pattern that encloses all the needed criteria (eg. Tasks/processors communications, shared resources access, etc) for a specific scheduling approach.
- Solution:
  The proposed design pattern aims at supporting the semi-partitioned scheduling while enclosing all the needed criteria in the same view.

4.2 Modeling

The proposed DP-ASPS is modeled through static and dynamic views annotated using the UML/MARTE profile.

4.2.1 Static View

This view (Figure 2) [14] is used to ensure an automatic choice of the semi-partitioned scheduling approach and algorithms. It is annotated through Non-Functional Properties (NFP), Software Resource Modeling (SRM), Hardware Resource Modeling (HRM) and Allocation Modeling (Alloc) sub-profiles.

The static view of a design pattern aims at modeling the various entities with their relationships and the following properties, which are required to solve a specified issue. The static view of the proposed design pattern (Figure 2) encloses a set of entities (such as «Task», «ExecutionHost», «Memory», «ExclRes», «CommRes», etc) that may be used to fulfill an automatic semi-partitioned scheduling.

In a semi-partitioned scheduling context, two types of schedulers are used; a global scheduler and a scheduler associated with every processor. The global scheduler is specified through the task «GlobalScheduler» annotated «gaExecHost». The class «ExecutionHost» specifies both the processor and its scheduler respectively through the stereotypes «hwComputingResource» and «scheduler». The «ExclRes» entity models the exclusion resources that are used for controlling the access for shared resources. A task class is annotated through

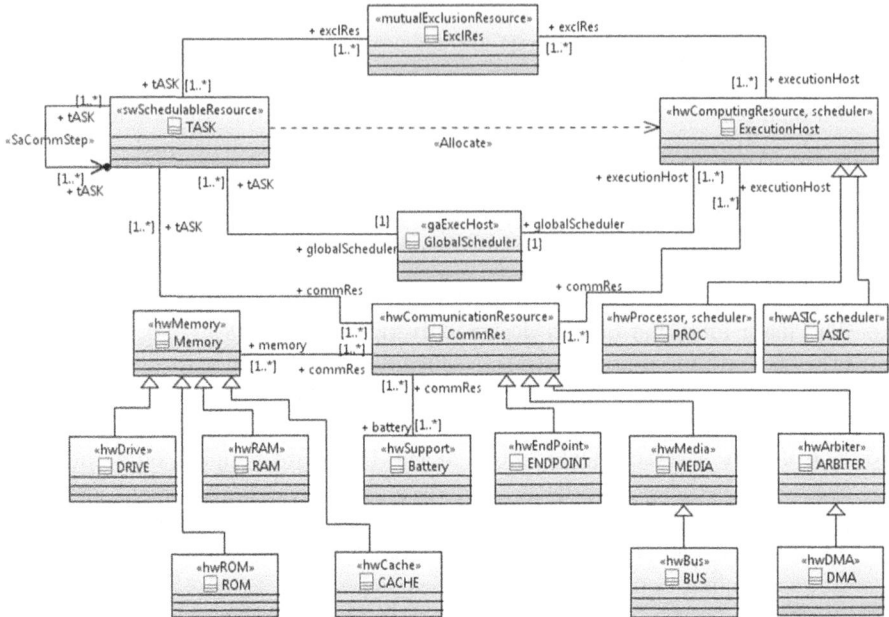

Fig. 2. The static view of the DP-ASPS

the stereotype «swSchedulableResource» that encloses a set of attributes such as «arrivalPattern», «isStaticSchedulingFeature», «isPreemptable», «DeadlineElements», «PeriodElements», «type: ArrivalPattern», etc. The attribute «isStaticSchedulingFeature» is used to specify whether the scheduling is static or dynamic. To specify whether a task may be preemptable during its execution, the attribute «isPreemptable» is used. The attributes «DeadlineElements» and «PeriodElements» aims at specifying, respectively, the task deadline and period.

What is worthwhile to note is that other entities (eg. memories) may be used while scheduling RTES to ensure an optimization of the scheduling. Other entities such as the communication resources ("CommRes") are used to link some components such as CPUs and memories.

4.2.2 Dynamic View

A dynamic view describes the invocations of methods between the specified entities. It can be modeled through a sequence diagram or an activity diagram. In this proposal, we opt for the use of a sequence diagram (Figure 3) annotated via the stereotypes of NFP and SAM (Schedulablity Analysis Modeling) sub-profiles.

This view is used mainly to ensure an early schedulability analysis. It is annotated through the operator «seq» to mention that the interactions order is not strict. «TASK», «GlobalScheduler», «ExecutionHost», «ExclRes» are the objects of the sequence diagram.

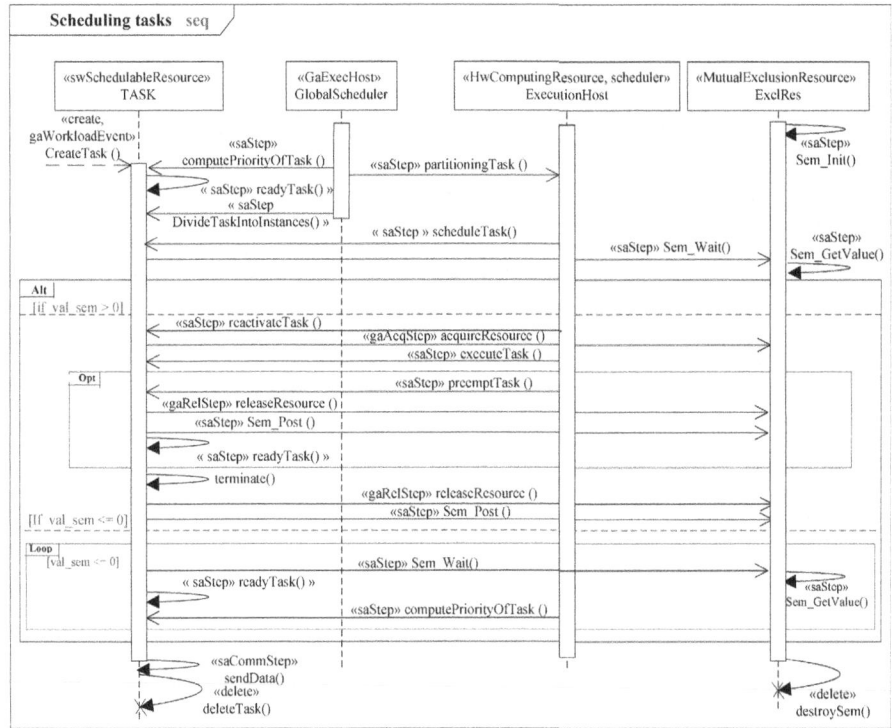

Fig. 3. The dynamic view of the DP-ASPS

Based on its priority, a created task is placed in the waiting queue of the global scheduler. This affects the waiting tasks on the available execution hosts. They will be then placed in the corresponding local schedulers queues. The tasks that are not affected by executions hosts have to be divided into instances («DivideTaskIntoInstances()») to be executed, not simultaneously, on different computing resources. This strategy improves the processors occupation rate.

A task that is placed in the scheduler queue is «ready». It is waiting for a shared resource access to be executed. In this case, the operation «Sem_Wait()» is launched to ask access for a shared resource.

This step is followed by a «Sem_GetValue» operation, which specifies the value associated with the shared resource. To lock the required shared resource, the condition «val_sem()>0» (Value of the semaphore) must be satisfied. Consequently, the task locks the shared resource («acquireResource()») and begins its execution on the corresponding computing resource. Once the time quantum of the task is finished, it unlocks the shared resource. If the condition «val_sem()>0» is not satisfied the task will be in a waiting state until the availability of a shared resource.

What is worthwhile to note is that a task/instance may lock, not simultaneously, different shared resources. The parallelism of a task instances is forbidden;

an instance (also named job) has to finish its execution before the lunching of the following instance.

The tasks preemption is allowed in the semi-partitioned scheduling. Indeed, a task may be preempted by the scheduler upon the activation of a higher priority task. Consequently, the preempted task has to unlock the shared resource and passes to be in the ready state «readyTask()» until the availability of another shared resource.

The communication between tasks is specified through the operation «sendData()» annotated through «saCommStep». As mentioned before, in the context of semi-partitioned scheduling, the task migration is allowed. This may be specified implicitly in the dynamic view of the design pattern. It is mentioned by the changing of the processors/schedulers as well as through the attributes indicating the context switching.

5 Design Pattern for Automatic Global Scheduling: DP-AGS

5.1 Specification

- Context:
  This design pattern aims at guiding designers to choose the appropriate scheduling algorithm regarding the global approach.
- Problem:
  The scheduling design patterns proposed in the literature support only partitioned scheduling algorithms. No attention has been given for global automatic scheduling.
- Solution:
  The DP-AGS seeks to support the automatic global scheduling while enclosing all the needed criteria in the same view.

5.2 Modeling

The proposed design pattern is modeled through two views annotated through MARTE profile.

5.2.1 Static View

Similarly to the static view of the DP-ASPS, the static view of the design pattern intended for global scheduling (DP-AGS) is based on a set of entities modeling the application, the architecture and the allocation of the software part on the target architecture. This view (Figure 4) is likewise annotated through NFP, SRM, HRM, SAM and alloc sub-profiles.

While the semi-partitioned scheduling uses two types of schedulers, the global scheduling is based only on the use of a global scheduler.

Consequently, the DP-AGS view encloses all the entities of the static view of the DP-ASPS except the local schedulers (schedulers associated with processors). Thus, the entities indicating the execution hosts («hwcomputingResource», «hwProcessor», «hwAsic») will not be annotated by the stereotype «scheduler».

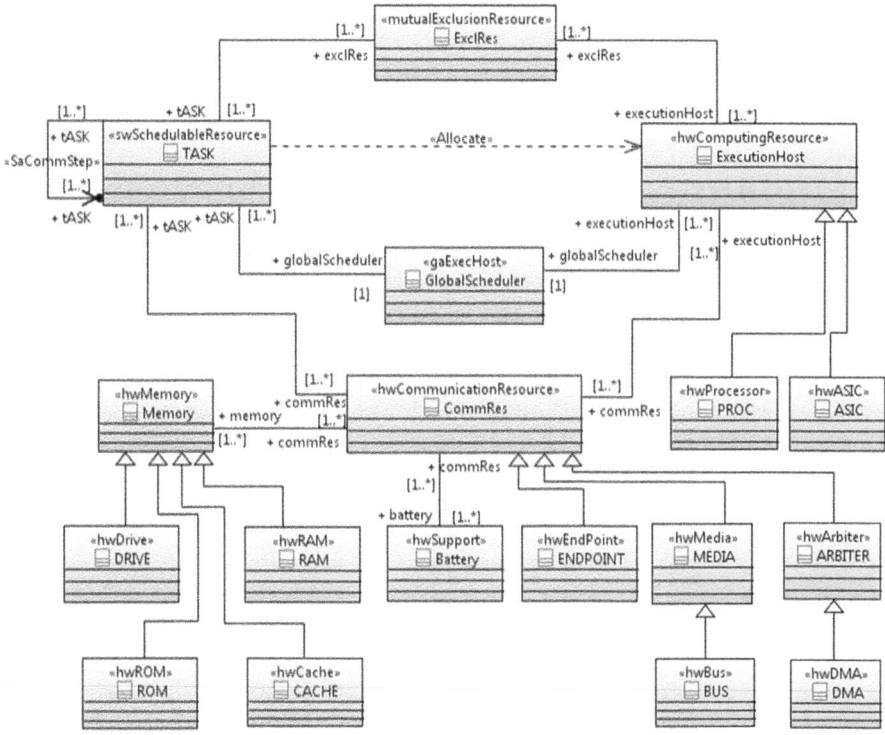

Fig. 4. The static view of the DP-AGS

5.2.2 Dynamic View

The sequence diagram displayed in Figure 5 illustrates the systems scheduling behavior while adopting a global approach.

The global scheduling is based mainly on the tasks set, the global scheduler, the execution resources and mutual exclusion resources.

Following its creation, a task is placed in the global scheduler queue. This placement is based on the task priority that is computed by the scheduler. The task is waiting for an available execution host («readyTask()»). The «sem_Wait()» operation has to be executed to check the availability of a shared resource. In this context, two cases arise:

– If the result value of «sem_wait()»<0 :
  In this case, the task remains in the waiting state until the availability of an execution host. The test of the availability of an execution task is repeated until the satisfaction of the request. This may be represented through the fragment «Loop».
– If the result of «sem_wait()»>0 :
  In this case, a shared resource is available. The scheduler then reactivates the most prior task in its queue («reactivateTask()») to lock the shared resource («acquireResource()»). The task begins its execution on the corresponding execution host («executeTask()»).

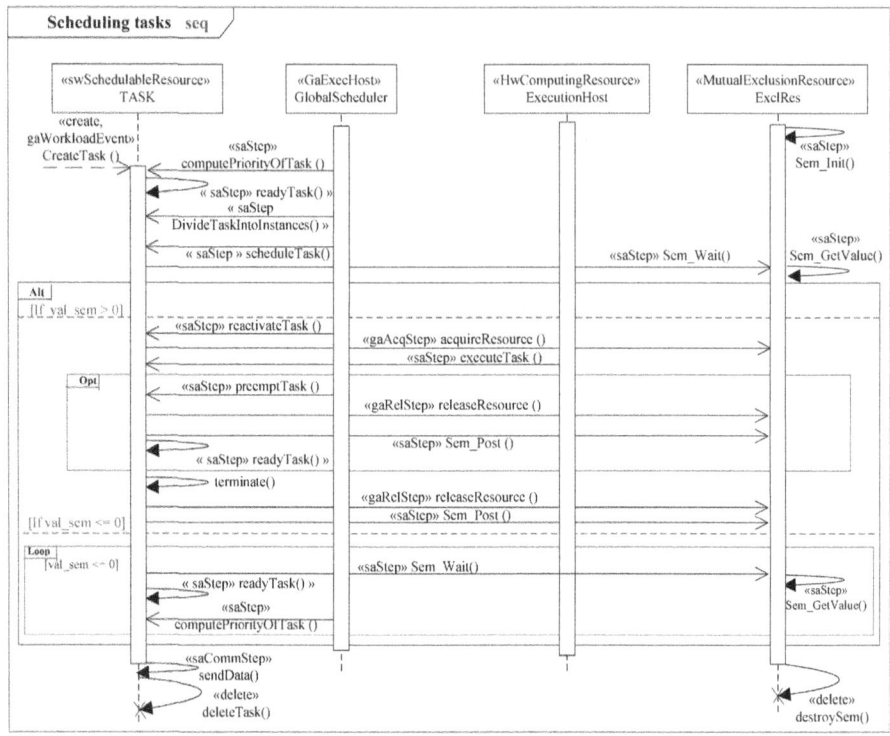

Fig. 5. The dynamic view of the DP-AGS

If the processor is unable to execute the whole task, this last will be subdivided into instances («DivideTaskIntoInstances()») and affected, not simultaneously, on different processors («scheduleTask()»).

Once the task execution is finished, the task releases («releaseResource()») the shared resource that will be locked through another ready task. Consequently, the «Sem_Post()» operation is launched to change the semaphore value and specify that the shared resource becomes available.

i.e. The two static views of the proposed design patterns for semi-partitioned and global scheduling seem to be substantially similar (except for the classes «ExecutionHost», «PROC» and «ASIC»), but the filling of the classes' properties vary depending on the adopted scheduling approach.

6 Case Study

To evaluate the proposed design pattern for global scheduling, we consider the networking router application described in [13] (i.e. a case study has been provided in [14] to validate the design pattern for semi-partitioned scheduling).

Table 1. Tasks parameters for the networking router application

Task	Priority	WCET	Deadline
IF1	1	200	201
IF2	2	100	101
IF3	3	400	401
IF4	4	200	201
IF5	5	200	201
IF6	6	400	401
Mem1	7	100	201
Mem2	8	100	201
Mem3	9	100	101
PktRx1/2	10	1442	1443
Chlk	11	450	451
HdrCal	12	2400	2401
PktRx3	13	1530	1531
PktRx4	14	670	671
Pkt_fwd	15	600	601
Proc1	16	2800	2801
Cnfg	17	1600	1601
Assm	18	800	2401
Recalc1	19	3000	3801
PktRx5/6	20	2100	2101
Enc	21	4900	4901
Recalc2	22	1750	6381
Proc2	23	5600	5601

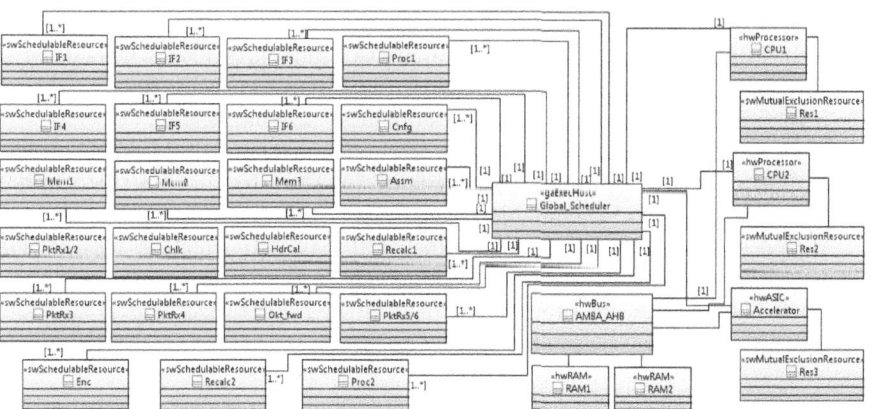

Fig. 6. An application of the DP-AGS (static view)

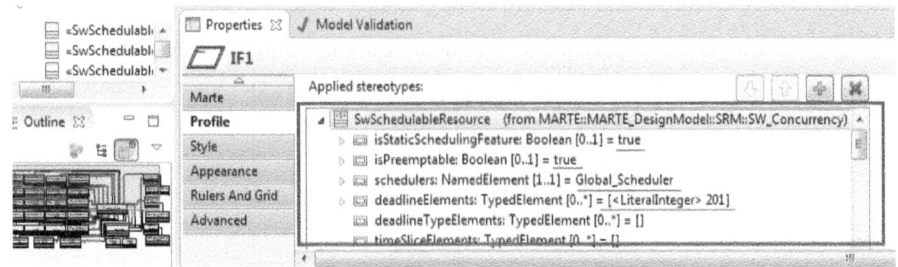

Fig. 7. A design pattern class filling using MARTE profile

The software architecture of the studied application is composed of twenty three tasks such that every task is characterized by a priority, a WCET (Worst Case Execution Time), and a deadline (Table 1).

The software architecture of the proposed case study is mapped to a preemptive execution platform composed of two embedded processors (CPU1 and CPU2), an accelerator (ASIC), two physical memories (RAM1 and RAM2), and a communication bus (AMBA AHB). A global scheduler has to be used to schedule the available tasks/jobs. Moreover, Exclusion resources are mandatory for shared resources access management. Indeed, every processor is protected by a mutual exclusion resource.

Parameters of both the system software and hardware architecture must be specified in the static view of the proposed design pattern application to guide the automatic scheduling approach and algorithm choice.

Figure 6 exposes a modeling of the studied system by the proposed design pattern intended for the global scheduling. As mentioned before, the proposed design pattern is modeled through the UML/MARTE profile, which supports a high-level modeling and encloses a big set of annotations. In this context, Papyrus UML2 modeler ([1]) is used to apply this modeling. Figure 6 is exported from papyrus (Under Eclipse Indigo).

In the exposed model, the properties of classes are hidden.

To clarify the specification of classes attributes under UML/MARTE profile, an example of one class filling is given (Figure 7). This class models the task IF1 annotated through «swSchedulableResource».

As specified in Table 1, the deadline of IF1 is 201, which is set through the attribute «deadlineElements: TypedElement[0..*]= [<LiteralInteger>201]». The task IF1 may be preemptable by other high prior tasks. This is specified through the attribute «isPreemptable», which set on «true» («isPreemptable: Boolean[0..1] =true»).

i.e. A class in the static design pattern may contain a big set of features, but while instantiating the pattern the designer has to specify only the attributes needed for the scheduling of his system (example Figure 7).

In fact, the studied system specification (XMI representation of the design pattern reuse) is entered to our tool for the scheduling algorithm and approach selection. Then, once the appropriate scheduling algorithm and approach are selected, the dynamic view of the proposed design pattern for global scheduling is reused to check the schedulability analysis of the studied system.

7 Conclusions

Throughout this paper we have proposed two UML/MARTE-based design patterns aiming at supporting semi-partitioned and global scheduling. The main advantage of these design patterns use is the automatic selection of scheduling approach and algorithm. Furthermore, by using design patterns the obtained results are efficient.

To achieve the automatic scheduling, the proposed design patterns have been integrated in an MDE-based approach. A case study has been performed in this paper to validate the use of the proposed design pattern intended for global scheduling.

References

1. http://www.papyrusuml.org
2. Dorin, F., Yomsi, P.M., Goossens, J., Richard, P.: Semi-partitioned hard real-time scheduling with restricted migrations upon identical multiprocessor platforms. CoRR Journal (2010)
3. Du, C., Sun, X.-H., Wu, M.: Dynamic scheduling with process migration. In: Seventh IEEE International Symposium on Cluster Computing and the Grid, CCGRID 2007, pp. 92–99, May 2007
4. Florescu, O., Voeten, J., Verhoef, M., Corporaal, H.: Reusing real-time systems design experience through modelling patterns. In: FDL, pp. 375–381. ECSI, Darmstadt, September 19–22, 2006
5. Fritzsche, R., Ristig, C., Siemers, C.: An approach and design pattern for intra-application scheduling. Technical Report IfI-10-11, Clausthal University of Technology (2010)
6. Gamma, E., Helm, R., Johnson, R., Vlissides, J.: Design Patterns: Elements of Reusable Object-oriented Software, 1st edn. Addison-Wesley Professional, Boston (1995)
7. Gaudel, V., Singhoff, F., Plantec, A., Rubini, S., Dissaux, P., Legrand, J.: An ada design pattern recognition tool for aadl performance analysis. In: Proceedings of the 2011 Annual International Conference on Special Interest Group on the Ada Programming Language, SIGAda 2011, pp. 61–68. ACM, Denver (2011)
8. Goossens, J.: Introduction à l'ordonnancement temps réel multiprocesseur. In: Ecole d'été Temps Réel, pp. 157–166 (2007)
9. OMG (Object Management Group). A uml profile for marte: Modeling and analysis of real-time embedded systems. standard, June 2008
10. HadjKacem, Y., Mahfoudhi, A., Magdich, A., Karamti, W., Abid, M.: Using mde and priority time petri nets for the schedulability analysis of embedded systems modeled by uml activity diagrams. In: The 19th Annual IEEE International Conference and Workshops on the Engineering of Computer Based Systems (ECBS), pp. 316–323 (2012)

11. Kacem, Y.H., Mahfoudhi, A., Tmar, H., Abid, M.: From UML/MARTE to RTDT: A model driven based method for scheduling analysis and HW/SW partitioning. In: The 8th ACS/IEEE International Conference on Computer Systems and Applications, AICCSA 2010, Hammamet, Tunisia, May 16–19, 2010, pp. 1–7 (2010)
12. Konrad, S., Cheng, B.H.C., Campbell, L.A.: Object analysis patterns for embedded systems. IEEE Transactions on Software Engineering **30**(12), 970–992 (2004)
13. Madl, G.: Model-based Analysis of Event-driven Distributed Real-time Embedded Systems. PhD thesis, Long Beach, CA, USA (2009)
14. Magdich, A., Kacem, Y.H., Mahfoudhi, A., Kerboeuf, M.: A uml/marte-based design pattern for semi-partitioned scheduling analysis. In: The 23th International Conference on Enabling Technologies: Infrastructure for Collaborative Enterprises (WETICE), Collaborative Software Processes track (CSP). IEEE Computer Society, Juin 2014
15. Magdich, A., Kacem, Y.H., Mahfoudhi, A.: Extending UML/MARTE-GRM for integrating tasks migrations in class diagrams. In: Lee, R. (ed.) SERA 2013. SCI, vol. 496, pp. 73–84. Springer, Heidelberg (2013)
16. Magdich, A., Kacem, Y.H., Mahfoudhi, A., Abid, M.: A MARTE extension for global scheduling analysis of multiprocessor systems. In: The 23th IEEE International Symposium on Software Reliability Engineering (ISSRE), pp. 371–379. IEEE Computer Society, November 2012
17. Olteanu, A., Pop, F., Dobre, C., Cristea, V.: An adaptive scheduling approach in distributed systems. In: IEEE International Conference on Computational Photography (ICCP), pp. 435–442 (2010)
18. Schmidt, D.C.: Model-driven engineering. IEEE Computer **39**, February 2006
19. Tsantalis, N., Chatzigeorgiou, A., Stephanides, G., Halkidis, S.T.: Design pattern detection using similarity scoring. IEEE Trans. Softw. Eng. **32**(11), 896–909 (2006)
20. Zamfirache, F., Frincu, M.: Automatic selection of scheduling algorithms based on classification models. In: International Conference on Knowledge Engineering: Principles and Techniques (2011)

A Generic Traceability Framework for Model Composition Operation

Youness Laghouaouta[1](✉), Adil Anwar[2], Mahmoud Nassar[1], and Jean-Michel Bruel[3]

[1] IMS-SIME, ENSIAS, Mohammed V University, Rabat, Morocco
y.laghouaouta@um5s.net.ma, nassar@ensias.ma
[2] Siweb, EMI, Mohammed V University, Rabat, Morocco
anwar@emi.ac.ma
[3] IRIT, University of Toulouse, Toulouse, France
bruel@irit.fr

Abstract. In order to handle complexity, model driven engineering aims at building systems by developing several models, where each model represents a specific concern of the system. In this context, designers need mechanisms to validate, synchronize and understand interactions between those perspectives. Model composition deals with these issues but remains a complex task. For these reasons, we believe that a strong traceability mechanism is a key factor to handle relationships between models and manage the complexity of the composition operation. This paper describes a generic approach to keep track of the model composition operation. We also define a traces generation process to adapt our proposal to any specific composition language. Finally, an example is presented to illustrate our contributions.

Keywords: Model traceability · Model composition · Aspect-oriented modeling · Graph transformations

1 Introduction

Model Driven Engineering (MDE) proposes models to represent all artifacts handled by a software development process. The principle is to raise the abstraction level by using models as first class entities in dedicated model management operations. Besides, systems are built based on various models that express particular viewpoints. This allows managing the system's complexity, but requires mechanisms to validate, synchronize and understand the interrelation between contributing models. The generation of models that cross separate views tackles these tasks.

Nevertheless, the composition of models remains a complex activity. We advocate that traceability mechanisms are key features to handle this complexity. Indeed, traceability information exposes the effects of executing this operation and helps to comprehend relationships existing among managed models. This kind of information supports validation tasks and offers a way to optimize the co-evolution of models and composition chains.

In previous papers [1][2], we treat the tracing of composition specifications written in the Epsilon Merging Language EML [3] and the ATLAS Transformation Language ATL [4]. The current extends these works and focuses on the generic nature of our proposal. Indeed, we aim to keep track of the composition effects regardless of its specification features. For this purpose, we specify a generic traceability framework that allows tracing the model composition operation, and provide a traces generation process for adapting our proposal to a specific composition language.

The remainder of this paper is structured as follows. Section 2 presents an example that motivates the need for tracing the model composition operation. Section 3 provides an overview of the traces generation process. Section 4 details the structuring and the generation of trace links. In Section 5, we demonstrate the soundness of our approach using a specialization case with the dedicated merging language EML [3]. In Section 6, we review the related approaches. Finally, Section 7 summaries this paper and presents future work.

2 Motivating Example

We illustrate the necessity to automatically keep track of the model composition operation using a Library Management System (LMS). The composition scenario we have chosen is the merging of structural models that express an extract of the librarian and head librarian activities. Fig. 1(a) shows an excerpt of the class diagram related to the head librarian, while Fig. 1(b) depicts the class diagram that models the specific librarian requirements. As a result of composing these models, we obtained the class diagram depicted in Fig. 1(c). The composed model contains two categories of elements: elements that originate from only one input model (e.g., the class named *Loan*) and those that existed in both of them (e.g., the class named *Book*). Therefore, two kinds of traceability links have to be captured: merging links and translation links. Essentially, this kind of information reveals the logical relation existing among the managed models and specifies how source models contribute to the production of the composed one. Fig. 1 depicts also extracts of traces that have to be captured. Trace links are represented by nodes labeled with 'M' for merging links and 'T' for translation links, while the dashed lines represent the left, right and target references. In the context of model composition, such information has many possible reuses:

- *Validation of the composition*: trace links provide a detailed view of the flow of execution. Indeed, they represent relationships between source model elements and their target equivalents. Through these links, we can verify the consistency and the completeness of the model composition.
- *Support the co-evolution of models*: those links are useful to analyze the impact of changing sub-models during the evolution of the system. For instance, as the class *Loan* is connected by a translation link with the class

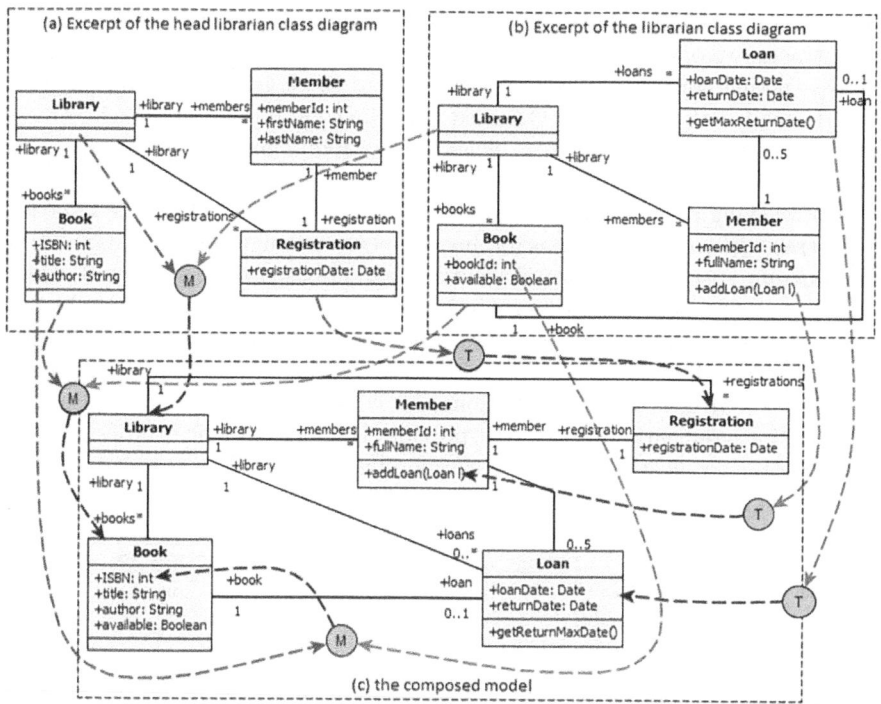

Fig. 1. Motivating example

Loan in the merged model, adding a new attribute to the source class will result in adding this attribute to the target one without reestablishing matching links between the source models elements.

– *Composition chain optimization*: as a part of a model composition chain, the restriction to source artifacts of a given stage of the overall chain confuses its management. Hence, the use of the trace model can broaden its scope, through the reuse of previous valuable links. As an example, the *bookId* and *ISBN* were considered as describing the same concept. This design decision has been held with the merging link that connects the *ISBN* and *bookId* attributes in the source models with the *ISBN* attribute in the resulting one. Consequently, exploiting this link in subsequent compositions will assist the matching step by specifying that the *ISBN* and *bookId* attribute's names are equivalent.

3 Overview of the Traces Generation Process

In this section, we provide an overview of the way traces are captured and structured. Drawing on the work presented by Jouault [5], we propose to generate the trace model as an additional target model of the specification to trace.

However, rather than using a higher-order transformation (HOT) to allow generating traces, we opted for aspect orientation and consider this concern as being a cross-cutting concern. Indeed, the weaving of the traces generation patterns is performed with an Aspect Oriented Modeling (AOM) [6] approach. Traces thus generated will conform to a generic traceability metamodel presented in Section 4.1.

The weaving process has been defined in such a way that allows the insertion of the traces generation patterns for any composition language while abstracting the concrete specification nature (textual, model-based or graph-based). This solution is organized around two parts (see Fig. 2): the bottom part which is generic and reusable and the top part which is specific for a given composition language. In dealing with the generation task, we simulate the aspect modeling orientation with graph transformations [7]. For this reason, it is necessary to implement a serialization service that provides language specific utilities to parse the concrete specification into the corresponding model (M1).

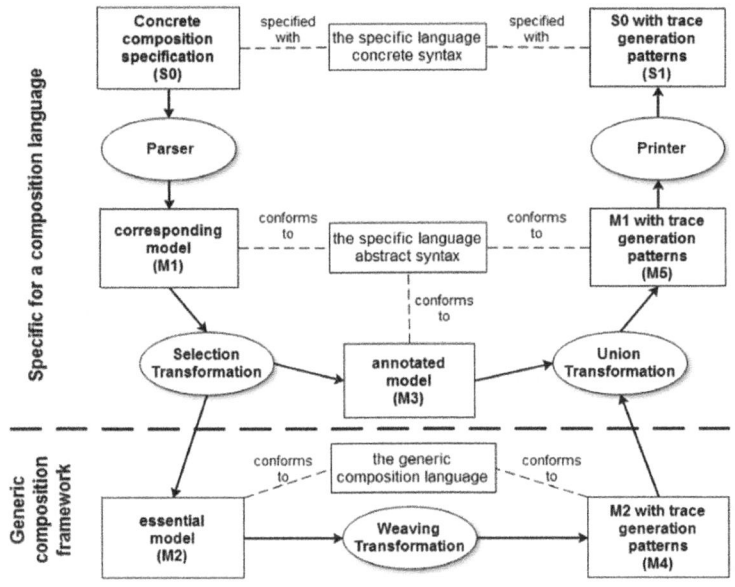

Fig. 2. The traceability weaving process

Besides, in order to specify the graph transformations which implement the weaving mechanism independently from a given composition language, we propose a generic composition language (c.f., Section 4.2). Essentially, this formalization describes the core elements with respect to the traceability perspective and does not take account of all the operational semantic of a composition framework. Accordingly, we have to select the relevant elements from the corresponding model (M1) and therefore we translate them to conform to our generic proposal. This extraction is performed using a specific graph transformation called *selection transformation*. Besides, the specific composition metamodel has to be augmented with traceability capabilities to allow injecting the omitted elements after performing the weaving. Indeed, we add the meta-class *TracedElement* that generalizes all concepts described in the host language. *TracedElement* contains two attributes: *traceID* to identify each element and *status* to precise whether the element has been selected or omitted. Hence, the *selection transformation* generates two models: the essential model (M2) containing the traceability relevant elements and the model annotated with the aforementioned markers (M3).

Once the essential model is generated, we apply the *weaving transformation* that inserts the traces generation patterns. Thereafter, the *union transformation* transforms the resulting model (M4) to conform to the specific language. On the other hand, it reuses the annotated model (M3) to weave the omitted elements at their relevant containers with respect to the presented markers. Finally, we reproduce the concrete specification which involves the traces generation patterns by using a specific language printer.

4 A Generic Framework for Model Composition Traceability Management

In this section, we detail each component of the generic part of the traces generation process. The traceability framework is based on a generic traceability metamodel accounting for structuring traces. Whereas, the traces generation concern is encapsulated on a traceability aspect which is defined around a generic metamodel that formalizes the core elements of a composition language.

4.1 A Generic Metamodel for Structuring Traces

In the literature, several metamodels for the model transformation traceability have been proposed [5][8][9]. The core concept in all of them is the trace link construct, which represents a relationship between a set of source elements and the targets ones. In our context, two categories of relationships have to be expressed: merging links and translation links. However, metamodels addressing the model transformation traceability do not support this case of categorization or poorly express it through the assignment of additional information. In fact, the way trace links are structured must take into account the composition mechanism.

Fig. 3 depicts our composition traceability metamodel [2]. It specifies two types of trace links: merging links and translation links. On the one hand, this categorization allows expressing the composition relationships kinds in a trivial manner. On the other hand, it guides the reuse of traces (e.g., matching correspondences can be deduced from merging links). A merging link connects the source artifacts (belonging to the left and right models) to their target equivalent. While a translation link represents a transition from a left or right element to the target one.

We represent rule invocation by a nesting of trace links which is expressed through parent-child relations among them. This structure provides a multi-scales character to the generated traces and allows the final user to configure the granularity level he desires. In addition, the *Context* concept brings another configuration mechanism. It provides us with the support to represent semantically rich traces by assigning further information to a subset of trace links. This configuration can be achieved through the definition of relevant expressiveness data (e.g., the composition rule name, the traceability intention ...), where a context attribute is tied to the additional information to be appended to a specific set of traces, and a context is as a well thought out combination of attributes.

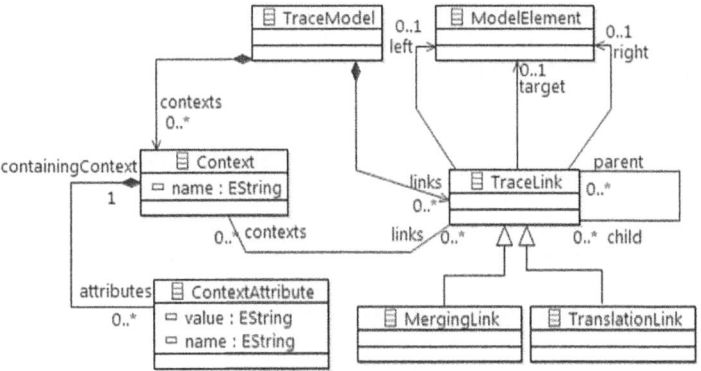

Fig. 3. The composition traceability metamodel

4.2 A Generic Model Composition Language

Several approaches addressing the model composition field have been proposed: AMW [10], EML [3], Kompose [11]. In order to provide a generic traceability solution, we have to abstract the composition operation from its syntax nature (textual or graphical) or the language used to express it. A typical composition process involves two major steps: matching and merging [12]. During the first step, correspondences between left and right model elements are calculated. Matching elements are merged while other elements may be transformed into target model elements. In addition, Bezivin *et al.* [12] set a list of requirements for model composition frameworks. We consider these aspects to identify the main elements that constitute the backbone of such an operation (see Fig. 4).

– *Source and target models*: the composition operator combines source models in order to produce the target ones. Hence, the composition specification has to be aware of the relevant information (models name, their metamodel...).
– *Merging rules*: they describe the behavior needed to combine two elements that match with respect to some correspondences criteria. These rules have a name, a statements block which specifies the merging mechanism, and a set of parameters referencing the contributing elements.

- *Translation rules*: their structure is similar to merging rules. However, they are applied to transform elements that have not been merged into the target model, and therefore have at most one source parameter.
- *Rule invocation*: without an adequate mechanism to call rules, the target model looks fragmented. This mechanism enables the weaving of structural relations between target elements (e.g., an attribute belonging to its class) by linking the result of a rule application to the elements previously created. We encapsulate this behavior in an abstract operation named *targetEquivalent*. It resolves the target element corresponding to a source one.

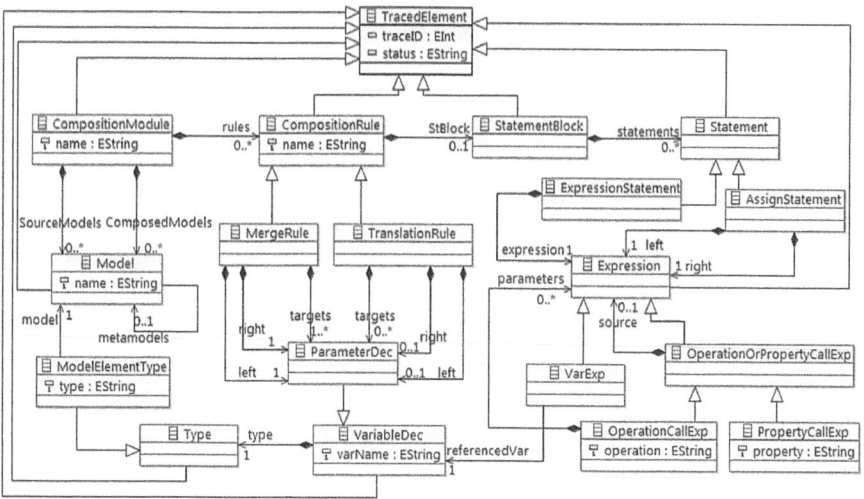

Fig. 4. Abstract syntax for a generic composition language

We formalize these requirements with the metamodel depicted in Fig. 4. Note that this formalization includes a set of concepts defined in the rule based languages EML [3] and ATL [4]. Furthermore, this metamodel does not structure all the operational part of a composition language. It is just a formalization that allows specifying the traceability aspect in a generic way. Essentially, it focuses on the core concepts with regards to the traceability perspective (managed models, merging rules, translation rules and rules invocation) and is restricted to elements that are used to specify the traces generation weaving, such as:

- *Types*: the *Type* meta-class is the basis of all types. It generalizes the *ModelElementType* meta-class which represents a meta-level classifier.
- *Expressions*: it provides expressions to navigate properties (*PropertyCallExp*) and invoke operations (*OperationCallExp*). The *VarExp* expression allows accessing a declared variable.
- *Statements*: the *AssignStatement* elements are used to assign the right value to the left expression, while an expression statement refers to one expression.

The section that follows is concerned implementing the weaving transformations, and clarifying the missing parts of this abstract syntax.

4.3 Graph Transformations for Traceability Weaving

The AOM focuses on modularizing and composing crosscutting concerns during the design phase of a software system. Indeed, both the aspects that encapsulate the crosscutting structure and the base model they crosscut are models [6]. Our objective is to build the trace model without manually encoding the generation patterns and regardless of the specification nature; this approach aligns perfectly with these tasks. It allows encapsulating these patterns in an aspect and abstracts the composition specification through the corresponding model.

Nevertheless, the AOM is a paradigm for conceptualizing aspects, which requires an implementation mechanism. We use graph transformation rules to simulate the aspect orientation. A graph rewriting rule consists of two parts: a left-hand side (LHS) and a right-hand side (RHS). A rule is applied by substituting the objects of the LHS with the objects of the RHS, only if the pattern of the LHS can be matched to a given graph [13]. Therefore, the LHS part is used to determine where the aspect should be applied (the pointcuts); whereas the RHS defines the crosscutting structure that replaces those points (the advice).

We employ a graph transformation unit to weave the traces generation patterns. Its first rule declares the trace model to be an additional target model of the composition to trace. Thereafter, it calls two loop sub-units to trace all the merging and translation rules. Finally, trace links are nested by applying a responsible rule. In what follows, we use the Henshin project [14] to express these graph transformations. Note that the Henshin representation of a rule does not explicit the description of the left and right hand sides. It is based on stereotyping edges to depict the rule application semantic instead.

Trace Model Declaration. We propose to generate the trace model like any other target model. For this purpose, we create two *Model* instances. The first is connected as a new target of the composition and references the trace model, while the second node corresponds to our generic traceability metamodel.

Trace a Merge Rule. Keeping track of merge rules consists of declaring the traceability link, which captures the relationship between the matched source elements and the target one, as being an extra output. For that, the rule depicted in Fig. 5 looks for a *MergeRule* node in the graph corresponding to the specification we wish to trace; subsequently, it appends a new *ParamDec* node of type *MergingLink* to the rule that have been matched as one its target parameters. The added parameter allows the generation of trace links while producing the target elements. Furthermore, this rule creates three assign statements referenced by the *AssignStatement* nodes. Each affects the reference of the corresponding element to the appropriate trace link property (*left*, *right*, and *target*).

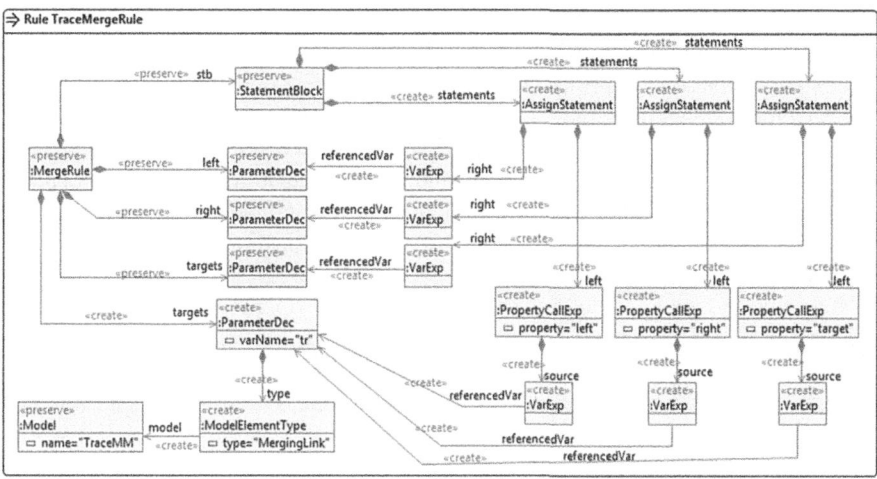

Fig. 5. Trace a merging rule

Trace a Translation Rule. The application of the rule shown in Fig. 6 inserts the pattern that keeps track of the transition from a source element to its target equivalent. This rule searches for a *TranslationRule* node and declares a new parameter of type *TranslationLink*. This link captures the correspondence between the managed source and target elements that are matched with the *ParamDec* nodes stereotyped with *preserve*. As with merging rules, we assign the traceability data to the generated link using *AssignStatement* nodes.

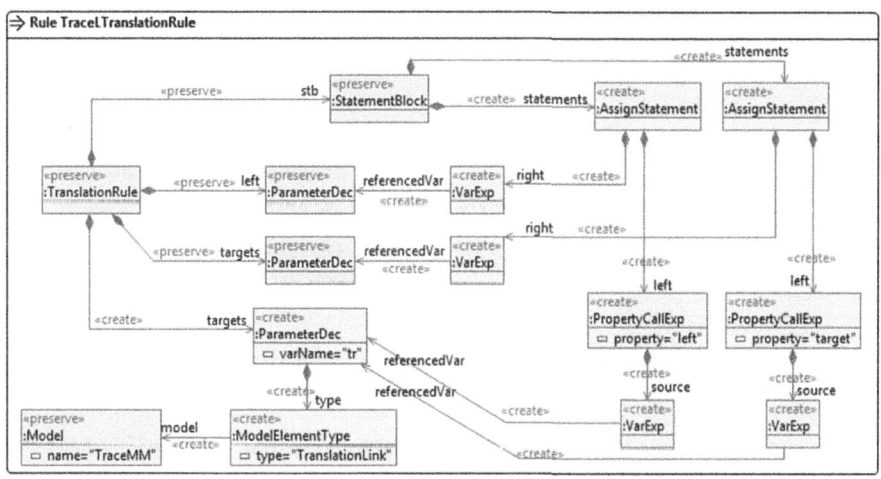

Fig. 6. Trace a translation rule

Trace Links Nesting. In Section 4.2, we discussed the use of rules invocation to structure the composed model. The operation *targetEquivalent* encapsulates this behavior by providing a mechanism for resolving the corresponding target element of a source one. This target equivalent is produced by an anterior application of a given composition rule. We propose to reuse a similar mechanism for structuring trace links. Actually, it follows from the application of the preceding graph transformations, the production of additional elements corresponding to the traceability links. Therefore, the *targetEquivalent* operation resolves these links as potential equivalents.

Accordingly, we have defined two other abstract operations: *traceEquivalent* and *defaultTargetEquivalent*. They provide a filtering mechanism to select trace links from other target elements. Hence, we can assign the resolved trace equivalent to be a child of the link produced by the current rule (which calls the *targetEquivalent* operation).

Fig. 7 depicts the rule that implements this nesting mechanism. It matches a composition rule involving an invocation of the *targetEquivalent* operation (referenced by the *OperationCallExp* node stereotyped with *delete*). Thereafter, it copies the reference of the element to resolve its target equivalent (which corresponds to the parameter of the *targetEquivalent* call) to the *resolvedElt* variable. Finally, the two other *Statement* nodes allow copying the original call of *targetEquivalent* (using the *defaultTargetEquivalent* operation) and binding the traceability element as a parent of the trace equivalent.

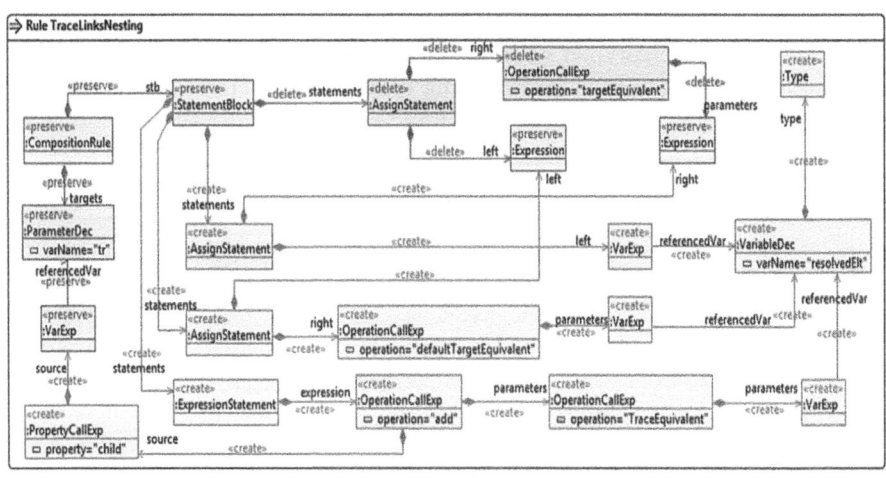

Fig. 7. Trace links nesting rule

5 Specializing the Generic Framework for a Dedicated Merging Language

In this section, we illustrate the specialization of our generic framework by tracing the merging scenario we have presented before as a motivating example (c.f., Section 2). The composition specification is specified by EML [3], which is a dedicated merging language. We aim through this language enforcing our generation process to reveal the generic character of our traceability framework.

The specialization work consists of identifying the meaning of the main composition elements (managed models, merging rule, translation rule and rule invocation) regarding a given composition approach (e.g., how the rule invocation is performed in EML?). Thereafter, we establish correspondences between the specific representation of each concept and its equivalent conforming to our generic composition language. Those correspondences underpin the selection and the union transformations of the traces generation process (c.f., Section 3).

5.1 Perform the Merging Scenario with EML

An EML specification is defined using three types of rules: match rules, merge rules, and transform rules. Match rules are applied on the source models to calculate correspondences between their elements. Subsequently, merge rules are used to combine elements that describe the same concept, while transformation rules allow transforming elements that have no corresponding element.

In order to apply our traceability weaving process, we have implemented an EMFText[1] parser that transforms the textual representation of any EML specification into a corresponding model that conforms to the EML abstract syntax[2]. Once the corresponding EML model is generated, the selection transformation is applied to annotate it with the *traceID* and *status* markers and produce the essential model. Table 1 summarizes the mapping between the main generic composition elements and the relevant concept in EML.

Table 1. The EML relevant concepts of the main generic composition elements

Generic concept	EML relevant concept
Composition module	Composition module element
Merging rule	Merge rule
Translation rule	Transformation rule
Rule invocation	A call of the *equivalent* operation

Fig. 8 depicts the graph transformation rule that allows selecting merging rules. It looks for a *MergeRule* node (belonging to the EML abstract syntax) with three *ParameterDeclaration* nodes which reference, respectively, the *left*, *right* and *target* elements. Thereafter, it creates the corresponding *MergeRule* node (belonging to our generic composition metamodel) with its connected elements. Note that this rule has two other effects. It duplicates the *traceID* value of the selected elements to the created ones. Also, it changes the value of the *status* attribute into *Selected* in order to annotate the corresponding model.

---
[1] See http://www.emftext.org.
[2] See http://www.eclipse.org/epsilon/doc/book/.

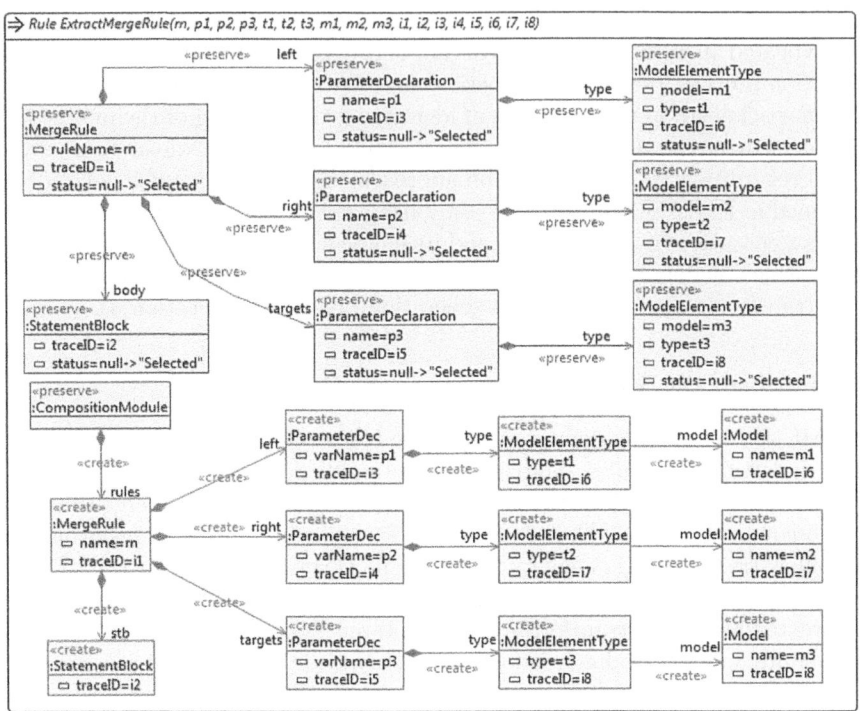

Fig. 8. Selection of merging rules in EML

The weaving transformation is applied on the essential model for weaving the traces generation patterns. Subsequently, the union transformation reinserts the omitted elements into their corresponding containers (i.e., each omitted assign statement must be nested on its corresponding statements block). The resolution of the relevant container is based on the *tracedID* attribute. Listing 1 depicts the EML rule that merges two matching classes, while Listing 2 represents the resulting modifications over this rule that generate traces.

Listing 1. EML rule to merge two classes

```
1 rule MergeClassWithClass
2 merge l : left!Class
3 with r : right!Class
4 into t : target!Class
5 {
6 t.name = l.name;
7 t.ownedAttribute = l.ownedAttribute.includingAll(r.ownedAttribute).equivalent();
8 }
```

As a result of applying the traces generation weaving process, the traceability parameter is declared as another target parameter and the traceability information is assigned to it (Listing 2: lines 8,11-13). In addition, the call of the *equivalent* operation (Listing 1: line 7) has been captured and replaced with the fragment that allows to nest traces (Listing 2: lines 14-16). Note that the generic operations *defaultTargetEquivalent* and *traceEquivalent* was translated into the host language using the EML *select* operation. Besides, the assignment of the *name* property value (Listing 1: line 6) was marked as omitted and has been injected in the resulting specification (Listing 2: line 10).

Listing 2. EML rule to merge two classes with traces generation

```
 1  pre
 2  {
 3  var resolvedElt : new Any ;
 4  }
 5  rule MergeClassWithClass
 6  merge l : left!Class
 7  with r : right!Class
 8  into t : target!Class , tr:trace!MergingLink
 9  {
10  t.name = l.name;
11  tr.left=l;
12  tr.right=r;
13  tr.target=t;
14  resolvedElt = l.ownedAttribute.includingAll(r.ownedAttribute);
15  t.ownedAttribute = element.equivalent().select(it | not it.isKindOf(trace!TraceLink));
16  tr.child.add(element.equivalent().select(it | it.isKindOf(trace!TraceLink)));
17  }
```

5.2 Results

In Fig. 9 we provide an extract of the trace model generated with our framework. Note that we have used the Emf2gv[3] project to provide a user friendly representation of traces. The trace model conforms to our composition traceability metamodel and captures relationships between the contributing models and the merged one. The dashed lines represent the left, right and target references; while the trace links nesting is represented with solid lines.

The trace model contains two types of trace links that are generated with respect to the composition relationships kinds. For instance, the first level merging link connects the composed model to the *Librarian* and *Head Librarian* class diagrams. The contained merging link connects the *Book* classes corresponding, respectively, to the librarian and head librarian concerns, and the composed model. Fig. 9 depicts also a translation link that represents the transition from the *Loan* class in the *Librarian* class diagram to its target equivalent. Furthermore, the nesting of traces is closely modeled on the rule invocation sequence.

6 Related Work

In the literature, several works dealing the traceability of model driven development operations are presented [5][8][9][15]. We focus here on two of them that are distinguished by their generic nature:

Grammel and Kastenholz [9] have defined a generic approach to trace various model transformation approaches. Their proposal is based on a generic interface that offers two mechanisms for augmenting a transformation engine with traceability. The first one consists of transforming the implicit trace model to conform to their generic traceability metamodel. Besides, in the case of a lack

---

[3] See http://sourceforge.net/projects/emf2gv.

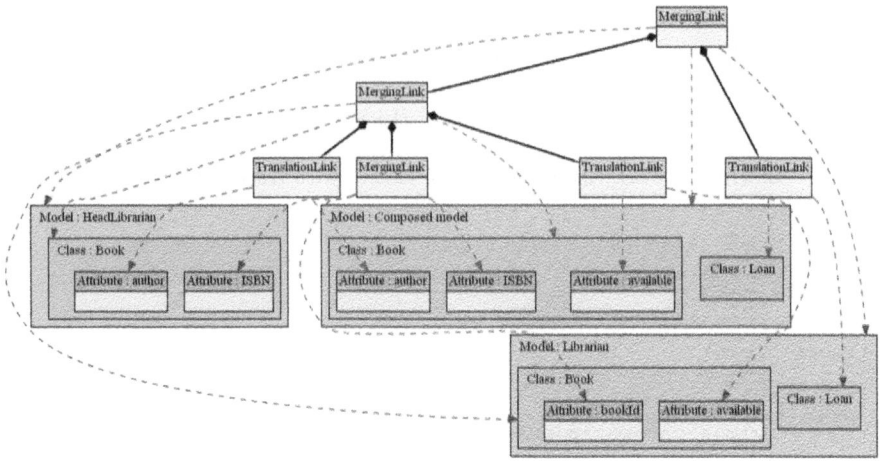

Fig. 9. Extract of the generated trace model

of an implicit traceability tool, they provide support to generate traces based on Aspect Oriented Programming (AOP).

In [15], Vara *et al.* propose a methodological framework that supports the model-driven development of model transformations and allows generating traces freely as a side effect. The extraction of traceability relationships is implemented by a HOT operating on the high-level specification of transformation. This specification is augmented with the traces generation patterns and a set of transformations are described for producing the lower-level transformation models. Thus, the executable transformation generates the target models accompanied by an additional model capturing traceability links. However, the applicability of the approach is restricted to newly model transformations developed by this framework, and no mechanism is defined to trace pre-existing transformations.

Our approach makes use of the benefits of the presented works while focusing on model composition. As for the approach of Grammel and Kastenholz [9], we adopt aspect orientation to generate traces. However, instead of using AOP, we have chosen an AOM solution to abstract the composition specification through the corresponding model. Thereby, we handle the plurality and diversity of the composition approaches.

7 Conclusion and Future Work

In this paper, we proposed a generic approach to automatically build traces of models composition. We consider the traceability management as a cross-cutting concern. Actually, we designed some graph transformation rules that encapsulate the trace links generation concern. Those rules have been defined around a generic composition metamodel which formalizes the core element of the composition operation. Moreover, a traces generation process, partly independent of the composition language, is introduced to specialize the application of our proposal.

We are currently exploring a pre-configuration support to provide the user with the tool to configure the application of the traceability aspect depending on its purpose. Besides, we intend to work on the traces reusability issue. We have set three possible reuses: validation in model composition, co-evolution of models and optimization of composition chain.

References

1. Laghouaouta, Y., Anwar, A., Nassar, M., Coulette, B.: A graph based approach to trace models composition. JSW **9**(11), 2813–2822 (2014)
2. Laghouaouta, Y., Anwar, A., Nassar, M., Bruel, J.M.: On the use of graph transformations for model composition traceability. In: IEEE International Conference on Research Challenges in Information Science (RCIS 2014), pp. 1–11. IEEE (2014)
3. Kolovos, D.S., Paige, R.F., Polack, F.A.C.: Merging models with the Epsilon Merging Language (EML). In: Wang, J., Whittle, J., Harel, D., Reggio, G. (eds.) MoDELS 2006. LNCS, vol. 4199, pp. 215–229. Springer, Heidelberg (2006)
4. Jouault, F., Kurtev, I.: Transforming models with ATL. In: Bruel, J.-M. (ed.) MoDELS 2005. LNCS, vol. 3844, pp. 128–138. Springer, Heidelberg (2006)
5. Jouault, F.: Loosely coupled traceability for ATL. In: Proceedings of the European Conference on Model Driven Architecture (ECMDA) workshop on traceability, Nuremberg, Germany, vol. 91. Citeseer (2005)
6. France, R., Ray, I., Georg, G., Ghosh, S.: Aspect-oriented approach to early design modelling. IEE Proceedings-Software **151**(4), 173–185 (2004)
7. Ehrig, H., Engels, G., Rozenberg, G.: Handbook of graph grammars and computing by graph transformation, vol. 2. world Scientific (1999)
8. Amar, B., Leblanc, H., Coulette, B.: A traceability engine dedicated to model transformation for software engineering. In: ECMDA Traceability Workshop (ECMDA-TW), pp. 7–16 (2008)
9. Grammel, B., Kastenholz, S.: A generic traceability framework for facet-based traceability data extraction in model-driven software development. In: Proceedings of the 6th ECMFA Traceability Workshop, pp. 7–14. ACM (2010)
10. Del Fabro, M.D., Bézivin, J., Jouault, F., Breton, E., Gueltas, G.: Amw: a generic model weaver. In: Proceedings of IDM 2005 (2005)
11. France, R., Fleurey, F., Reddy, R., Baudry, B., Ghosh, S.: Providing support for model composition in metamodels. In: 11th IEEE International Enterprise Distributed Object Computing Conference, EDOC 2007, pp. 253–253. IEEE (2007)
12. Bézivin, J., Bouzitouna, S., Del Fabro, M.D., Gervais, M.-P., Jouault, F., Kolovos, D.S., Kurtev, I., Paige, R.F.: A canonical scheme for model composition. In: Rensink, A., Warmer, J. (eds.) ECMDA-FA 2006. LNCS, vol. 4066, pp. 346–360. Springer, Heidelberg (2006)
13. Lambers, L., Ehrig, H., Orejas, F.: Conflict detection for graph transformation with negative application conditions. In: Corradini, A., Ehrig, H., Montanari, U., Ribeiro, L., Rozenberg, G. (eds.) ICGT 2006. LNCS, vol. 4178, pp. 61–76. Springer, Heidelberg (2006)
14. Arendt, T., Biermann, E., Jurack, S., Krause, C., Taentzer, G.: Henshin: advanced concepts and tools for in-place EMF model transformations. In: Petriu, D.C., Rouquette, N., Haugen, Ø. (eds.) MODELS 2010, Part I. LNCS, vol. 6394, pp. 121–135. Springer, Heidelberg (2010)
15. Vara, J.M., Bollati, V.A., Jiménez, Á., Marcos, E.: Dealing with traceability in the mddof model transformations. IEEE Trans. Software Eng. **40**(6), 555–583 (2014)

Novel Applications of Modeling
(Short Papers)

Applying Predicate Abstraction to Abstract State Machines

Alessandro Bianchi[✉], Sebastiano Pizzutilo, and Gennaro Vessio

Department of Informatics, University of Bari, 70125, Bari, Italy
{alessandro.bianchi,sebastiano.pizzutilo,
gennaro.vessio}@uniba.it

Abstract. Abstract State Machines (ASMs) represent a general model of computation which subsumes all other classic computational models. Since the notion of ASM state naturally captures the classic notion of program state, ASMs are suitable to be verified through a *predicate abstraction* approach. The aim of this paper is to discuss how predicates over ASM states can support the formal verification of ASM-based models. The proposal can overcome the main limitations that penalize traditional model checking techniques applied to ASMs.

1 Introduction

Abstract State Machines (ASMs) represent a general model of computation which subsumes all other classic computational models [27], [12]. Indeed, they suffice to capture the behavior of wide classes of sequential [20], parallel [8] and distributed algorithms [18]. The origin of this generality lies in the notion of ASM *state*. In classic formalisms, such as finite state machines and Turing machines, states are symbolically represented by (sequence of) symbols belonging to finite alphabets [21]. Conversely, ASM states are syntactically and semantically represented by algebraic structures defined over finite signatures. Therefore, ASM states can model any object of arbitrary complexity [20].

Thanks to their generality, ASMs have been successfully applied for modeling critical and complex systems in a wide range of application domains, and for analyzing their computationally interesting properties, both domain-independent (e.g. the *termination* of the execution, *deadlock-* and *starvation-freedom*, and so on) and domain-specific (e.g. security issues, the movement of robotic arms, synchronization issues of real-time controllers, and so on) [9]. However, despite the advantages they provide, the computational power and the arbitrary complexity of the formalism cause an unavoidable drawback: several computationally interesting properties are undecidable, so the formal verification of ASM-based models cannot be fully automated [29].

Traditional model checking approaches to the problem of verifying properties typically model systems with finite state machines (or variants) and express the properties to be verified using some temporal logic [5]. Analogously, when model checking techniques are applied to ASMs, the ASM-based model under study is transformed into the input required by the adopted model checker [14], [3], which is in general less expressive. Therefore, this approach suffers from two main limitations: the loss of expressive power due to the translation of the ASM specifications, and the difficulty in using the declarative notations of temporal logics, considered less comfortable for practitioners with respect to operational specifications of the properties (e.g. [4]).

Our long term research is aimed at providing a theoretical framework for treating properties analysis entirely within the ASM framework, i.e. without translating ASMs, and without using temporal logics. To this end, the present paper takes a step towards the application of a *predicate abstraction* approach for formally analyzing ASMs. Predicate abstraction is a popular and widely used technique for automatizing the verification of programs [19], [11]. It consists in the approximation of the program states into a finite number of predicates defined over these states. In this context, the goal of the present paper is to support the verification of ASM properties through predicates over the states. In this way, the goal of formally verifying ASM models entirely within the ASM framework can be achieved.

The rest of this paper is organized as follows. Section 2 is about related work. Section 3 provides background knowledge about the ASM formalism. Section 4 deals with predicate abstraction for ASMs. Section 5 depicts some illustrative examples. Finally, Section 6 concludes the paper and sketches future developments.

2 Related Work

The ASM formalism allows both manual and automated formal verification of systems. In [9] numerous proofs are provided to illustrate how a modeler can verify properties of a given ASM. These proofs range from simple to complex and, since ASMs are executable machines which lend themselves to traditional inductive proofs, they are often formulated in a mathematical way. For example, in [15] a manual verification calculus based on the Hoare logic is proposed. Conversely, in [14] and [3] the authors use an automatic model checking approach for verifying ASMs. However, the translation of the ASM into the input required by the model checkers may cause a loss of expressive power. Moreover, in all these cases, properties are expressed in some temporal logic [5]. But, the effectiveness of this hybrid approach is not unanimously recognized: several authors emphasize the need of an entirely operational specification of properties within the same formalism, e.g. [22], [4].

ASMs have been successfully used for modeling several systems, often concurrent and distributed, and for investigating their properties. For example, an ASM specification to model concurrency in a Web browser and to prove some consistency properties has been proposed in [17]; ASMs have also been used to model and validate vision-based robot control applications in [25]; and they have been applied for studying Grid systems in [6]. However, all these works are characterized by the lack of a theoretical framework for systematically treating the analyzed properties.

Concerning the application of predicate abstraction to formal methods, other formalisms, such as Petri nets [10], already employs predicates on states whenever properties are to be analyzed. However, Petri nets typically provide only few levels of abstraction, so they are not able to support refinements till to implementation details. Conversely, the expressive power of ASMs provides a way to describe algorithmic issues in a simple abstract pseudo-code, which can be translated into a high level programming language source code in a quite simple manner [9]. Furthermore, predicate abstraction has been already used within the ASM formalism in [16]; however, the authors use what they call *test predicates* in a way which is different from ours. Indeed, they use predicates on states in order to generate test sequences.

3   Background on ASMs

Abstract State Machines are finite sets of so-called *rules* of the form **if** *condition* **then** *updates* (possibly with the **else** clause) which transform *abstract* states [9]. An ASM state is an algebraic structure, i.e. a domain of objects with functions and relations defined on them. Partial functions are turned into total functions by using the special value *undef*. Moreover, without loss of generality, relations are treated as particular functions that evaluate to *true* or *false*. On the other hand, the concept of rule reflects the notion of transition occurring in traditional transition systems: *condition* is a first-order formula whose interpretation can be *true* or *false*, whereas *updates* is a finite set of assignments of the form $f(t_1, ..., t_n) := t$, whose execution consists in changing in parallel the value of the specified functions to the indicated value.

Pairs of function names, fixed by a signature, together with values for their arguments are called *locations*: they abstract the notion of memory unit. Therefore, a state can be viewed as a function that maps locations to their values: the current configuration of locations together with their values determines the current state of the ASM. As usual in computational models, an ASM *step* is a pair (s, s') of states: in a given state, all conditions are checked, so that all updates in rules whose conditions evaluate to *true* are simultaneously executed, and the result is a transition of the machine from that state to another. Note that for the unambiguous determination of the next state, updates must be *consistent*, i.e. no pair of updates must refer to the same location.

A generalization of basic ASMs is represented by Distributed ASMs (DASMs) [9], [18], capable to capture the formalization of multiple agents acting in a distributed environment. Essentially, a DASM is intended as an arbitrary but finite number of independent agents, each executing its own underlying ASM. In a DASM the keyword **self** is used for supporting the relation between local and global states and for denoting the specific agent which is executing a rule.

4   Predicates Over ASM States

Classic computational models, such as finite state machines and Turing machines, represent the current state of the computation with (sequence of) symbols belonging to finite alphabets [21]. This poses a limitation: the representation of states is restricted to a specific data structure. Instead, as explained in Section 3, ASMs allow any algebraic structure to serve as representation of states. This results in a great amount of details specifying the states, so making the analysis of the properties of the whole system more difficult, mainly for what concerns the comprehension of the semantics of each state, with respect to the computational behavior of the modeled system.

For better explaining this issue, the next section will elaborate two examples, both concerning distributed systems: the analysis of *starvation-freedom* [2] and *deadlock-freedom* [28], and the analysis of the computation of an agent capable to play two or more roles at the same time. In the former case, the simple execution of one or more updates does not necessarily involve the change of the locations values in such a way that the process makes real computational progress, so driving to starvation. In fact, an ASM could starve even if the computation continues to evolve through different states. In other words, it is difficult to recognize effective progress. As an extreme case, this computational behavior can lead to deadlock. On the other hand, the second case concerns, for example, the case of a process that acts both as a client with respect to a service, and, simultaneously, as a server with respect to another service. In this case, ASMs easily capture in a same state different computational activities to be run in parallel. However, it is difficult for the modeler to distinguish, inside the same state, what computational branches have been entered or not.

In order to overcome these problems, the need of an abstraction framework capable to capture the semantics of the ASM states arises. More precisely, there is the need to partition the set of locations into subsets and extract from them the locations specifically interesting for the verification purposes. To this end, we apply a predicate abstraction approach. Predicate abstraction is a popular and widely used technique proposed to analyze programs [19], [11]. It aims at generating an abstract model from the concrete system to be verified, so checking the former instead of the latter. Briefly speaking, the system states are mapped to model states according to their evaluation with respect to a finite set of predicates defined over the system states. The model has the same control flow of the original program but it concerns only the predicates over the states. The model can then be used in the place of the original program when performing model checking, theorem proving, or other kinds of verification techniques.

Literature agrees that a program state coincides with the configuration of program variables and their current values, e.g. [24]. Analogously, an ASM state coincides with the configuration of ASM locations and values. So, since there exists a natural parallelism between classic program states and ASM states, predicate abstraction can be applied to ASMs as much as to programs of traditional programming languages. In this context we can apply predicate abstraction to ASMs through the following:

Definition. *A* predicate ϕ over an *ASM* state s *is a first-order formula defined over the locations in* s, *such that* s ⊨ ϕ.

Predicates over the states serve to represent the semantics of each state, i.e. the properties locally satisfied, and can be regarded as a non-injective labeling function that maps predicates to each state. An ASM model can then be equipped with a set of predicates $\Phi = \{\phi_1, ..., \phi_n\}$, such that, in the current state, each ϕ_i can be satisfied or not. In this way, the ASM control flow can be represented by the truth value of the predicates over the states, i.e. by composing the local properties of the various states. So, global properties to be verified can be analyzed by focusing on this composition.

Note that our use of predicate abstraction is quite different with respect to the traditional way: instead of extracting abstract models from the ASMs to be verified, our aim is to use predicates over the states in order to support the verification of ASM models. In particular, applying predicate abstraction to ASMs induces the partition of locations we need for expressing the semantics of the states.

5 Two Examples

In order to show the application of predicate abstraction over ASM states two examples are discussed: the Dining Philosophers problem, and the Ad-hoc On-demand Distance Vector (AODV) routing protocol for Mobile Ad-hoc NETworks (MANETs). The first example deals with starvation-freedom and deadlock-freedom analysis: here the same value for a predicate holds for several states. The second example discusses the case of several predicates holding over the same state in the context of an agent playing several different roles within the same system.

5.1 Dining Philosophers

The Dining Philosophers problem, due to Dijkstra [13], is a well-known metaphor for discussing concurrent processes. Five philosophers are sitting around a table with a bowl of spaghetti in the middle. For them, life consists only of two moments: thinking and eating, rigorously using two forks. Since each philosopher has a right fork and a left fork, (s)he thinks till both forks become available, eats for a certain amount of time, then stops eating (putting back both forks on the table), and starts thinking again. The problem is that in between two neighboring philosophers there is only one fork: each one shares a fork with a neighbor. The ASM-based model of Dining Philosophers is in [9]: it is a DASM with a set of *philosophers* = $\{p_1, ..., p_5\}$, i.e. the agents of the system, and a set of *forks* = $\{f_1, ..., f_5\}$, i.e. their shared resources. The computation evolves through the states characterized by the following predicates:

- thinking: ¬(*owner*(*rightFork*(**self**)) = **self** ∨ *owner*(*leftFork*(**self**)) = **self**). The philosopher is thinking, so (s)he is waiting for both forks to become available;
- eating: *owner*(*rightFork*(**self**)) = **self** ∧ *owner*(*leftFork*(**self**)) = **self**. The philosopher is eating, so (s)he has obtained both forks,

  where:

- *rightFork*: *philosophers* → *forks* indicates a philosopher's right fork;
- *leftFork*: *philosophers* → *forks* indicates a philosopher's left fork;
- *owner*: *forks* → *philosophers* denotes the current user of a fork.

Initially, each philosopher p_i thinks, and has fork f_i on the right and fork f_{i-1} on the left, except for p_1 that has fork f_5 on the left. The ASM program for p_i is shown below:

PhilosopherProgram(p_i) =
   if *owner*(*rightFork*(**self**)) = *undef* ∧ *ower*(*leftFork*(**self**)) = *undef* **then** {
      owner(*rightFork*(**self**)) := **self**
      owner(*leftFork*(**self**)) := **self**
   }
   if *owner*(*rightFork*(**self**)) = **self** ∧ *owner*(*leftFork*(**self**)) = **self** **then** {
      owner(*rightFork*(**self**)) := *undef*
      owner(*leftFork*(**self**)) := *undef*
   }

Ideally, p_i, denoted by **self**, would like to execute alternatively the two rules above to get and later to release the desired forks. Indeed, even if ASM rules are executed in parallel by definition, the second rule (i.e. the second **if-then** statement in the program above) can be performed if and only if the first rule has been previously executed.

During the waiting for both forks, the computation of p_i can go through four states:

1. (*owner*(*rightFork*(**self**))=philosopher on the right) ∧ (*owner*(*leftFork*(**self**))=*undef*);
2. (*owner*(*rightFork*(**self**))=*undef*) ∧ (*owner*(*leftFork*(**self**)) = philosopher on the left);
3. (*owner*(*rightFork*(**self**)) = philosopher on the right) ∧ (*owner*(*leftFork*(**self**)) = philosopher on the left);
4. (*owner*(*rightFork*(**self**)) = *undef*) ∧ (*owner*(*leftFork*(**self**)) = *undef*).

In all four states, the `thinking` predicate holds. The state changes whenever an update is executed by the neighboring philosophers over the shared locations; however, the ASM could not make a real computational step towards the state characterized by the `eating` predicate. In fact, only state (4) allows the first rule to be executed, so the desired state can be reached. In this particular case, even if the ASM state changes, the computation could not make a real progress, i.e. the process risks to starve.

For verification purposes, predicate abstraction is very suitable for capturing starvation. In particular, starvation could arise if there are rules: (*i*) whose condition concerns functions which represent the dependency of the agent from external resources; and (*ii*) whose execution/non-execution could have effects that does not change the value of the predicate over the states which represents the "waiting for something" issue. In our case, the first rule of the *PhilosopherProgram* shows these issues.

Finally, it is worth noting that, if resource holding holds, the scenario above is affected by the risk of deadlock: each philosopher picks up his/her right fork and waits for the left fork to become available. Thanks to predicate abstraction, this issue can be captured by the following predicate: *owner*(*rightFork*(*p*)) = *p*, ∀ *p* ∈ *Philosophers*. Therefore, the model is deadlock-free if, during the DASM computation, it is not possible that its global state fulfill the predicate above, i.e. at any moment there must be at least one ASM whose state fulfills ¬(*owner*(*rightFork*(**self**)) = **self**).

5.2 A MANET Routing Protocol

Mobile Ad-hoc NETworks (MANETs) [1] are wireless networks designed for communications among nomadic hosts, in absence of fixed physical infrastructure. Each node plays a twofold role: end-point of a communication session and router supporting other communications. Both activities evolve concurrently. A MANET that adopts the AODV routing protocol [26] can be modeled by a DASM including a homogeneous set of *hosts* = {h_1, ..., h_n}. Each ASM can be in one of different states, which are characterized by the following predicates over the states:

- `idle`: the host is inactive. Its formula is given by: *wishToInitiate*(**self**, *dest*) = *false* ∧ *receivedRREQ*(**self**, *dest*) = *false* ∧ *isEmpty*(*replies*(**self**)) = *true*, ∀ *dest* ∈ *hosts*;
- `router`: the host has received a control packet directed to it. It is characterized by *receivedRREQ*(**self**, *dest*) = *true*;

- `initiator`: the host has to start a new communication session. It is characterized by *wishToInitiate*(**self**, *dest*) = *true*;
- `forwarding`: the host is forwarding a control packet to another recipient. It is characterized by *isEmpty*(*replies*(**self**)) = *false*.

where:

- *wishToInitiate*: *hosts* × *hosts* → *boolean* indicates whether a new communication session to a destination is required;
- *receivedRREQ*: *hosts* × *hosts* → *boolean* indicates whether an RREQ packet has been received;
- *isEmpty*: *queues* → *boolean* states if a queue of messages is empty or not.

In fact, in order to model broadcasting and unicasting, each host is associated with two queues of messages: *requests* and *replies*, including: RREQ (Route REQuest) packets for requesting a route to a desired destination, and RREP (Route REPly) packets for replying this request, respectively. This allows us to model sending/receiving of packets by means of enqueuing/dequeuing abstract messages into the corresponding queue. In addition, each ASM includes the following functions:

- *routingTable*: *hosts* → PowerSet(*records*), which represents the information about the hosts recorded into the host's routing table;
- *hostsInRT*: PowerSet(*records*) → PowerSet(*hosts*), which returns the set of the hosts stored in a given routing table, for checking information about hosts.

The ASM pseudo-code of the i-th host is shown below.

HostProgram(h_i) =
  if ¬*isEmpty*(*requests*(**self**)) **then** {
    RREQ = *top*(*requests*(**self**))
    nextHop = sender of *top*(*requests*(**self**))
    updateRoutingTable(**self**, *RREQ*)
    receivedRREQ(**self**, *dest*) := *true*
    Router(*RREQ*, *nextHop*)
  }
  if *wishToInitiate*(**self**, *dest*) = true **then**
    Initiator(*dest*)
  if ¬*isEmpty*(*replies*(**self**)) {
    RREP = *top*(*replies*(**self**))
    if *RREP.init* ≠ **self then** {
      nextHop = select *c.nextHop* ∈ *hostsInRT*(*routingTable*(**self**))
        with *RREP.init* = *c.dest*
      updateRoutingTable(**self**, *RREP*)
      UnicastRREP(*RREP*, *nextHop*)
      dequeue RREP from *replies*(**self**)
    }
  }

It is worth noting that the activation of a host unfolds different computational branches: two of them lead to the execution of the *Router* or *Initiator* submachine, respectively; in the third case, the forwarding of RREPs is executed. In particular, the *Router* submachine models the behavior of the node when it supports communications between other end-points; instead, the *Initiator* submachine models the behavior of the node when it acts as the initiator of a new communication session. Note that if initiator does not know a route to reach destination, then it starts a *route discovery* process aimed at discovering this route. For clarity, an ASM *submachine* is a parameterized rule [9]: it allows the declaration of *local* functions, so that each call of a submachine works with its own instantiation of its local functions.

For verification purposes, predicate abstraction can help in expressing the node's behavior. In our case, the simultaneous fulfillment of different predicates over the same ASM state is very suitable for capturing the intrinsic concurrency of the nodes' computation. Indeed, when the MANET starts operating, each host is idle. But, during the normal execution of the MANET, a host can, for example, fulfill the router predicate with respect to a destination, but at the same time it can fulfill other predicates, e.g. initiator, for what concerns other destinations. The values of the arguments help in distinguishing the various cases.

Moreover, predicate abstraction can help in investigating some specific properties for MANETs: the correctness of the activities of sending/receiving packets, the starvation-freedom of initiator when it starts a route discovery process, and so forth. For example, concerning the starvation issue, the presence of a timeout in the *Initiator* submachine allows initiator to escape the waiting for RREPs if a route to destination cannot be found. So, for that specific destination, after the timeout expiration, the ASM state does not fulfill the initiator predicate but the idle predicate.

For the purposes of the present work, it is not necessary to further detail the *updateRoutingTable* and *UnicastRREP* rules, as well the *Router* and *Initiator* submachines. The interested reader can find the full specification of the model and the proof of its correctness in [7].

6 Conclusion

This paper proposes predicate abstraction over ASM states. The proposed approach can support the verification of ASM-based models by overcoming the main limitations that penalize classic model checking techniques applied to ASMs. In fact, applying predicate abstraction to ASMs enables the analysis of the global properties of the system to be verified through a representation of its local properties through predicates over the states. In this way, the analysis is executed entirely within the ASM framework, without the need of less expressive models and without the burden of temporal logics. Possible applications are represented by various kinds of critical and complex systems that can benefit from a formal approach: Internet-based services, security protocols, Cloud, Grid and mobile systems, and so on.

It is worth specifying that researchers usually distinguish between two classes of properties [23]. *Safety* properties specify that "something bad never happens", e.g. deadlock-freedom. Instead, *liveness* properties stipulate that "something good eventually happens", e.g. starvation-freedom. From this point of view, safety properties require that certain predicates over the states, which represent a "bad thing", must never be satisfied, or, alternatively, their negation must always hold during the computation. Conversely, liveness properties require that certain predicates over the states, which represent a "good thing", must eventually be satisfied during the computation. Future directions of this research should investigate specific features of predicate abstraction with respect to the specific kinds of properties to be analyzed.

Acknowledgements. This work has been partially funded by the Italian Ministry of Education, University and Research, within the "Piano Operativo Nazionale" PON02_00563_3489339.

References

1. Agrawal, D.P., Zeng, Q.A.: Introduction to Wireless and Mobile Systems. Thomson Brooks/Cole (2003)
2. Alpern, B., Schneider, F.B.: Defining Liveness. Information Processing Letters **21**(4), 181–185 (1985)
3. Arcaini, P., Gargantini, A., Riccobene, E.: AsmetaSMV: a way to link high-level ASM models to low-level NuSMV specifications. In: Frappier, M., Glässer, U., Khurshid, S., Laleau, R., Reeves, S. (eds.) ABZ 2010. LNCS, vol. 5977, pp. 61–74. Springer, Heidelberg (2010)
4. Arcaini, P., Gargantini, A., Riccobene, E.: CoMA: conformance monitoring of java programs by abstract state machines. In: Khurshid, S., Sen, K. (eds.) RV 2011. LNCS, vol. 7186, pp. 223–238. Springer, Heidelberg (2012)
5. Baier, C., Katoen, J.P.: Principles of Model Checking. The MIT Press (2008)
6. Bianchi, A., Manelli, L., Pizzutilo, S.: An ASM-based Model for Grid Job Management. Informatica (Slovenia) **37**(3), 295–306 (2013)
7. Bianchi, A., Pizzutilo, S., Vessio, G.: Suitability of Abstract State Machines for Discussing Mobile Ad-hoc Networks. Global Journal of Advanced Software Engineering **1**, 29–38 (2014)
8. Blass, A., Gurevich, Y.: Abstract State Machines Capture Parallel Algorithms. ACM Transactions on Computational Logic **4**(4), 578–651 (2003)
9. Börger, E., Stärk, R.: Abstract State Machines: A Method for High-Level System Design and Analysis. Springer (2003)
10. Chen, Z., Zhou, C., Ding, D.: Automatic abstraction refinement for petri nets verification. In: 10th Int. Workshop on High-Level Design, Validation and Test, pp. 168–174 (2005)
11. Das, S., Dill, D.L., Park, S.: Experience with predicate abstraction. In: Halbwachs, N., Peled, D.A. (eds.) CAV 1999. LNCS, vol. 1633, pp. 160–171. Springer, Heidelberg (1999)
12. Dershowitz, N.: The Generic Model of Computation. Electronic Proceedings in Theoretical Computer Science (2013)
13. Dijkstra, E.W.: Hierarchical Ordering of Sequential Processes. ACTA Informatica **1**(2), 115–138 (1971)
14. Farahbod, R., Glässer, U., Ma, G.: Model checking CoreASM specifications. In: 14th International ASM Workshop (2007)
15. Gabrisch, W.: A hoare-style verification calculus for control state ASMs. In: 5th Balkan Conference on Informatics, pp. 205–210 (2012)

16. Gargantini, A., Riccobene, E., Rinzivillo, S.: Using spin to generate testsfrom ASM specifications. In: Börger, E., Gargantini, A., Riccobene, E. (eds.) ASM 2003. LNCS, vol. 2589, pp. 263–277. Springer, Heidelberg (2003)
17. Gervasi, V.: An ASM model of concurrency in a web browser. In: Derrick, J., Fitzgerald, J., Gnesi, S., Khurshid, S., Leuschel, M., Reeves, S., Riccobene, E. (eds.) ABZ 2012. LNCS, vol. 7316, pp. 79–93. Springer, Heidelberg (2012)
18. Glausch, A., Reisig, W.: An ASM-characterization of a class of distributed algorithms. In: Abrial, J.-R., Glässer, U. (eds.) Rigorous Methods for Software Construction and Analysis. LNCS, vol. 5115, pp. 50–64. Springer, Heidelberg (2009)
19. Graf, S., Saidi, H.: Construction of abstract state graphs with PVS. In: Grumberg, O. (ed.) CAV 1997. LNCS, vol. 1254, pp. 72–83. Springer, Heidelberg (1997)
20. Gurevich, Y.: Sequential Abstract State Machines Capture Sequential Algorithms. ACM Transactions on Computational Logic **1**(1), 77–111 (2000)
21. Hopcroft, J.E., Ullman, J.D.: Introduction to Automata Theory, Languages, and Computation. Addison-Wesley (1979)
22. Klai, K., Desel, J.: Checking soundness of business processes compositionally using symbolic observation graphs. In: Giese, H., Rosu, G. (eds.) FMOODS/FORTE 2012. LNCS, vol. 7273, pp. 67–83. Springer, Heidelberg (2012)
23. Kindler, E.: Safety and Liveness Properties: A Survey. EATCS Bulletin **53**, 268–272 (1994)
24. Laplante, P.: Dictionary of Computer Science. Engineering and Technology. CRC Press (2000)
25. Luzzana, A., Rossetti, M., Righettini, P., Scandurra, P.: Modeling synchronization/communication patterns in vision-based robot control applications using ASMs. In: Derrick, J., Fitzgerald, J., Gnesi, S., Khurshid, S., Leuschel, M., Reeves, S., Riccobene, E. (eds.) ABZ 2012. LNCS, vol. 7316, pp. 331–335. Springer, Heidelberg (2012)
26. Perkins, C.E., Belding-Royer, E.M., Das, S.R.: Ad hoc On-Demand Distance Vector (AODV) Routing. RFC 3561 (2003). http://tools.ietf.org/html/rfc3561
27. Reisig, W.: The Expressive Power of Abstract State Machines. Computing and Informatics **22**, 209–219 (2003)
28. Singhal, M.: Deadlock Detection in Distributed Systems. IEEE Computer **22**(11), 37–48 (1989)
29. Spielmann, M.: Automatic verification of abstract state machines. In: Halbwachs, N., Peled, D.A. (eds.) CAV 1999. LNCS, vol. 1633, pp. 431–442. Springer, Heidelberg (1999)

A Workaround Design System for Anticipating, Designing, and/or Preventing Workarounds

Steven Alter[✉]

University of San Francisco, San Francisco, USA
alter@usfca.edu

Abstract. Idealized system design produces requirements reflecting management intentions and "best practices." This paper proposes a workaround design system (WDS) for anticipating, designing, and/or preventing workarounds that bypass systems as designed. A WDS includes a process and an interactive "workaround design tool" (WDT) for identifying and evaluating foreseeable workarounds based on work system theory and a theory of workarounds. This paper summarizes the conceptual background and explains the form, use, and implications of the proposed WDS and WDT.

The idea of WDS addresses significant gaps in practice and research. Designers should have methods for identifying likely obstacles and anticipating and evaluating a non-trivial percentage of plausible workarounds. Methods for identifying workarounds might help in training work system participants. Researchers might use WDS to explore why specific responses to obstacles did or did not occur. The lack of methods related to anticipating, designing or preventing workarounds implies that WDS may prove fruitful even though it is impossible to anticipate all possible workarounds.

Keywords: Workaround · Systems analysis and design · Business process management · Emergent change

1 Augmenting Design by Placing Workarounds in the Foreground

As a contribution to EMMSAD 2015 (Exploring Modeling Methods for Systems Analysis and Design), this paper[1] introduces a way to explore workarounds, an important topic that is ignored or barely mentioned in modeling methods and systems analysis. Workarounds are widely recognized in everyday business life. Some workarounds address unanticipated obstacles; others bypass cumbersome processes or technologies; yet others involve taking personal advantage of incomplete management oversight. Some authors view workarounds as essential occurrences in everyday work and as sources of innovation; others view them as hazardous activities, noncompliance, or opportunistic behavior that undermines management intentions.

---

[1] This is an abbreviated version of a longer paper that is available at www.stevenalter.com. The longer version provides more background, more complete explanations, and many literature references that could not be included in this abbreviated version.

Ignoring workarounds has negative consequences for systems analysis and design. It places systems at greater jeopardy of being undermined by inappropriate workarounds. It also increases the probability of creating cumbersome features that work system participants will view as indications that the designers did not understand the nature and details of the work being done.

This paper questions the common assumption that application software captures and enforces best practices. Many examples in the management, operations, and sociotechnical literature demonstrate that work system participants with behavioral discretion may perform activities in ways that were not prescribed by the software or by management, and may act in ways that conflict directly with management expectations. Systems analysis and design and related modeling methods should address those issues if the goal is to build realistic systems that will achieve business goals.

Augmenting Established Methods. This paper's new idea is a workaround design system (WDS) focusing on what work system participants might do when current or proposed specifications of routines, processes, best practices, or methods do not fit realities that they encounter. This involves imagining exceptions or obstacles and identifying appropriate responses. Those responses may involve workarounds or may address exceptions and obstacles in other ways. Thus, a WDS is quite different from typical approaches such as identifying alternate paths in typical use case narratives.

The idea of a WDS starts with a broadly defined process that includes identifying the work system, identifying foreseeable exceptions or obstacles that might call for a workaround, identifying plausible workarounds, evaluating those workarounds, and deciding how to adjust the design, if necessary. The WDS uses a workaround design tool (WDT) that provides knowledge-based support for each step through templates and compilations of available knowledge in forms such as lists of typical workaround drivers, lists of typical design moves and characteristics that might change, and possibly even workaround design patterns that resemble design patterns for software.

A systematic approach for imagining and evaluating foreseeable workarounds could have many uses that augment established design approaches:

- A WDS might help in identifying circumstances under which aspects of a work system might be bypassed or undermined. Anticipation of exceptions or obstacles and the resulting workarounds might help in designing the work to encourage appropriate behavioral discretion while also blocking inappropriate workarounds.
- An effective WDS might lead to more complete and useful instructions about what to do when foreseeable exceptions and obstacles occur in real world practice.
- An effective WDS might be incorporated into training during and after implementation of new or improved work systems. Use of the WDS might help work system participants understand their own work system in a deeper way.
- An effective WDS might sensitize managers and designers to be more realistic about how work systems and software will be developed, implemented, and used.
- An effective WDS might encourage more effective participation in analysis and design efforts by giving work system participants a way to contribute that engages both their imagination and their knowledge of their own work settings.
- An effective WDS might make likely workarounds visible as a contribution to future improvements in the work system or information system that supports it.

Source of Ideas About Workarounds. This paper applies a theory of workarounds that was developed as part of research about unplanned change [1]. The theory attempted to encompass workarounds discussed in "300+ articles" that mentioned examples of workarounds or ideas about workarounds. The articles were found through Google Scholar searches such as "workaround + nursing" or "workaround + bureaucracy" or "workaround + hazard." Those articles covered disciplines ranging from IS and medical informatics to organization behavior and public administration.

Related Methods and Research. The idea of using a WDS for anticipating foreseeable workarounds is a new variation on a long established practice of enriching design and planning processes by identifying events whose occurrence might have a significant impact. Strategy studies have used many related methods (e.g., scenario analysis, cross impact analysis, and Delphi studies). At an operational level, various forms of risk analysis and crisis management planning have been used (e. g., failure modes and effects analysis (FMEA) in Six Sigma and Monte Carlo simulation).

Organization. This paper's explanation of the proposed WDS and WDT starts by summarizing the theory of workarounds that forms the basis of a WDS, a temporary work system for devising workarounds, supported by a software based tool called a workaround design tool (WDT). It explains a WDT's capabilities and sources of knowledge. Overall, it explains an approach for taking workarounds seriously enough to incorporate the anticipation of workarounds into assumptions about how systems in organizations operate and how to analyze and design systems. A multi-year research project could build on these ideas by creating a WDT and testing the entire WDS/WDT approach in experimental or real world design situations.

2 Background about Workarounds

The proposed WDS and WDT are based on a theory of workarounds that is explained in depth in [1], which includes a lengthy literature review covering previous discussions of workarounds along with many examples from various disciplines. Due to page limitations, only a brief summary of the theory of workarounds is included here.

Definition of Workaround and Related Preconditions. Workarounds occur for a variety of reasons. In some cases, people need to respond in some way to unanticipated obstacles or exceptions encountered while doing work. In other cases, workarounds bypass cumbersome or inefficient process steps. Workarounds may bypass organizational routines that emerged over time without an explicit design and fail to consider important contingencies. Activities may deviate from expectations due to a lack of knowledge or training, personal opportunism, or other reasons. The theory of workarounds is based on the following definition of workaround:

> A workaround is a goal-driven adaptation, improvisation, or other change to one or more aspects of an existing work system in order to overcome, bypass, or minimize the impact of obstacles, exceptions, anomalies, mishaps, established practices, management expectations, or structural constraints that are perceived as preventing that work system or its participants from achieving a desired level of efficiency, effectiveness, or other organizational or personal goals.

This definition is broader and more encompassing than 12 other definitions in the literature [1]. Workarounds include adaptations in any part of a work system rather than just in processes or technologies. A work system is a system in which human participants and/or machines perform work using information, technology, and other resources to produce products/services for internal or external customers [2].

Workarounds affect details of a work system's operation, either temporarily or over an extended period, but do not change its overall identity, purpose, and high-level architecture. Workarounds may be totally ethical, ethically questionable, or fraudulent. Decisions related to creating and executing workarounds may or may not consider ethics and legality along with many other factors.

Preconditions for the occurrence of a workaround include: 1) a specific process, policy, or set of practices within an existing work system; 2) organizational and/or personal goals related to that situation; 3) an obstacle, exception, anomaly, or even an established practice that might be perceived as something to bypass or overcome; 4) an ability to imagine and execute a workaround.

Goals and Effects of Workarounds. The literature review found a wide range of goals for workarounds. These included: overcoming inadequate IT functionality, bypassing obstacles built into existing routines, overcoming transient obstacles due to anomalies or mishaps, responding to mishaps with quick fixes, augmenting existing routines without developing new resources, substituting for unavailable or inadequate resources, implementing new resources, preventing mishaps, pretending to comply, and other actions for personal or mutual benefit such as cheating, stealing and colluding. Direct effects of workarounds included: continuation of work despite obstacles, mishaps, or anomalies; creation of hazards, inefficiencies or errors; impacts on subsequent activities; and compliance or non-compliance with management intentions. Perspectives on workarounds ranged from quite positive to quite negative: necessary activities in everyday life, creative acts, sources of future improvements, quick fixes that don't go away, add-ons, shadow systems, or feral systems, inefficiencies or hazards, resistance, and distortions or subterfuge.

Theory of Workarounds. As represented in Figure 1, the theory of workarounds identifies steps in designing and executing workarounds along with common factors that affect perceived needs for workarounds and decisions about which workarounds will be designed and executed. It encompasses the descriptions of workarounds that were found in the literature review mentioned earlier, ranging from small, localized workarounds that are forgotten quickly through software add-ons, or shadow systems designed to address work flow or software shortcomings over long time spans. It also covers all of the goals and perspectives on workarounds that were mentioned above.

The factors included in Figure 1 have significant impact in some situations and minimal impact in others because the theory spans a wide range of situations. For example, monitoring systems and ethical considerations usually are more important for workarounds that affect activities, information, or results elsewhere and usually are less unimportant for workarounds of temporary, local conditions that have no impact elsewhere. Italicized terms on the left side of Figure 1 identify generic steps in perceiving the need for a workaround and then creating it. The sequence reflects a rationalist view in which work system participants create workarounds by identifying obstacles and deciding what to do about them. The theory combines ideas from the theory of planned behavior, improvisation and bricolage, and agency theory.

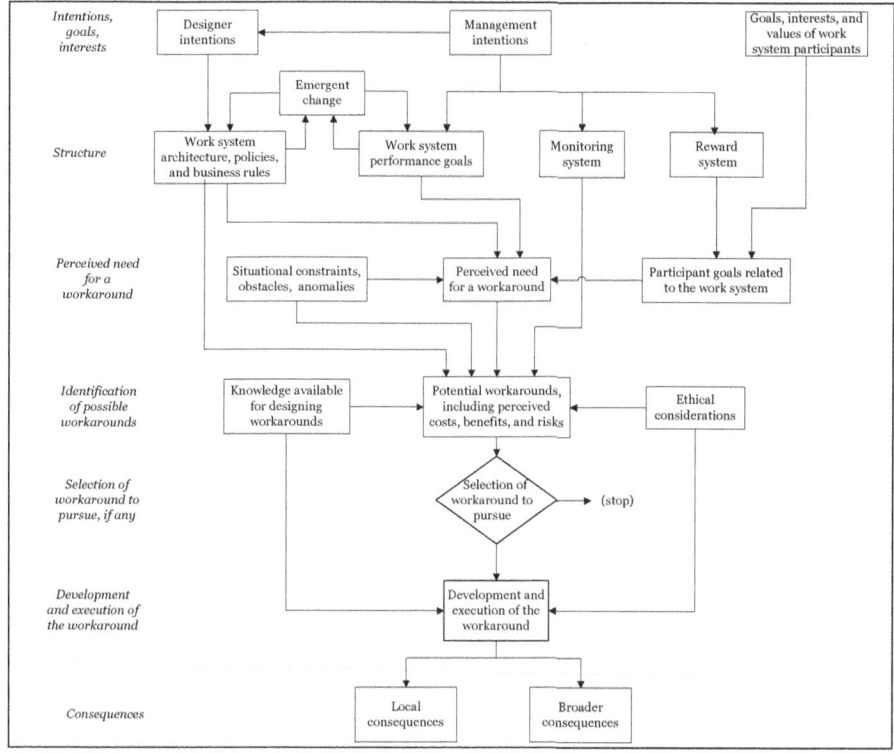

Fig. 1. Theory of workarounds [1]

3 A Workaround Design System

The proposed WDS is a temporary work system whose participants use an interactive workaround design tool (WDT) to design workarounds related to a proposed or existing work system in an organization. Potential applications and benefits of a WDS were mentioned at the outset, e.g., anticipation of foreseeable workarounds, insight for managers and system designers, training and assistance for work system participants, and more effective user participation in analysis and design efforts.

Design Assumptions for a WDS. Table 1 shows important differences between assumptions underlying the proposed WDS and assumptions for textbook descriptions of systems analysis and design. WDS challenges the assumption that systems in organizations will operate as designed or intended. It challenges the assumption that business processes and information systems represent best practices that remain appropriate even as the surrounding context changes over time and as occasional exceptions and obstacles prove awkward or insurmountable with established practices. It also challenges the assumption that designers are capable of designing processes and related software that will encompass every possible situation that the work system will encounter. Overall, considering only best case assumptions and most likely cases is myopic and increases the probability of surprise responses to conditions that could have been anticipated.

Table 1. Comparing assumptions for a WDS versus Typical Systems Analysis and Design

Topic	Typical analysis and design methods	A workaround design system
Unit of analysis	Information system	IT-reliant work system
Hardware/software usage	Requirements specify usage patterns	Formal requirements may not exist. If they exist, they may not be followed completely.
Responsibility of work system participants	Get work done using the prescribed methods	Get done work using prescribed methods when practical and applying workarounds if appropriate
Nature of official business process	It represents best practices, i.e., the right way to perform the task.	The official business process may not describe current practices and may or may not represent best practices.
Expectations about compliance	Work system participants will comply with business processes.	Work system participants may or may not comply with official business processes.
Nature of requirements	Requirements are rational and are based on management goals	Requirements may or may not exist. Any existing requirements may not be appropriate, especially when exceptions occur.
Alignment of goals and incentives	Organization's goals are aligned with participants' incentives	Goals of the organization may or may not be aligned with incentives of work system participants.
Expected mastery and knowledge levels	Work system participants know how to do the job.	Work system participants know how to do the job and also know enough to create appropriate workarounds when they encounter obstacles.
View of workarounds	Workarounds are inappropriate. Work practices should conform to the design of the work system.	Workarounds are appropriate when participants encounter exception conditions or obstacles, except when negative consequences would occur or when explicitly prohibited for an understandable reason.

Structure of a WDS. The structure of a WDS is based on the rationalist assumptions built into Figure 1. Table 2 uses the format of a work system snapshot [2] to summarize a WDS as a work system that uses a WDT. WDS participants are called designers to avoid confusion with participants in the work system in which the workaround will occur. Notice that the proposed WDT is included as a technology that is used within the WDS. The following discussion focuses on anticipating foreseeable obstacles and workarounds for a proposed work system that contains or uses an information system. Other uses such as producing a workaround to a currently operational work system would call for changes in some of the steps.

Table 2. Summary of a WDS in the Format of a Work System Snapshot

Customers	Products/ Services
• Managers and designers who attain insights • Trainers who use plausible workarounds in training • Work system participants who use insights and information the WDS generates	• List of anticipated exceptions or obstacles and plausible workarounds, if any • Documentation, suggestions, and warnings for each plausible workaround that is identified
Major Processes or Activities	
• Designers identify the work system in which workarounds might occur. • Designers summarize the work system using a work system snapshot • Designers identify foreseeable exceptions or obstacles that call for a workaround. • Designers identify plausible workarounds for specific exceptions or obstacles. • Designers evaluate plausible workarounds. • Designers decide how to adjust the design, if necessary.	

Participants	Information	Technologies
• Designers of workarounds, who may be managers, work system designers, or work system participants	• Description of work system • Description of exceptions, obstacles, other relevant factors • Description of workarounds	• Workaround design tool (WDT)

Each of the major processes or activities in Table 2 calls for a bit of elaboration.

Identify the Work System. Designers name the work system using a verb phrase (e.g., invoicing for construction work, answering customer queries, producing month-end financial statements, finding and fixing bugs in operational software).

Summarize the Work System using a Work System Snapshot. Identifying possible workarounds requires a summary description of a work system The format of Table 2 shows that a work system snapshot is a one-page summary of a work system's customers, product/services, processes and activities, participants, information, and technologies. Work system snapshots have been used by many hundreds of MBA and Executive MBA students [2, 3].

Identify Foreseeable Exceptions or Obstacles that might Call for a Workaround. To facilitate this task, a computerized version of Table 3 could be presented to suggest common obstacles that might be considered.

Table 3 is organized around the six elements included in a work system snapshot plus the three other elements of the work system framework and "work system as a whole." Due to page limitations entries for only three cells are shown, but these suffice to illustrate the content of this type of table. A version of Table 3 was proposed as a sort of negative design space consisting of things to be avoided in a new or existing work system [5]. Other versions of this type of table can be developed, e.g., for specific types of situations, such as purchasing or manufacturing systems.

Table 3. Common Stumbling Blocks and Risk Factors [4, p. 65], abbreviated

Customers	Product/Services
• Unrealistic expectations • Unmet customer needs or concerns • Customer segments with contradictory needs • Unsatisfying customer experience • Lack of customers or customer interest	• Unfamiliar products or service • Products/ services are difficult to use • High cost of ownership • Incompatibility with other aspects of the customer's work environment

Process and Activities	
• Inadequate resources or capacity • Inadequate quality controls • Uncertainty about work methods • Excessive variability in work practices • Over-structured work practices • Excessive interruptions • Excessive complexity • Inadequate security	• Omission of important functions • Built-in delays • Unnecessary hand-offs, authorizations • Steps that don't add value • Unnecessary constraints • Low value variations • Inadequate scheduling of work • Large fluctuations in workload

Participants	Information	Technologies
• (entries omitted)	• (entries omitted)	• (entries omitted)

Infrastructure	• (entries omitted)
Environment	• (entries omitted)
Strategies	• (entries omitted)
Work System as a Whole	• (entries omitted)

Identify Plausible Workarounds Related to Specific Exceptions or Obstacles. A similarly formatted table could display common design "moves" that describe the form of a workaround. For example, a workaround might skip a business process step, might not conform to a business rule, might use different information, or might be performed by a substitute. As above, different versions of this type of table could be developed for different types of systems.

A second type of support for identifying foreseeable workarounds could be an organized checklist of frequently used workarounds. That could come from empirical studies of workarounds and/or analysis of hundreds of examples in the literature. For example, [6] studied workarounds in healthcare, accounting, and automotive and found generic types of workarounds including the following, among many others: download and save protected data (that is inconvenient to access when it is protected); provide a vague reason for accessing data (when a formal reason is required); post a password in view (when passwords are hard to remember); share one person's password (when sharing a device); and split transactions to avoid alarms (with a 5000 euro limit, convert one 6000 euro transaction into two 3000 euro transactions).

A third type of checklist or script for identifying workarounds could be based on design characteristics that might be viewed as design dimensions. [5, p. 9] identifies characteristics that might be included in a script or checklist to help designers consider how a workaround might make the process more structured or less structured, more complex or less complex, more collaborative or less collaborative, and so on.

Evaluate Plausible Workarounds. Evaluating plausible workarounds could be merged, at least partially, with the previous step if WDS participants believe that approach would minimize redundant effort. In either case, the designers would consider only workarounds that have a plausible rationale and that work system participants might actually consider. The WDT could facilitate this evaluation by providing a checklist or script based on issues implied by factors in Figure 1.

4   Discussion and Conclusion

This paper proposed a theory-based workaround design system (WDS) that includes a workaround design tool (WDT). It illustrated how existing knowledge related to common problems and issues, design moves, and work system characteristics could be incorporated into a WDT in the form of checklists, scripts, or sliders. Extensions for specific types of situations would display selected, domain-specific examples that occurred in the past, such as common ways to work around log-on procedures, authorizations, and controls in service and order entry systems.

WDS provides a perspective for looking at many topics and issues in new and different ways. It provides a system and tool for making accumulated knowledge about workarounds visible and useful for managers, designers, and trainers. It augments established pedagogy, practice, and research and has potential value in many areas:

Value to Managers. A WDS could help in anticipating workarounds that might occur, thereby leading to tactics and strategies for facilitating appropriate workarounds and preventing inappropriate workarounds. It might help managers attain more realistic views of capabilities and limitations of computerized systems that try to embody and enforce best practices. It might support better communication about what is expected, both in terms of following rules and in terms of behavioral discretion.

Value to System Designers and Developers. A WDS would augment established systems analysis and design methods by providing a practical way to deal with workarounds, a topic that does not appear in the glossaries of most systems analysis and design textbooks. A WDS would complement established methods that generate UML or BPMN documentation of system structure and operation. The WDS would explore possible uses of workarounds to address practical limits of any particular idealized view of how work should be performed. Especially useful applications of WDS might occur during ERP configuration processes, which often encounter misfits between the situational needs within a department and the options offered by the ERP software. Like many formal methods and tools, WDS and WDT might be especially useful for less sophisticated designers, implementers, and users who had not yet honed their ability to anticipate, design, and evaluate possible workarounds.

Value to Trainers and Trainees. After initial training on a work system's basic structure and operation and the related software, trainees could use WDS and WDT to suggest possible workarounds and to evaluate possible consequences for themselves, for their work groups, and for others in the organization. That level of engagement might generate deeper learning than current training approaches. It might overcome some of the common inadequacies of training on ERP and other complex software.

Value to Business Students and Their Instructors. Current introductory courses often contain hands-on exercises using tools such as transaction processing software, databases, spreadsheets, and search engines. While that hands-on experience is useful, most of it is about details and concepts of a specific information system or software tool. It is not about more challenging questions such as how to propose and evaluate workarounds when confronted with anomalies that call for workarounds. Use of a WDS in classroom settings could generate a more creative atmosphere in IS courses and could lead to deeper understandings that help students become more productive employees. Classroom applications could use crowd sourcing and gamification approaches to make the entire experience more engaging.

Value for Future Development of Specific Work Systems. Use of WDS and WDT might help in setting paths for future improvements of IT-enabled work systems. This possibility is consistent with research concluding that workarounds often are the springboard for future improvements.

Supplementing Traditional Approaches and Assumptions. The idea of WDS reflects an unconventional stance toward systems analysis and design. The system is an IT-enabled work system, not a technical artifact. Workarounds likely will occur and unanticipated obstacles and contingencies will arise even if requirements capture management intentions and "best practices." Work system design should consider foreseeable workarounds and should anticipate their consequences. This paper suggested the idea of WDS, explained its theoretical basis, and explained how concepts can be built into tools that support a WDS. The next step is to produce and test working prototypes to understand how these ideas can be used effectively in practice.

References

1. Alter, S.: Theory of Workarounds. Communications of the Association for Information Systems **34**, 1041–1066 (2014)
2. Alter, S.: Work System Theory: Overview of Core Concepts, Extensions, and Challenges for the Future. Journal of the Association for Information Systems **14**, 72–121 (2013)
3. Truex, D., Alter, S., Long, C.: Systems analysis for everyone else: empowering business professionals through a systems analysis method that fits their needs. In: ECIS (2010)
4. Alter, S.: The Work System Method: Connecting People, Processes, and IT for Business Results. Work System Press, Larkspur (2006)
5. Alter, S.: Design spaces for sociotechnical systems. In: Proceedings of ECIS (2010)
6. Röder, N., Wiesche, M., Schermann, M.: Why managers tolerate workarounds – the role of information systems. In: Proceedings of ECIS (2014)

An Evaluation of an Enhanced Model Driven Approach for Computer Game Creation

Hong Guo[✉], Shang Gao, John Krogstie, and Hallvard Trætteberg

Department of Computer and Information Science,
Norwegian University of Science and Technology, Trondheim, Norway
{guohong,shanggao,krogstie,hal}@idi.ntnu.no

Abstract. Various game authoring tools have been used to ease the game creation. However, these pre-defined tools may not be suitable for some emerging domains. We proposed an approach (GCCT) to create tools for certain domains first, and then create games with these tools. GCCT is based on the Model Driven Development (MDD) approach which has been successfully applied in many domains to fulfill similar requirements. But MDD also has drawbacks and as a result, persons often have concerns to adopt MDD. To alleviate this, some enhancements were made in GCCT and a user survey was performed to probe the user attitude towards this enhanced MDD approach. 46 persons responded to the survey and the result indicated that in general, GCCT was useful and easy to use. The participants intended to use GCCT primarily because they thought that GCCT was useful. Users' familiarity with computer games and MDD did not have a strong impact on users' understanding and adoption of GCCT.

Keywords: Model driven development · Domain specific modeling · Computer game development · TAM · Survey

1 Introduction

Various authoring tools have been adopted to ease computer games creation [1, 2] by providing easy user interfaces and code automation. However, such tools usually do not address the complexity required by sophisticated games [3]. On the other hand, game engine tools are more powerful and they are the mainstream tools to create commercial games. However, they are usually huge, complex, and lack usability. This makes the learning curve steep and the using cost expensive. What is more, both authoring tools and engine tools target at pre-defined domains. Some emerging or innovative domains like pervasive games [4] and education games may not be able to benefit from them. To ease the game creation especially for such specific domains, we propose an approach named Game Creation with Customized Tools (GCCT). By using GCCT, developers create tools according to specific domain requirements first, and then create games with these tools. GCCT is derived from the general Model Driven Development (MDD) approach which has been proven effective in many domains [5-7] to achieve higher productivity, shortened development cycle, and better software quality. The success of MDD was primarily achieved by providing high level and domain-specific abstractions (as the base of easy user interfaces) as well as code automation. Despite the apparent fitness and strength, involving MDD may also bring risks. For instance, MDD is not easy [8] and it imposes high upfront cost. As a result, managers and developers often have concerns and feel reluctant to adopt MDD [9]. To alleviate this, some enhancements were made in GCCT and a user survey was performed to probe the user attitude towards this enhanced MDD approach.

The remainder of this paper is organized as below. We illustrate the GCCT approach in Section 2. We introduce the evaluation model and the research hypotheses in Section 3. Then in Section 4 and Section 5, we present the settings and results of the evaluation. We discuss the result in Section 6 and conclude the paper in Section 7.

2 GCCT Overview

GCCT contains tools creation and games creation. The tools creation in GCCT involves three tasks: game feature customization, game editor customization, and game code generator customization (task 1-3 in Table 1). While game feature customization defines which game features should be supported in the tools and needs to be done according to the domain requirements firstly, game editor and code generator customization can be done afterwards based on the decided game feature set. For game editors, we select the style (textual, tree-based, diagrammatical, or graphical). Based on the editor style, we connect editor elements to the feature elements. For code generators, we define the rules regarding how code snippets should be generated.

Table 1. GCCT Tasks

GCCT tasks		Corresponding MDD tasks
Tools Customization	1. Game feature Customization	1.1 Domain analysis according to project requirements
		1.2 DSL meta-model/ abstract syntax definition based on domain analysis results
	2. Game Editor Customization	2.1 Style selection according to project requirements
		2.2 DSL concrete syntax definition based on the DSL meta-model and the editor style
	3. Game Code Generator Customization	3.1 Code template definition based on the DSL meta-model and codes of a working prototype
4. Game Creation		4.1 Model creation based on the DSL
		4.2 Code generation according to the model

In addition to tools customization/creation, another part of GCCT is games creation (task 4 in Table 1). Tasks involved in both parts are formalized according to MDD traditions (as shown in Table 1). In order to fully play the strengths of MDD, more specialization and adaptation are done in GCCT to lower the technical barrier and the upfront and general cost. The main improvements are introduced briefly here, and more details were reported in [10]. *Firstly*, instead of introducing a new domain analysis task, we propose to structure existing game design and level design tasks/documents to produce the domain analysis outputs. *Secondly*, GCCT regulates and accelerates the domain analysis task based on predefined domain vocabularies/ontologies. *Thirdly*, GCCT also proposes to reuse existing working prototypes to construct the template of code generator. *Lastly*, GCCT utilizes the state of the art language workbench tools to automate general tasks [11].

Both the tools creation and the games creation are highly iterative. The necessity of intensive iterations in GCCT comes from the requirements of both the game domain and the MDD domain. Due to the difficulties to judge whether a game play is really interesting [12], games are often developed in an iterative way [13]. On the other hand, MDD approaches often start with some basic artifacts, and the number of artifacts grows increasingly within necessary iterations. When requirements change, MDD are supposed to reflect these changes in terms of artifacts/ models [14]. Although both parts are highly iterative, different parts are focused on when the projects progress. In the earlier stage, usually more iterations of tools customization are performed, while several iterations of game creation may be used in order to validate the tools and provide feedback to improve the tools. In the latter stage of the projects, fewer iterations of tools creation may be performed because the tools have grown to be quite mature. And more games are developed in an iterative way.

3 Evaluation Model and Hypotheses

We have made enhancements in GCCT to alleviate some of the MDD drawbacks. And we realized that, the risks of using such an MDD based approach should be evaluated before really applying it in practice. We have evaluated the cost and other major technical issues internally in [10]. In this article, we present the evaluation on potential users' acceptance of this enhanced MDD approach.

User acceptance has received fairly extensive attention in the evaluation of information systems. Among those research efforts, the Technology Acceptance Model (TAM) [15] is considered the most influential and widely applied theoretical model describing an individual's acceptance of information systems [16]. Three main elements are defined in TAM: Perceived Usefulness (PU) as the extent to which potential users believes that using the technology will improve the job performance, Perceived Ease of Use (PEOU) as the extent to which potential users believes using the technology will be free of effort, and Intention to Use (IU) as the strongest predictor of usage behavior. TAM claims that IU can be explained by PU and PEOU, and PU is determined by PEOU. Figure 1 presents the overview of the main TAM variables and expected relationships. In addition, a number of external variables have been introduced into TAM which may affect PU, PEOU, or IU, and their relationships. As GCCT in now in its early stage of development, the focus of evaluating user acceptance is more on probing the general attitude than identifying the affecting factors. Therefore, we choose the original constructs of PU, PEOU and IU as the main research objects in our evaluation.

Based on TAM model, we derive three hypotheses (H1-H3). In addition, as MDD requires high technical skills of the development team, we also explore whether corresponding technical background of users may bring impact on their attitudes towards GCCT. We explore whether experienced game developers find GCCT more useful than amateurs or non-developers (H4). And we want to find out if lack of MDD expertise would hinder the understanding and perceived ease of use of GCCT (H5).

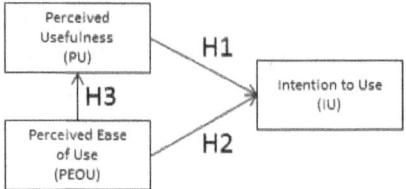

Fig. 1. Technology Acceptance Model [15]

- H1 For GCCT, PU will have a significant influence on IU
- H2 For GCCT, PEOU will have a significant influence on IU
- H3 For GCCT, PEOU will have a significant influence on PU
- H4 Those familiar with computer game development perceive GCCT to be more useful
- H5 Those familiar with MDD perceive GCCT to be easier of use

4 Evaluation Settings

Participants Recruitment
The selection criteria for finding proper test persons are important. As GCCT is not designed for amateurs, understanding and using them may be difficult for persons without necessary expertise. We intended to recruit students with at least some basic knowledge about programming languages & modeling languages, and preferably to recruit those with experiences of either MDD or game development. As most of the recruitments were from a Norwegian university, the educational level of participants is comparatively high (at least bachelor student). As a result, 60 persons were recruited to perform our survey, and 46 of them answered all questions.

Instrument Development
The survey consisted of two parts. The first part aimed to know the participants' personal backgrounds. The second part was intended to examine users' thoughts on GCCT. Validated instrument measures from the previous study [15] were used as the foundation to create the instrument for this study. In order to ensure that the instrument better fit the testing of GCCT, some minor changes in wording were made to ensure easy interpretation and comprehension of the questions. As a result, 14 measurement items (see Table 2) were included in the second part of the survey. A 5-point Likert scale, with 1 being the negative end of the scale (strongly disagree) and 5 being the positive end of the scale (strongly agree), was used to examine participants' responses to all items in this part. Furthermore, we added another two items to test participants' knowledge in information technology.

Process
Because GCCT is not a ready-to-use tool, instead of allowing participants trying it out, we developed two videos (15 minutes in total) introducing GCCT and necessary background for participants to understand it before answering questions. The actual person recruitment and data collection procedure consisted of three parts. *Firstly*, we

tested the videos and questions among several colleagues with different technical background. The testers did not answer the questions formally. Instead they provided feedbacks which were used to revise the videos for smoother illustration and easier understanding. *Secondly*, we performed a small scale survey in a model driven development course given at the Norwegian University. We assumed students in the course should have a comparatively better knowledge of MDD than programmers in general. We played the two videos on the spot and 14 volunteers filled out a paper-based survey. Finally, we recruited more persons from various channels. Both videos and questions were provided online so that participants were able to watch the videos and answer the questions when they were available. 32 persons answered the survey. Thus in total, 46 responses were included in the data analysis.

Table 2. Measurement Items

Perceived Usefulness	PU1. Using GCCT would enable me to create series of pervasive games more quickly.
	PU2. Using GCCT would improve my performance to create series of pervasive games.
	PU3. Using GCCT to create series of pervasive games would increase my productivity.
	PU4. Using GCCT would enhance my effectiveness on creating series of pervasive games
	PU5. Using GCCT would make it easier to create series of pervasive games.
	PU6. I find GCCT useful to create series of pervasive games.
Perceived Usefulness	PEOU1. Learning to use GCCT would be easy for me.
	PEOU2. I find it easy to use GCCT to create pervasive games that I want to create.
	PEOU3. I find the approach of GCCT clear and understandable.
	PEOU4. I find GCCT to be flexible to use.
	PEOU5. It would be easy for me to become skillful at using GCCT.
	PEOU6. I find GCCT easy to use.
Intention to Use	IU1. Assuming I need to create pervasive games and I have access to the tools, I intend to use GCCT.
	IU2. Given that I need to create pervasive games and I have access to the tools, I predict that I would use GCCT.
Technical Background	TB1. How familiar are you with Computer Game Development?
	TB2. How familiar are you with DSL/ DSM/ MDSD?

5 Evaluation Results

Demographics of the Respondents

46 out of the overall 60 participants responded with valid answers to the survey. The responding rate is 76.6%. The majority of the respondents are in the age groups 20-29 (43%) and 30-39 years (46%). More than half of the respondents are Ph.D. students or with higher education. The gender distribution is a little dominant of men (61%). And among all the people, more than half (65%) indicated that they plan or probably plan to develop games. Among the total 46 respondents, 35 indicated that they knew something about how to develop computer games, and another 2 were very familiar with it. Regarding pervasive games knowledge, around half of the respondents knew something about (19) or were very familiar with (3) it. All respondents knew UML and most of them (34) were very familiar with it. The distribution of MDD knowledge is comparatively even. 12 respondents were very familiar with MDD, 19 had some knowledge, while another 12 knew nothing about MDD.

Descriptive Result and Test of Measures

The average values for TAM variables are summarized in Table 3. From the table, we can see that in general, GCCT is acceptable with mean value of all key constructs (PU, PEOU, and IU) above 3.0. We noticed that the value of PU is very close to that of IU (3.79 for PU and 3.75 for IU). But the average value of PEOU is lower than both PU and IU.

Table 3. Mean Value of TAM Constructs

	N	Minimum	Maximum	Mean	Std.Deviation
PU	46	1.0	5.0	3.79	.65869
PEOU	46	1.0	5.0	3.45	.68301
IU	46	1.0	5.0	3.75	.90523

According to [17], data collected from the individual items in a Likert scale are strictly speaking ordinal data. However, when multiple Likert items are combined to form a scale, parametric procedures can be used in the statistical analysis of the data if the scales pass the Cronbach's alpha or the Kappa test of inter-correlation and validity. Reliability statistics on Cronbach's Alpha coefficient were conducted on our survey items used to measure each scale. The reliabilities of all the three scales were above the 0.90 level (0.926 for 6 items of PU, 0.925 for 6 items of PEOU, and 0.954 for 2 items of IU). In [18], the author recommended the minimum requirement for internal consistency as 0.70. The results indicated that all the three scales were with good reliabilities.

Test of TAM Related Hypotheses

To validate TAM related hypotheses (H1-H3), we first used correlation to measure linear relationship between variables as shown in Table 4. In the table, the correlation value of an item with other items indicates to what extent the common entity is. For GCCT, we can see high correlation between each pair of TAM constructs.

We further used regression to investigate the relationship among PU, PEOU and IU for GCCT. Figure 2 summarizes coefficients (labelled on the lines), t-statistics significance (as the stars), and R2 values (in the boxes of IU and PU). From the data, we found that for GCCT, PU is a predictor of IU as we expected, but PEOU was excluded as a predictor of IU. Regarding to PU, PEOU was found to be predictor as expected. The model showed in Figure 2 explains around 70% of variance for both IU and PU of GCCT. As a result, we can see that, hypotheses H1 and H3 were supported. But H2 was not supported with statistics significance.

Table 4. Correlation of TAM Constructs for GCCT

		PU	PEOU	IU
PU	Pearson Correlation	1	.857**	.820**
	Sig. (1-tailed)		.000	.000
PEOU	Pearson Correlation	.857**	1	.782**
	Sig. (1-tailed)	.000		.000
IU	Pearson Correlation	.820**	.782**	1
	Sig. (1-tailed)	.000	.000	
*. Correlation is significant at the 0.05 level (1-tailed).				
**. Correlation is significant at the 0.01 level (1-tailed).				

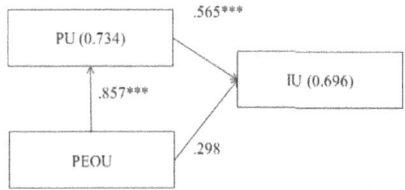

Fig. 2. Structured TAM Model for GCCT

Test of Additional Hypotheses

We assumed that computer game development experience may probably lead to different understanding of usefulness of GCCT. We calculated correlation between game development experience and PU. But the correlation was not found significant (Pearson Correlation is .110, and 1-tailed Significance is .234). This indicates that computer game development experience did not bring obvious impact to how participants perceived the usefulness of GCCT. Lack of general domain knowledge and development experience may not make people be resistant to adopt GCCT. As GCCT is derived from general MDD approaches, we have had concerns about whether lack of MDD knowledge will hinder persons to accept GCCT as it appears difficult to use. We removed complex terminologies in the introduction video, and tried to explain our approach from purely game development perspective. Further, we tested correlation between MDD knowledge and PEOU for GCCT. The correlation is not significant as well (Pearson Correlation is .148, and 1-tailed Significance is .163).

6 Discussion

General Results

The results presented in Section 5 indicate that in overall, GCCT is acceptable by the participants. GCCT was perceived as useful and an average score of 3.79 was given to PU of GCCT. Although ease of use was perceived a little bit lower (3.45), the final result of intention to use was not bad (3.75). In our survey, there is also an open question for general comments. Thirteen participants left some words in this area which may help us explain the scores better. Among them, six respondents considered this approach as "interesting", "cool", "good", or they "like" the approach. Especially, GCCT was thought to be a choice for "efficiently creating lightweight casual games" (compared with other heavy tools that is available at present). It indicates that GCCT is a promising approach and deserves further investigation.

For the TAM related hypotheses, both H1 and H3 were supported by the evaluation result. The perceived usefulness supported the majority of final intention of use. This indicates that people may adopt GCCT mainly because they think it is useful for their work. On the other hand, PEOU did not have a significant influence on IU in our case. Similar results on this were also found in previous research. As introduced in [16], among 101 papers reviewed by the research, only 58 studies showed a significant relationship between PEOU and IU indicating PEOU is "an unstable measure" in predicting IU. In [19], the author noted that "no amount of PEOU will compensate for low usefulness" and questioned the overall effects of PEOU in TAM. This does not matter much as the aim of this evaluation is to probe users' attitude in a roughly quantitative way. We want to know the usefulness of GCCT, the easy to use of GCCT, and to what extend users intent to adopt GCCT for game creation when needed. Identifying potential impact factors of users' attitude to this approach may also be useful, but is not critical at present.

Further, we found computer game development experience does not correlate to PU of GCCT significantly. This indicates that GCCT provides opportunities for amateurs to develop games easily, so that person with no game development experience would find the approach useful in the same way with those who have such experiences. On the other hand, lack of model driven expertise seems not to make people feel difficult to understand and use GCCT. Persons who do not have any MDD knowledge perceived similar PEOU of GCCT. This indicates that with proper specialization and re-explaining, there may not be extra difficulty to involve persons without MDD background in a GCCT process. Most traditional computer game developers do not have an MDD background. Therefore, given a proper working environment and support, those persons may transit to use GCCT approach without much effort. This primarily applies to the game development however. For the tools development, more MDD knowledge is required apparently.

Threats to Validity
We discuss about the most important threats to the validity of this evaluation. *Construct validity* concerns is about whether the sampling particulars can be defended as measures of general constructs [20]. In our survey, we based on TAM model to evaluate PU, PEOU, and IU. TAM is a widely used model and the constructs have been rigorously validated. However, when we developed our scales for PU and PEOU, we adapted the usage scenario and subjects to fit our study. After adaption, some scales may not work as well as original ones. This may lead to the uncertainty of answers from correspondents. *Internal validity* is about "the validity of inferences about whether observed co-variation between A (the presumed treatment) and B (the presumed outcome) reflects a causal relationship from A to B as those variables were manipulated or measured [20]". In our study, we calculated Cronbach Alpha to test internal consistency reliability. We used correlation and regression to validate the TAM models. Although these statistic methods are effective in general, using them on small samples may have threats to some degree (the sample size of our evaluation is 46). *External validity* [20] is about to what extend the study result can be generalized to other situations and other people. Most of the respondents in this survey are students with preliminary IT knowledge like programming and modelling. In addition, most of the respondents do not have real game development experiences in industry. Thus the study result may not apply well for persons who do not have programming background or real game developers in companies.

7 Future Work and Conclusion

In this paper we have introduced GCCT, an enhanced MDD approach for computer game creation. Furthermore, we evaluated users' thoughts on this enhanced MDD approach through a user survey. The result indicated that the majority of the participants found GCCT to be useful, easy to use, and promising. PU has a significant impact on the IU, while PEOU does not. What is more, the result also showed that lack of computer game development experience and MDD knowledge might not hinder the participants to understand and adopt GCCT.

Consequently, we identified some future work for further GCCT development and evaluation: 1) we will investigate more aspects of the approach in addition to the technical aspect. For instance, how to reuse existing documents in traditional computer game development, and how the traditional game creation participants participate in the new process; 2) when further evaluation is considered, we will evaluate on more specific and concrete aspect of the GCCT approach, trying to involving more participants in industry, and allowing them to trying out the approach in the IDE to create benchmark applications.

References

1. Wang, Y., Langlotz, T., Billinghurst, M., Bell, T.: An authoring tool for mobile phone AR environments. In: Proceedings of New Zealand Computer Science Research Student Conference. Citeseer (2009)
2. Gandy, M., MacIntyre, B.: Designer's augmented reality toolkit, ten years later: implications for new media authoring tools. In: Proceedings of the 27th Annual ACM Symposium on User interface Software and Technology. ACM (2014)
3. Furtado, A.W., Santos, A.L.: Using domain-specific modeling towards computer games development industrialization. In: The 6th OOPSLA Workshop on Domain-Specific Modeling (DSM 2006). Citeseer (2006)
4. Hong, G., Trætteberg, H., Wang, A.I., Meng, Z.: TeMPS: a conceptual framework for pervasive and social games. In: 2010 IEEE 3rd International Conference on Digital Game and Intelligent Toy Enhanced Learning (DIGITEL 2010) (2010)
5. Hernandez, F.E., Ortega, F.R.: Eberos GML2D: a graphical domain-specific language for modeling 2D video games. In: Proceedings of the 10th Workshop on Domain-Specific Modeling, p. 1. ACM, Reno (2010)
6. Völter, M., Stahl, T., Bettin, J., Haase, A., Helsen, S.: Model-driven software development: technology, engineering, management. John Wiley & Sons (2013)
7. Hutchinson, J., Whittle, J., Rouncefield, M.: Model-driven engineering practices in industry: Social, organizational and managerial factors that lead to success or failure. Science of Computer Programming **89**, 144–161 (2014)
8. Mernik, M., Heering, J., Sloane, A.M.: When and how to develop domain-specific languages. ACM Comput. Surv. **37**(4), 316–344 (2005)
9. Selic, B.: The pragmatics of model-driven development. IEEE Software **20**(5), 19–25 (2003)
10. Hong, G., Hallvard, T., Alf Inge, W., Shang, G.: RealCoins: A Case Study of Enhanced Model Driven Development for Pervasive Game. International Journal of Multimedia and Ubiquitous Engineering **10**(5) (2015)
11. Maier, S., Volk, D.: Facilitating language-oriented game development by the help of language workbenches. In: Proceedings of the 2008 Conference on Future Play: Research, Play, Share, pp. 224–227. ACM, Toronto (2008)
12. Adams, E.: Fundamentals of game design. New Riders (2010)
13. Gal, V., Le Prado, C., Natkin, S., Vega, L.: Writing for video games. In: Proceedings Laval Virtual (IVRC) (2002)

14. Kelly, S., Tolvanen, J.-P.: Domain-Specific Modeling Enabling Full Code Generation. John Wiley & Sons, Inc. (2008)
15. Davis, F.D.: Perceived usefulness, perceived ease of use, and user acceptance of information technology, pp. 319–340. MIS Quarterly (1989)
16. Lee, Y., Kozar, K.A., Larsen, K.R.: The technology acceptance model: Past, present, and future. Communications of the Association for Information Systems **12**(1), 50 (2003)
17. Allen, I.E., Seaman, C.A.: Likert scales and data analyses. Quality Progress **40**(7), 64–65 (2007)
18. Kline, P.: Handbook of psychological testing. Routledge (2013)
19. Keil, M., Beranek, P.M., Konsynski, B.R.: Usefulness and ease of use: field study evidence regarding task considerations. Decision Support Systems **13**(1), 75–91 (1995)
20. Shadish, W.R., Cook, T.D., Campbell, D.T.: Experimental and quasi-experimental designs for generalized causal inference. Wadsworth Cengage Learning (2002)

Author Index

Abid, Mohamed 447
Alter, Steven 489
Andersson, Alexander 293
Anwar, Adil 461
Asadi, Mohsen 184

Bagheri, Ebrahim 184
Baier, Thomas 119
Barforoush, Ahmad Abdollahzadeh 429
Bernstein, Vered 200
Bianchi, Alessandro 283, 479
Bider, Ilia 19, 169
Bolt, Alfredo 102
Brucker, Achim D. 246
Bruel, Jean-Michel 461

Cabanillas, Cristina 37
Calegari, Daniel 53
Comuzzi, Marco 68

Delgado, Andrea 53
Di Ciccio, Claudio 119

Fahland, Dirk 85

Gailly, Frederik 151
Gao, Shang 499
Gharib, Mohamad 231, 379
Giorgini, Paolo 231, 246, 379
Guo, Hong 499

Haisjackl, Cornelia 217
Heggset, Merethe 395
Huber, Steffen 265

Jablonski, Stefan 37
Jalali, Amin 19
Josefsson, Magnus 169

Kacem, Yessine Hadj 447
Kahani, Mohsen 184
Kammerer, Klaus 135

Karagiannis, Dimitris 413
Karakostas, Bill 68
Kemelman, Mark 311
Kerboeuf, Mickaël 447
Kholkar, Deepali 326
Kolb, Jens 135
Krogstie, John 293, 395, 499
Kulkarni, Vinay 326

Laghouaouta, Youness 461
Leemans, Sander J.J. 85
Leopold, Henrik 3
Lim, Shao Yi 217

Magdich, Amina 447
Mahfoudhi, Adel 447
Mannhardt, Felix 3
Mendling, Jan 37, 119
Mertens, Steven 151
Mohammadi, Reza Gorgan 429

Nassar, Mahmoud 461

Pinggera, Jakob 217
Pizzutilo, Sebastiano 283, 479
Poels, Geert 151
Pourmasoumi, Asef 184

Ramos, Andrés 345
Rangiha, Mohammad Ehson 68
Reichert, Manfred 135
Reijers, Hajo A. 3
Reinhartz-Berger, Iris 311

Sadi, Mahsa H. 360
Sáenz, Juan Pablo 345
Salnitri, Mattia 246
Sánchez, Mario 345
Schlegel, Thomas 265
Schönig, Stefan 37
Seiger, Ronny 265
Söderström, David 19

Soffer, Pnina 200, 217
Sunkle, Sagar 326

Trætteberg, Hallvard 499

van der Aa, Han 3
van der Aalst, Wil M.P. 85, 102
Vessio, Gennaro 283, 479
Villalobos, Jorge 345

Weber, Barbara 217
Wesenberg, Harald 395
Weske, Mathias 119
Widman, Kim 169

Yu, Eric 360

Zamansky, Anna 311
Živković, Srđan 413
Zugal, Stefan 217

The manufacturer's authorised representative in the EU is Springer Nature Customer Service Centre GmbH, Europaplatz 3, 69115 Heidelberg, Germany. If you have any concerns regarding our products, please contact ProductSafety@springernature.com

Printed and bound by CPI Group (UK) Ltd, Croydon, CR0 4YY

23/03/2026

02076672-0020